Geoheritage and Geotourism Resources

Geoheritage and Geotourism Resources

Editors

Nicoletta Santangelo
Ettore Valente

MDPI • Basel • Beijing • Wuhan • Barcelona • Belgrade • Manchester • Tokyo • Cluj • Tianjin

Editors
Nicoletta Santangelo
Department of Earth,
Environmental and Resources
Sciences, University of Naples
Federico II
Italy

Ettore Valente
Department of Earth,
Environmental and Resources
Sciences, University of Naples
Federico II
Italy

Editorial Office
MDPI
St. Alban-Anlage 66
4052 Basel, Switzerland

This is a reprint of articles from the Special Issue published online in the open access journal *Resources* (ISSN 2079-9276) (available at: https://www.mdpi.com/journal/resources/special_issues/ Geoheritage_Geotourism).

For citation purposes, cite each article independently as indicated on the article page online and as indicated below:

LastName, A.A.; LastName, B.B.; LastName, C.C. Article Title. *Journal Name* **Year**, *Article Number*, Page Range.

ISBN 978-3-03936-788-7 (Hbk)
ISBN 978-3-03936-789-4 (PDF)

Cover image courtesy of Ettore Valente.

Contents

About the Editors

Nicoletta Santangelo (Prof.) Ph.D. in Earth Science at University of Naples Federico II (1991), Assistant Professor at University of Naples Federico II (1992), Associate Professor in Geomorphology at University of Naples Federico II (1998).

Ettore Valente (Dott.). Ph.D. in Earth Science at University of Naples Federico II (2010), Research Fellow at University of Naples Federico II (2017).

Editorial

Geoheritage and Geotourism Resources

Nicoletta Santangelo and Ettore Valente *

Department of Earth, Environmental and Resources Sciences, University of Naples Federico II, via Cinthia 21, 80126 Naples, Italy; nicsanta@unina.it
* Correspondence: ettore.valente@unina.it

Received: 22 June 2020; Accepted: 26 June 2020; Published: 28 June 2020

Abstract: This Special Issue wants to outline the role of Geoheritage and Geotourism as potential touristic resources of a region. The term "Geoheritage" refers to a peculiar type of natural resources represented by sites of special geological significance, rarity or beauty that are representative of a region and of its geological history, events and processes. These sites are also known as "geosites" and, as well as archaeological, architectonic and historical ones, they can be considered as part of the cultural estate of a country. "Geotourism" is an emerging type of sustainable tourism, which concentrates on geosites, furnishing to visitors knowledge, environmental education and amusement. In this meaning, Geotourism may be very useful for geological Sciences divulgation and may furnish additional opportunities for the development of rural areas, generally not included among the main touristic attractions. The collected papers focused on these main topics with different methods and approaches and can be grouped as follows: (i) papers dealing with geosites promotion and valorisation in protected areas; (ii) papers dealing with geosites promotion and valorisation in non-protected areas; (iii) papers dealing with geosites promotion by exhibition, remote sensing analysis and apps; (iv) papers investigating geotourism and geoheritage from the tourists' perspective.

Keywords: geoheritage; geotourism; geosite; geomorphosite; geoitinerary; geological science divulgation

---

1. Editorial for the Special Issue Geoheritage and Geotourism Resources

The Earth, the planet we live on, can be seen as a dynamic mixture of water and rocks forming the landscape we look at and we walk on every day. It is usual to admire landscapes in different parts of the world, which means in areas with a different climate and different geological history, that may inspire a sense of wonder and that let the people to travel for very long distances to be in touch with these amazing portions of the globe. Typical examples are the Yellowstone National Park and the Niagara Falls in the USA, the Iguazu Falls in South America, the fjords in Scandinavia, the Vesuvius in Italy, the Hymalaya in Asia, the Sahara Desert in Africa and the Ayers Rock in Australia. These fascinating rock outcrops and landforms are an example of the so-called Geoheritage [1–3]. They are places (generally named geosites/geomorphosites) where the processes responsible for Earth's dynamic and evolution are so well represented that they may be considered as temples or open air museum to explain the history of our planet. The concept of Geoheritage find its full application in the context of protected areas and of Geoparks in particular, but there many geosites/geomorphosites widespread all over the world and they are not always included in protected areas. Rocks and landforms may be used to divulge the knowledge of Earth's history to a wide audience and not only to experts, thus acting as a resource not only from an economic point of view [4]. To use the rocks and the landforms as a resource, their scientific, educational and touristic values, among all, must be defined thus enhancing their role as geosites and geomorphosites [5–10]. Once the role of either a geosite or a geomorphosite as a potential geotouristic attraction is established, a Geotourism plan may be diffused.

The Geotourism [11] is a double-faced concept because the suffix Geo may be interpreted either as Geological or Geographical [12]. A tentative to pass over the dualism between the Geological and the Geographical approach to Geotourism has been proposed by the Arouca Declaration [13]. The Geological and the Geographical approaches to Geotourism may be seen as two working ideas that run parallel each other, such as they are never meeting each other in analysis. In contrast, they are complementary, and their combination may provide a comprehensive analysis of the Geotourism potential of some areas, showing what nature offers to humans and how humans are able to manage nature in a sustainable way.

2. The Special Issue

The main aim of this Special Issue is to outline the role of Geoheritage and Geotourism as potential tools for touristic development and educational activity, especially for geological sciences divulgation. In this Special Issue we have collected 14 papers dealing with Geoheritage and Geotourism in Europe (moreover in Italy, seven papers; Poland, three papers; Malta, one paper; Iceland, one paper; and Czech Republic, one paper) and South America (in Ecuador, one paper). Both the Geological and Geographical approaches to Geotourism are present in the papers published in the Special Issue (Table 1), with study areas that comprehend both portions of Geoparks and areas not included in a Geopark.

Table 1. List of papers published in the Special Issue grouped according to the type of approach, the main aim and the contents.

Paper	Approach to Geotourism	Main Aims/Geographic Context	Contents
- Valente et al. [14]; - Perotti et al. [15]; - Santangelo et al. [16];	Geological	- Promotion of Geoheritage and Geotourism - Educational purposes - Areas already included in Natural park or Geopark	- Proposal of a geo-itinerary and 3D models (Valente et al. [14]); - Proposal of fieldtrips and virtual tours (Perotti et al. [15]); - Proposal of a geoitinerary with educational purposes (Santangelo et al. [16]);
- Piacentini et al. [17]; - Sansò, [18]; - Selmi et al. [19]; - Kubalíková, [20]; - Carrión-Mero et al. [21]	Geological	- Promotion of Geoheritage and Geotourism - Areas not included in a Geopark	- Proposal of a Geo-archaeological itinerary (Piacentini et al., [17]); - Recognition, selection and Promotion of a single Geomorphosite (Sansò, [18]); - Recognition, selection and quantitative assessment of geosites (Selmi et al., [19]); - Assessment of tourist's potential of geomorphosites and extended SWOT analysis (Kubalíková, [20]) - Qualitative and quantitative assessment of a potential geosite (Carrión-Mero et al. [21])
- Melelli, [22]; - Pasquarè-Mariotto et al. [23]; - Filocamo et al. [24]	Geological	- Promotion of Geoheritage and Geotourism by exhibition, remote sensing analysis and apps - Areas not included in a Geopark	- Proposal of a multimedia exhibition (Melelli, [22]); - Field photos, UAV images and 3D models (Pasquarè-Mariotto et al. [23]); - Creation of an app for electronic devices (Filocamo et al. [24])
- Rozenkiewicz et al. [25]; - Widawski et al. [26]; - Zgłobicki et al. [27];	Geographical	- Investigate Geotourism and Geoheritage from the tourists' perspective	- Online information availability on georesources (Rozenkiewicz et al. [25]); - Opinion of weekend tourists (Widawski et al. [26]) - Geosite assessments and questionnaire to tourists (Zgłobicki et al. [27]);

The general approach includes geosites/geomorphosites identification and evaluation by means of several methodologies [5–10]. This step is generally followed by the proposition of geo-itineraries that, in the case of areas already included in protected areas [14–16,26], aims at implementing the valorizations of the sites and at focusing on their potential didactic/touristic values. In particular, Perotti et al. [15] explained how different methods and approaches (fieldtrips and virtual tours together with more traditional ones like explicative panels and in situ museum) helped in outlining the geodiversity of the Valsesia Geopark. Widawsky et al. [26] instead analyzed the tourist potential in terms of Geotourism of the Gorczański National Park in Poland. They focused on tourist trails and didactic routes already existing in the park and investigated the tourists' opinion on the geotourist attractiveness of the Park. Given the peculiar park location near urbanized areas, the Authors suggested it as an attractive field for research on weekend tourism development.

Most of the other papers deals with territories that are not included in protected areas [17–21,27] presenting studies that aim at promoting Geotourism and/or at assess the Geotourism potential of these areas.

Selmi et al. [19] presented a quantitative assessment of geosites and geomorphosites in the island of Malta, in central Mediterranean Sea, showing the links existing between Geoheritage and other values of the sites (natural and cultural). One of the aim of the paper is collecting data useful for the local authorities in promoting a sustainable and responsible tourism development.

Kubalikova [20] carried out the assessment of the Geotourism potential of two areas in Southern Moravia located outside of protected areas and not far from a big city. This paper stresses how these areas can be considered as important resource for local and regional development.

Piacentini et al. [17] focused on a mountain area of the Apennines (central Italy) combining different types of cultural tourism, such as the archaeological and the geological ones. They proposed an integrated geological–archeological itinerary called the "Fan of the Terre Peligne", with the aim of enhancing the natural and cultural heritage of a poorly visited area of the inner Apennines.

There are two papers dealing with the characterization and valorization of one geosite [18,21]. The case of Monte del Diavolo [18] in Southern Italy is a proposal of valorizations of a site with archaeological, cultural and natural value that has been unexploited, notwithstanding the remarkable touristic flow along the nearby coastline. The case of El Sexmo gold mine [21] in Ecuador aims at highlighting the mining identity of Ecuador and to promote this site within the "Ruta del Oro" (Gold Route) Geopark project. The Authors demonstrated that the valorisation of this site will boost tourism, knowledge, and economic activity, favoring the sustainable development of Geotourism. Zgłobicki et al. [27] analysed the potential for Geotourism development in the Central Roztocze region (SE Poland) with the aim of demonstrating the importance of including this area in the proposed Geopark "Stone Forest in Roztocze". Collecting both data on geosites and tourists' opinions, the Authors demonstrate a high potential for Geotourism and consistency between scientific assessments and ratings from tourists. Despite that, the idea of geopark and Geotourism development is not supported by the state yet, whether institutionally or financially.

The paper from Rozenkiewicz et al. [25] faces the problem of Geotourism from a different point of view with respect to all the other papers. It considers the online information availability on georesources, presented on the official websites of the National Tourism Organizations (NTOs) of the Czech Republic, Poland and Slovakia. Unfortunately, the Authors found out that the information on Geotourism resources available online is rather dispersed and outlined the need of performing coordinate actions to solve this problem.

We finally want to highlight some papers [22–24] that introduce new, original and interesting approaches in the valorization of geoheritage at different levels. These papers deal with the utility of new visualization techniques as well as of devoted mobile applications in the promotion and popularization of geosites. Pasquarè Mariotto et al. [23] choose an iconic place for the geology of our planet, Iceland, to show how the main volcano–tectonic geosites may be explained and illustrated through 3D models derived from field photographs and unmanned aerial vehicle (UAV)-captured images. These models are also an example of interactive, navigable Virtual Outcrops, which are also available online. This kind of approach is surely interesting and novel and may serve as an example on how to promote Geoheritage online to reach a wide audience, with special reference to the younger generation. It could be a suggestion for all the promoters of protected areas to add to their websites' virtual reconstructions of single geosites/geomorphosites or virtual tours to catch the curiosity of a wider audience and increase the number of tourists.

Filocamo et al. [24] have implemented a mobile devise application for the Molise region, in central Italy, with the aim of helping the promotion of sustainable and eco-friendly tourism development in rural and inner areas of the Apennine. This application, called MoGeo App, aims at providing diversified Geotourism information that combines geologic attractions (geosites and geologic itineraries) with other possible tourist attractions (other sites of natural and cultural interest), to reach an audience as wide as possible and to respond to different interests and needs.

The paper of Melelli [22] is a unique and interesting example of Geotourism developed in an urban area, which is the city of Perugia (Umbria, central Italy). Panels, interactive tools, laboratories inside a museum and trekking tours outside allowed the tourists to come close to Geological science as well as favored the increase of visitors in a few months, with important consequences in terms of visibility and financial return. This paper, together with those of Valente et al. [14] and Santangelo et al. [16] belong to the group of papers dedicated to explaining the educational role of Geotourism in the divulgation of Geological sciences. These studies outline the idea that the improvement of geological knowledge is a basic resource for the social and economic development of any community. Instead, increasing the sensitivity of citizens regarding the environmental estate and in the awareness of natural disasters, are fundamental steps for territorial planning policies, aiming at controlling and mitigating the risks associated with natural disasters.

Author Contributions: All authors contributed equally to writing this editorial. All authors have read and agreed to the published version of the manuscript.

Funding: This research received no external funding.

Acknowledgments: We thank our informants and supporting staff for their contribution to the Special Issue.

Conflicts of Interest: The authors declare no conflict of interest.

References

1. Newsome, D.; Dowling, R. Geoheritage and geotourism. In *Geoheritage. Assessment, Protection, and Management*; Reynard, E., Brilha, J., Eds.; Elsevier: Amsterdam, The Netherlands, 2018; pp. 305–321. [CrossRef]
2. Reynard, E.; Brilha, J. Geoheritage: A Multidisciplinary and Applied Research Topic. In *Geoheritage. Assessment, Protection, and Management*; Reynard, E., Brilha, J., Eds.; Elsevier: Amsterdam, The Netherlands, 2018; pp. 3–9. [CrossRef]
3. Brilha, J. Geoheritage and geoparks. In *Geoheritage. Assessment, Protection, and Management*; Reynard, E., Brilha, J., Eds.; Elsevier: Amsterdam, The Netherlands, 2018; pp. 323–334. [CrossRef]
4. Bruschi, V.M.; Coratza, P. Geoheritage and environmental impact assessment (eia). In *Geoheritage. Assessment, Protection, and Management*; Reynard, E., Brilha, J., Eds.; Elsevier: Amsterdam, The Netherlands, 2018; pp. 251–264. [CrossRef]
5. Pereira, P.; Pereira, D.; Caetano Alves, M.I. Geomorphosite assessment in Montesinho Natural Park (Portugal). *Geogr. Helv.* **2007**, *62*, 159–168. [CrossRef]
6. Reynard, E.; Fontana, G.; Kozlik, L.; Scapozza, C. A method for assessing the scientific and additional values of geomorphosites. *Geogr. Helv.* **2007**, *62*, 148–158. [CrossRef]
7. Vujičić, M.D.; Vasiljević, D.A.; Marković, S.B.; Hose, T.A.; Lukić, T.; Hadžić, O.; Janićević, S. Preliminary geosite assessment model (GAM) and its application on Fruška Gora Mountain, potential Geotourism destination of Serbia. *Acta Geogr. Slov.* **2011**, *51*, 361–376. [CrossRef]
8. Kubalíková, L. Geomorphosite assessment for geotourism purposes. *Czech J. Tour.* **2013**, *2*, 80–104. [CrossRef]
9. Brilha, J. Inventory and Quantitative Assessment of Geosites and Geodiversity Sites: A Review. *Geoheritage* **2016**, *8*, 119–134. [CrossRef]
10. Mucivuna, V.C.; Reynard, E.; Garcia, M.D.G.M. Geomorphosites Assessment Methods: Comparative Analysis and Typology. *Geoheritage* **2019**, *11*, 1799–1815. [CrossRef]
11. Ólafsdóttir, R.; Tverijonaite, E. Geotourism: A Systematic Literature Review. *Geosciences* **2018**, *8*, 234. [CrossRef]
12. Dowling, R.; Newsome, D. Geotourism: Definition, characteristics and international perspectives. In *Handbook of Geotourism*; Dowling, R., Newsome, D., Eds.; Edward Elgar: Cheltenham, UK, 2018; pp. 1–22.

13. The Arouca Declaration. Available online: http://www.unesco.org/new/fileadmin/MULTIMEDIA/HQ/SC/pdf/Geopark_Arouca_Declaration_EGN_2012.pdf (accessed on 12 June 2020).

14. Valente, E.; Santo, A.; Guida, D.; Santangelo, N. Geotourism in the Cilento, Vallo di Diano and Alburni UNESCO Global Geopark (Southern Italy): The Middle Bussento Karst System. *Resources* **2020**, *9*, 52. [CrossRef]

15. Perotti, L.; Bollati, I.M.; Viani, C.; Zanoletti, E.; Caironi, V.; Pelfini, M.; Giardino, M. Fieldtrips and Virtual Tours as Geotourism Resources: Examples from the Sesia Val Grande UNESCO Global Geopark (NW Italy). *Resources* **2020**, *9*, 63. [CrossRef]

16. Santangelo, N.; Amato, V.; Ascione, A.; Russo Ermolli, E.; Valente, E. GEOTOURISM as a Tool for Learning: A Geoitinerary in the Cilento, Vallo di Diano and Alburni Geopark (Southern Italy). *Resources* **2020**, *9*, 67. [CrossRef]

17. Piacentini, T.; Somma, M.C.; Antonelli, S.; Buccolini, M.; Esposito, G.; Mancinelli, V.; Miccadei, E. The "Fan of the Terre Peligne": Integrated Enhancement and Valorization of the Archeological and Geological Heritage of an Inner-Mountain Area (Abruzzo, Central Apennines, Italy). *Resources* **2019**, *8*, 118. [CrossRef]

18. Sansò, P. Devil Landforms as Resources for Geotourism Development: An Example from Southern Apulia (Italy). *Resources* **2019**, *8*, 131. [CrossRef]

19. Selmi, L.; Coratza, P.; Gauci, R.; Soldati, M. Geoheritage as a Tool for Environmental Management: A Case Study in Northern Malta (Central Mediterranean Sea). *Resources* **2019**, *8*, 168. [CrossRef]

20. Kubalíková, L. Assessing Geotourism Resources on a Local Level: A Case Study from Southern Moravia (Czech Republic). *Resources* **2019**, *8*, 150. [CrossRef]

21. Carrión-Mero, P.; Loor-Oporto, O.; Andrade-Ríos, H.; Herrera-Franco, G.; Morante-Carballo, F.; Jaya-Montalvo, M.; Aguilar-Aguilar, M.; Torres-Peña, K.; Berrezueta, E. Quantitative and Qualitative Assessment of the "El Sexmo" Tourist Gold Mine (Zaruma, Ecuador) as A Geosite and Mining Site. *Resources* **2020**, *9*, 28. [CrossRef]

22. Melelli, L. "Perugia Upside-Down": A Multimedia Exhibition in Umbria (Central Italy) for Improving Geoheritage and Geotourism in Urban Areas. *Resources* **2019**, *8*, 148. [CrossRef]

23. Pasquaré Mariotto, F.; Bonali, F.L.; Venturini, C. Iceland, an Open-Air Museum for Geoheritage and Earth Science Communication Purposes. *Resources* **2020**, *9*, 14. [CrossRef]

24. Filocamo, F.; Di Paola, G.; Mastrobuono, L.; Rosskopf, C.M. MoGeo, a Mobile Application to Promote Geotourism in Molise Region (Southern Italy). *Resources* **2020**, *9*, 31. [CrossRef]

25. Rozenkiewicz, A.; Widawski, K.; Jary, Z. Geotourism and the 21st Century–NTOs' Website Information Availability on Geotourism Resources in Selected Central European Countries: International Perspective. *Resources* **2020**, *9*, 4. [CrossRef]

26. Widawski, K.; Oleśniewicz, P.; Rozenkiewicz, A.; Zaręba, A.; Jandová, S. Protected Areas. Geotourist Attractiveness for Weekend Tourists Based on the Example of Gorczański National Park in Poland. *Resources* **2020**, *9*, 35. [CrossRef]

27. Zgłobicki, W.; Kukiełka, S.; Baran-Zgłobicka, B. Regional Geotourist Resources—Assessment and Management (A Case Study in SE Poland). *Resources* **2020**, *9*, 18. [CrossRef]

Article

Geotourism in the Cilento, Vallo di Diano and Alburni UNESCO Global Geopark (Southern Italy): The Middle Bussento Karst System

Ettore Valente [1,*], Antonio Santo [2], Domenico Guida [3] and Nicoletta Santangelo [1]

[1] Department of Earth, Environmental and Resource Science, DISTAR, University of Naples Federico II, via Cinthia 21, 80126 Naples, Italy; nicsanta@unina.it

[2] Dipartimento di Ingegneria Civile, Edile e Ambientale, DICEA, University of Naples Federico II, piazzale Tecchio 80, 80125 Naples, Italy; antonio.santo@unina.it

[3] Department of Civil Engineering and Consorzio inter-Universitario per la previsione e prevenzione dei Grandi Rischi (CUGRI), University of Salerno, Campus of Fisciano, Via Giovanni Paolo II, 132–84084 Fisciano, Italy; dguida@unisa.it

* Correspondence: ettore.valente@unina.it

Received: 27 March 2020; Accepted: 17 April 2020; Published: 26 April 2020

Abstract: In this paper we want to stress the role of geotourism as a means to promote environmental education and, on occasion, as a way to increase the touristic interest of an area. Geoparks are certainly the territory where geotourism can be best exploited. We propose a geoitinerary to discover the amazing, but poorly known, Middle Bussento Karst System, with the blind valley of the Bussento River, in the southeast of the Cilento, Vallo di Diano and Alburni United Nations Educational, Scientific and Cultural Organization (UNESCO) Global Geopark. This is the only example, in Southern Italy, of a stream sinking underground and it is the second longest subsurface river path in Italy, making this a core area of the Geopark. We combined field surveys and literature data to create a geoitinerary that can be useful in helping to promote this site. This geoitinerary is applicable to both simple generic visitors and geo-tourists and has an educational purpose, especially in explaining the significance and the fragility of karst areas in terms of environmental protection. Moreover, it may represent a sort of stimulus for the growth of touristic activity in this inner area of the Geopark.

Keywords: geotourism; geomorphosites; environmental education; Cilento, Vallo di Diano and Alburni Geopark; Middle Bussento Karst System

---

1. Introduction

1.1. Geoparks and Geotourism: An Overview

In recent decades, the growing awareness to protect nature from human impact has led to the diffusion of several legislative and social actions, both at international and national levels. The protection of nature is, for example, clearly expressed in article no. 9 of the Italian Constitution [1], from December 1947. In Italy, one of the first laws to address the protection of nature is the Galasso Law (law no. 431, 22 August 1985). This law introduced the concept of "protected areas" in Italian territory, namely seas, rivers, mountain areas from 1200 m above sea level (m a.s.l.), volcanoes, forests, glaciers, national parks, and archaeological areas.

At the international level, in 1972, the United Nations Educational, Scientific and Cultural Organization (UNESCO) adopted the convention "Concerning the Protection of the World Cultural and Natural Heritage". This convention identified 1500 sites all over the world to be included in the World Heritage List, due to outstanding values of their natural and cultural features ([2,3] and references therein). Currently, the sites included in the World Heritage List are mainly cultural sites (869 of the 1121 sites), whereas a lower number of natural sites with a high geological and geomorphological scientific and scenic value are included in the list (213 of the 1121 sites; [4]). To overpass this low number of natural sites included in the World Heritage List and considering the large number of sites in the world with high geo-scientific significance, in 1997 the Division of Earth Science at UNESCO proposed a new program called the "UNESCO's Geoparks Programme" [5,6]. Prior to the creation of this program, the idea of establishing a Geoparks network was settled for the first time during the 30th International Geological Congress in Beijing in 1996 [6]. A Geopark is defined as "a territory with well-defined limits that has a large enough surface area for it to serve local economic development. The Geopark comprises a number of geological-paleontological heritage sites of special scientific importance, rarity or beauty; it may not be solely of geological-paleontological significance but also of archaeological, ecological, historical or cultural value" [2]. The first 25 Geoparks were established in Europe and China, and in 2004 they formed the UNESCO Global Geoparks Network [2,6]. It is worth mentioning that the first protected area in the world was established at Yellowstone National Park in North America in 1872 [7]. Currently, only three Canadian national parks have gained the title of 'Geoparks' in North America [8]. UNESCO Global Geoparks are defined as "single, unified geographical areas where sites and landscapes of international geological significance are managed with a holistic concept of protection, education and sustainable development" [9]. UNESCO Global Geoparks focus their activities on raising awareness in the local community about the geological heritage of the area, promoting the concept that the landscape is a dynamic element, stimulating sustainable Geotourism, and encouraging the protection of geological resources [9].

The definition of Geopark includes a clear reference to local economic development, so geological and geomorphological features within a Geopark must be considered as crucial geological resources ([10] and references therein) forming the so-called geoheritage [11–13]. Growth of the local economy must be practiced through sustainable management strategies that seek to develop geotourism by attracting an increasing number of visitors. The concept of geotourism has widely diffused in the last decades [14–19] among international papers covering many countries throughout the world ([20] and references therein). Moreover, Dowling and Newsome [21] highlighted that geotourism can be defined from both a geological and a geographical point of view. The geological definition of geotourism was first proposed by Hose [14,22] who defined it as "the provision of interpretive and service facilities to enable tourists to acquire knowledge and understanding of the geology and geomorphology of a site beyond the level of a mere aesthetic appreciation". This definition has since been refined by the same author [23,24]. In 2006 another definition was proposed by Dowling and Newsome [16] who introduced the concept of scale, suggesting that geotourism focuses on both small geological and paleontological sites and large landforms and landscapes. Newsome and Dowling [17] pointed out that geotourism is a form of tourism focused on geology and landscape that can be carried out either by independent visits or guided tours. Hose [15] suggested that geotourism is underpinned by the so-called 3G's, namely geoconservation, geohistory, and geo-interpretation. In contrast, in 2000, the National Geographic Society of the United States of America defined geotourism as "tourism that sustains or enhances the geographical character of a place-its environment, culture, aesthetics, heritage, and the well-being of its residents" ([21] and references therein). In 2006 a new definition was set by Pralong [25] that highlighted the emotional aspects of geotourism and introduced the concept of geomarketing, thus placing geotourism as a component of the regional economy. Subsequently, in 2011, during the International Congress on Geotourism at Arouca (Portugal) a new definition was proposed that included the term "geology" in the geographical definition of geotourism [26]. These two points of view on what is geotourism suggest considering it either as a "type or form" of tourism (geological definition) or as an "approach" to tourism (geographical definition) [21]. In this paper we adopt the geological definition of geotourism.

To discuss geoheritage and geotourism it is necessary to recognize geosites and geomorphosites. The former are "portions of the geosphere that present a particular importance for the comprehension and reconstruction of the history of the Earth, climate and life" [27]. The latter are "geomorphological landforms that have acquired a scientific, cultural/historical, aesthetic and/or social/economic value due to human perception or exploitation" [28]. Moreover, the assessment methods of geomorphosites and their importance in geotourism and geoheritage have been widely discussed in international papers [29–45].

The concept of geotourism has been widely diffused in Italy. With its ten UNESCO Geoparks, is the third country in the world with the highest number of Geoparks, being preceded only by China and Spain [8]. Addressing Geotourism and Earth Science education in Italy, numerous scientific papers have been produced which deal with geosites and geomorphosites inventory in many areas of the national territory [46–53]. Most of these papers focus on Central and Northern Italy, and less international papers have been published which deal with geosites and geotourism in Southern Italy [54–58]. In particular, few international papers [59] and conference proceedings [60] have addressed Cilento, Vallo di Diano and Alburni Geopark and only some national papers [61–63] deal with geosites inventory and geotourism in the territory of the Geopark.

1.2. The Cilento, Vallo di Diano and Alburni UNESCO Global Geopark

The Cilento, Vallo di Diano and Alburni Geopark (Figure 1) is among the largest Italian Geoparks, covering an area of 181,048 ha that includes 80 municipalities in the province of Salerno, with a population of ~280,000 inhabitants. The Geopark was firstly established as a national park, namely the National Park of Cilento and Vallo di Diano, in 1991 under the law 394/91. In 1997 the national park was included in the Man and the Biosphere Program of UNESCO and it became part of UNESCO's World Heritage List in 1998. Then in 2010 during the 9th European Conference of Geoparks in Lesvos (Greece), it gained the title of Geopark and became part of the European Geopark Network. The Geopark finally gained the title of UNESCO Global Geopark in 2015.

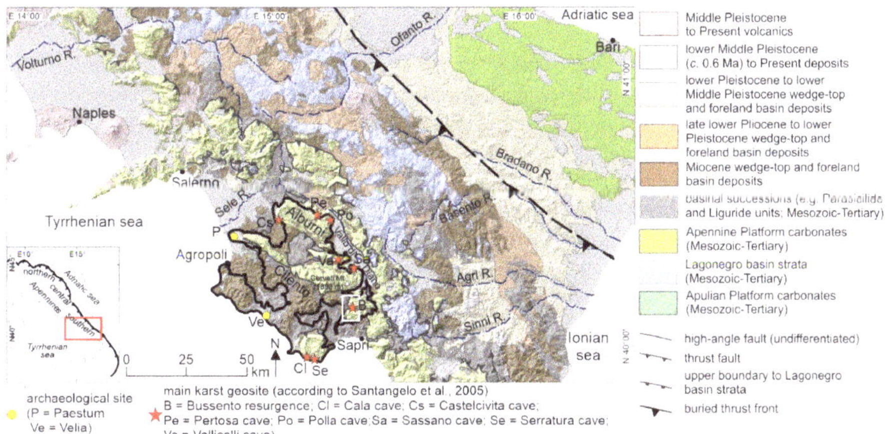

Figure 1. Geological map of the Southern Apennines (modified from [64]). Thick black line indicates the boundary of the Cilento, Vallo di Diano and Alburni Geopark (from [65]). White box in the southest (SE) corner of the Geopark indicates the location of Figure 2.

The Geopark of Cilento, Vallo di Diano and Alburni is located in the southern part of the Campania region. It extends from the coastal areas between the towns of Agropoli, to the north, and Sapri, to the south, up to the mountain ridges of the Southern Apennines chain, towards the east, reaching a maximum elevation of 1898 m above sea level (m a.s.l.) at the peak of Mt. Cervati (Figure 1). The Geopark is characterized by a complex geological setting, with large outcrops of permeable carbonate units in the inner mountainous landscape and less permeable or even impermeable wedge-top and basinal units towards the coast (Figure 1). Its complex geological setting results in a large variety of stratigraphical and geomorphological features, whose scientific importance and beautiful scenery are undisputed [59]. While the coastal area with its beautiful beaches and the well-known archeological sites of Paestum and Velia (Figure 1) is a continuously growing touristic attraction, with thousands of visitors per year, the inner portion of the park is less visited and there are few initiatives promoted to invert this trend.

Inventory, evaluation, and protection of the wide geological estate of the Geopark have been carried out by the competent authorities and are listed in the official catalogue promoted by national [66] and regional institutions [67], as well as by the Geopark itself. Moreover, Santangelo et al. [59] identified a total of 263 sites of geological interest, 32% of which are geomorphosites.

In this paper, we investigate a poorly known portion of the Geopark that is placed along its southeastern border. This portion of the Geopark is mainly made up of Mesozoic carbonate platform successions (Figure 1) which are strongly karstified and host some of the most important water reservoirs of the region with springs that have a total discharge of more than 20 m^3/s [68,69]. Karst processes are so well represented in the area both at the surface (doline, polje, gorges, ponors, karst springs) and underground (horizontal and vertical cave/karst systems [70]). Many of these karst landforms are listed in the official catalogue of the Geopark [71] as geomorphosites according to different perspectives: some cave, for instance, preserve important archaeological records [72], while others represent the longest or the deepest karst systems of the area; meanwhile others are exemplary or have a particular didactic value [59].

Unfortunately, up until now, few initiatives have been carried out to promote this significant geological estate. The foundation of the Musei Integrati dell'Ambiente (MIDA) museum [73], which is part of the touristic Pertosa cave management system (see Figure 1 for location), has, among others, a sector dedicated to the explanation of karst processes. At the same time, the proposition of geotouristic itineraries in this portion of the Geopark is still at an early stage with only some papers addressed to both Italian (itineraries n. 11, 12, and 13 in [61–63]) and international tourists [54]. Moreover, Aloia and colleagues [61] briefly discuss the importance of correct management strategies of the Geopark as a tool to direct touristic flows from the coastal area to the inner, hilly and mountainous areas.

Our study aims to focus attention on the karst geomorphosite of the middle Bussento river system. This site is the only example of stream sinking underground in Southern Italy [59] and is the second longest underground river in Italy, being preceded only by the Timavo River in North-East Italy [74]. This geomorphosite includes a system of ponors, the largest of which is the La Rupe ponor, where the Bussento River sinks, and the Bussento Resurgence, where the Bussento River reemerges after a ~4 km long subsurface path. The area is already the object of some, local scale, touristic promotion activities such as the foundation of the Museo Virtuale (MU.VI.), a virtual museum conceived as an educational and scientific center, managed by the Caselle in Pittari administration, where teaching materials are organized for presentation by means of visual technologies (multi-touch screen, 3D room for virtual reconstruction). Yet despite the high scientific and educational values of this portion of the Bussento river valley, an adequate, comprehensive geotourism policy has not been assessed by the local administration.

In this context, we attempt to contribute to an increase in the knowledge of this fascinating portion of the Geopark by the promotion of a comprehensive geoitinerary, which should serve both as a scientific and educational instrument at inter-municipality scale. The main aims of this work are: (i) contributing to explain how karst processes can make a river disappear underground; (ii) discussing the importance of karst aquifer and the main environmental implications connected to the communication between surface and underground waters; (iii) increasing curiosity about this site, helping to promote the integrate management of this inner portion of the Geopark as a touristic attraction and thus helping to grow the local and the district economy.

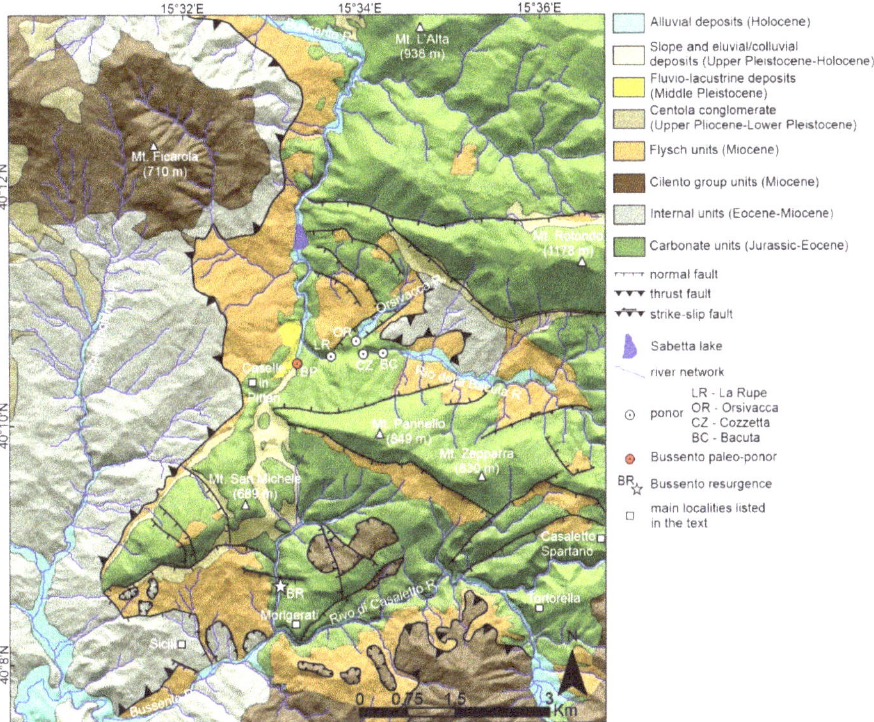

Figure 2. Geological map of the Middle Bussento Karst System (modified from [75]).

2. Study Area: The Middle Bussento Karst System

The study area is characterized by the occurrence, in a very restricted area, of Jurassic to Eocene carbonate units, of Eocene to Miocene internal units, and of Miocene flysch and Cilento units (Figure 2). Carbonate units consist of inner platform limestone with high fossils content (rudists and gasteropods) passing upward to open shelf limestone with interlayered marls and clays. Internal units consist of clays with interbedded quartz-rich and feldspar-rich sandstones passing upwards through calcilutite, calcarenite, and calcirudite with cherty lists and nodules and with interbedded quartz-arenite and marls strata. Miocene flysch units consist of a fining-upward sequence of thrust-top deposits whereas the Miocene Cilento group consists of a coarsening upward sequence of a wedge-top basin [75].

The actual structural setting of the area derives from its complex tectonic evolution with internal units thrusted over the carbonate units that were covered by the flysch units. The tectonic setting is completed by the occurrence of E-W, NW-SE, and NW-SE normal faults that articulate the topography. Moreover, flysch units have been eroded from the high carbonate peaks and are preserved only along the main valleys [75]. The complex tectonic evolution of the area is discussed in other studies [76–82].

The main river flowing in the study area is the Bussento River (Figure 2). This river originates from Mt. Cervati and Mt. Ficarola, following a mainly NW-SE orientation in the upper portion of its basin where it carves impermeable units. At the base of Mt. L'Alta (Figure 2) the Bussento River flows along a N-S trend and starts carving into the Mesozoic carbonate units. Then, it suddenly disappears near the village of Caselle in Pittari in the so called La Rupe ponor to flow out again ~4 km to the south, in the so-called Bussento Resurgence, near the village of Morigerati (Figure 2). To the east of the middle Bussento River there are two more rivers—Orsivacca River and Rio della Bacuta River—that, after flowing in the impermeable flysch and internal units, suddenly disappear underground when they carve the carbonate units. Moreover, the Orsivacca River ends in the Orsivacca ponor whereas the Rio della Bacuta River has both a fossil and an active ponor, respectively named the Cozzetta and Bacuta ponors (Figure 2). D'Elia et al. [83] suggested that the Orsivacca and the Rio della Bacuta rivers were left tributaries of the Bussento River and that they were separated from it following the ponor regression mechanism [84,85]. The separation of the Orsivacca and the Rio della Bacuta rivers from the Bussento River is also testified by a paleo-valley placed in-between the relative ponors. When these three river segments were connected, the Bussento River should have had a higher discharge that could justify the presence of a large fossil ponor, named the Bussento paleo-ponor (Figure 2), which is placed to the southwest of the active La Rupe ponor [83].

It is worth noting that the Bussento River is intercepted by the Sabetta reservoir, an artificial basin created in 1958, which serves a hydroelectrical powerplant. Therefore, the Bussento River discharge downstream of the Sabetta reservoir depends on the amount of water released for operational activities and the amount of water coming from the residual drainage basin between the Sabetta Lake and the La Rupe ponor [86].

Regarding the hydrogeological setting of the area, the main aquifer is part of the carbonate structure of the Salice-Coccovello Mts, which drains towards the basal spring systems of Morigerati, located at about 90 m a.s.l., which has a total discharge of 1.5 m^3/s. Hydrogeological analysis has proved a subsurface connection between the La Rupe, Orsivacca, Cozzetta, and Bacuta ponors with the Morigerati springs, that are all part of the so-called Middle Bussento Karst System [87].

Recent scientific interdisciplinary research has studied the Middle Bussento Karst System as one of the most interesting karst systems in Southern Italy, spanning from the singular karst back-flooding to the karst pulse floods, suggesting it as experimental UNESCO karst basin [86] for geodiversity conservation, protection, and promotion.

3. Materials and Methods

As we already discussed in Section 1, Geoparks are intended to not only protect the geoheritage of an area but also to promote a holistic concept of education and sustainable development. Within a Geopark, geotourism represents the most useful tool to enable tourists to acquire knowledge and understanding of the geology and geomorphology of a site. With these concepts in mind, we focused our attention on the inner area of the Cilento, Alburni and Vallo di Diano Geopark, and chose one of the geosites already listed in the official catalogue by the competent authorities. We used this site as an example of how geotourism may be a useful tool to increase the touristic attraction of an area and promote environmental education.

The inner part of the Geopark is mainly made of carbonate rocks and for this reason our choice fell among karstic geomorphosites. We selected the Middle Bussento Karst System because of its uniqueness as the sole example of an underground river in Southern Italy. Moreover, it also has a high didactic value because it is representative of the extreme significance and sensitivity of the karst environment as water reservoirs.

We collected and revised all literature data about the geological, geomorphological, hydrogeological, and speleological setting of the Middle Bussento Karst System and planned the geoitinerary.

Field surveys (5 days in total) were carried out to detail the technical issues of the geoitinerary. It is 17 km long and can be done entirely by walking or it is possible to move by car from one stop to the next. For those who prefer the latter option, we determined the path length for each stop. It indicates the distance from the place where tourists can leave the car until they reach the ponor. In addition, the duration (in hours) of both the entire geoitinerary and every single path is indicated, together with the differences in elevation of every path and the main geological and geomorphological features that can be admired along each path.

To emphasize the scientific and educational importance of the proposed itinerary we prepared some sketches to provide tourists with a complete overview of both the surface and subsurface setting of the Middle Bussento Karst System. These sketches include a 10 m digital terrain model (DTM) of the investigated area and a 3D view of the topography (both from the south and from the east). The 10 m DTM was created from elevation data (both contour lines and elevation points) derived from a detailed scale topographic map (Technical Map of the Campania Region, at scale 1:5000). Elevation data were imported in a Geographic Information System (GIS) software (ArcGis 10.7©, Redlands, CA, USA) and interpolated by means of the Topo to Raster tool to obtain the DTM that, successively, was used to derive the hillshade map (by means of the 3D Analyst tool in ArcGis) and the river network (by means of the Hydrology tool in ArcGis). Both the hillshade map and the river network were then imported in the ArcScene module of ArcGis to obtain two 3D views of the topography that focused on the area between the Sabetta Lake, to the north, and the Bussento Resurgence, to the south. These two 3D views, respectively from the south and from the west, were then imported in Corel Draw© (Ottawa, MI, Canada) to produce the final sketch.

To analyze the potential of the proposed geoitinerary, we carried out a classical SWOT (strengths, weaknesses, opportunities and threats) analysis. Finally, to assess the geotouristic value of the Middle Bussento Karst System geomorphosite, we calculated the index proposed by Pica et al. [88] named the value of a site for geotourism (VSG). It results from the following equation:

$$VSG = RP + RR + SCE + SAC + AC \tag{1}$$

RP is the representativeness index, RR is the rarity index, SCE is the scenic-aesthetic value, SAC is the historical-archaeological-cultural value, and AC is the accessibility index. Each index has a maximum score of 5 so that the highest value of the VSG is 25. The scores derive from the geosites' characteristics reported in Table 1. According to Pica et al. [88], VSG values lower than 8 indicate that the site has low touristic potential, VSG values between 9 and 16 indicate a medium touristic potential, and values between 17 and 25 indicate a high touristic potential.

Table 1. Scores of the indexes used for the evaluation of the value of a site for geotourism (VSG) index (from Pica et al. [88]). RP = representativeness index; RR = rarity index; SCE = scenic-aesthetic value; SAC = historical-archaeological-cultural value; AC = accessibility index.

Value of a Site for Geotourism		
VSG = RP + RR + SCE + SAC + AC, VSG max = 25		
Attributes	**Values**	
Representativeness RP	**0, 1, 3, 5**	
Ideal model correspondence	5, 3, 3, 1, 0	
Peculiarity (lithostratigraphy, carsism, hydrology, paleontology, geomorphology, structural geology, mineralogy)	5, 3, 3, 1, 0	
Typicality	5, 3, 3, 1, 0	
Interest peculiarity	5, 3, 3, 1, 0	
Rareness RR	**0, 1, 3, 5**	
Geographical range	local, regional, national, international	Two-way table
Frequency	5, 4, 3, 1, 0	

Table 1. *Cont.*

Value of a Site for Geotourism		
VSG = RP + RR + SCE + SAC + AC, VSG max = 25		
Attributes	**Values**	
Scenic-aesthetic SCE	**0, 1, 3, 5**	
Viewpoints	5, 3, 1, 0	
Cromatic contrast	5, 3, 1, 0	summarize intervals
Landform queerness	5, 0	
Historical-archeological-cultural SAC	**0, 1, 3, 5**	
National restriction	3, 5 (area, geosite)	
Regional/local restriction	1, 3 (area, geosite)	summarize intervals
Protected area	3, 5, 1	
Other (archeological, monumental, architectural values; legends, stories, tradition; toponym)	2, 2, 1	
Accessibility AC	**0, 1, 3, 5**	
Ways to approach the site	5, 3, 1	
Difficulty to approach the site	5, 4, 3, 1	summarize intervals
Services	5, 4, 3, 1, 0	

4. Results

4.1. Karst Landforms (Ponors, Blind Valleys, Resurgences, and Karst Springs): Importance for Environmental Education

Karst processes occur on Earth's surface in any place where soluble rocks like limestones or gypsum crop out and are exposed to the action of meteoric waters. The latter have a natural content of CO_2 that increases during percolation in the soil. For this reason, they can dissolve soluble rocks, creating spectacular morphologies both at the surface and underground [89] (Figure 3). Dissolution makes the rocks highly permeable, allowing for the circulation and accumulation of water underground. A mountain made up of permeable and soluble rock, like limestone, behaves as a sponge, adsorbing all the meteoric waters dropping above its surface. For this reason, at depth, a subterranean water body originates, called by geologists as a "water table". The mountain containing this water body is called "aquifer". Karst water aquifers represent the most important water reservoirs on our planet, furnishing a high percentage of the drinkable water feeding our aqueducts [90].

Karst processes are governed by the following simple chemical equation:

$$CaCO_3 + CO_2 + H_2O \leftrightarrow Ca(HCO_3)_2 \tag{2}$$

In this work, we describe some peculiar karst morphologies that originate when a karst system is fed not only by meteoric waters (autogenic karst), but also by runoff waters coming from a non-karst area (allogenic karst; [86]). This phenomenon is possible when limestone successions crop out in association with other less permeable or impermeable rocks. During meteoric events, the running waters collect on impermeable rocks and create a drainage network that flows on the surface until it meets soluble rocks. At this point, water may attack the more fractured rocks by chemical dissolution where it starts to create a concentrate infiltration point, the so-called ponor. Thus, what are ponors? They are simply holes on the Earth's surface where streams disappear underground. They are also called swallow holes or stream-sinks. The stream waters, during geological time, are able to "dig" these holes in the carbonate rocks because they are soluble and susceptible to karst processes. This kind of ponor, located at the contact between permeable and impermeable rocks, is referred to by geologists as a "contact ponor" (Figure 4; [70]). The largest is the upstream valley, and the biggest will be the hole that the water may dig in the carbonate rocks. The drainage basin located upstream of a ponor is called the "blind valley" because it has no continuation on the surface, but it disappears underground. In association with ponors there are always cave systems (Figure 5A) that, depending on the difference in altitude between the ponor and the basal water table, may transfer the surface water towards a resurgence or may directly feed the water table (Figure 5B). This condition makes karst aquifers highly vulnerable regarding possible contamination between surface waters and underground waters [91]. Anything spilled anywhere in the catchment of a blind valley may directly reach an underground water table, even if it is very far from the spilling point.

Figure 3. Examples of epikarst and endokarst morphologies in the Cilento, Vallo di Diano and Alburni Geopark: (**A**) karren; (**B**) doline at the summit of the Mt. Cervati peak; (**C**) carbonate concretion, stalactites, in a cave.

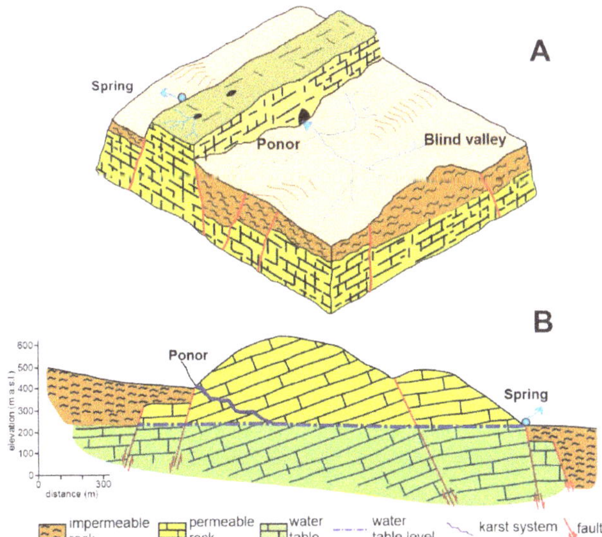

Figure 4. A 3D sketch (**A**) showing the formation of contact ponor and (**B**) cross-section view showing the communication between the stream waters sinking in the ponor and the basal water table.

Figure 5. (**A**,**B**) Internal views of karst systems associated with ponors (Cilento, Vallo di Diano and Alburni Geopark).

In the southern Apennines, the main springs that are intercepted for drinkable use are placed at the base of the carbonate massifs, with single discharge that often exceeds 2000–3000 L/s [68,92]. Most of these water resources are fed by the karst system for which a communication between ponor at the surface and subsurface water table has been proven [93]. Amazing examples of these features are the endoreic karst basin of the Matese Massif and the Picentini Mts. [64,93,94] and the carbonate slopes of the Cilento, Vallo di Diano and Alburni Geopark [70,95]. In some cases, the contamination of the water sinking in the ponor has been proven [95,96]. This is what makes karst areas, where stream sinks and blind valleys are present, extremely sensitive environments, where surface waters may come directly in contact with underground water reservoirs. All human activities insisting on a drainage basin located upstream of a ponor must take into account the possibility of interfering with the basal water table. These simple concepts on how a karst aquifer functions should be basic components in the environmental education of every citizen and administrator. Geotourism offers the possibility of coming directly in contact with these problems, not only by studying but also in such activities as taking a walk. Therefore, we propose a geoitinerary in the area of the middle Bussento River system since it is a good case to understand this karst environment.

4.2. The Karst Ponors and the Blind Valley of the Bussento River: Proposed Geoitinerary

In the middle Bussento river valley karst processes are well developed and mostly represented by contact ponors with associated cave systems and blind valleys. The Bussento River originated from Mt. Cervati and Mt. Figarola (Figures 1 and 2). Its drainage basin is carved on impermeable and soft rocks belonging to Miocene flysch deposits (Cilento group) and to Eocene-Miocene clayey basinal successions (internal units). The river flows for about 10 km in a N-S direction on this soft rock then it reaches the northern slope of the carbonate ridge of Mt. Pannello and abruptly disappears, sinking into a big hole called the La Rupe ponor (Figure 6).

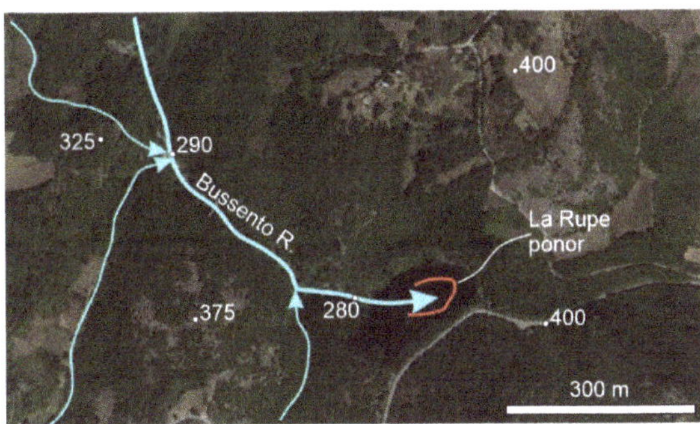

Figure 6. Google Earth image of the La Rupe ponor area. Arrows indicate flow direction. Red line indicates the La Rupe ponor. White points are elevation points.

The proposed geoitinerary is shown in Figure 7, whereas its technical issues are reported in Table 2.

Table 2. Technical issues of the proposed itinerary.

The Middle Bussento Karst System Geoitinerary				
Length	Duration	Number of Stops	How to Move between Stops	Type of Itinerary (According to Italian Excursionist Federation classification, [97])
~17 km	It could be completed in a one-day itinerary (expected duration is 8 h) or split in a two-day itinerary	5	by walking or by car	excursionist
Stop no. 1: La Rupe ponor (3 km far from the urban center of Caselle in Pittari)				
Path Length	Duration	Difference in elevation	Main geological and geomorphological features	
1 km	2 h	180 m	La Rupe ponor; Bussento River blind valley; dissolution pans; karren	
Stop no. 2: Orsivacca ponor (0.8 km far from stop no. 1)				
Path Length	Duration	Difference in elevation	Main geological and geomorphological features	
300 m	1 h	40 m	Orsivacca ponor; Orsivacca River blind valley	
Stop no. 3: Rio della Bacuta valley (1.6 km far from stop no. 2)				
Length	Duration	Difference in elevation	Main geological and geomorphological features	
500 m	1 h	60 m	Cozzetta ponor; Bacuta ponor; Rio della Bacuta River blind valley	
Stop no. 4: MU.VI. (3.2 km far from stop no. 3)				
Duration of the visit		Main geological and geomorphological features		
1 h		Virtual tour of the Middle Bussento Karst System		
Stop no. 5: Bussento Resurgence (6.5 km far from stop no. 5)				
Length	Duration	Difference in elevation	Main geological and geomorphological features	
700 m	2 h	120 m	Bussento Resurgence; Bussento gorge; fossils of rudists; Mulino spring	

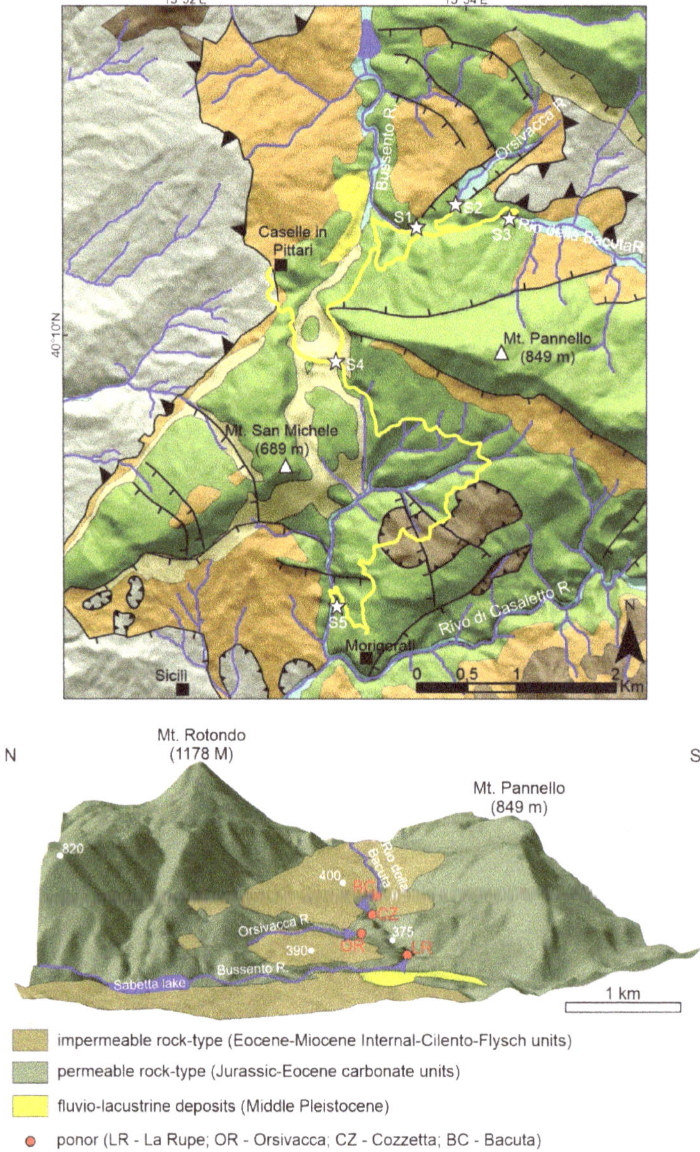

Figure 7. Upper panel: proposed geoitinerary in the discover of the Middle Bussento Karst System (yellow line). Stars indicate geoitinerary stops. S1: La Rupe ponor; S2: Orsivacca ponor; S3: Rio della Bacuta valley with its active (Bacuta) and fossil (Cozzetta) ponors; S4: MU. VI.; S5: Bussento Resurgence. See Figure 2 for legend of the geological units. Lower panel: 3D view, from west, of the area between the Bussento River, the Orsivacca River, and the Rio della Bacuta River. Red points are ponors. Blue lines indicate the river network and arrows indicate the flow direction. White points are elevation points.

The map and the 3D model of Figure 7 show that two little streams occur to the left of the Bussento River, named the Rio della Bacuta River and the Orsivacca River, which disappear underground and may be classified as blind valleys. Their drainage basins upstream of the ponors are carved into soft rocks and the streams disappear when they meet the carbonate rocks.

The starting point of the path that bring tourists to stop no. 1, the La Rupe ponor (Figure 8A), is 3 km far from the urban area of Caselle in Pittari and it is near a gentle surface placed at the foot of the northern slope of Mt. Pannello. During the drop towards the La Rupe ponor, tourists may admire the stratified carbonate bedrock with several small-scale epikarst features (e.g., dissolution pans, karren). Tourists will reach the Bussento River valley bottom after having travelled for 900 m. The water level is very low because it is regulated by the Sabetta Lake hydroelectric powerplant placed about 2.5 km upstream of the Bussento ponor. Before the dam construction in 1958, the river waters entered the ponor with a higher flow rate, carrying out a significant debris load, as indicated by the large (cubic decimeters to cubic meters), rounded to sub-rounded boulders covering the valley floor. During extreme floods, this material obstructed the entrance of the ponor causing the formation of a wide lake. Local inhabitants called this phenomenon "Votamare" or "Ultimare", which means the area looked like the sea. After a 100 m walk along the Bussento River valley floor, tourists will reach the La Rupe ponor. The entrance, which is about 30 m high and 10 m large, is very spectacular and represents the point where the Bussento river starts to flow underground (Figure 8B,C). During geological time the river waters created an underground channel, at least 4 km long, that cut across the Mt. Pannello ridge and resurged at the surface at the foot of its southern slope, near the village of Morigerati, in the so-called Bussento Resurgence (Figure 7). Only tourists with either a speleological background or under the guide of an expert speleologist may visit the initial part of this underground channel, which is explored for only 566 m. It develops mainly following SW-NE and NW-SE trends and ends in a siphon lake.

From the La Rupe ponor, tourists will return to the main road to reach the Orsivacca, Cozzetta, and Bacuta ponors (stops S2 and S3 in Figure 7). The Orsivacca valley floor is dry for most of the year but during autumn and winter seasons it is possible to admire the water falling within the ponor. As shown in Figure 9, the ponor's dimensions are very little in respect to those of "La Rupe", being smaller their catchments. These ponors are in connection with caves and, in these cases, only persons with a good speleological background and under the guide of an expert speleologist have access.

After visiting the Rio della Bacuta and Orsivacca blind valleys, tourists will return to the main road to reach the fourth stop of the geoitinerary, the MU.VI. (Museo Virtuale, "Virtual Museum", stop S4 in Figure 7). The MU.VI. is a building (Figure 10) where it is possible to admire a permanent virtual exhibition about the Middle Bussento Karst System under the guide of experts either from local associations addressed to the promotion of the territory or from speleologists of the Italian Alpine Club (a speleological group). The visit to the MU.VI. is 1 h long and here it is possible to organize other outdoor activities (e.g., trekking, canyoning, pedal cars, mountain biking) that are not included in the proposed itinerary, but that can be provided by local associations. In addition, for those that enjoy outdoor activities, there is an area where it is possible to practice fitness exercises right in front of the MU.VI. entrance.

Tourists will then move from the MU.VI. towards the last stop of the geoitinerary, the Bussento Resurgence (stop S5 in Figure 7), which is 6.5 km away and can be reached by car (15 min) or by walking (~1 h).

Figure 8. Stop no. 1: La Rupe ponor. (**A**) Panoramic view of the Bussento River valley and the entrance of the La Rupe ponor; (**B**) the La Rupe ponor entrance; (**C**) close view of the La Rupe ponor entrance.

Figure 9. (**A**) Stop no. 2: the Orsivacca ponor; (**B**) the fossil Cozzetta ponor; (**C**) the active Bacuta ponor.

Figure 10. (**A**) Stop no. 4: the entrance of the MU. VI. (virtual museum); (**B**) lateral view of the MU.VI. with the eastern slope of Mt. San Michele in the background.

This stop visits both the Old Mill springs and the Bussento River resurgence. Along the drop towards the valley floor, tourists can admire amazing rudists fossils (Figure 11A) in the carbonate bedrock. A small railway is also present to facilitate the tour either for tourists not used to trekking or for people that do not want to become tired. Once tourists reach the valley floor, they may admire the amazing Old Mill spring (Figure 11B) which is one of the basal springs of the Salice-Coccovello carbonate aquifer. This spring has a mean discharge of 50 L/s. Then, walking along a woody path next to the water level, they may reach the Bussento Resurgence, the point where the Bussento River rises again to the surface. The cave system associated with the resurgence (Figure 11C) is mostly horizontal and expert speleologists explore it for 462 m. Hydrogeological analysis (tracing tests) has proved a subsurface connection between the La Rupe, Orsivacca, Cozzetta, and Bacuta ponors with the Bussento Resurgence, as explained in Figure 12. The waters sinking in these ponors connect and travel underground at least for 4 km, making this case the only example of an underground river in

Southern Italy. Considering the ponors and the resurgence altitude in respect of the water table level, it is possible to understand that there is the possibility that the stream waters during their underground path may feed the basal water table (see blue arrows in Figure 12).

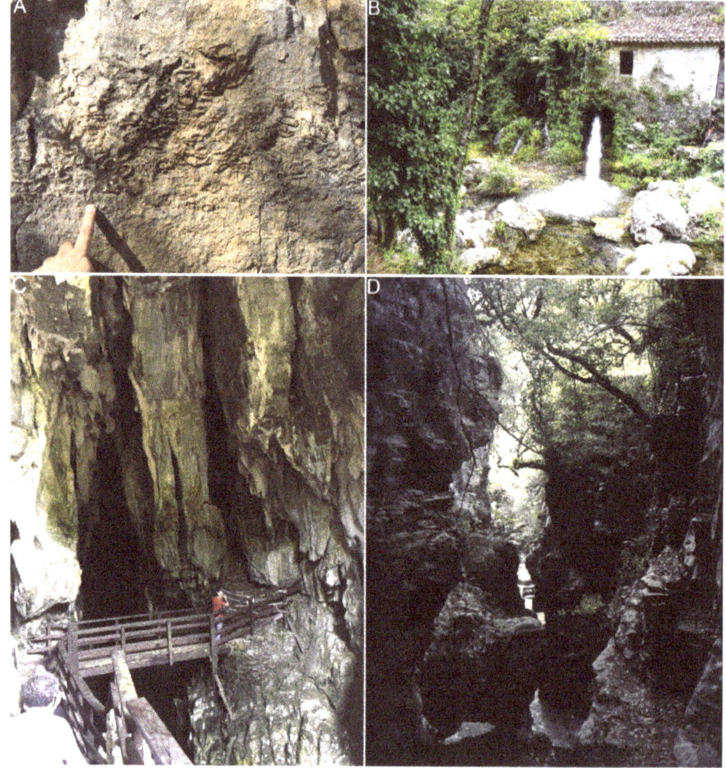

Figure 11. Stop no. 5: the Bussento Risurgence near Morigerati. (**A**) Detail of rudists fossils in the carbonatic bedrock; (**B**) the Old Mill spring; (**C**) the visiting tour in the interior of the Morigerati Resurgence; (**D**) view, from the inner side of the Bussento Resurgence, of the steep valley flanks.

The Bussento Resurgence, in addition to being an important geomorphosite, it is also a World Wildlife Foundation (WWF) Oasis, with an educational and international center, where it is possible to gain information about both the abiotic and biotic components of the environment.

Moreover, from the resurgence down-valley, the Bussento River created a gorge (Figure 11D), another peculiar karst morphology. Gorges are formed when a river flows on karst carbonate rocks and tends to dissolve them, leading to the formation of deep and narrow valleys, which often host a natural habitat of high environmental quality. Gorge flanks are high and very steep, making the gorges inaccessible areas except for expert speleologists or for those practicing canyoning.

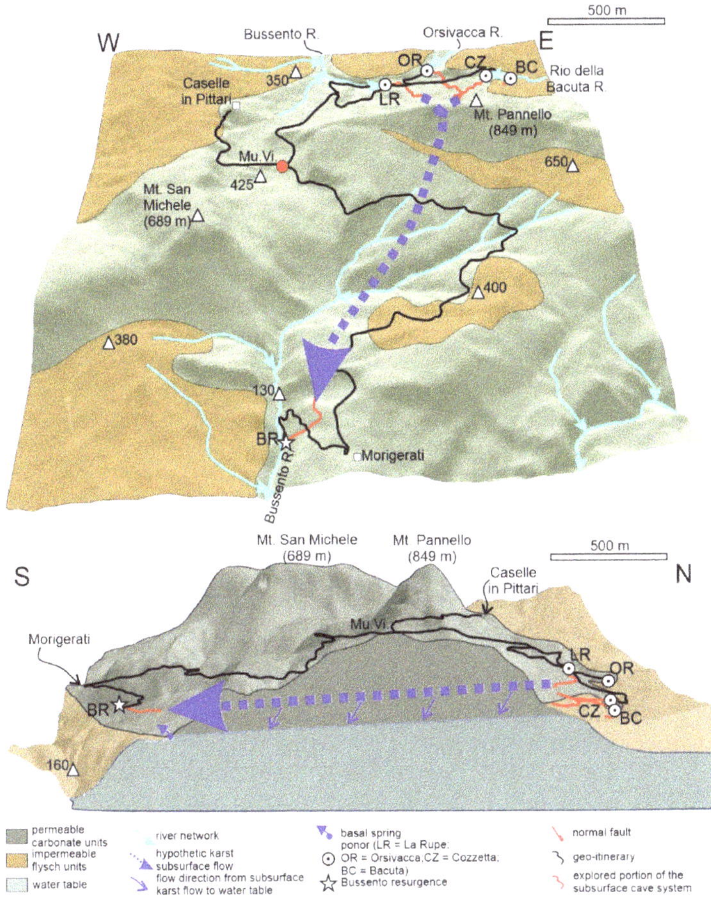

Figure 12. Frontal (from south, upper panel) and lateral (from east, lower panel) 3D views of the Middle Bussento Karst System. Red lines indicate the explored subsurface portion of the karst system (from [98]). White triangles in the upper panel are elevation points (expressed in meters above sea level).

5. Discussion

5.1. Geotourism in Karst Area

Karst terrains are widespread all over the world and cover about 17% of the Earth's surface [99], a percentage that increased up to 21.6% in Europe [100]. These data highlight the importance of karst terrains and landforms as crucial geological and geomorphological features to promote geotourism; in particular considering that 37% of Geoparks in the world exhibit karst phenomenon [101]. Ruban ([101] and references therein) highlighted the main features that make karst area prone to geotourism. According to Ruban, karst areas are, in fact, either unusual and rare landforms or windows into the dynamic geological environment: they attracted tourists before the actual concept of geotourism developed; they have aesthetic attractiveness; they are strongly related to the archaeological, historical, and ethnocultural peculiarities of some areas; they are of socio-economic importance because they are the main water reservoirs. Among karst landforms, the ones that play a major role in attracting tourists are caves [102]. Kim et al. [103] highlighted that cave tourism within geotourism gained large popularity in Korea. Accordingly, Calaforra and Cortes [99] strengthened the importance of karst caves in geotourism as an instrument to favor local economic growth, indicating that karst caves in Spain receive more than 2.5 million visitors per year, with an economic return for the related municipalities (which often have less than 5000 inhabitants) in excess of 15 million euros [104]. Doorne [105] also highlighted the importance of karst caves in the local economy by discussing the Waitomo Glowworm Cave in New Zealand which serves a village of 500 inhabitants with a tourist population of around 450,000 international visitors per year. Allan et al. [106] focused on the motivations that lead tourists to visit places of geological importance by providing a questionnaire to tourists that visited the Cristal Cave in Australia. They found that the motivations of the visit include relaxation, escape from daily routine, sense of wonder, and knowledge. Hurtado et al. [107] also investigated Cristal Cave tourist traffic. Kiernar [108] discussed nature protection and geotourism in Laos where several karst caves receive tourist traffic related to religious motivation.

It is important that tourist traffic in karst caves avoid alteration of the natural environment, as proven for the Cave of Marvels in Spain [109]. Moreover, Baker and Genty [110] highlighted that tourist traffic in karst caves where ventilation is poor may alter CO_2 concentrations and increase temperatures by 3°. Calaforra et al. [111] carried out experimental measurements of caves' temperature at the Cueva del Agua de Iznalloz, Granada, Spain before the cave was opened to tourists, to determine the impact of human presence within the cave. Eagles et al. [112] pointed out that good planning and appropriate practices can lead to sustainable management of tourism.

Williams [113] stressed the importance of preserving the integrity of karst areas as a mandatory task because karst systems are complex systems that develop both at the surface and subsurface. Moreover, karst environments are an example of a fragile ecosystem whose balance depends on several factors such as the energy and quality of water flow. Williams [113] also remarked that environmental conditions in the recharge areas, both in allogenic and autogenic karst systems, have a strong influence on environmental conditions in the subsurface, thus suggesting that correct management of these areas is mandatory to avoid pollution in karst caves with dramatic consequences for plant and animals living in there. The previous point is fundamental considering that surface and subsurface water divides do not necessarily correspond, thus, karst drainage areas are not easy to delimit. The problem of preserving the integrity of a karst area is relevant also in the Middle Bussento Karst System. In fact, the Bussento River suffers both from low discharge due to the activity of the Sabetta Lake hydroelectrical powerplant and the organic pollution. This could cause severe problem to the 43 species found during speleological exploration at the La Rupe ponor [114,115]. Fortunately, subsurface water auto-depuration, due to a still poorly known subsurface karst path, dilutes the organic contaminants, making the water flowing out at the Bussento Resurgence in Morigerati the cleanest and able to host many fluvial organisms [114].

For what we have discussed up to now, it is evident that for geotourism in karst areas there is a fundamental task to diffuse the concept of environmental protection in these fragile ecosystems. The geoitinerary we propose intends to make people aware about the importance of karst areas as the main source for drinkable water by the combination of outdoor (trekking), indoor (speleological exploration under the guide of expert speleologists), and educational (visit to the MU.VI.) activities. It should also serve as a promoter of the territory by increasing curiosity about this poorly known portion of the Geopark and thus contributing to local economic growth.

5.2. Promoting Geotourism in the Middle Bussento Karst System: What Has Been Done and What Can Be Done

The scientific and educational values of the Middle Bussento Karst System are well known since the first exploration of the La Rupe ponor in the early 1950s [116]. Explorations were carried out in a discontinuous way until 2007 when speleologists from the Italian Alpine Club, section of Naples, reached the farthest point of the subsurface path [116]. Due to the high scientific value of the area, the Speleological Group of the Italian Alpine Club, section of Naples, in combination with other speleological groups from the Campania Region, organized the 2nd Regional Conference on Speleology in Caselle in Pittari, from 3–6 June 2010 [117]. This conference has been, up to now, the most important scientific event during which the Middle Bussento Karst System has been at the center of discussions of hundreds of researchers and persons interested in speleology coming from all over Italy. During this conference, an intense educational activity was carried out at the MU.VI. by speleologists of the Campania Speleological Federation and the Speleological Group of the Italian Alpine Club, section of Naples [118].

Since then, educational activities geared towards knowledge and promotion of the territory have been carried out in a discontinuous way by local administrations and associations. These activities include outdoor sports such as canyoning, trekking, mountain biking, and pedal cars mainly concentrated during the spring and summer. In addition, local associations carry out environmental education activities in elementary, middle, and high schools in the surroundings of Caselle in Pittari. This often includes a visit of scholars to the MU.VI., where a permanent virtual exhibition on karst is present the entire year.

Further activities addressed to discover the Middle Bussento Karst System include personal initiatives by environmental guides often coming from areas outside the Geopark. These activities include a one-day trek to visit either the La Rupe ponor or the Bussento Resurgence. They usually do not include a one-night stay in the area to enjoy the hospitality of the locals and the amazing local food; thus, they do not contribute significantly in the growth of the local economy.

In our opinion, what is done to promote the territory, help the growth of the local economy, and emphasize the unicity of the Middle Bussento Karst System is good, but it is not enough. We must always have in mind that the Bussento River is the second longest subsurface path in Italy and this point should be a crucial promoting element to bring visitors to this area. The sporadic initiatives carried out either by local administrations and associations or environmental guides enhance the lack of coordination between these groups that, together, could do a lot for the local community. There is also a scarce amount of information on the Internet. For example, the local administration created a website to promote the area [119] but the amount of data available on the Internet about the Middle Bussento Karst System is sharply lower compared to other karst systems in the same Geopark (e.g., the Pertosa and Castelcivita caves).

We think that more convincing advertising actions are necessary if local administrations really want to help in the growth of local economy by attracting more tourists. The proposed geoitinerary emphasizes the fascinating subsurface world that characterizes the area, and tries to increase awareness among tourists, local people, and local administrations about the importance of karst areas as resources of potable water. In addition, the geoitinerary must be accompanied by a more effective and pervasive presence on the Internet through social media, and the production of attractive educational material that could make the Middle Bussento Karst System easily understood by non-geologists. An example could be the 3D reconstruction in Figure 12 that aims to describe the karst system in a simple way, helping tourists figure out the connection between the La Rupe, Orsivacca, Cozzetta, and Bacuta ponors and the Bussento Resurgence.

5.3. SWOT Analysis and VSG Index

To analyze the potential of the area and to highlight possible activities addressed to the efficient and effective use of the proposed geoitinerary, we carried out an analysis to define the strengths, weaknesses, opportunities, and threats (SWOT analysis). Results of the SWOT analysis are reported in Table 3.

Table 3. Results of the SWOT (strengths, weaknesses, opportunities and threats) analysis.

Strengths	Weaknesses
1. Discovery of a poorly known portion of the Cilento, Vallo di Diano and Alburni United Nations Educational, Scientific and Cultural Organization (UNESCO) Global Geopark. 2. The area is easily reachable by public transport. 3. The Bussento River is the second longest subsurface river in Italy and the longest one in Southern Italy. 4. High didactic value to explain the relation between surface and subsurface flows in karst areas. 5. Karst landforms are visible and comprehensible to tourists, even if they have no geological background.	1. The geoitinerary is at an initial stage and deserves the development of further activities, such as explanatory panels for each stop. 2. The geoitinerary is a personal initiative by the Authors and lacks, up to now, enough of an intense collaboration with local administration. 3. The full development of the geoitinerary needs funding and a management policy. 4. Local people and local administration have not yet fully understood the high touristic and didactic potential of the area. 5. Tourists reach the study area because of the amazing local food but they have no cognition of the karst system in the surroundings. 6. Local accommodation facilities can host very little tourist traffic.
Opportunities	**Threats**
1. The geoitinerary could be integrated in a wider tour because of its closeness with the main touristic attraction of the Geopark. 2. The geoitinerary could lead many tourists to move from the touristic coastal areas of Cilento to the poorly known inner areas of the Geopark. 3. This tourist flow could help the growth of the local economy (e.g., by increasing the number of accommodation facilities). 4. The growth of the local economy could provide work for young people and so it could reduce the abandonment of small mountain villages. 5. The geoitinerary could be split in two days thus providing enough time for the tourists to enjoy the hospitality of local people and food of excellent quality.	1. The high and steep carbonate walls near the La Rupe ponor may be affected by rock falls and deserves mitigation actions. 2. Some mule track and wooden paths need maintenance. 3. Visits to the La Rupe ponor is strongly influenced by the activities of the Sabetta Lake hydroelectrical powerplant and is not allowed during basin emptying. 4. The Bussento River at the La Rupe ponor suffers both from the low discharge due to the activity of the Sabetta Lake hydroelectrical powerplant and the organic pollution. 5. This could cause severe problems for the 43 species discovered during speleological exploration at the La Rupe ponor.

We then calculated the VSG (value of a site for geotourism) index by applying the Pica et al. [88] method reported in Table 1. Moreover, the Middle Bussento Karst System has a representativeness (RP) value of 5 because it is an amazing example of a subsurface river path with contact ponors representative of the ponor retreat mechanism, and its interest falls in many disciplines, including geomorphology, hydrology, and structural geology. The rareness (RR) index has a value of 5 because subsurface river paths are not so common in Italian territory and the Middle Bussento Karst System is the second longest subsurface river in Italy. The scenic-aesthetic (SCE) value is 5 because viewpoints are common along the entire geoitinerary and cromatic contrast is excellent, thus allowing a full appreciation of the karst landforms. The historical-archeological-cultural index (SAC) has a value of 3 because of restriction laws related to the protected area and to a poor connection with local tradition. The accessibility index (AC) has a value of 3 because the site is easily accessible by car and public transport, but it lacks services close to the stops of the geoitinerary.

The resulting value of the VSG index is 21 suggesting that the Middle Bussento Karst System has a high potential for geotourism.

6. Conclusions

Karst systems are sensitive environments that deserve accurate management and promotion strategies to avoid contamination of water resources used for drinkable needs and to avoid alteration of the environment, with drastic consequences for the flora and fauna that live in these areas and also for the karst landforms as well. Furthermore, karst landforms are among the more fascinating landforms to attract tourists.

To highlight the natural and socio-economic role of karst areas, we investigated the Middle Bussento Karst System, in the Cilento, Vallo di Diano and Alburni UNESCO Global Geopark by carrying out a comprehensive analysis of both its surface and subsurface karst landforms. The peculiarity of the area is that the Bussento River sinks at the La Rupe ponor, near the village of Caselle in Pittari, and remerges, after 4 km of subsurface path, at the Bussento Resurgence, near the village of Morigerati, making the Bussento River subsurface path the second longest one in Italy and the longest one in Southern Italy. Selected karst landforms include the La Rupe, Orsivacca, Cozzetta, and Bacuta ponors, whose subsurface paths are connected and end at the Bussento Resurgence. In addition, a visit to a local virtual museum (MU.VI.) is included in the geoitinerary. The SWOT analysis also enhances the touristic potential of the geoitinerary, highlighting the high potential of this area as a possible touristic attraction in the Cilento, Vallo di Diano and Alburni UNESCO Global Geopark. This is also testified by the high value of the VSG index, whose score of 21 places the geomorphosite among the areas with high potential for geotourism. Some touristic promotion activity has already been carried out, such as the foundation of the MU.VI., however more actions are necessary to strengthen the role of the Middle Bussento Karst System as a significant geo-touristic attraction. The proposed geoitinerary has high scientific and educational values that may help in the field of environmental education, making people aware about the importance of karst areas as containers of water resources. The combination of the proposed geoitinerary with educational materials could help visitors gain detailed knowledge about what a karst environment is and how it works. Future campaigns of touristic promotion in the Geopark should emphasize the role of the blind valley of the Bussento River as the second longest subsurface river path of Italy and the longest in Southern Italy. Moreover, the addition of possible outdoor activities carried out by local associations could help the growth of tourism.

We tried to contribute to an increase in the curiosity of this site, in order to promote this fascinating portion of the Geopark as a touristic attraction and aid in the growth of the local economy.

Author Contributions: Conceptualization, E.V. and N.S.; methodology, E.V. and N.S.; software, E.V.; validation, A.S., D.G., and N.S.; formal analysis, D.G.; investigation, E.V.; data curation, E.V.; writing—original draft preparation, E.V., A.S., and N.S.; writing—review and editing, E.V., A.S., and N.S.; visualization, D.G.; supervision, A.S. and N.S. All authors have read and agreed to the published version of the manuscript.

Funding: The research was funded by department research (DISTAR), Nicoletta Santangelo.

Acknowledgments: We wish to thank two anonymous reviewers whose suggestions helped to increase the quality of the paper.

Conflicts of Interest: The authors declare no conflicts of interest.

References

1. Senato Della Repubblica Italiana. Available online: https://www.senato.it/documenti/repository/istituzione/costituzione.pdf (accessed on 3 March 2020).

2. Eder, F.W.; Patzak, M. Geoparks—Geological attractions: A tool for public education, recreation and sustainable economic development. *Episodes* **2004**, *27*, 162–164. [CrossRef]

3. Brilha, J. Geoheritage and geoparks. In *Geoheritage. Assessment, Protection, and Management*; Reynard, E., Brilha, J., Eds.; Elsevier: Amsterdam, The Netherlands, 2018; pp. 323–334. [CrossRef]

4. UNESCO Website. Available online: https://whc.unesco.org/en/list/ (accessed on 3 March 2020).

5. Patzak, M.; Eder, W. "UNESCO GEOPARK" A new Programme—A new UNESCO label. *Geol. Balc.* **1998**, *28*, 33–35.

6. Zouros, N. The European Geoparks Network Geological heritage protection and local development. *Episodes* **2004**, *27*, 165–171. [CrossRef]

7. Nowlan, G.S.; Bobrowsky, P.; Clague, J. Protection of geological heritage: A North American perspective on Geoparks. *Episodes* **2004**, *27*, 172–176. [CrossRef] [PubMed]

8. Global Geopark Website. Available online: http://www.globalgeopark.org/aboutGGN/list/index.htm (accessed on 3 March 2020).

9. UNESCO Global Geopark Definition. Available online: http://www.globalgeopark.org/aboutGGN/6398.htm (accessed on 1 April 2020).

10. Bruschi, V.M.; Coratza, P. Geoheritage and environmental impact assessment (eia). In *Geoheritage. Assessment, Protection, and Management*; Reynard, E., Brilha, J., Eds.; Elsevier: Amsterdam, The Netherlands, 2018; pp. 251–264. [CrossRef]

11. Brilha, J. Inventory and Quantitative Assessment of Geosites and Geodiversity Sites: A Review. *Geoheritage* **2016**, *8*, 119–134. [CrossRef]

12. Newsome, D.; Dowling, R. Geoheritage and geotourism. In *Geoheritage. Assessment, Protection, and Management*; Reynard, E., Brilha, J., Eds.; Elsevier: Amsterdam, The Netherlands, 2018; pp. 305–321. [CrossRef]

13. Reynard, E.; Brilha, J. Geoheritage: A Multidisciplinary and Applied Research Topic. In *Geoheritage. Assessment, Protection, and Management*; Reynard, E., Brilha, J., Eds.; Elsevier: Amsterdam, The Netherlands, 2018; pp. 3–9. [CrossRef]

14. Hose, T.A. Selling the Story of Britain's Stone. *Environ. Interpret.* **1995**, *10*, 16–17.

15. Hose, T.A. 3G's for modern geotourism. *Geoheritage* **2012**, *4*, 7–24. [CrossRef]

16. Dowling, R.; Newsome, D. The scope and nature of geotourism. In *Geotourism*; Dowling, R., Newsome, D., Eds.; Routledge: Oxford, UK, 2006; pp. 31–53.

17. Newsome, D.; Dowling, R.K. Setting an agenda for geotourism. In *Geotourism: The Tourism of Geology and Landscape*; Newsome, D., Dowling, R., Eds.; Goodfellow Publishers Limited: Oxford, UK, 2010; pp. 1–12.

18. Migoń, P.; Różycka, M.; Michniewicz, A. Conservation and Geotourism Perspectives at Granite Geoheritage Sites of Waldviertel, Austria. *Geoheritage* **2018**, *10*, 11–21. [CrossRef]

19. Bentivenga, M.; Cavalcante, F.; Mastronuzzi, G.; Palladino, G.; Prosser, G. Geoheritage: The Foundation for Sustainable Geotourism. *Geoheritage* **2019**, *11*, 1367–1369. [CrossRef]

20. Ólafsdóttir, R.; Tverijonaite, E. Geotourism: A Systematic Literature Review. *Geosciences* **2018**, *8*, 234. [CrossRef]

21. Dowling, R.; Newsome, D. Geotourism: Definition, characteristics and international perspectives. In *Handbook of Geotourism*; Dowling, R., Newsome, D., Eds.; Edward Elgar: Cheltenham, UK, 2018; pp. 1–22.

22. Hose, T.A. Geotourism, or can tourists become casual rock hounds? In *Geology on Your Doorstep*; Bennett, M.R., Doyle, P., Larwood, J.G., Prosser, C.D., Eds.; The Geological Society: London, UK, 1996; pp. 207–228.

23. Hose, T.A. European geotourism: Geological interpretation and geoconservation promotion for tourists. In *Geological Heritage: Its Conservation and Management*; Barettino, D., Wimbledon, W.A.P., Gallego, E., Eds.; Instituto Tecnologico GeoMinero de Espana: Madrid, Spain, 2000; pp. 127–146.

24. Hose, T.A. Towards a history of Geotourism: Definitions, antecedents and the future. In *The History of Geoconservation*; Burek, C.V., Prosser, C.D., Eds.; The Geological Society: London, UK, 2008; pp. 37–60.

25. Pralong, J.P. Geotourism: A new form of tourism utilising natural landscapes and based on imagination and emotion. *Tour. Rev.* **2006**, *61*, 20–25. [CrossRef]

26. The Arouca Declaration. Available online: http://www.unesco.org/new/fileadmin/MULTIMEDIA/HQ/SC/pdf/Geopark_Arouca_Declaration_EGN_2012.pdf (accessed on 2 April 2020).

27. Reynard, E. Protecting Stones: Conservation of erratic blocks in Switzerland. In *Dimension Stone 2004. New Perspectives for a Traditional Building Material*; Prykril, R., Ed.; A.A. Balkema Publishers: Leiden, The Netherlands, 2004; pp. 3–7.

28. Panizza, M. Geomorphosites: Concepts, methods and examples of geomorphological survey. *Chin. Sci. Bull.* **2001**, *46*, 4–6. [CrossRef]

29. Reynard, E.; Panizza, M. Geomorphosites: Definition, assessment and mapping. *Géomorphologie* **2005**, *11*, 177–180. [CrossRef]

30. Reynard, E.; Coratza, P. Geomorphosites and geodiversity: A new domain of research. *Geogr. Helv.* **2007**, *62*, 138–139. [CrossRef]

31. Reynard, E.; Coratza, P. Scientific research on geomorphosites. A review of the activities of the IAG working group on geomorphosites over the last twelve years. *Geogr. Fis. Dinam. Quat.* **2013**, *36*, 159–168. [CrossRef]

32. Reynard, E.; Fontana, G.; Kozlik, L.; Scapozza, C. A method for assessing the scientific and additional values of geomorphosites. *Geogr. Helv.* **2007**, *62*, 148–158. [CrossRef]

33. Reynard, E.; Coratza, P.; Regolini-Bissig, G. (Eds.) *Geomorphosites*; Pfeil: München, Germany, 2009; p. 240.

34. Serrano, E.; González Trueba, J.J. Assessment of geomorphosites in natural protected areas: The Picos de Europa National Park (Spain). *Géomorphologie* **2005**, *3*, 197–208. [CrossRef]

35. Zouros, N. Geomorphosite assessment and management in protected areas of Greece. Case study of the Lesvos Island-coastal geomorphosites. *Geogr. Helv.* **2007**, *62*, 69–180. [CrossRef]

36. Miccadei, E.; Piacentini, T.; Esposito, G. Geomorphosites and geotourism in the parks of the Abruzzo region (Central Italy). *Geoheritage* **2011**, *3*, 233–251. [CrossRef]

37. Pralong, J.P. A method for assessing the tourist potential and use of geomorphological sites. *Géomorphologie* **2005**, *3*, 189–196. [CrossRef]

38. Piacente, S.; Coratza, P. (Eds.) Geomorphological sites and geodiversity. *Il Quat.* **2005**, *18*, 332.

39. Pereira, P.; Pereira, D.; Caetano Alves, M.I. Geomorphosite assessment in Montesinho Natural Park (Portugal). *Geogr. Helv.* **2007**, *62*, 159–168. [CrossRef]

40. Gregori, L.; Melelli, L. Geotourism & Geomorphosites: The G.I.S. Solution. *Il Quat.* **2005**, *18*, 285–292.

41. Coratza, P.; Hoblea, F. The Specificities of Geomorphological Heritage. In *Geoheritage. Assessment, Protection, and Management*; Reynard, E., Brilha, J., Eds.; Elsevier: Amsterdam, The Netherlands, 2018; pp. 87–106. [CrossRef]

42. Coratza, P.; De Waele, J. Geomorphosites and natural hazards: Teaching the Importance of Geomorphology in Society. *Geoheritage* **2012**, *4*, 195–203. [CrossRef]

43. Mucivuna, V.C.; Reynard, E.; Garcia, M.d.G.M. Geomorphosites Assessment Methods: Comparative Analysis and Typology. *Geoheritage* **2019**, *11*, 1799–1815. [CrossRef]

44. Pescatore, E.; Bentivenga, M.; Giano, S.I.; Siervo, V. Geomorphosites: Versatile Tools in Geoheritage Cultural Dissemination. *Geoheritage* **2019**, *11*, 1583–1601. [CrossRef]

45. Santos, D.S.; Reynard, E.; Mansur, K.L.; Seoane, J.C.S. The specificities of geomorphosites and their influence on assessment procedures: A methodological comparison. *Geoheritage* **2019**, *11*, 2045–2064. [CrossRef]

46. Bollati, I.; Zucali, M.; Giovenco, C.; Pelfini, M. Sport climbing sites as a new approach for education in Earth Science: Scientific representativeness of Montestrutto area (Austroalpine Domain, Piemonte, Italy). *Ital. J. Geosci.* **2014**, *133*, 187–199. [CrossRef]

47. Bollati, I.; Coratza, P.; Panizza, V.; Pelfini, M. Lithological and structural control on italian mountain geoheritage: Opportunities for tourism, outdoor and educational activities. *Quaest. Geogr.* **2018**, *37*, 53–73. [CrossRef]

48. Panizza, V.; Mennella, M. Assessing geomorphosites used for rock climbing: The example of Monteleone Rocca Doria (Sardinia, Italy). *Geogr. Helv.* **2007**, *62*, 181–191. [CrossRef]

49. Piacentini, T.; Castaldini, D.; Coratza, P.; Farabollini, P.; Miccadei, E. Geotourism: Some examples in northern-central Italy. *GeoJ. Tour. Geosites* **2011**, *8*, 240–262.

50. Aringoli, D.; Gentili, B.; Pambianchi, G.; Piscitelli, A.M. The contribution of the "Sibilla appenninica" legend to karst knowledge in the Sibillini Mountains (central Apennines, Italy). In *Myth and Geology*; Piccardi, L., Masse, W.B., Eds.; Geological Society, London, Special Publications: London, UK, 2007; Volume 273, pp. 329–340. [CrossRef]

51. Aringoli, D.; Farabollini, P.; Gentili, B.; Materazzi, M.; Pambianchi, G. Geoparks and geoconservation: Examples in the southern Umb.ro-Marchean Apennines (central Italy). *Geoacta Spec. Publ.* **2008**, *3*, 1–14.

52. Coratza, P.; Ghinoi, A.; Piacentini, D.; Valdati, J. Management of geomorphosites in high tourist vocation area: An example of Geo-Hiking maps in the Alpe di Fanes (Natural Park of Fanes-Senes_Braies, Italian Dolomites). *GeoJ. Tour. Geosites* **2008**, *2*, 106–117.

53. Venturini, C.; Pasquarè Mariotto, F. Geoheritage Promotion Through an Interactive Exhibition: A Case Study from the Carnic Alps, NE Italy. *Geoheritage* **2019**, *11*, 450–469. [CrossRef]

54. Santangelo, N.; Romano, P.; Santo, A. Geo-itineraries in the Cilento Vallo di Diano Geopark: A Tool for Tourism Development in Southern Italy. *Geoheritage* **2015**, *7*, 319–335. [CrossRef]

55. Sisto, M.; Di Lisio, A.; Russo, F. The Mefite in the Ansanto Valley (Southern Italy): A Geoarchaeosite to Promote the Geotourism and Geoconservation of the Irpinian Cultural Landscape. *Geoheritage* **2020**, *12*, 29. [CrossRef]

56. Margiotta, S.; Sansò, P. Abandoned Quarries and Geotourism: An Opportunity for the Salento Quarry District (Apulia, Southern Italy). *Geoheritage* **2017**, *9*, 463–477. [CrossRef]

57. Pilogallo, A.; Nolè, G.; Amato, F.; Saganeiti, L.; Bentivenga, M.; Palladino, G.; Scorza, F.; Murgante, B.; Las Casas, G. Geotourism as a Specialization in the Territorial Context of the Basilicata Region (Southern Italy). *Geoheritage* **2019**, *11*, 1435–1445. [CrossRef]

58. Bucci, F.; Tavarnelli, E.; Novellino, R.; Palladino, G.; Guglielmi, P.; Laurita, S.; Prosser, G.; Bentivenga, M. The History of the Southern Apennines of Italy Preserved in the Geosites Along a Geological Itinerary in the High Agri Valley. *Geoheritage* **2019**, *11*, 1489–1508. [CrossRef]

59. Santangelo, N.; Santo, A.; Guida, D.; Lanzara, R.; Siervo, V. The geosites of the Cilento-Vallo di Diano national park (Campania region, southern Italy). *Il Quat.* **2005**, *18*, 101–112.

60. Aloia, A.; De Vita, A.; Guida, D.; Toni, A.; Valente, A. National Park of Cilento and Vallo di Diano: Geodiversity, geotourism, geoarchaeology and historical tradition. In Proceedings of the 9th Europena Geopark Conference—European Geopark Network Mytilene Lesvos, Lesvos, Greece, 1–5 October 2010; p. 41.

61. Aloia, A.; Guida, D.; Valente, A. Geodiversity in the Geopark of Cilento and Vallo di Diano as heritage and resource development. *Rend. Online Soc. Geol. Ital.* **2012**, *21*, 688–690.

62. Aloia, A.; Guida, D. *The Geosites: Geopark's Gaia Synphony in the Cilento, Vallo Diano and Alburni Geopark, Geopark Book n. 1*; Cilento, Vallo di Diano and Alburni Geopark: Vallo della Lucania, Italy, 2014.

63. Calcaterra, D.; D'Argenio, B.; Ferranti, L.; Pappone, G.; Petrosino, P. *Guide Geologiche Regionali, Campania e Molise*; Società Geologica Italiana: Rome, Italy, 2016; p. 276.

64. Valente, E.; Ascione, A.; Ciotoli, G.; Cozzolino, M.; Porfido, S.; Sciarra, A. Do moderate magnitude earthquakes generate seismically induced ground effects? The case study of the M w = 5.16, 29th December 2013 Matese earthquake (southern Apennines, Italy). *Int. J. Earth Sci.* **2018**, *107*, 517–537. [CrossRef]

65. Italian National Parks Website. Available online: http://www.parks.it/parco.nazionale.cilento/map.php (accessed on 6 March 2020).

66. ISPRA Geosites Database. Available online: http://sgi.isprambiente.it/GeositiWeb/default.aspx?ReturnUrl=%2fgeositiweb%2f (accessed on 5 March 2020).

67. Cilento, Vallo di Diano and Alburni Geopark Geosites Database. Available online: http://www.cilentoediano.it/it/geositi-gli-ambiti-paesaggio (accessed on 5 March 2020).

68. Allocca, V.; Celico, F.; Celico, P.; De Vita, P.; Fabbrocino, S.; Mattia, C.; Monacelli, G.; Musilli, I.; Piscopo, V.; Scalise, A.R.; et al. *Note Illustrative Della Carta Idrogeologica Dell'italia Meridonale*; Istituto Poligrafico e Zecca dello Stato: Rome, Italy, 2007; p. 217.

69. Guida, D.; Guida, M.; Luise, D.; Salzano, G.; Vallario, A. Idrogeologia del Cilento. *Geol. Romana* **1980**, *19*, 349–369.

70. Santangelo, N.; Santo, A. Endokarst processes in the Alburni massif (Campania, southern Italy): Evolution of ponors and hydrogeological implications. *Z. Geomorphol.* **1997**, *41*, 229–246.

71. Geosites Census in the Bussento River Area by the Geopark. Available online: http://www.cilentoediano.it/it/node/1112 (accessed on 6 March 2020).

72. Piperno, M. *La Preistoria alle Falde del M. Cervati. Parco Nazionale del Cilento e Vallo di Diano*; Cilento, Vallo di Diano and Alburni Geopark: Vallo della Lucania, Italy, 2001; p. 12.

73. Fondazione MIDA Website. Available online: http://fondazionemida.com/ (accessed on 5 March 2020).

74. Braitenberg, C.; Pivetta, T.; Rossi, G.; Ventura, P.; Betic, A. Karst caves and hydrology between geodesy and archeology: Field trip notes. *Geodesy Geodynamics* **2018**, *9*, 262–269. [CrossRef]

75. Sheet 520—Sapri of the Geological Map of Italy at Scale 1:50000, Carg Project. Available online: http://www.isprambiente.gov.it/Media/carg/520_SAPRI/Foglio.html (accessed on 6 March 2020).

76. Cello, G.; Mazzoli, S. Apennine Tectonics in Southern Italy: A review. *J. Geodyn.* **1998**, *27*, 191–211. [CrossRef]

77. Menardi Noguera, A.; Rea, G. Deep structure of the Campanian–Lucanian Arc (Southern Apennine, Italy). *Tectonophysics* **2000**, *324*, 239–265. [CrossRef]

78. Vitale, S.; Ciarcia, S. Tectono-stratigraphic and kinematic evolution of the southern Apennines/Calabria–Peloritani Terrane system (Italy). *Tectonophysics* **2013**, *583*, 164–182. [CrossRef]

79. Ciarcia, S.; Vitale, S.; Di Staso, A.; Iannace, A.; Mazzoli, S.; Torre, M. Stratigraphy and tectonics of an Internal Unit of the southern Apennines: Implications for the geodynamic evolution of the peri-Tyrrhenian mountain belt. *Terra Nova* **2009**, *21*, 88–96. [CrossRef]

80. Ciarcia, S.; Mazzoli, S.; Vitale, S.; Zattin, M. On the tectonic evolution of the Ligurian accretionary complex in southern Italy. *Geol. Soc. Am. Bull.* **2012**, *124*, 463–483. [CrossRef]

81. Vitale, S.; Ciarcia, S.; Mazzoli, S.; Iannace, A.; Torre, M. Structural analysis of the 'Internal' Units of Cilento, Italy: New constraints on the Miocene tectonic evolution of the southern Apennine accretionary wedge. *CR Geosci.* **2010**, *342*, 475–482. [CrossRef]

82. Vitale, S.; Ciarcia, S.; Mazzoli, S.; Zaghloul, M.N. Tectonic evolution of the 'Liguride' accretionary wedge in the Cilento area, southern Italy: A record of early Apennine geodynamics. *J. Geodyn.* **2011**, *51*, 25–36. [CrossRef]

83. D'Elia, G.; Guida, M.; Terranova, C. Osservazioni sui fenomeni di deformazione gravitativa profonda nel bacino del fiume Bussento (Campania). *Boll. Soc. Geol. Ital.* **1987**, *106*, 253–264.

84. Cvijc, J. La geographie des terrains calcaires. *Monographies Acad. Serbe Sc. et Arts* **1960**, *26*, 141–207.

85. Ford, D.; Williams, P. *Karst Geomorphology and Geology*; Unwin Hyman: London, UK, 1989.

86. Bovolin, V.; Cuomo, A.; Guida, D. Hydraulic modeling of flood pulses in the Middle Bussento Karst System (MBSKS), UNESCO Cilento Global Geopark, southern Italy. *Hydrol. Process.* **2017**, *31*, 639–653. [CrossRef]

87. Guida, M.; Guida, D.; Guadagnuolo, D.; Cuomo, A.; Siervo, V. Using Radon-222 as a Naturally Occurring Tracer to investigate the streamflow-groundwater interactions in a typical Mediterranean fluvial-karst landscape: The interdisciplinary case study of the Bussento river (Campania region, Southern Italy). *WSEAS Trans. Syst.* **2013**, *12*, 85–104.

88. Pica, A.; Fredi, P.; Del Monte, M. The Ernici Mountains Geoheritage (Central Apennines, Italy): Assessment of the geosites for Geotourism development. *GeoJ. Tour. Geosites* **2014**, *14*, 193–206.

89. Santo, A.; Budetta, P.; Forte, G.; Marino, E.; Pignalosa, A. Karst collapse susceptibility assessment: A case study on the Amalfi Coast (Southern Italy). *Geomorphology* **2017**, *285*, 247–259. [CrossRef]

90. Ford, D.; Williams, P.D. *Karst Hydrogeology and Geomorphology*; John Wiley & Sons: Hoboken, NJ, USA, 2013.

91. De Waele, J.; Gutiérrez, F.; Parise, M.; Plan, L. Geomorphology and natural hazards in karst areas: A review. *Geomorphology* **2011**, *134*, 1–8. [CrossRef]

92. Celico, F. Idrogeologia dei massicci carbonatici delle piane quaternarie e delle aree vulcaniche dell'Italia centro-meridionale (Marche e Lazio meridionale, Abruzzo, Molise e Campania). *Quad. Cassa Mezzog.* **1983**, *4*, 1–225.

93. Santo, A. Karst processes and potential vulnerability of the Campanian carbonatic aquifers: The state of the knowledge. In Proceedings of the International Conference on Enviromental Changes in Karst areas. I.G.U.-U.I.S.-University of Padua, Padua, Italy, 15–27 September 1991; pp. 95–107.

94. Russo, N.; Del Prete, S.; Giulivo, I.; Santo, A. *Grotte e Speleologia della Campania*; Elio Sellino Editore: Avellino, Italy, 2005; pp. 363–396.

95. Santo, A.; Liguori, M.; Aquino, S.; Galasso, M. Problemi di geologia ambientale nei polje appenninici: l'esempio della Piana di Forino (Campania). *Il Quat.* **1998**, *11*, 233–245.

96. Parise, M.; Santo, A. Tracer Tests History in the Alburni Massif (Southern Italy). *IOP Conf. Ser. Earth Environ. Sci.* **2017**, *95*, 062006. [CrossRef]

97. Italian Excursionist Federation Website, Classification of Muletrack. Available online: https://www.ficitalia.com/fie/cosa-facciamo/sport/escursionismo/classificazione-dei-sentieri/ (accessed on 13 March 2020).

98. Campania Region Cave Database. Available online: http://sit.regione.campania.it/catastogrotte/ (accessed on 13 March 2020).

99. Calaforra, J.M.; Fernandez-Cortes, A. Geotourism in Spain: Resources, environmental management. In *Geotourism. Sustainabilty, Impact and Management*; Dowling, R., Newsome, D., Eds.; Elsevier: Oxford, UK, 2006; pp. 199–220.

100. Chen, Z.; Auler, A.S.; Bakalowicz, M.; Drew, D.; Griger, F.; Hartmann, J.; Jiang, G.; Moosdorf, N.; Richts, A.; Stevanovic, Z.; et al. The World Karst Aquifer Mapping project: Concept, mapping procedure and map of Europe. *Hydrogeol. J.* **2017**, *25*, 771–785. [CrossRef]

101. Ruban, D. Karst as Important Resource for Geopark-Based Tourism: Current State and Biases. *Resources* **2018**, *7*, 82. [CrossRef]

102. Cigna, A.A.; Forti, P. Caves: The Most Important Geotouristic Feature in the World. *Tour. Karst Areas* **2013**, *6*, 9–26.

103. Kim, S.S.; Kim, M.; Park, J.; Guo, Y. Cave tourism: Tourists' characteristics, motivations to visit, and the segmentation of their behavior. *Asia Pac. J. Tour. Res.* **2008**, *13*, 299–318. [CrossRef]

104. Rivas, A.; Durán, J.J.; López-Martínez, J. Spanish show caves as elements of the geological heritage. In Proceedings of the 4th International Show Caves Association Congress, Use of Modern Technologies in the Development of Caves for Tourism, Postojna, Slovenia, 21–27 October 2002; pp. 155–158.

105. Doorne, S. Caves, cultures and crowds: Carrying capacity meets consumer sovereignty. *J. Sustain. Tour.* **2000**, *8*, 116–130. [CrossRef]

106. Allan, M.; Dowling, R.K.; Sanders, D. The motivation for visiting geosites: The case of Cristal Cave, Western Australia. *GeoJ. Tour. Geosites* **2015**, *16*, 141–152.

107. Hurtado, H.; Downling, R.K.; Sanders, D. An explanatory study to develop a Geotourism typology model. *Int. J. Tour. Res.* **2014**, *16*, 608–613. [CrossRef]

108. Kiernan, K. The nature conservation, Geotourism and poverty reduction nexus in developing countries: A case study from the Lao PDR. *Geoheritage* **2013**, *5*, 207–225. [CrossRef]

109. Pulido-Bosch, A.; Martin-Rosales, W.; Lopez-Chicano, M.; Rodriguez-Navarro, C.M.; Vallejos, A. Human impact in a tourist karstic cave (Aracena, Spain). *Environ. Geol.* **1997**, *31*, 142–149. [CrossRef]

110. Baker, A.; Genty, D. Environmental pressures on conserving cave speleothems: Effects of changing surface land use and increased cave tourism. *J. Environ. Manag.* **1998**, *53*, 165–175. [CrossRef]

111. Calaforra, J.M.; Fernandez-Cortes, A.; Sanchez-Martos, F.; Gisbert, J.; Pulido-Bosch, A. Environmental control for determining human impact and permanent visitor capacity in a potential show cave before tourist use. *Environ. Conserv.* **2003**, *30*, 160–167. [CrossRef]

112. Eagles, P.F.J.; McCool, S.F.; Haynes, C.D. *Sustainable Tourism in Protected Areas Guidelines for Planning and Management*; IUCN: Gland, Switzerland, 2002.

113. Williams, P. *World Heritage Caves and Karst*; IUCN: Gland, Switzerland, 2008; p. 57.

114. Del Vecchio, U.; Mitrano, T.; Ruocco, M. Recenti esplorazioni al corso sotterraneo del fiume Bussento. In Proceedings of the Atti del II Convegno Regionale di Speleologia, Caselle in Pittari, Italy, 3–4 June 2010; pp. 21–29.

115. Guida, D.; Ragone, G. Valutazione parametrica della vulnerabilità del sistema carsico del medio Bussento (parco nazionale del Cilento e Vallo di Diano). In Proceedings of the Atti del II Convegno Regionale di Speleologia, Caselle in Pittari, Italy, 3–4 June 2010; pp. 263–276.

116. Mitrano, T. Inghiottitoio del Bussento (CP18). *L'Appennino Meridionale* **2008**, *5*, 252–257.

117. Mitrano, T. *Atti del II Convegno Regionale di Speleologia*; Federazione Speleologica Campana: Caselle in Pittari, Italy, 2010; p. 281.

118. Ruocco, M.; Avitabile, M.; Catuogno, F. Mostre ed attività didattiche al II convegno regionale di speleologia, 3–4 Guigno 2010. In Proceedings of the Atti del II Convegno Regionale di Speleologia, Caselle in Pittari, Italy, 3–4 June 2010; pp. 277–281.

119. Middle Bussento Ponors Website. Available online: http://www.inghiottitoiodelbussento.it/ (accessed on 15 March 2020).

Article

Fieldtrips and Virtual Tours as Geotourism Resources: Examples from the Sesia Val Grande UNESCO Global Geopark (NW Italy)

Luigi Perotti [1], Irene Maria Bollati [2,*], Cristina Viani [1], Enrico Zanoletti [3], Valeria Caironi [2], Manuela Pelfini [2] and Marco Giardino [1]

[1] Earth Science Department, University of Torino, Via Valperga Caluso, 35, 10125 Torino, Italy; luigi.perotti@unito.it (L.P.); cristina.viani@unito.it (C.V.); marco.giardino@unito.it (M.G.)
[2] Earth Science Department "A. Desio", University of Milan, Via Mangiagalli, 34, 20133 Milano, Italy; valeria.caironi@unimi.it (V.C.); manuela.pelfini@unimi.it (M.P.)
[3] Geoexplora, Geologia & Outdoor, V. Mussi 5, 28831 Baveno (VB), Italy; enrico.zanoletti@geoexplora.net
* Correspondence: irene.bollati@unimi.it; Tel.: +39-02-50315516

Received: 30 April 2020; Accepted: 27 May 2020; Published: 29 May 2020

Abstract: In the 20th anniversary year of the European Geopark Network, and 5 years on from the receipt of the UNESCO label for the geoparks, this research focuses on geotourism contents and solutions within one of the most recently designated geoparks, admitted for membership in 2013: the Sesia Val Grande UNESCO Global Geopark (Western Italian Alps). The main aim of this paper is to corroborate the use of fieldtrips and virtual tours as resources for geotourism. The analysis is developed according to: (i) geodiversity and geoheritage of the geopark territory; (ii) different approaches for planning fieldtrip and virtual tours. The lists of 18 geotrails, 68 geosites and 13 off-site geoheritage elements (e.g., museums, geolabs) are provided. Then, seven trails were selected as a mirror of the geodiversity and as container of on-site and off-site geoheritage within the geopark. They were described to highlight the different approaches that were implemented for their valorization. Most of the geotrails are equipped with panels, and supported by the presence of thematic laboratories or sections in museums. A multidisciplinary approach (e.g., history, ecology) is applied to some geotrails, and a few of them are translated into virtual tours. The variety of geosciences contents of the geopark territory is hence viewed as richness, in term of high geodiversity, but also in term of diversification for its valorization.

Keywords: Sesia Val Grande UNESCO Global Geopark; geodiversity; geoheritage; fieldtrips; virtual tours; multidisciplinary approach; Italian NW Alps

---

1. Introduction

Geotourism has been defined as "tourism that sustains or enhances the distinctive geographical character of a place - its environment, heritage, aesthetics, culture, and the well-being of its residents" (National Geographic Society, https://www.nationalgeographic.com/maps/geotourism/). In its updated concept, geotourism is rapidly emerging as a form of urban and regional sustainable development [1]. As Earth scientists involved in Alpine geological and geomorphological studies, we focused the diversity of physical characters in a geographical region of the Western Italian Alps for understanding their significance in terms of natural and cultural heritage, and their possible contributions to the development of a sustainable, environmental-friendly geotourism.

From this perspective, we analyzed several sites and areas of geological and geomorphological interests, particularly those with significant scientific, educational, cultural, or aesthetic value, which are collectively named geoheritage sites [2]. This term has been used widely to define geological features valuable to the society because of their uniqueness [3–5] either in terms of scientific value or educational purposes and tourism [6,7].

Even if geoheritage is a generic but descriptive term, it is based on an advanced and inclusive consideration of Earth's landforms, materials, and processes, recalling the geodiversity concept. According to Gray's [8] definition, geodiversity is not just a matter of different features, but also of their assemblages, structures, systems, and contribution to landscapes. The complexity of geodiversity is a challenge for its study, but also an opportunity to be explored for the possible recognition of geoheritage sites and the establishment of tourist destinations for providing local and regional economic benefits [9,10].

As geodiversity represents a basis for the geotourism, it can be considered an important resource for the local and regional development [11]. Therefore, for the effective enhancement of geotourism, we considered both its geological and territorial dimensions:

1. "landscape, interaction and time": the geodiversity deeply contributes not only to the structure of a geographical area, but also to its cultural meaning and to its perception by people, either within natural or urban areas [12,13]. As a result, the landscape character derives from the action and interaction of natural and/or human factors (European Landscape Convention of Firenze; [14]) and their historical changes [15];

2. "territorial dimensions of geodiversity and geoheritage": within a territory, legal and economic issues related to the protection, preservation, and exploitation of geoheritage also have to be considered [8,16]. A comprehensive, integrated approach to the management of any natural heritage should be addressed, for combining directions from international, national, and regional laws and for stimulating a balance between the need to protect and enhance the natural heritage and the legitimate needs of local populations or visitors [17].

Such a comprehensive approach can offer relevant contributions to both geotourism activities and sustainable use of georesources (including hydro-georesources, sensu Perotti et al. [18]). Significant, long-lasting, international experiences are available for enhancing geotourism. A European Working Group for Earth Sciences Conservation (EWGES) was created in 1988, then (1993) transformed in the ProGEO association (http://www.progeo.se), devoted to diffusion of activities on Earth Sciences and to the establishment of an international network for geosite inventory and conservation. A similar working group was created within the International Union of Geoscience (IUGS), whose outreach activities were the "Geosites" (global inventory of sites of geological interest) and "Geoparks" (a UNESCO partnership for promoting territories including geosites). This site-based approach has worked extremely well for geoconservation purposes [19,20], although geosites cannot by themselves maintain and enhance geodiversity. If complemented by a "regional geothematic" approach, however, the benefits to geoconservation can be enormous. This approach encompasses a targeted selection of sites to create areas of outstanding value within a certain region, based on the attractiveness of scientific knowledge and the possibility of the sustainable fruition of both cultural and natural heritage [21]. The targeted selection of sites within geothematic areas makes easier efforts of regional geodiversity enhancement through careful planning of investments for local activities such as geotourism. In fact, as schematized by Brilha [22], *geodiversity sites (on-site; in-situ)* and *geodiversity elements (off-site; ex-situ)* could be selected as representative of the geodiversity of a region, and when they are recognized as being of relevant scientific value, they become *geosites (on-site; in-situ)* or *geoheritage elements (off-site; ex-situ)* within the wider context of the geoheritage.

An ideal methodological framework for making geotourism effective could be one devoted to the creation of material and virtual field trips aimed to raise awareness of the importance of our geoheritage, and the need to conserve it, amongst different actors: teachers and students, decision makers and stakeholders, entrepreneurs and the general public [23].

The importance of fieldtrips has been underlined since a long time having the function of a "direct experience with concrete phenomena and materials" [24]. When fieldtrips are then intended for schools, moreover, a learning-by-doing approach is favored, including the activity of "observation, identification, measurements and comparison" [24–26]. Bollati et al. [27], for example, proposed a multidisciplinary approach to physical landscape reading, based on the use of vegetation, to reconstruct the evolution of landforms (i.e., dendrogeomorphology) along a geotouristic itinerary. This kind of approach towards physical landscape evolution under geomorphic processes action, and its different responses according to geodiversity, could also be useful to people for hoping to gain awareness of Earth as a complex system, whose dynamics may induce hazards and risks [27–29]. If these experiences are then proposed in iconic sites (i.e., geosites of national or international relevance; [30]) they acquire even more efficacy.

Digital tools—geo-information, geo-visualization, digital monitoring, and Geographic Information Systems (GIS)—are allowing new approaches to geoheritage assessment and mapping [31,32], and geotourism communication and education [28,33]. Direct interactions between institutions and users, and the general public or schools, are enhanced, and favored on a worldwide scale [34]. Digital tools applications for key geoheritage areas are rapidly evolving, and there are several examples [32,34–37].

An important innovation is that many areas are expanding their reach to a global audience by the use of mobile apps, that are set for running on small, wireless devices, such as smartphones and tablets, rather than desktop or laptop computers. A mobile app can provide navigation aid in general, or can be made specific to a park or other protected area, to provide users with real-time guidance and knowledge as they either explore in the field, or virtually navigate the area at home [38]. Some protected areas provide physical or digital visitors with specialized apps equipped with a virtual park ranger or specialist as storyteller, offering particular views on trails. For example, the PROGEO Piemonte project [23,39] has developed two mobile apps devoted to exploring specific areas in the Western Italian Alps [40]. In the last few decades, moreover, the possibility to instantly share events and news and cross-posting between many platforms at the same time [34], is the favored dissemination of information among people, as it allows them to share their opinions and rate their experiences visiting different sites.

The aims of the paper are, hence, to: (i) illustrate the geodiversity of the territory of the Sesia-Val Grande UNESCO Global Geopark (Western Italian Alps) by means of on-site and off-site geoheritage (sensu Brilha [22]); (ii) to propose different approaches to geoheritage valorization, depending on the features of each specific area, using examples of fieldtrips and virtual tours as container of on-site and off-site geoheritage from the Sesia-Val Grande UNESCO Global Geopark.

2. Study Area

2.1. Location

The Sesia-Val Grande UNESCO Global Geopark (SVUGG) is located on the north-east of the Piemonte Region (NW Italy), and encompasses areas of the Verbano-Cusio-Ossola, Biella, Novara and Vercelli Provinces (Figure 1). It is bordered to the west by the Monte Rosa massif, to the north by the Ossola and Vigezzo Valleys towards the Swiss border, to the south-east by Lake Maggiore and, to the south, by an area degrading towards the Po plain. The SVUGG includes, in the north, the whole Val Grande National Park and surrounding territories and, in the south, most of the mountain range of the Sesia river basin, including the whole Sesia Valley, and portions of neighboring territories, such as Valsessera, and Strona Valley.

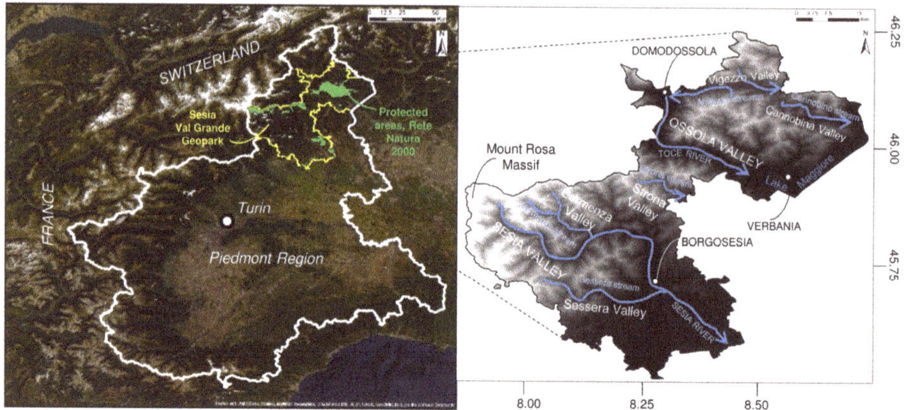

Figure 1. Geographical location of the Sesia-Val Grande UNESCO Global Geopark. On the left, the position of the SVUGG within the Piemonte Region with the protected area's location; on the right, the Digital Terrain Model (5 m resolution, source Geoportale Regione Piemonte; http://www.geoportale. piemonte.it/geocatalogorp/?sezione=catalogo) highlighting the articulation of the relief in the geopark with the main streams, lakes, and peaks.

2.2. The Sesia-Val Grande UNESCO Geopark History

The Sesia Val Grande UNESCO Global Geopark (SVUGG) is member of the European Geopark Network (EGN) since 2013, and of the UNESCO Global Geoparks Program-UGGP since November 2015. It covers an area of about 2202 km^2 and has a perimeter of around 423 km.

The Val Grande National Park (www.parcovalgrande.it), the first partner of the geopark and entirely located within the geopark, was established in 1992 by the decree of the Italian Environment Ministry. It is recognized as the largest wilderness area in Italy, but also in Europe, and since its creation it protects both habitats and endangered animal and botanical species. The National Park was also established as a Site of Community Importance (SCI) and a Special Protection Zone (SPZ) of the Natura 2000 network, because it preserves 10 priority habitats in its territory. Since 2007, the geological heritage also became a strategic target of the activities in the Park, finally leading to its candidature for the European Geopark Network (EGN).

The second partner of the geopark is the Geotouristic Association of the Valsesia Supervolcano (www.supervulcano.it/home.html), which represents the area of the Sesia Magmatic System (middle and lower Sesia Valley), and includes among its members two Natural Parks: Monte Fenera and Alta Valsesia.

The SVUGG comprises other protected areas as the Natural Park of the Alta Val Strona, the Natural Reserves of Baragge and Fondo Toce, the Special Reserves (UNESCO Heritage Sites) of Ghiffa, Varallo, and Domodossola Sacri Monti, as well as other protected areas (Oasi Zegna, Oasi Bosco Tenso, and Pian dei Sali).

The SVUGG has a website (http://www.sesiavalgrandegeopark.it/) and a social network page where updates on events and initiatives are available (https://www.facebook.com/pg/AssociazioneSesiaValGrandeGeopark/posts/).

As a whole, the territory of the geopark offers its visitors the opportunity to observe the effects of geologic processes, which formed the continental crust at different depths. It also introduces them to the concepts of global plate tectonics, as it is located astride the Insubric Line, representing a major alpine lineament that marks the boundary between the Central Alps, consisting of intricate refolded basement nappes, and the Southern Alps with S-vergent thrusts [41].

Since the geopark extends from the Monte Rosa massif to the northern boundary of the Po Plain, it also shows the record of past climate changes and of the glacial, periglacial, water- and gravity-related processes, which continuously shape the landscape [42].

Last but not least, the geopark territory is also an open-air museum of the ancient civilization of the Alps, since it preserves the traces of a "stone culture" of different ages: from the Paleolithic human settlements in the Monte Fenera caverns, up to the historical use of local georesources, and the construction of defense works in World War I, by taking advantage of the landscape morphology.

Some research projects devoted to the dissemination of Earth Sciences among schools and general public had the SVUGG as focus area. The most important ones are:

- The ERASMUS+ Project "GEOclimHOME: Geoheritage and climate change discovering the secrets of home" (https://geoclimhomeblog.wordpress.com/). This is a three-year project funded in 2015 by the Erasmus+ Programme within the Key Action 2 (cooperation for innovation and the exchange of good practices which promote strategic partnership for school education). It supports high school teaching and scientific research between Rokua (Finland) and Sesia-Val Grande (Italy) UNESCO Geoparks. It aims to improve both the general perception of climate and environmental changes in Europe and the appraisal of geoheritage. The project was implemented with the participation of the Chablais UNESCO Geopark (France).
- The PROGEO-Piemonte Project (http://www.progeopiemonte.it/). This offers an innovative approach for the management and enhancement of the geological heritage of the Piemonte Region. Within the project, nine "geothematic areas" have been identified for representing the regional geodiversity of the Piemonte Region. Geological sites identification, the enhancement of museum collections, the activation of educational projects with the schools, the installation of exhibitions, and nature trails designed to promote the geotourism, by means of virtual tours and fieldtrips, are among its main aims. Its main actions concern: (i) the progress of scientific knowledge; (ii) land development, education, and communication through innovative methodologies; (iii) collaboration with the local communities in order to involve them and provide them with benefits.
- SITINET – Geological and archeological sites of the Insubria Region. This was a project devoted to the inventory of the geological and archeological spots in the Insubria region, including the Ossola area (http://www.sitinet.org/). It ended in 2013, just during the acceptance of the study area in the EGN, which was also a starting point for the inventory of the geoheritage of the geopark.

Another relevant project currently ongoing in the SVUGG, and particularly related to the Val Grande National park territory, is the COMUNITERRAE - Maps of Cultural Communities of Alpine Landscapes in the Val Grande National Park (http://www.comuniterrae.it/). It is a project by the Associazione Ars.Uni.Vco and Val Grande National Park, included in the European Chart for Sustainable Tourism. It aims to promote innovative methods to ensure sustainable features, and to clearly communicate the project benefits both to the local and to a wider public. The valorization of assets, places, and components of the material and immaterial heritage of a territory along centuries is the main focus. In 2019 the project was awarded with the European Heritage Award 2019 in the category "Education, Training and Awareness-Raising".

The different aspects of the geopark are then described in the following paragraphs.

2.3. Geology

The geologic context exposed in the SVUGG territory is of high scientific interest, as witnessed by thousands of papers published in the last 50 years by researchers from all over the world.

The geodiversity which characterizes the geopark results from processes lasted over 500 million years, whose effects are still recognizable in the field.

The SVUGG (Figure 2) is stretched along the Canavese Line (CL), a segment of the Insubric Line, a major tectonic boundary separating the N-vergent nappes of the Central Alps (Austroalpine and

Pennine domains), affected by the Alpine metamorphism, from the S-vergent South Alpine domain [43] (to the SE), a pre-Alpine metamorphosed basement with its sedimentary coverage.

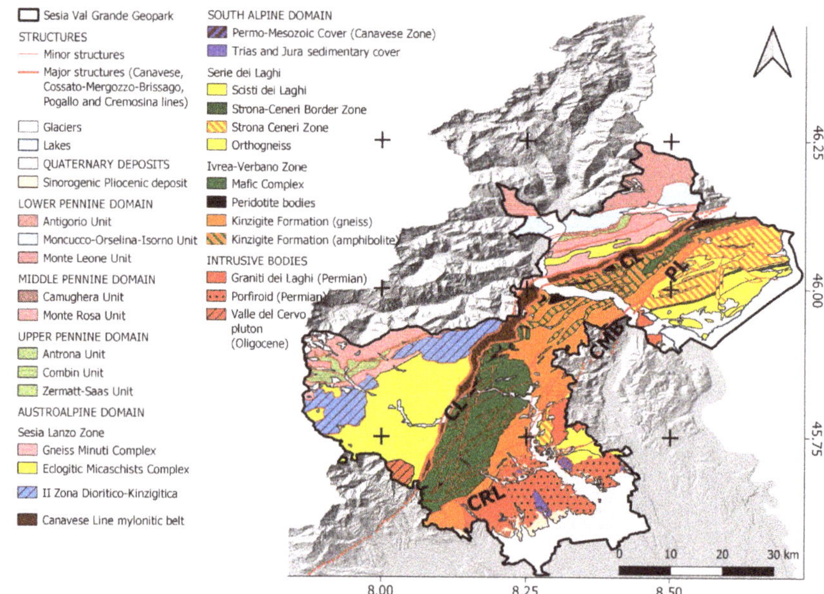

Figure 2. Regional Geological setting of the Sesia-Val Grande UNESCO Global Geopark from the ARPA Piemonte Geological map (1:250000) (modified from https://webgis.arpa.piemonte.it/geoportale/).

Most of the territory of the geopark (Figure 2) belongs to the South Alpine domain. Along the CL, which crosses the area in SW–NE direction, the alpine units are represented by the Austroalpine domain, whereas the other nappes (Lower, Middle and Upper Pennine domains) crop out only along the north-western boundary of the geopark.

Therefore, the South Alpine domain and its relations with the Austroalpine domain along the CL are the focus of most of the geotouristic activities proposed to the general public along geotrails, in exemplary geosites (see Sections 4.1 and 4.2), as well as in different thematic museums and laboratories (see Section 4.3). The main geological themes will be described in the following paragraphs.

2.3.1. The Canavese Line

The Canavese Line (CL) (sensu Schmid et al. [44] and Steck et al. [41]) here represents the contact between the Austroalpine domain, to the north-west, involved in the Alpine metamorphism, and the South Alpine domain, to the south-east, which preserves much older structures, despite having experienced some Alpine tectonic deformation in greenschist to anchizonal facies [45].

The Austroalpine domain is here represented by the Sesia-Lanzo Zone, a composite unit (Gneiss minuti and Eclogitic Micaschists Complexes) with a polyphase deformation history (HP/LT; mainly blue-schist to eclogite facies conditions; [46,47]) related to specific phases of the Alpine orogeny (Late Cretaceous–Early Tertiary), and by the II Dioritic-Kinzigite Zone, characterized by micaschists, gneisses, and metabasites in granulite to amphibolite facies [48].

The CL is visible on the field as a greenschist facies mylonite belt [41], up to 1 km thick, which may involve rocks belonging to:

i.   The northern border of the Ivrea-Verbano Zone (South Alpine basement), with its Permo-Mesozoic cover;

ii.   The Canavese zone (South Alpine domain; [49]), consisting of amphibolite facies basement rocks overlain by Permian silicoclastic sediments and Triassic-Liassic carbonate rocks (sedimentary and metamorphic); the latter were accreted to the Ivrea-Verbano Zone margin, before the Alpine metamorphic greenschists facies overprint and are regarded as the distal continental margin of the Adria plate facing the Ligurian–Piedmont Ocean [49];

iii.   The southern border of the Sesia-Lanzo Zone (Austroalpine domain) [44], well exposed in the Ossola Valley.

In the Loana Valley, near the northern boundary of the Val Grande National Park, the mylonite occurrence (i.e., the Scaredi Formation [41]) is particularly meaningful [44,50–52]. There, mylonites derived from the Permo-Mesozoic cover rocks are dismembered, and often imbricated with or folded into the Ivrea-Verbano-derived mylonites; Sesia-Lanzo mylonitized gneiss also occur along restricted bands [49].

2.3.2. The South Alpine Domain

In the SVUGG area, the South Alpine domain is represented by the Massiccio dei Laghi [53,54], which comprises two main lithotectonic units: The Ivrea-Verbano Zone and the Serie dei Laghi. After Ferrando et al. [49], it also comprises the Canavese Zone (see before).

The Massiccio dei Laghi exposes a spectacular cross section from the lower crustal levels of the Ivrea-Verbano Zone to the middle and upper crustal levels of the Serie dei Laghi. It is considered a model for a magmatically underplated and extended crustal section [55,56].

The Ivrea-Verbano Zone mainly consists of two portions:

i.   The Kinzigite Formation: a metamorphosed volcano-sedimentary sequence, composed of dominant metapelites, with minor quartzites, thin meta-carbonate horizons and interlayered metabasites [57]. Mantle peridotite lenses are tectonically interfingered with the metasedimentary rocks [58], especially in the north-western part, near the CL (Balmuccia in the Sesia Valley, Premosello in the Ossola Valley and Finero in the Cannobina Valley: [59] and references therein). The metamorphic grade decreases from the granulite facies in the northwest to the upper amphibolite facies in the southeast [60].

ii.   The Mafic Complex: gabbroic to dioritic intrusive rocks, representing the deepest level of the Sesia Magmatic System (described below).

The scientific importance of the Ivrea-Verbano Zone is once more underlined by the recent Project Drilling the Ivrea-Verbano Zone (DIVE), which, through four drilling operations in the Sesia and Ossola Valleysaims to unravel the physico–chemical properties and architecture of the lower continental crust towards the crust–mantle (Moho) transition [61].

The Serie dei Laghi [54] consists of different units (from NW to SE):

i.   Strona-Ceneri Zone: this consists of two types of paragneisses. The Gneiss Minuti, fine-grained metasandstones still preserving relicts of sedimentary structures, and the Cenerigneiss, coarse-grained to conglomeratic gneisses containing a variety of enclaves (quartzite pebbles, nodules rich in aluminium silicates, fragments of metamorphic rocks). Both gneisses contain calc-silicate enclaves, deriving from calcareous concretions frequent in arenaceous deposits. Gneiss Minuti and Cenerigneiss are respectively interpreted as well sorted deposits from turbidity currents and as mass flow turbidites, deposited in an accretionary prism along an active continental margin [62,63].

ii.   Strona-Ceneri Border Zone [64]: a continuous horizon, one to several hundreds of meters thick, between the Strona-Ceneri Zone and the Scisti dei Laghi. It mainly consists of banded amphibolites, with lenses of ultramafites, metagabbros, garnet bearing amphibolites (retrogressed eclogites) and minor paragneisses. The banded amphibolites are an example of the Leptynite-Amphibolite Group (LAG), an association widespread throughout the Hercynian belt in Europe. The LAG is formed by tuffites of alternate mafic and acidic composition deposited in a marine environment [65].

iii. Scisti dei Laghi: mainly garnet and staurolite and kyanite micaschists, with minor paragneiss intercalations.

Thick Orthogneiss lenses are intercalated in all these units, but mainly within or close to the Strona-Ceneri Border Zone. They are metaluminous tonalites to granites with calcalkaline affinity ([66,67] and references therein) and an Ordovician intrusion age around 466 Ma (Rb-Sr whole rock isochron; [68]).

The intrusives and their sedimentary host rocks suffered together the Variscan orogenic metamorphism, mainly in amphibolite facies conditions, recorded by mineral ages of 311–325 Ma [67].

The original contact between the Ivrea-Verbano Zone and Serie dei Laghi is the Cossato-Mergozzo-Brissago Line (CMB; [69]), an important subvertical tectonic lineament characterized by the simultaneous occurrence of three distinctive features [53]: high-T mylonites, migmatites, and mafic to intermediate dykes and stocks (called the "Appinite Suite" [69]), mostly concordant with the CMB mylonitic foliation. The best estimate of the intrusion age of the Appinites is a U-Pb age of 285 ± 5 My [70] on a monazite from a dyke near Mergozzo. The CMB is cut at low angle and dislocated by the Pogallo Line [53], characterized by amphibolite to greenschist facies mylonites and by the lack of Appinite intrusions. In the Sesia sector, the CMB has been reactivated by a younger fault (Cremosina Line; CRL) [53].

The last large-scale event in the Serie dei Laghi was the intrusion of granitic magmas forming different plutons (Graniti dei Laghi) outcropping along the southeastern border of the SVUGG. The most famous of them are the Montorfano and Mottarone–Baveno plutons (dated at 275 Ma [67]).

2.3.3. The Sesia Magmatic System

The Sesia Magmatic System is part of a large Late Carboniferous to Early Permian igneous province [71], a bimodal suite of basic and silicic volcanic and plutonic rocks outcropping across Europe from Spain to Scandinavia in association with extensive crustal rifting. From the lower to the upper levels, it consists of:

i. The Mafic Complex (part of the Ivrea-Verbano Zone): this is an 8-km-thick composite layered intrusion (peridotites, pyroxenites, norites, and the main gabbro [72–74]), which intruded the deep crust around 288 Ma ago [75]. Along the intrusive contacts, partial melting of the kinzigites produced migmatites within 1 to 2 km from the intrusion [76]. Residual melt from the Mafic Complex and silicic melt generated by anatexis migrated to higher crustal levels.

ii. The Valle Mosso granite: this is a compositionally zoned pluton [77] grading upwards into a fine-grained to granophyric facies with miarolitic cavities. It also contains some basaltic to andesitic dykes and intrudes the base of the overlying caldera.

ii. The Sesia Supervolcano: this forms the upper part of the system, together with relics of a bimodal volcanic field of basaltic andesite and rhyolite. The supervolcano, partially covered by younger sedimentary deposits, is a huge rhyolitic caldera with a diameter exceeding 15 km and an estimated volume of ignimbrite erupted above 300 km^3 [78]. The caldera-forming events are well documented along the Sesia Valley and its hydrographic network, with beautiful exposures of volcanic megabreccia within the welded rhyolitic ignimbrite that fills the caldera, and huge blocks of country rocks (Scisti dei Laghi) slided into the caldera during the eruption. After Quick et al. [78], volcanism lasted approximately 6 My, beginning about 288 Ma and culminating in the caldera-forming eruption at about 282 Ma. The karstic Triassic marine carbonate of Monte Fenera is deposited on the caldera ignimbrite.

2.4. Geomorphology

From a geomorphological point of view, the Sesia Val Grande UNESCO Global Geopark (SVUGG) preserves several geomorphological landscapes tracing back to the long-term modelling history of the Alpine relief. A diversity of landforms recalls ancient and present surficial processes, which shaped the landscapes by means of their dynamic interactions with geological and tectonic conditioning factors. As a result, the altitudinal range of the whole SVUGG area is articulated and extreme: from the 4634 m a.s.l. of Monte Rosa to less than 200 m a.s.l. of the Po plain, down to the bottom of Lake Maggiore, a crypto-depression at −179 m. The Geomorphodiversity (sensu Panizza [79]) of the geopark is presented below, with references to both the geological constrains and the morphodynamic processes affecting the geomorphological landscape.

2.4.1. Lithostructural Constrains and Long-term Geomorphological History

The SVUGG area belongs to the Western Alps, an arch-shaped mountain chain which shows, at the regional scale, an asymmetrical transversal profile: the inner SE-facing side is shorter and steeper than the outer NW-facing side. Within the inner side of the Alps (Figure 1), the SE-NW altitudinal profile of the geopark shows morphological steps with distinctive mean elevation: from the upper plain (altitude between 200–350 m a.s.l.), to the foothills (350–1000 m a.s.l.) through all to the mountain relief (1000–4634 m a.s.l.) up the current alpine watershed [42].

At the regional scale, these distinctive morphological steps correspond roughly to the distribution areas of major geological complexes of the Western Alps, namely (from SE to NW): The Quaternary deposits of the Po plain and synorogenic Pliocene deposits, the sedimentary, magmatic, and metamorphic units of the South Alpine domain and the metamorphic units of the Austroalpine and Pennine domains (Figure 2) [80,81].

At the local scale, a diversity of peculiar relationships can be recognized between landforms and lithotypes. Some examples include: enhanced effects of differential erosion along the valleys, where schist units outcrop between massive magmatic or metamorphic rocks, (e.g., the morphological change corresponds to the narrow, deeply incised gorge of the Mastallone river within the hard diorites, while upstream of Fobello, a large valley developed, due to most erodible metamorphic schists); microscale competence contrasts between hard rock inclusions within pyroclastic breccias; subsurface karst landforms for dissolution phenomena of carbonatic rocks.

Moreover, strong conditioning factors to the geomorphological landscape are due to the geometrical setting of regional schistosity, to local geological structures, and to major tectonic discontinuities. At a regional scale, this is particularly evident along the shear zones related to the Canavese, Cossato-Mergozzo-Brissago, Pogallo and Cremosina Lines (Figure 2). As an example, the whole NE sector of the SVUGG is dominated by marked NE-SW (within Serie dei Laghi Unit) and NNE-SSW (within Ivrea-Verbano Unit) trends of morphostructures, either represented by deep incised tributary valleys (Figure 3b) and linear segments of the hydrographic network (Figure 3c).

The alpine valleys across the Geopark develop radially from the Po Valley towards the watershed. The two major valleys (Sesia and Toce valleys) are deeply incised in the bedrock and their slopes sometimes exceed 3000 m a.s.l.. The engravings of these alpine valleys can be dated to the Messinian [82] or to the Pliocene [83]. This long-term geomorphological history is witnessed by landforms and deposits within the Geopark, such as the Lake Maggiore cryptodepression, the deep valleys and the foothills sector where marine sediments were deposited during the Pliocene [84]. This ancient modeling of the mountain chain was followed by the deposition of a continental regressive sedimentary sequence between the middle Pliocene and the lower Pleistocene (Villafranchiano; [84]).

Thereafter, following the climatic changes of Quaternary period, the entire Alpine sector of Sesia Val Grande was repeatedly occupied by important glacial masses. The glacial pulsations have shaped the main valleys of the area, strongly influencing the current geomorphological and hydrographic regional structure.

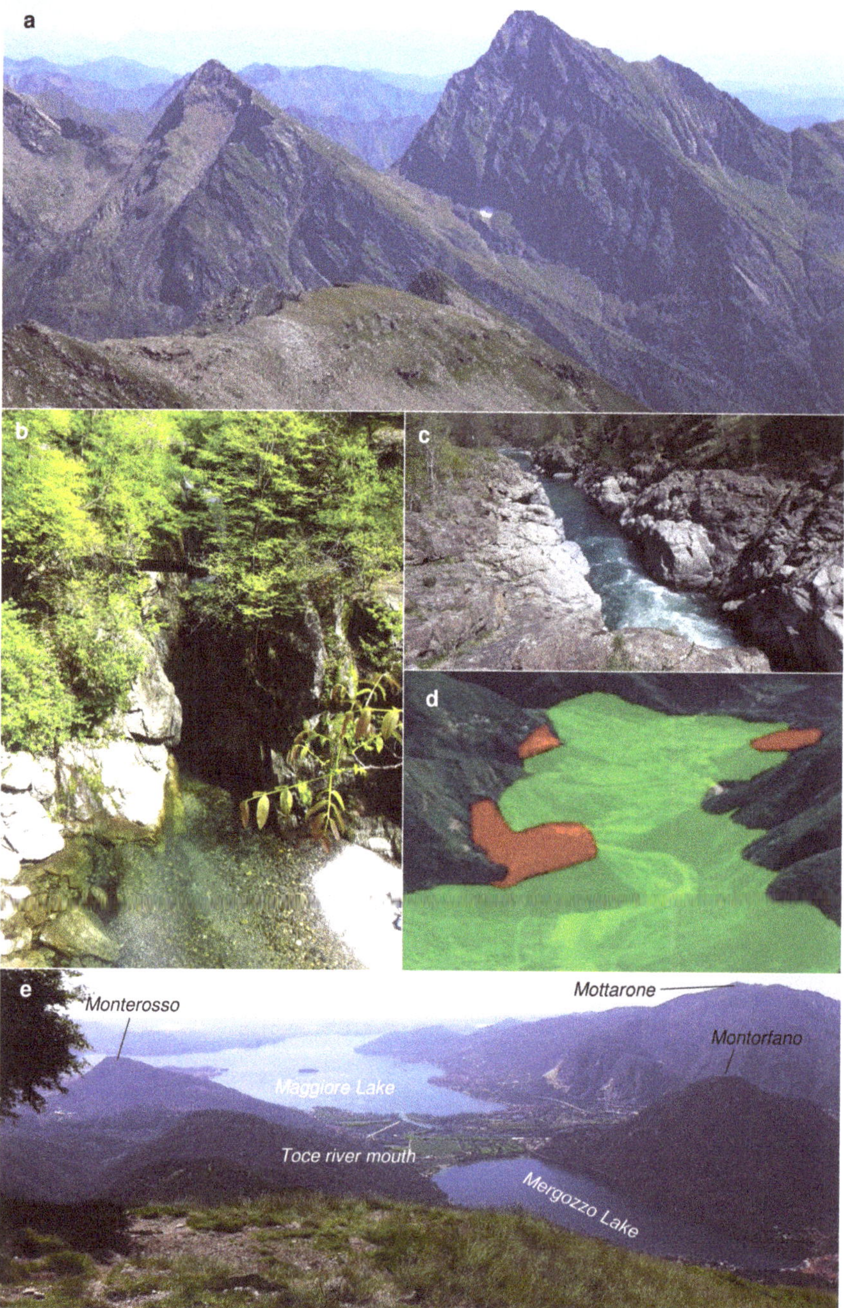

Figure 3. Example of geomorphological features characterizing the Sesia Val Grande UNESCO Global Geopark: (**a**) Tagliaferro peak in Sesia Valley; (**b**) incised tributary stream of the Pogallo Valley; (**c**) linear stream at Balmuccia (Sesia Valley); (**d**) Unipiano and Sacro Monte di Varallo glacial terraces; (**e**) Toce river mouth in the Maggiore Lake, with the Montorfano peak on the right.

2.4.2. "Recent" and Present-day Geomorphological Landforms and Processes

Within an area of large geomorphodiversity, the SVUGG offers both a live demonstration of active glacial processes and a window on past climate changes recorded in the Pleistocene landforms, marked by repeated glacial advances and retreats.

The onset of Quaternary glaciations in the Western Alps led to the formation of a large ice sheet, and major valley glaciers [85]. Within the geopark, both erosional and depositional landforms witness the extent of major Pleistocene glacial modelling phases from the Monte Rosa massif to the Sesia and Ossola Valleys. Starting from the higher elevation areas, erosional landforms such as magnificent nunatak-like mountains are visible (e.g., Mud Horn and Tagliaferro peak, high Sesia Valley, Figure 3a), large and steep U-shaped valleys (e.g., Toce Valley, Figure 3e), as well as trimlines and in-valley sequences of erosional and depositional landforms (e.g., the Unipiano and Sacro Monte di Varallo glacial terraces, examples of further valley deepening after glacial erosional modelling, Figure 3d).

Distinctive depositional features all around the Maggiore Lake witness the Toce and Ticino glaciers extension to the upper Po plain through the Verbano lobe and towards the Orta lake [86] leaving few areas free of ice cover, such as Monte Mottarone, SW of Verbania. According to geomorphological evidences, radiocarbon datings, and numerical models [87–89]; also, the Pogallo Valley was supposedly indicated as of incomplete glacial coverage during the Late Glacial Maximum (LGM, around 24ky BP, in the geopark area), thus making possible survival of older structural and fluvial landforms.

Successively, during the Little Ice Age (LIA, XIV-XIX century), favorable climatic conditions allowed local glacier expansion. Later, SVUGG glaciers experienced a strong retreat; from the end of the LIA until now, they lost about the 50% of their area (from about 7 km^2 to about 3.5 km^2, [90,91]). Currently, according to the New Italian Glacier Inventory [92] only seven glaciological units exists on the slopes of the Monte Rosa southeast side: four glaciers and three snowfields. As a result of the ongoing climate change, their shrinkage is still underway.

The possible appraisal of several environmental changes through time as a consequence of glacial, periglacial, water- and gravity-related processes, is among the most attractive characters of the geomorphological heritage in the geopark. A thematic representation of the SVUGG geomorphological landscapes is offered in Figure 4, which is a land-systems map created as a factor map for regional assessment of geodiversity within the geopark. GIS synthesis and interpretation of regional datasets with geomorphological contents (hydrography, glaciers, glacial cirques and periglacial features, landslides, debris-flows, and alluvial fans) allowed the recognition of areas characterized by prevalence of a certain type of landforms and related processes. These are useful for regional enhancement of geomorphodiversity and framing of local geosites.

As shown in the map, a large area of the geopark is characterized by the hydrographic basins of two main rivers (Sesia and Toce) and by slopes providing several insights on the effects of water-related processes. The selection of geosites and geotrails concerning fluvial and water related landforms and processes, that could also affect their evolution (e.g., [93,94]), is of great importance for the enhancement of SVUGG geoheritage. This particular aspect of geodiversity (i.e., hydro-geodiversity sensu Perotti et al. [18]), in fact, includes several abiotic ecosystem services that have been recognized of high relevance for the local community, either for their material contents, for the provisioning character (water for human and agricultural consumption and for renewable energy), or for their valuable contribution to cultural and leisure activities (environmental education, tourism, sports).

To complete this geomorphological framework, examples of karst morphologies (i.e., mainly hypogean caves) are also present where carbonatic rocks occur, such as in the Monte Fenera area.

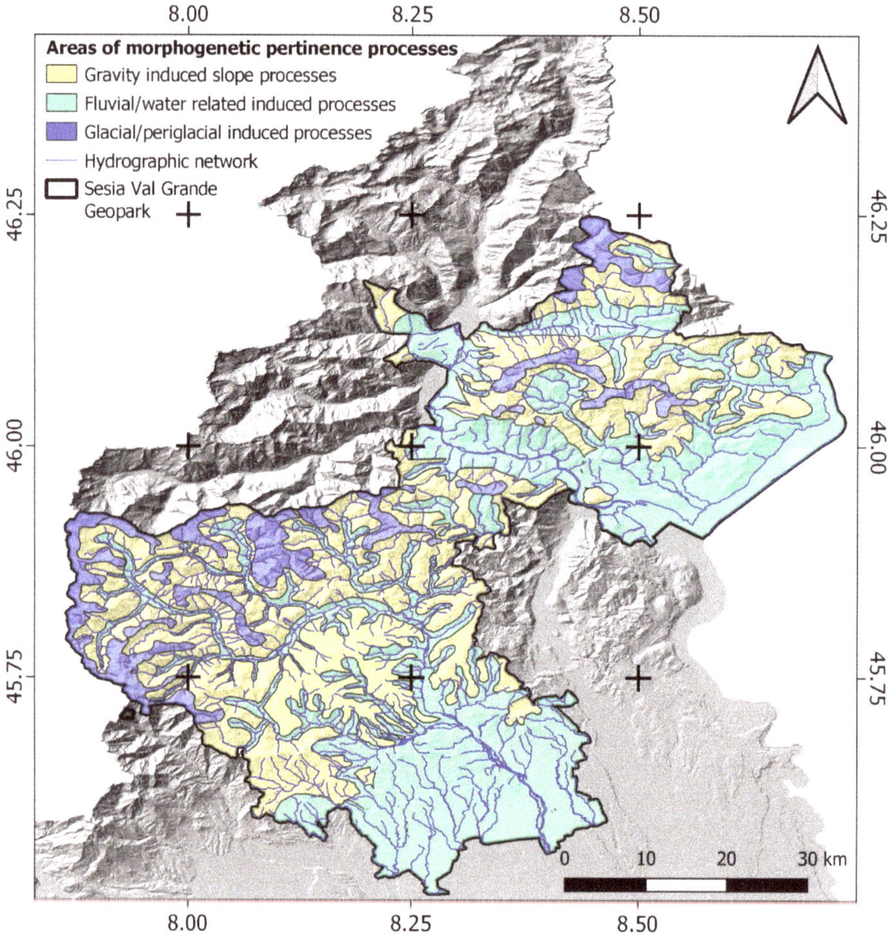

Figure 4. A land-systems map created as a factor map for regional assessment of geodiversity within the Sesia Val Grande UNESCO Global Geopark.

2.5. Georesources and Ancient Human Settlements

Georesources within the geopark territory are strictly related to its high lithological geodiversity and morphological traits. These features are related to the abiotic ecosystem services (such as provisioning and cultural services, as outlined by Gray [8]).

Quarrying and mining were important economic activities in the geopark area and its vicinity for centuries, whereas, at present, all of the mines and most of the quarries are closed and the land rehabilitated, in some cases through geotouristic valorisation activities (see Section 4.3).

The Ossola Valley is one of the most important quarrying areas of the Italian Alps [95–97], materials are used since a long time all around the national territory and also abroad. Several web resources are available to explore this richness (http://www.pietredelvco.it/; http://pietredelcusio.weebly.com/).

The most relevant georesources for the local communities, in the portion of the Ossola Valley within the SVUGG territory, are:

- The Permian Graniti dei Laghi: Mottarone-Baveno (Figure 5a), Montorfano, and Mergozzo quarries were active since the XVI century, and their pink, white, and green granites were extensively used in architecture outside the region [97,98].

- Carbonate metasedimentary rocks, dated back to Permian-Mesozoic, and marbles. The metacarbonates have a stripe-like distribution along the Canavese Line, within the Ivrea-Verbano and Canavese Zones. Especially in the Ossola Valley, they were used for lime production within the lime kilns (e.g., the Loana Valley [99]; SF 4-I). Marbles occurring as lenses within the Ivrea-Verbano kinzigites were extensively quarried as ornamental rocks since ancient times: the most famous quarry is the Cava Madre in Candoglia, exclusively reserved since 1387 AD for the Milan Cathedral (Duomo di Milano) (Figure 5b) [100].

- Talc and/or chlorite-rich metamorphic rocks, deriving from mafic—ultramafic protoliths within the Austroalpine domain, locally known as "Pietra laugera" or "Pietra ollare". They are "soft", easily workable stones typically used for jars, pots, and pipes (Figure 5c) [101].

Figure 5. Georesources providing abiotic ecosystem services to the community within and outside the Sesia Val Grande UNESCO Global Geopark: (**a**) Mottarone and Montorfano granite: on the left, a rocky chain in Rome city made of Montorfano white granite and the columns in Baveno pink granite; on the right, a view on the Baveno quarry with Montorfano quarry in the background; (**b**) the Cava Madre in Candoglia (on the left), exclusively reserved since 1387 AD for the Milan Cathedral on the right); (**c**) an outcrop with the traditional signs of Pietra ollare extraction (above) and a pot made of the same rock and conserved in the Ecomuseo ed leuzerie e di scherpelit (below, courtesy of Archivio Ecomuseo by Riccardo Rapini).

Concerning the human settlements in the area, both in the Ossola Valley and in the Sesia Valley, there is evidence of human habitation dating back to the Paleolithic, as it would have had favorable environmental conditions. In the lower Sesia Valley, within the karstified Triassic marine carbonate complex of Monte Fenera, caverns utilized by Paleolithic inhabitants can be found [102]. Moreover, famous petroglyphs, carved into different lithologies, probably related to religious ceremonies, are located at the Alpe Sassoledo (Figure 6a), Alpe Prà, Alpe Pianzà (Figure 6b), and Malesco villages, in the Ossola Valley. In particular, the petroglyphs of the Alpe Sassoledo (Figure 6a) were the source of inspiration for the creation of the Val Grande National Park logo. Finally, remnants of an ancient necropolis ascribed to the Leponti population (II-I century BC) were found near Ornavasso [103].

Figure 6. The (pre)historical background signs retrieved in different localities withint the Sesia Val Grande UNESCO Geopark: (**a**) Alpe Sassoledo; (**b**) Alpe Pianzà in the Vigezzo Valley; (**c**) Walser tytpical architecture in the Otro Valley.

According to the geographical concept of geotourism we also analysed further development of the use of stone resources in other SVUGG sectors, namely at Alagna ("Im Land" in the Walser German language) an alpine town of Upper Valsesia. This is the access point to the North face of Monte Rosa. It was settled by Walser colonist from Valais, Switzerland in the 14th century: since then it has preserved its alemanic language, culture and architecture (Figure 7c). The present day permanent resident population is about 600 inhabitants, while during winter season, over 5000 tourists per day are present at Alagna Valsesia. It has preserved its pristine character of typical alpine stone village.

Concerning the adaptation of defensive strategies to the morphology of the territory, the geomorphodiversity played an important role during human history. An important historical feature within the geopark is the presence of the Northern defensive border of the Italian territory towards North during the First World War, known as the Linea Cadorna (Figure 7a,b) In the Verbano area and in the lower Ossola Valley, this artificial path makes it possible to reach, relatively easily, high altitude spots in the SVUGG, and has locally become part of geotrails, as described in the results. In the specific case of the Toce Valley bottom, the line passes in correspondence of the narrower part (Stretta di Bara), where the Ossola Valley bottom reaches it lowest width value (about 700 m, (Figure 7c)).

Other examples of this relation are represented by the Special Reserves (UNESCO Heritage Sites) of the Sacri Monti. Among them, the Varallo (Figure 7d) and Domodossola Sacri Monti in particular are located on isolated hills along the Sesia and Toce rivers, respectively, while the Orta Sacro Monte occupy the top of a hill on a peninsula in the Orta Lake (Figure 7e).

Figure 7. Military defense and religious buildings withint the Sesia Val Grande UNESCO Geopark;
(**a,b**) Strada Cadorna at Pian Vadà (**a**) and Monte Spalavera (**b**); (**c**) Strada Cadorna nearby the Stretta
di Bara; (**d,e**) UNESCO Sacri Monti, located in peculiar geomorphological contexts: Varallo (**d**, photo
courtesy of Carlo Pozzoni) and Orta (**e**, Google Earth 3D view).

3. Materials and Methods

3.1. Geoheritage Analysis

In the framework of this research, a complete inventory of the on-site geoheritage (i.e., geosites
and geotrails; in situ sensu Brilha [22]) and off-site geoheritage (museums, geo-laboratories; ex situ
sensu Brilha [22]) has been performed, as potential resources for geotourism.

Concerning the on-site geoheritage, and geosites in the specific, the lists of the geosites included
in the area of the SVUGG, retrieved from the geopark documentation, was examined and revised: they
were related to both the application and the revalidation dossiers of the geopark. In the first phase
of application (2013, http://www.sesiavalgrandegeopark.it/dossier-candidatura.html), during which
the greater part of geosites were identified, they were assessed according to a matrix considering their
scientific value, educational contents, and relevance as geotouristic destinations. In addition, their
potentials for geoconservation, economic valorization, sustainable management, and conscious usage
were carefully considered. This qualitative-quantitative method was applied according to the territory
administrative boundaries, counting the presence of certain parameters (i.e., 0 or 1), and quantifying
some of them (e.g., vulnerability from 1 to 4).

This initial list was then implemented considering the more recent adds related to the UNESCO
Revalidation Dossier (2017). These new geosites derive from the most recent researches carried
out on the SVUGG territory concerning mainly geosites meaning in terms of cultural geology and
geomorphological landscape.

Hence, in the framework of this research, a shapefile containing all these inventoried geosites,
with the respective location (WGS84 coordinates), the geometric properties (point, line, area) and
a brief description, was created. The elevation of the geosites was extracted from the last Digital
Terrain Model of the Piemonte Region (2009–2011; 5 m resolution). In order to provide a homogeneous
classification, the geosites were assigned of:

(i)   A primary and a secondary scientific interest, according to the topics characterizing each
      one of them: GM = geomorphology; GRS = georesources; HYD = hydrogeology; M =
      mineralogy; P = petrography; PAL = paleontology; SD = sedimentology; SS = soil science;
      ST = structural. Moreover, the primary interest was classified according to its level of importance
      at the international, national, regional or local scale (sensu Panizza [104]).

(ii)   Additional interests: A = aesthetic; H = history/archaelogy; S = sport, e.g., [21,105].

All these considerations were made by widening the scope beyond the boundaries of the SVUGG, taking into consideration the scientific researches carried out at the international, national and regional levels, on these geosites and on other similar ones across the Alps.

Concerning geotrails, they were finally inventoried and grouped according to specific topics (GM = geomorphology; GRS = georesources; HYD = hydrogeology; M = mineralogy; P = petrography; PAL = paleontology; SD = sedimentology; SS = soil science; ST = structural). They were described according to type of installations (panels) and support materials (virtual, apps, paper guides). Moreover, they have been put in relation with geosites (related strictly to the geotrails or satellite) and with off-site geoheritage.

The off-site geoheritage has been also inventoried providing a brief description.

These 3 categories (geosites, geotrails and off-site geoheritage) were put in relation each other in specific tables (SF 1–3).

3.2. Methods for Implementing Geotrails

The second main goal of this work, which was to present solutions for enhanced geoheritage within the Sesia-Val Grande UNESCO Global Geopark, implies the popularization of geodiversity by "translating" the complexity of Earth system contents with simple languages [23], thus allowing a knowledgeable approach not only for people involved in the field of geosciences, but also for the general public and professionals involved in educational activities [27,106]. Currently, it is necessary to start cultural growth based on a process of communication and interpretation of our geological heritage, leading people to observe the processes affecting the physical world with greater awareness [28].

Hence, the trails proposed have been implemented with specific tools. Hence, among the listed geotrails, the most meaningful and representative ones mirroring the SVUGG geodiversity and offering a particular approach to landscape view were described in detail.

In the following subsections, the tools used at the selected geotrails, according to the specificity of each trail, are described.

3.2.1. MobileApp and Websites Tools for Enhancement of Virtual Tours within Geosites

In the last few years, the advancement of digital technologies and the large diffusion of Internet facilities have favored, in the field of cartography, but also for educational purposes [107], the increasing use of electronic devices and the development of dedicated software. In addition to the well-established appeal of videos available on the web [107], examples of these progresses are offered by webmap applications: GIS functionality is combined with Internet technology, allowing the publication of cartographical data integrated with other information, including hyperlinks to images and information [31,108]. These solutions can be valuable and comprehensive instruments to present results of Geosciences researches to the general public. In order to reach these goals and to promote the knowledge and the exploitation of the geosites in Piemonte region, a webmap application and mobile app (Figure 8) have been developed, through which it is possible to reach a large number of people. Moreover, this solution is economical and easy for users: whatever hardware or software configuration they are using for internet connection, and with only elementary computer knowledge, users can access the data shared by the webmap with a classic internet browser.

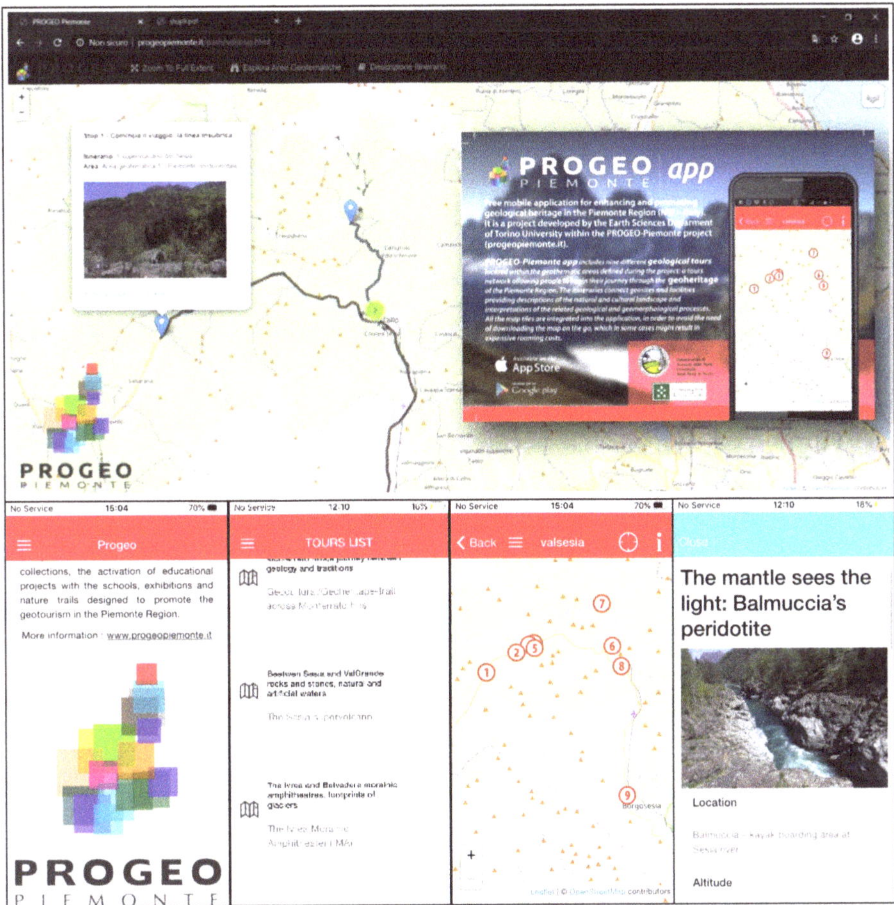

Figure 8. Virtual tours inside the Sesia Val Grande UNESCO Global Geopark–Progeo Piemonte. Virtual App available in IOS and Google Play app stores.

The tools developed in the SVUGG context were largely based on the innovative approach introduced by the multidisciplinary project PROGEO-Piemonte, [23,40], further developed and tested within cooperative research and educational activities of the University of Torino and the SVUGG (H2020-COFUND "Tech4Culture" project). In order to select the most popular routes, popular products, educational initiatives, and public engagement events within the SVUGG were inventoried following the PROGEO-Piemonte standards, thus implementing both the related websites (http://www.progeopiemonte.it/en/aree/piemonte-nordorientale/; http://www.sesiavalgrandegeopark.it/). Each tour connects geosites and facilities ("stops") providing descriptions of the natural and cultural landscape and interpretations of the related geological and geomorphological processes. The mobile app and webpage include nine different geotouristical tours located within the geothematic areas defined during the project: a tour network allowing people to begin their journey through the geoheritage of the Piemonte Region, considering both geological and cultural aspects. All the map tiles are integrated into the application, in order to avoid the need of downloading the map on the go, which in some cases might result with expensive roaming costs.

3.2.2. Multidisciplinary Educational Fieldworks for Understanding Spatio-Temporal Evolution of the Alpine Landscape

According to the principle of learning-by-doing [109], specific educational tools (learning aid sensu Orion [24]) were set for being expendable by students [27] and useful for teachers, who were often not familiar with Earth Sciences on the field and laboratory works [110]. The aim was to allow users to observe, measure and compare [24,111]. The geotrail along which these tools were implemented was, in particular, focused on the relation between geomorphological processes, climate change, vegetation response, and human settlements (i.e., Earth as a System [112]). Hence, simplified version of the geomorphological map (i.e., geomorphological boxes (Figure 9c) [21]) and simple exercises (Figure 9c), putting in relation geomorphology, dendrochronology, and dendrogeomorphology, were thought to make students work with methodologies of investigations applied by researchers for reconstructing the evolution of the Alpine physical landscape. These activities have been already tested in the framework, among others, of the Erasmus+ Project (see Section 2.2), are available on the panels along the trail, and are going to be freely downloadable from the website of the trail, together with dedicated web-based videos (Figure 9a).

Figure 9. Fieldtrips activities inside the Sesia Val Grande UNESCO Global Geopark: (**a**) multimedia video available on the web on the lime kilns history in the Loana Valley for supporting field activities; (**b**) practical demonstration of the lime production in the Loana Valley; (**c**) simple exercises on the activity of geomorphological processes for students in the Loana Valley (modified from [27]); (**d**) fieldtrips with explanation to students of glacial modeling along the Sentiero Azzurro.

4. Results

The results of the inventory and implementation for geotrails are described here. Supplementary Files (SF 1–4) are available, including tables with the data and images of the on-site and off-site geoheritage.

4.1. Geosite Inventory

Figure 10a shows the distribution of the geosites in the SVUGG and the complete information is included in SF 1. From a geographical point of view, they spread over the entire area of the geopark with a concentration along the existing geotrails. The elevation range of the geosites varies from about 200 m a.sl., in correspondence of the bottom of the Toce Valley, to more than 3200 m a.s.l. for the Monte Rosa massif.

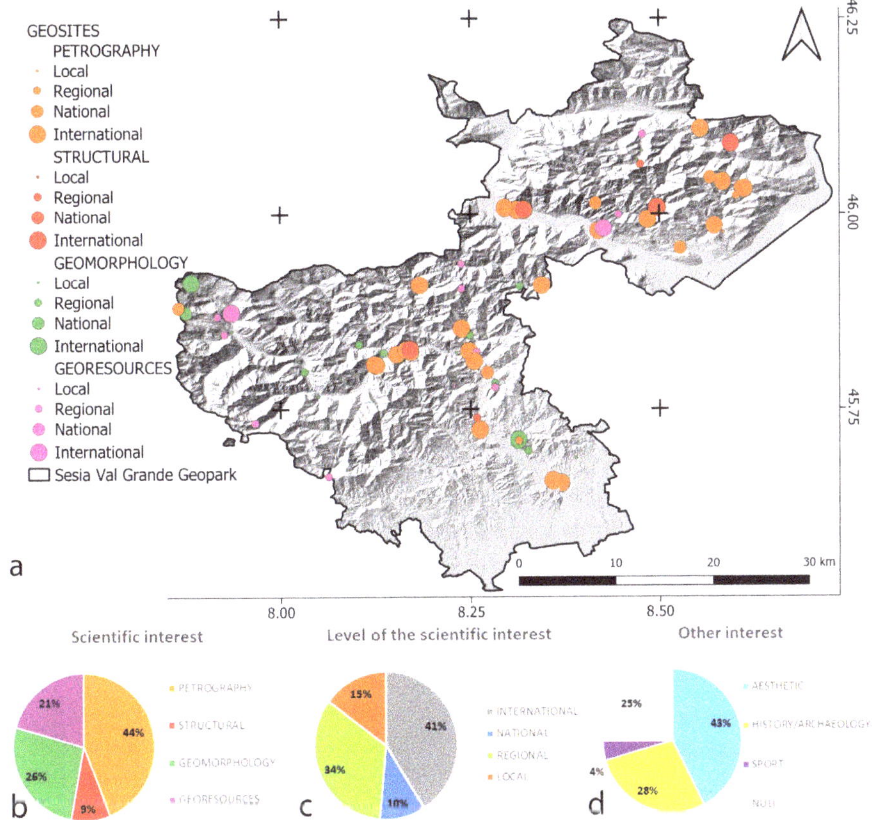

Figure 10. Distribution of the geosites in the Sesia Val Grande UNESCO Global Geopark (**a**), with the relative abundance of geosites according to the Scientific interest (**b**), level of Scientific interest (**c**), and other interest (**d**). Refer also to SF 1.

Considering their primary scientific value (Figure 10a,b), four main geological interests were identified: petrography (P), structural geology (ST), geomorphology (GM), and georesources (GRS). Some of the categories used in the inventory and applied to both geosites and geotrails were not associated with being of primary interest to any of the geosites (HYD = hydrogeology; M = mineralogy; PAL = paleontology; SD = sedimentology; SS = soil science). They are associated with some of the geosites as secondary interest. About half of the geosites (44%) are related to petrographic topics. Geomorphology, including glaciology, and georesources interests are quite equally represented (26% and 21% respectively), while the topic concerning structural geology can be identified in the 9% of the geosites as primary interest. Concerning the level of the scientific interest of the geosites (Figure 10a,c), the majority of them have an international importance (41%) and mainly correspond to the petrographic and structural categories. They are followed, in number, by the geosites with regional interest (34%). Geosites of national and local importance are less represented (10% and 15% respectively). Finally, regarding other interest (Figure 10d), many geosites are characterized by attributes related to their aesthetic value (43%). Some of them present an historical and/or archaeological interest (28%), and the minority are exploited for sport activities (4%). For 25% of the geosites, none of the mentioned additional interests can be identified.

4.2. Geotrails Analysis

In the following paragraphs, a selection of seven geotrails, over a total of 18 within the geopark territory reported in Figure 11, is proposed, to show geodiversity of the SVUGG (Figure 12) and the different approaches adopted to involve users (Figures 8 and 9), be they the general public (tourists, amateurs) or students in schools of different orders.

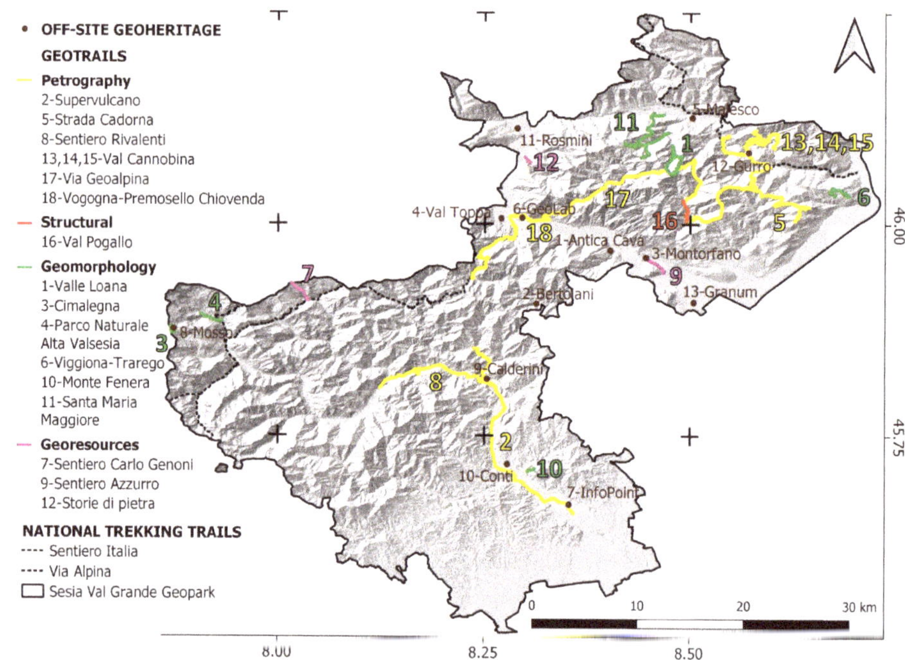

Figure 11. Distribution of the 18 geotrails and 13 off-site geoheritage sites in the Sesia Val Grande UNESCO Global Geopark. Refer also to SF 2 and SF 4.

The selection (grey in SF 2) is presented, starting from the petrographic topic (4.2.1; 4.2.2; yellow in Figure 11), through geomorphology and glaciology topics (4.2.3; 4.2.4; 4.2.5; 4.2.6; green in Figure 11), and concluding with a geotrail specifically focused on georesources (4.2.7; violet in Figure 11). The data of each geotrail are reported in SF 2, and more images are available in SF 4. In SF 2 and in Figure 11, the Via Geoalpina is inserted (17, SF 2): it is a complex, regional relevant trail that includes parts of other geotrails described and/or reported in SF 2 (1, 5, 16, 18; h, i, l, SF 4-II). Hence, portions of this regional trail are equipped, but the whole Via Geoalpina is herein marked. Moreover, the Via Alpina and the Sentiero Italia are included in the map too, since they represent non-thematic trails of national relevance. Variations of trails conditions are communicated to visitors near real-time, as, for example, in the case of interruption of trails due to hydrogeological instability processes, such as the one very recently occurred along the Pogallo Valley trail (16 in Figure 11) during May 2020.

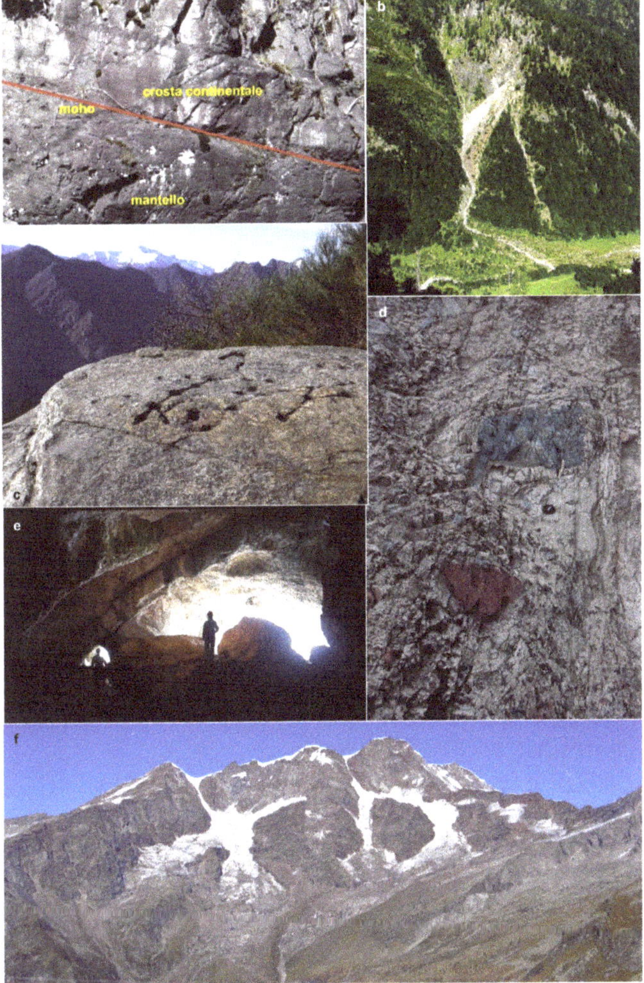

Figure 12. Geodiversity along the geotrails in the Sesia Val Grande UNESCO Global Geopark: (**a**) Moho surface in outcrop at Premosello Chiovenda (51, SF 1; 18, SF 2); (**b**) Pizzo Stagno landslide (18, SF 1; 1, SF 2); (**c**) Alpe Prà petroglyphs in the Val Grande National Park; (**d**) Caldera megabreccia at Prato Sesia (50, SF 1; 2, SF 2); (**e**) Ciota Ciara cave in the Monte Fenera karst complex (22, SF 1; 10, SF 2); (**f**) panoramic view on the Monte Rosa from the Sesia Valley (31, SF 1; 4, SF 2). For other views on geodiversity along geotrails of the SVUGG, refer also to the SF 4.

4.2.1. "Viaggio Spazio-Temporale Nelle Profondità Della Terra" (A Space-Time Journey Inside the Earth depths; 18, SF 2)

This trail was inaugurated in April 2013. It runs along the mountain slope behind the Vogogna and Premosello-Chiovenda villages and is relatively simple, but with some exposed stretches. It is equipped with 10 thematic panels located on significant outcrops, available on the web, plus one introductory panel at the beginning of the itinerary. It is focused on three main geologic themes, which may stimulate the curiosity of the general public:

i.   The boundary between the Central and the Southern Alps (c, SF 4-II) is along the Insubric Line. The attention is focused on the juxtaposition between the Central Alps, consisting of refolded basement nappes affected by the Alpine metamorphism, and the Southern Alps, which are little affected by that phase and preserving much older structures. The visitors may observe the phyllonites (8, SF 1; b, SF 4-II), produced by the deformation of Austroalpine rocks along the fault zone, and their contact with the mafic granulites of the Ivrea-Verbano Zone (Southern Alps).

ii. A journey from the upper to the lower crust: Along the path, the visitor ideally walks deeper and deeper inside the crust, reaching rocks formed at a depth of over 30 km.

iii. The Mohorovicic discontinuity (MOHO; i.e., the mantle–crust transition): this surface, normally located at a depth > 35 km, is here represented by the contact between serpentinized peridotites and mafic granulites (Premosello outcrop, Figure 12a; 51, SF 1).

The itinerary also allows observations on the geomorphology of the area (glacial, gravity- and water-related landforms). The trail is illustrated in detail in the Geolab "Luigi Burlini" (6, SF 2; e, f, SF 4-II), located in the Vogogna village, near the beginning of the itinerary and in a web-based video (SF 2; g, SF 4-II). Unfortunately, a section of the trail is currently closed for a landslide and the path indicated in Figure 11 and in SF 2 is the part allowed at this moment.

4.2.2. "Il Supervulcano della Valsesia" (The Sesia Supervolcano; 2, SF 2)

The Sesia Supervolcano geotrail was realized within the network of PROGEO-Piemonte geothematic virtual tours and published as a book and online by PROGEO mobile app and Progeo online webmap. The free mobile application (iOS and Google Play versions) and the free webmap page (www.progeopiemonte.it/path/valsesia.html) were published to enhance and promote geological heritage in the Piemonte Region, and specifically in the Sesia Valley. The itinerary proposed consists of 10 stops reachable by car or by short walking trails, and it is supported by the Supervolcano Infopoint in Prato Sesia (7, SF 3). The Valsesia itinerary permits observations on what was going on around 280 Ma ago, in and below an active supervolcano, which extended for at least 25 km deep in the Earth's crust. Today this area is an open-air laboratory: by observing different evidences (1, 4, 6, 16, 20, 23, 24, 25, 26, 33, 41, 47, 50, 58, SF 1), geologists can study the processes that lead an active supervolcano to collapse in a caldera (Figure 12d), after a major eruption (SF 4-IV). The wealth of scientific data and interpretations presented through the stops of the itinerary allows also not expert visitors to reconstruct accurately the history of magmatic processes of the Sesia Supervolcano. Approximately 295 Ma ago, partial melting of the mantle produced magmas that were introduced into the deeper part of the crust, forming the so-called Mafic Complex. About 288 Ma ago, heat from this deep magmatic body melted the upper crust, forming granitic bodies known as the Graniti dei laghi. In both lower and upper crust, hybrid magmas formed as well, due to mixing of magmas of different origin. In the same period, the magmatic activity reached the Earth's surface. Later, about 280 Ma ago, a super-eruption collapsed the volcanic system, forming a caldera with a diameter of at least 13 km: it is estimated that more than 500 km^3 of magma were erupted. It is one of the most violent known magmatic events, which evidences are still preserved along Sesia Valley despite the successive geological events. Then, 90 Ma ago, the Earth's crust slowly opened, creating the Tethys Ocean. Only in the last 30 Ma, during the formation of the Alps, the collision between Africa and Europe exposed a slice of the African crust containing the whole magmatic system of the supervolcano (c, SF 4-IV).

4.2.3. L' Anello Geoturistico della Valle Loana (The Loana Valley Geotouristic Ring; 1, SF 2)

In 2019, a geotouristic trail was equipped along the Loana Valley in the Malesco Municipality, thanks to GAL funding from Fondo Europeo Agricolo per lo Sviluppo Rurale (FEAR). The trail is articulated in two parts: (i) the first one runs along the valley bottom; it is a touristic, easily accessible; (ii) the second one reaches the head of the valley as far as the northern border of the Val Grande National Park. Both the proposals were set starting from the existing excursionist trail network and they have both a ring pattern. Six geostops were identified along the trails and equipped with panels (e, SF 4-I,) containing essential information, and most of them are enriched with simplified mapping tools [21]. The trail is focused on the evolution of the Alpine landscape under glacial, snow-, water- and gravity-related processes, in relation to vegetation and human settlements. Considering these topics, the trail is recommended only in safe conditions. The trail is inserted in the offer of the "Ecomuseo ed leuzerie e di scherpelit - Museo del Parco Nazionale della Val Grande" of the Malesco Municipality (5, SF 3), from which, in 2020 a pdf-format guide to the trail will be made also available, together with virtual videos (f, SF 4-I), in the specific session of the thematic trails. The trail is focused on three main topics:

i. Lithological and structural control on geomorphological modelling—the Loana Valley is a S-N oriented valley that was carved by glaciers during the Pleistocene. The head of the valley is characterized by a glacio-structural saddle due to the presence of the Canavese Line (a, SF 4-I). The area represents a noteworthy place for the study of deformations related to the CL (i.e., Scaredi Formation; [41]), as testified by several authors [44,50–52]. The influence of the deformation pattern on the surficial modelling is particularly evident where different lithotypes, including marbles (27, SF 1), which crop out in structural contact, producing different relief features.

ii. Ecologic support role on landforms—the geomorphic processes interesting currently and during the past the Loana valley deeply affect the vegetation distribution in the area. Bollati et al. [99,113] detected specific patterns of disturbance on vegetation growing on the geosites located along the valley bottom (18, SF 1; b, c, SF 4-I), and on vegetation growing on carbonate isolated reliefs modelled by glaciers in the past, near the head of the valley (17, SF 1; Figure 12b). These results were proposed in form of simple exercises, addressed to general public and schools, allowing them to relate the vegetation behavior with the geomorphological activity (Figure 9c).

iii. Cultural value of georesources—in addition to the most famous "Pietra ollare" (x, SF 1), along this trail, the attention is focused on carbonate outcrops located in the valley and quarried in the past to produce lime (Figure 9a). Old lime kilns, located in the valley bottom, have been recently restored with the possibility, for tourists and schools, of experimenting lime production (Figure 9b; d, SF 4-I). The carbonate outcrops (17, SF 1) represent georesources, a particularly relevant consideration in a geopark where the link with local populations and resources usages is extremely important.

4.2.4. L'Itinerario Glaciologico del Parco Naturale Alta Valsesia (Upper Sesia Valley Natural Park Glaciological trail, 4, SF 2)

At the beginning of the 1990s, a glaciological trail was equipped in the Upper Sesia Valley Natural Park in the Alagna Valsesia Municipality. Starting from the existing network of local hiking tracks, the trail was set up as a round-trip route. Eight geostops were identified along the trail, each one equipped with display panels. The southeastern side of Monte Rosa belonging to Sesia Valley is the less glacierized among the five sides of the massif. There are no active valley glaciers similar to Gorner (northwestern side), Verra and Lys glaciers (southwestern side), neither large debris-covered glaciers like Belvedere Glacier (northeastern). Nevertheless, thanks to the glaciological trail of the Upper Sesia Valley Natural Park, it is possible to observe magnificent glacial landscapes of high environmental and scientific interest. The aim of the trail is to guide hikers in the observation and recognition of the most evident and significant landforms shaped by glaciers, during their expansion and recessional phases. For this purpose, the first geostop describes past climatic fluctuations: particularly those dating back to the last million year related to Pleistocene glaciations. The seven geostops illustrate three main topics:

i. The process of glacial erosion and related landforms, from the micro to the macro scale—at the beginning of the trail, it is possible to observe the "Caldaie del Sesia" landform system (e, SF 4-V) modelled by subglacial water of the ancient Sesia Glacier. The system is composed by a gorge where the Sesia River is channeled through a rock step, producing a high waterfall, and a huge kettle. Another example is a roche moutonnée with potholes, good example of landforms related to abrasion and/or plucking processes. Finally, some geostops illustrate large scale erosion processes, particularly those related to the Bors Valley hanging valley and glacial cirque.

ii.   The processes and landforms related to glacial accumulation—hikers walk along the ridge of a glacial deposit, the "Fondecco moraine" (d, SF 4-V). This shows the dimensions reached by the ancient Sesia Glacier during the late glacial advances, during the end of Pleistocene.

iii.   Glaciers and their dynamics—glaciers of the Sesia Valley side of the Monte Rosa (31, SF 1; Figure 12f; g, h, SF 4-V) are described with information on their past areal dimensions. The trail ends in proximity of the current glacier snout, where it is possible to observe crevasses and seracs; moreover, it is also possible to observe moraines and roche mountonnée (f, SF 4-V) shaped by glaciers during the Little Ice Age.

Thanks to the information acquired from the display panels during the ascending route, the hikers are invited, during the descent, to identify and recognize glacial landforms by themselves. The glaciological trail of the Upper Sesia Valley Natural Park has been recently included among the glaciological itineraries on the Italian glacial mountains [90,91]. The contents of the trail drive the attention of the geotourists to some of the most important topics of the present-day debate on environmental changes of the glacial environments [114].

4.2.5. Geological-Pedological Trail of the Cimalegna Plateau (Itinerario geologico-pedologico dell'altopiano di Cimalegna; 3, SF 2)

The Cimalegna glaciological-pedological trail is located in the high-altitude plateau (2800–3000 m a.s.l.) at the Western border of the geopark (Alagna Valsesia municipality). For its structural geology context, the Cimalegna plateau (3, SF 1) is an ideal place to examine the geological history of the North-Western Alps, with particular regard to the geological dynamics of the last 200 Ma [115]. Moreover, here, glacial and periglacial features and soils, show a high variety, and the typical pedogenetic processes of this high-altitude cold environment are well expressed. The Cimalegna plateau is also a relevant location for scientific studies promoted by University of Torino. The "Angelo Mosso Scientific Institute" (8, SF 3; a, SF 4-V) was established here on 1907 for physiological studies at high altitude [116], then upgraded by the Geophysical Observatory conducted by Umberto Monterin since 1927 [117] and, later, by the Snow and Alpine Soils Laboratory. Here, the NatRisk research Team (www.natrisk.unito.it) established a base for studies on glacial and periglacial environments [118,119]), and contributed to the recognition of the Cimalegna site as the Alpine site of the Italian Network for Long-Term Ecological Research (www.lteritalia.it), part of the LTER International Network (ILTER; www.ilternet.edu/). The Cimalegna trail was established by the Ente di Gestione delle Aree Protette della Valle Sesia in 2008 [120], to offer public engagement activities for high school students, naturalistic and environmental associations, such as training courses relating to hiking and mountaineering (c, SF 4-V). The circular route of the Cimalegna path starts from the Passo dei Salati, descends to the "Angelo Mosso Scientific Institute", near Bodwitch Lake, continues east to the Col d'Olen (2881 m), climbs up the Corno del Camoscio horn (3024 m a.s.l.) for a 360° viewpoint on the southern slope of the Monte Rosa, then it descends back to the Passo dei Salati. Along the trail, eight display panels were placed, illustrating the geological history of this area with photographs and diagrams, starting from an ancient ocean, the Tethys, up to the formation of the Alps, also dwelling on the soils (b, SF 4-V), which are formed here in particular conditions, due to the presence of an almost flat area, of high-altitudes, and extreme climatic conditions. Further engaged research activities are possible at the "Angelo Mosso Scientific Institute", with audiovisuals projections on specific topics [36], and the exhibition of pedoliths of the soils of Cimalegna (reconstruction in the laboratory of an entire soil profile, made using the material taken in the field; [121]) and a collection of lithotypes from the area [120].

4.2.6. Monte Fenera Caves Trail (Sentiero delle Grotte del Monte Fenera; 10, SF 2)

The peculiarity of Monte Fenera is the presence of many caves (22, SF 1; f, g, SF 4-III) which, over the course of thousands of years, have been occupied by living beings of various animal species, including some extinct ones, such as *Ursus Spelaeus, Merk's rhinoceros*, and even by *Homo neanderthalensis*, whose finds retrieved at Fenera give the mountain the primacy of the oldest prehistoric site in Piemonte [102,122]. An articulated karst system has developed inside the carbonate rocks (limestone and dolomites) of the Monte Fenera. The preservation of these kinds of rocks in Valsesia is the consequence of the tectonics of the area: in particular, of those faults, the most important is the Cremosina Line, that, causing the movement of large rocky masses, allowed the preservation of Mesozoic rocks from erosion as happened in the neighboring portions. The geotrail was set up with 12 explanatory display boards. The first seven panels focus on geology, describing the rocks that constitute the Monte Fenera (from the oldest dating back to 280 Ma ago, to the more recent ones that emerge from the slopes of the mountain relief), and on the structural assets of the area, characterized by important tectonic lines. Two panels are dedicated to the description of the karst system with concretions. The most important caves (22, SF 1) are: "Grotta delle Arnarie" (3500 m of development), "Buco della Bondaccia" (500 m of development), "Ciota Ciara" (200 m; Figure 12e) and "Il Ciutarùn" (about 70 m). At the entrances of the last three caves, a descriptive panel of each individual underground cavity was placed: the development, the speleological and biological aspects, with images of the interior of the caves and of the species adapted to the underground life that live in them or find a temporary shelter. A panel relating to the archaeological excavations made by the University of Ferrara was also placed at the "Ciota Ciara".

4.2.7. Sentiero Azzurro (The Azzurro Trail; 9, SF 2)

The Mont'Orfano (Figures 3e and 5b) is an isolated peak at the end of Ossola Valley, facing both the Mergozzo and Maggiore lakes. Its lithological composition, mainly of white and green granite, together with the presence of local faults, made it to survive to erosion by the Toce Glacier during the Ice Ages, and thus, standing alone, separated from the Mottarone massif, the Massone massif, and the mountains belonging to Val Grande National Park. Due to the presence of granite, it had (and still has, even if reduced) significant importance as a georesource for local communities. On its slopes, for many centuries, about thirty quarries operated to extract big and small rock blocks, mainly used in local buildings, but in the 19th and 20th centuries, lots of sculptures and monuments were made all over the world with the Montorfano white granite. The trail, belonging to the cultural offer of Ecomuseo del Granito (3, SF 3), develops on an easy and historical path, starting from the main village of Mergozzo to reach the small village of Montorfano, just in the middle of the largest quarries, of which only one is still active. Along the path, some panels explain, with the aid of historical pictures, the techniques used in extracting, moving and working the granite blocks (a, b, SF 4-III). Thus, the trail has not only a geological importance, but even an anthropological one, because the quarries have been the only economic resource for decades for the local communities of Mergozzo and the surrounding area. The trail ends at Belvedere, a scenic point of view on Mergozzo and Maggiore lakes (x, SF 1), with an explanatory panel about lake geomorphology and the formation of the first one, in relation to sediment transport rates that have characterized the Toce River (Figures 3e and 9d).

4.3. The Off-Site Geoheritage within the Sesia-Val Grande UNESCO Global Geopark (Museum, Geo-Laboratories)

In the wide area of the SVUGG there are 13 science museums and Park visitor centers closely linked to the activities and themes of the geopark (SF 3, SF 4). Some of them are directly managed by the Regional Natural Park or the National Natural Park and others are managed by four official ecomuseums (recognized by Piemonte regional law), together with local Municipalities. The Ecomuseum is a tool for the participatory management of the natural and cultural heritage of a territory [123], thus it does not focus only on the scientific aspect and the static element (i.e., a rock collection), but enhances that element in the context of the territory and its community. Moreover, it is not an open-air museum nor a diffuse museum, but it is a cultural container. Only few museums are privately owned, but are usually open to the public during tourist season or on reservation.

This off-site geoheritage represents a very important element for the Earth Sciences knowledge of the area, because each of these entities focuses its attention on a specific topic, sometimes with substantial collections that also have a historical value (i.e., Museo di Scienze Naturali "Mellerio Rosmini", 11, SF 3, and Museo "Pietro Calderini" 9, SF 3). The off-site geoheritage is only not represented by collections of rocks and minerals, but also by specific explanatory sections, inside museums and visitor centers, about the georesources and their use in past and present times. In some cases, there is a strong connection between the archaeological heritage and the geological one (i.e., Ecomuseo ed leuzerie e di scherpelit - Museo del Parco Nazionale della Val Grande, 5, SF 3; f, SF 4-I; Museo Archeologico e Paleontologico "Carlo Conti", 10, SF 3). Many of them are also connected with the on-site geoheritage, are near-by geosites or geotrails, and offer lots of information for the outdoor visits. In two cases, Ecomuseo della Val Toppa (4, SF 3) and Antica Cava (1, SF 3) are actually on-site geoheritage elements (an old gold mine and an old marble quarry), but they are managed as museums, with opening times and guided visits. Some structures offer educational services both for students of all ages and for adults, with the aid of well and specific trained educators and guides. These kinds of activities can join both laboratories and excursions, not only for working on general themes, but also to make the territory and its peculiarities known especially by students in schools. In particular, the Geolab "Luigi Burlini" in Vogogna (6, SF 3; e, f, SF 4-II), was thought as an area for the study and the teaching of themes related to Earth Sciences. It is equipped with a stereo microscope and a polarized light microscope, featuring high-definition video cameras and video devices, and also provides a collection of thin sections of the main rocks in the area, a mineralogical and lithological collection, and interactive instruments with pictures and animations of the main themes related to the geology of the area of the Val Grande National Park.

5. Discussions and Conclusions

The results of our research indicate that the Sesia Val Grande UGG is an area of high geodiversity, rich geoheritage, and multifold geotourism opportunities. Within the UGG Network, several geoparks are related to a specific thematic, frequently inferable from the name of the geopark itself (e.g., the petrified forest of Lesvos Geopark, Greece; Tuscan Mining Park, Italy; Marble Arch Caves, Ireland and UK). In the present case, the name of the geopark refers to its geographic hubs, the Sesia and Val Grande valleys. Regarding the original idea of the SVUGG, it centers around a "deep" geological focus: The South Alpine domain, where several relevant spots of international scientific value have been selected as geosites with a prevalent interest on petrography of crustal rocks. By also carefully considering the landscape and territorial dimensions of geodiversity and geoheritage, other focal points of the geopark were outlined by this study, here below discussed as contributions for integrating fieldtrips and virtual tours in the perspective of enhanced geotourism.

According to our results, it emerges that geosites are lacking in some areas (e.g., north-eastern sectors like Vigezzo Valley (Figure 1)), along some geotrails no geosites are up to now identified (e.g., Sentiero 117-Carlo Genoni, 7 in SF 2, with its ancient gold mining sites), or some geosites are not even linked through specific geotrails (e.g., geosites of the Cava Madre, 12 in SF, Chiesa di Albo; 13 in SF 1, and the off-site geoheritage of Antica cava, one in SF 3). Considering the absence of geosites mainly within satellite geological units (i.e., the Austroalpine domain and other domains north to the Canavese Line) and their potential significance as cultural geoheritage, they are of fundamental importance for the geotourism of the SVUGG. In fact: (i) they contribute even more to the geodiversity of the geopark; (ii) they are useful to contextualize the geopark reality in the framework of a regional evolution; and, hence, (iii) they are potential new resources for the territories and local communities. The same considerations apply in particular to the geomorphological evidences in such parts of the geopark, which in some cases are very relevant.

As an example, the Vigezzo Valley represents an important geological and geomorphological spot for the geopark where it is possible to retrieve witnesses on Earth evolution at different spatio-temporal scales: (i) the effects of one of the most disastrous hydrogeological instability events in the Western

Italian Alps occurred on 7th August 1978 [113]; (ii) the evidences of the tectonic activity of the Centovalli Line, related to the Simplon Line tectonic system and affecting paleoenvironmental settings, as well as current slope instabilities [124]; (iii) the mineralogical richness that allowed the identification of new mineralogical species (e.g., Vigezzite [125]).

Moreover, it emerges that in some areas (e.g., south-western regions of the geopark), the lateral valleys are less valorized from a geodiversity point of view despite important evidences. For example, the Sermenza and Sessera Valleys well represent, from a geomorphological point of view at the small scale, the relevance of the action of fluvial process, which could be compared with that of the glacial processes dominating the major valleys (Sesia and Toce Valleys). The possibility of comparing these main geomorphological processes in the field, just by moving between valleys of different orders is of high relevance from an educational point of view. Finally, the valorization of the remote lateral valleys would help local communities in the enhancement of tourist activities and, in some cases, to help new ones emerge, which corresponds to one of the auspices of UNESCO Global Geoparks.

In the present context of climate change, high-altitude geotrails are essential for raising awareness among the general public of ongoing environmental changes, and the glacial and periglacial areas offer some of the best evidences of these developments (e.g., glacier shrinkage, new glacier lakes formation, permafrost degradation [91,114,118]). Moreover, high altitude geotrails could also contribute to the explanation of the indirect effect of climate change on human settlements in mountain areas (eg., ski resorts, water supply, mountain huts access), and allow to test the fruitor perception about the environmental resources related to climate changes [126]. In fact, essential abiotic ecosystem services for human activities are being severely impacted by the ongoing changes [93]. The effects of the ongoing deglaciations are well perceptible along this kind of trail, where the recolonization of vegetation characterizes increasingly widening proglacial areas [127], where geosites are affected by paraglacial-type transformation [94]. Therefore, the availability of high altitude geotrails represents a strength in areas such as the SVUGG, for bringing attention to the abiotic ecosystem services (sensu Gray [8]). In this framework, a particular focus should be also driven towards hydrogeodiversity [18] and future geosites for enhancing water resources, among which high altitude environments host water towers (i.e., glaciers). For example, according to the results and the comparison with the current state of geoconservation of the Val Grande territory, there are also important areas in terms of hydro-geosystemic services, as they are directly related to the withdrawal and consumption of water (e.g., drinking water, for agriculture, for breeding). There are also areas in which human impact is deeper, and where there are no instances of hydrogeological protection sufficient for a good preservation. Therefore, more studies and insights about these issues are needed.

Hence, among the perspectives for the future there is, on the whole, the implementation of valorization initiatives by means of upgrading the geosites inventory and geotrails proposals, considering the easy accessibility and high educational value of potential resources in those areas, that are under evaluation. All these analyses will be accompanied by more and more detailed investigations on the evolution of alpine landscapes, in areas more sensitive to climate change: for example, at the Indren Glacier and Cimalegna Plateau, within the Monterosa Ski resort, on the southern face of Monte Rosa massif, where the LTER site "Mosso, Passo dei Salati-Col d'Olen" is also present in the area and it is part of the Italian Long Term Ecological Research Network. As proposed in other areas [126], the monitoring of tourists' perception and appreciation about valorization initiatives could be a further aim to pursue.

Finally, more multimedia applications and multidisciplinary activities could be implemented for the already existing geotrails, and for the future-planned ones, after their application in pilot areas, as described herein. More suggestions could also derive from the framework of international exchanges related to ERASMUS+ Projects, as those are being performed in the studied area up to now.

Supplementary Materials: The following are available online at http://www.mdpi.com/2079-9276/9/6/63/s1, Supplementary file 1 (SF 1): Table of complete data on geosites, Supplementary file 2 (SF 2): Table of complete data on geotrails, Supplementary file 3 (SF 3): Table of complete data on off-site geoheritage, Supplementary file 4 (SF 4): Collection of pictures of geosites, geotrails and off-site geoheritage.

Author Contributions: Conceptualization, L.P., I.M.B., M.G. and M.P.; Methodology, L.P., I.M.B., C.V., M.G. and M.P.; Software, L.P. and C.V.; Validation, M.G. and V.C.; Formal analysis, L.P.; Investigation, I.M.B., V.C., C.V., M.G. and M.P.; Resources, I.M.B., C.V., E.Z., V.C. and M.G.; Data curation, L.P., I.M.B., C.V., E.Z. and V.C.; Writing—original draft preparation, L.P., I.M.B., M.G., C.V., E.Z. and V.C.; Writing—review and editing, I.M.B., M.G., C.V., M.P., E.Z. and V.C.; Visualization, I.M.B., C.V. and E.Z.; Supervision, M.G. and M.P.; Project administration, L.P. and E.Z.; Funding acquisition, M.G. and E.Z. All authors have read and agreed to the published version of the manuscript.

Funding: ERASMUS+ project GEOclimHOME-PRO - Geoheritage and climate change for highlighting the professional perspective. Grant 2018-1-FI01-KA201-047206 (European fundings) Ministry of Education, University and Research (MIUR), Project Dipartimenti di Eccellenza 2018–2022, Dipartimento di Scienze della Terra 'A. Desio' of the University of Milano (National fundings) MIUR PRIN 2010–2011 project "Response of morphoclimatic system dynamics to global changes and related geomorphological hazards" (grant number 2010AYKTAB_0069) (National Fundings); Fondi Potenziamento della Ricerca – Linea 2 – 2015 (entrusted to D. Zanoni), 2016 and 2017 (entrusted to A. Zerboni) (National fundings) FEASR – Fondo Europeo Agricolo per lo Sviluppo Rurale – L'Europa investe nelle zone rurali (European fundings) PSR 2014–2020 della Regione Piemonte – MISURA 19 – Sostegno allo sviluppo locale LEADER (CLLD - Community Led Local Development LEADER) (Regional fundings) GEODIVE – Compagnia San Paolo bank foudation in partnership with University of Torino, under the funding program 2016; the project "GeoDIVE—From rocks to stones, from landforms to landscapes" was funded with grant #CSTO169034 (Regional fundings).

Acknowledgments: The Authors are very grateful to all the municipalities, managers of the protected areas, associations, local population and funders as mentioned along the text. Moreover, they are grateful to the students working inside the geopark territory, the visiting groups and to everybody contributing in enhancing the valorisation of geopark resources.

Conflicts of Interest: The authors declare no conflict of interest.

References

1. Dowling, R.K.; Newsome, D. *Handbook of Geotourism*; Edward Elgar Publishing: Cheltenham, UK, 2018; Volume 520.

2. O'Halloran, D.; Green, C.; Harley, M.; Knill, J. (Eds.) Geological and landscape conservation. In *Proceedings of the Malvern International Conference 1993*; The Geological Society, Geological Society Publishing House: London, UK, 1994; Volume 530.

3. Prosser, C.; Murphy, M.; Larwood, J. *Geological Conservation: A Guide to Good Practice*; English Nature: Peterborough, UK, 2006; Volume 145.

4. Reynard, E.; Brilha, J. *Geoheritage: Assessment, Protection, and Management*; Elsevier: Amsterdam, The Netherlands, 2017; Volume 482.

5. Sallam, E.S.; Abd El-Aal, A.K.; Fedorov, Y.A.; Bobrysheva, O.R.; Ruban, D.A. Geological heritage as a new kind of natural resource in the Siwa Oasis, Egypt: The first assessment, comparison to the Russian South, and sustainable development issues. *J. Afr. Earth Sci.* **2018**, *144*, 151–160. [CrossRef]

6. Dowling, R.K. Geotourism's global growth. *Geoheritage* **2011**, *3*, 1–13. [CrossRef]

7. Hose, T.A. 3G's for modern geotourism. *Geoheritage* **2012**, *4*, 7–24. [CrossRef]

8. Gray, J.M. *Geodiversity: Valuing and Conserving Abiotic Nature*, 2nd ed.; Wiley Blackwell: Chichester, West Sussex, UK, 2013; Volume 508.

9. Knudsen, D.C.; Metro-Roland, M.M.; Soper, A.K.; Greer, C.E. *Landscape, Tourism, and Meaning*; Ashgate Publishing: Burlington, VT, USA, 2008; Volume 176.

10. Bollati, I.; Coratza, P.; Panizza, V.; Pelfini, M. Lithological and structural control on Italian mountain geoheritage: Opportunities for tourism, outdoor and educational activities. *Quaest. Geog.* **2018**, *37*, 53–73. [CrossRef]

11. Kubalíková, L. Assessing geotourism resources on a local level: A case study from Southern Moravia (Czech Republic). *Resources* **2019**, *8*, 150. [CrossRef]

12. Borghi, A.; d'Atri, A.; Martire, L.; Castelli, D.; Costa, E.; Dino, G.; Favero Longo, S.E.; Ferrando, S.; Gallo, L.M.; Giardino, M.; et al. Fragments of the Western Alpine Chain as historic ornamental stones in Turin (Italy): Enhancement of urban geological heritage through geotourism. *Geoheritage* **2014**, *6*, 41–55. [CrossRef]

13. Pelfini, M.; Brandolini, F.; D'Archi, S.; Pellegrini, L.; Bollati, I. Papia civitas gloriosa: Urban geomorphology for a thematic itinerary on geocultural heritage in Pavia (Central Po Plain, N Italy). *J. Maps* **2020**, 1–9. [CrossRef]

14. Council of Europe. Explanatory Report to the European Landscape Convention. Available online: http://conventions.coe.int/Treaty/en/Reports/Html/176.htm (accessed on 18 April 2020).

15. Giardino, M.; Mortara, G.; Borgatti, L.; Nesci, O.; Guerra, C.; Lucchetti, C.C. Dynamic geomorphology and historical iconography. Contributions to the knowledge of environmental changes and slope instabilities in the Apennines and the Alps. In *Engineering Geology for Society and Territory*; Lollino, G., Giordan, D., Marunteanu, C., Christaras, B., Yoshinori, I., Margottini, C., Eds.; Springer: Berlin, Germany, 2015; Volume 8, pp. 463–468. [CrossRef]

16. Dauvin, J.C.; Lozachmeur, O.; Capet, Y.; Dubrulle, J.B.; Ghezali, M.; Mesnard, A.H. Legal tools for preserving France's natural heritage through integrated coastal zone management. *Ocean. Coast. Manag.* **2004**, *47*, 463–477. [CrossRef]

17. Lucchesi, S.; Giardino, M. The role of geoscientists in human progress. *Ann. Geophys.* **2012**, *55*, 355–359. [CrossRef]

18. Perotti, L.; Carraro, G.; Giardino, M.; De Luca, D.A.; Lasagna, M. Geodiversity evaluation and water resources in the Sesia Val Grande UNESCO Geopark (Italy). *Water* **2019**, *11*, 2102. [CrossRef]

19. Wimbledon, W.A.P.; Benton, M.J.; Bevins, R.E.; Black, G.P.; Bridgland, D.R.; Cleal, C.J.; Cooper, R.G.; May, V.J. The development of a methodology for the selection of British geological sites for conservation: Part 1. *Mod. Geol.* **1995**, *20*, 159–202.

20. Wimbledon, W.A.P. GEOSITES-a new conservation initiative. *Episodes* **1996**, *19*, 87–88. [CrossRef]

21. Bollati, I.; Crosa Lenz, B.; Zanoletti, E.; Pelfini, M. Geomorphological mapping for the valorization of the alpine environment. A methodological proposal tested in the Loana Valley (Sesia Val Grande Geopark, Western Italian Alps). *J. Mt. Sci.* **2017**, *14*, 1023–1038. [CrossRef]

22. Brilha, J. Inventory and quantitative assessment of geosites and geodiversity sites: A review. *Geoheritage* **2016**, *8*, 119–134. [CrossRef]

23. Ferrero, E.; Giardino, M.; Lozar, F.; Giordano, E.; Belluso, E.; Perotti, L. Geodiversity action plans for the enhancement of geoheritage in the Piemonte region (north-western Italy). *Ann. Geophys.* **2012**, *55*, 487–495. [CrossRef]

24. Orion, N. A Model for the development and implementation of field trips as an integral part of the Science curriculum. *Sch. Sci. Math.* **1993**, *93*, 325–331. [CrossRef]

25. Garavaglia, V.; Pelfini, M. Glacial geomorphosites and related landforms: A proposal for a dendrogeomorphological approach and educational trails. *Geoheritage* **2011**, *3*, 15–25. [CrossRef]

26. Magagna, A.; Ferrero, E.; Giardino, M.; Lozar, F.; Perotti, L. A selection of geological tours for promoting the Italian geological heritage in the secondary schools. *Geoheritage* **2013**, *5*, 265–273. [CrossRef]

27. Bollati, I.M.; Crosa Lenz, B.; Zanoletti, E. A procedure to structure multidisciplinary educational fieldworks for understanding spatio-temporal evolution of the Alpine landscape. *Rend. Online Soc. Geol. Ital.* **2019**, *49*, 11–19. [CrossRef]

28. Giardino, M.; Lombardo, V.; Lozar, F.; Magagna, A.; Perotti, L. GeoMedia-web: Multimedia and networks for dissemination of knowledge on geoheritage and natural risk. In *Engineering Geology for Society and Territory*; Lollino, G., Arattano, M., Giardino, M., Oliveira, R., Peppoloni, S., Eds.; Springer: Berlin, Germany, 2014; Volume 7, pp. 147–150. [CrossRef]

29. Bertotto, S.; Perotti, L.; Bacenetti, M.; Damiano, E.; Marta, C.; Giardino, M. Integrated geomatic techniques for assessing morphodynamic processes and related hazards in glacial and periglacial areas (Western Italian Alps) in a context of climate change. In *Engineering Geology for Society and Territory*; Lollino, G., Manconi, A., Clague, J., Shan, W., Chiarle, M., Eds.; Springer: Berlin, Germany, 2015; Volume 1, pp. 173–176. [CrossRef]

30. Coratza, P.; De Waele, J. Geomorphosites and natural hazards: Teaching the importance of Geomorphology in society. *Geoheritage* **2012**, *4*, 195–203. [CrossRef]

31. Ghiraldi, L.; Giordano, E.; Perotti, L.; Giardino, M. Digital tools for collection, promotion and visualisation of geoscientific data: Case study of Seguret Valley (Piemonte, NW Italy). *Geoheritage* **2014**, *6*, 103–112. [CrossRef]

32. Santos, I.; Henriques, R.; Mariano, G.; Pereira, D.I. Methodologies to represent and promote the geoheritage using Unmanned Aerial Vehicles, multimedia technologies, and augmented reality. *Geoheritage* **2018**, *10*, 143–155. [CrossRef]

33. Magagna, A.; Palomba, M.; Bovio, A.; Ferrero, E.; Gianotti, F.; Giardino, M.; Judica, L.; Perotti, L.; Tonon, M.D. GeoDidaLab: A laboratory for environmental education and research within the Ivrea Morainic Amphitheatre (Turin, NW Italy). *Rend. Online Soc. Geol. Ital.* **2018**, *45*, 68–76. [CrossRef]

34. Tormey, D. New approaches to communication and education through geoheritage. *Int. J. Geoheritage Parks* **2019**, *7*, 192–198. [CrossRef]

35. Cayla, N.; Hobléa, F.; Reynard, E. New digital technologies applied to the management of geoheritage. *Geoheritage* **2014**, *6*, 89–90. [CrossRef]

36. Giordano, E.; Magagna, A.; Ghiraldi, L.; Bertok, C.; Lozar, F.; d'Atri, A.; Dela Pierre, F.; Giardino, M.; Natalicchio, M.; Martire, L.; et al. Multimedia and virtual reality for imaging the climate and environment changes through Earth history: Examples from the Piemonte (NW Italy) geoheritage (PROGEO-Piemonte Project). In *Engineering Geology for Society and Territory*; Lollino, G., Giordan, D., Marunteanu, C., Christaras, B., Yoshinori, I., Margottini, C., Eds.; Springer: Berlin/Heidelberg, Germany, 2015; Volume 8, pp. 257–260. [CrossRef]

37. Lozar, F.; Clari, P.; Dela Pierre, F.; Natalicchio, M.; Bernardi, E.; Violanti, D.; Costa, E.; Giardino, M. Virtual tour of past environmental and climate change: The Messinian Succession of the Tertiary Piedmont Basin (Italy). *Geoheritage* **2015**, *7*, 47–56. [CrossRef]

38. Bacenetti, M.; Bertotto, S.; Ghiraldi, L.; Giardino, M.; Magagna, A.; Palomba, M.; Perotti, L. Geomathics tools and web-based open-source GIScience public participation field approach for field data collection: Supports for a Civil Protection exercise in a hydrological risk scenario. In *Engineering Geology for Society and Territory*; Lollino, G., Arattano, M., Giardino, M., Oliveira, R., Peppoloni, S., Eds.; Springer: Berlin, Germany, 2014; Volume 7, pp. 181–184. [CrossRef]

39. Perotti, L.; Palomba, M.; D'Atri, A.; Lozar, F.; Giardino, M. *Tutti I Colori Di Una Regione—Tredici Itinerari Geologici in Piemonte*; Museo di Scienze Naturali: Torino, Italy, 2019; Volume 372.

40. Gambino, F.; Borghi, A.; d'Atri, A.; Gallo, L.M.; Ghiraldi, L.; Giardino, M.; Martire, L.; Palomba, M.; Perotti, L.; Macadam, J. TOURinSTONES: A free mobile application for promoting geological heritage in the city of Torino (NW Italy). *Geoheritage* **2019**, *11*, 3–17. [CrossRef]

41. Steck, A.; Della Torre, F.; Keller, F.; Pfeifer, H.-R.; Hunziker, J.; Masson, H. Tectonics of the Lepontine Alps: Ductile thrusting and folding in the deepest tectonic levels of the Central Alps. *Swiss J. Geosci.* **2013**, *106*, 427–450. [CrossRef]

42. Giardino, M.; Mortara, G.; Chiarle, M. The glaciers of the Valle d'Aosta and Piemonte regions. Records of present and past environmental and climate changes. In *Landscapes and Landforms of Italy. World Geomorphological Landscapes*; Soldati, M., Marchetti, M., Eds.; Springer: Berlin, Germany, 2017; pp. 77–88. [CrossRef]

43. Laubscher, H.P. Large-scale, thin-skinned thrusting in the Southern Alps: Kinematic models. *Geol. Soc. Am. Bull.* **1985**, *96*, 710–718. [CrossRef]

44. Schmid, S.M.; Zingg, A.; Handy, M. The kinematics of movements along the Insubric Line and the emplacement of the Ivrea Zone. *Tectonophysics* **1987**, *135*, 47–66. [CrossRef]

45. Steck, A. Tectonics of the Simplon massif and Lepontine gneiss dome: Deformation structures due to collision between the underthrusting European plate and the Adriatic indenter. *Swiss J. Geosci.* **2008**, *101*, 515–546. [CrossRef]

46. Manzotti, P.; Ballèvre, M.; Zucali, M.; Robyr, M.; Engi, M. The tectonometamorphic evolution of the Sesia–Dent Blanche nappes (internal Western Alps): Review and synthesis. *Swiss J. Geosci.* **2014**, *107*, 309–336. [CrossRef]

47. Festa, A.; Balestro, G.; Borghi, A.; De Caroli, S.; Succo, A. The role of structural inheritance in continental break-up and exhumation of Alpine Tethyan mantle (Canavese Zone, Western Alps). *Geosci. Front.* **2020**, *11*, 167–188. [CrossRef]

48. Zanoni, D.; Bado, L.; Spalla, M.I.; Zucali, M.; Gosso, G. Evoluzione strutturale e metamorfica pre-sin-e post-intrusiva nell'incassante dei plutoni oligocenici di Biella e Traversella. *Rend. Online Soc. Geol. Ital.* **2006**, *2*, 196–198.

49. Ferrando, S.; Bernoulli, D.; Compagnoni, R. The Canavese zone (internal Western Alps): A distal margin of Adria. *Schweiz. Mineral. Petrogr. Mitt.* **2004**, *84*, 237–256.

50. Reinhardt, B. Geologie und Petrographie der Monte Rosa-Zone der Sesia-Zone und des Canavese im gebiet zwischen Valle d'Ossola und Valle Loana (Prov. di Novara, Italien). *Schweiz. Mineral. Petrogr. Mitt.* **1966**, *46*, 533–678.

51. Kruhl, J.H.; Voll, G. Fabrics and metamorphism from the Monte Rosa Root Zone into the Ivrea Zone near Finero, southern margin of the Alps. *Schweiz. Mineral. Petrogr. Mitt.* **1976**, *56*, 627–633.

52. Altenberger, U.; Hamm, N.; Kruhl, J.H. Movements and metamorphism North of the Insubric Line between Val Loana and Val d'Ossola (Italy). *Jb. Geol. B.-A.* **1987**, *130*, 365–374. Available online: https://www.zobodat.at/pdf/JbGeolReichsanst_130_0365.pdf (accessed on 28 April 2020).

53. Boriani, A.; Burlini, L.; Sacchi, R. The Cossato-Mergozzo-Brissago Line and the Pogallo Line (Southern Alps, Northern Italy) and their relationships with the late-Hercynian magmatic and metamorphic events. *Tectonophysics* **1990**, *182*, 91–102. [CrossRef]

54. oriani, A.; Giobbi Origoni, E.; Borghi, A.; Caironi, V. The evolution of the "Serie dei Laghi" (Strona-Ceneri and Scisti dei Laghi): The upper component of the Ivrea-Verbano crustal section; Southern Alps, North Italy and Ticino, Switzerland. *Tectonophysics* **1990**, *182*, 103–118. [CrossRef]

55. Rutter, E.H.; Brodie, K.H.; Evans, P.J. Structural geometry, lower crustal magmatic underplating and lithospheric stretching in the Ivrea-Verbano zone, northern Italy. *J. Struct. Geol.* **1993**, *15*, 647–662. [CrossRef]

56. Quick, J.E.; Sinigoi, S.; Mayer, A. Emplacement dynamics of a large mafic intrusion in the lower crust, Ivrea-Verbano Zone, northern Italy. *J. Geophys. Res.* **1994**, *99*, 21559–21573. [CrossRef]

57. Sills, J.D.; Tarney, J. Petrogenesis and tectonic significance of amphibolites interlayered with metasedimentary gneisses in the Ivrea Zone, Southern Alps, northwest Italy. *Tectonophysics* **1984**, *107*, 187–206. [CrossRef]

58. Quick, J.E.; Sinigoi, S.; Mayer, A. Emplacement of mantle peridotite in the lower continental crust, Ivrea-Verbano zone, northwest Italy. *Geology* **1995**, *23*, 739–742. [CrossRef]

59. Zanetti, A.; Mazzucchelli, M.; Sinigoi, S.; Giovanardi, T.; Peressini, G.; Fanning, M. SHRIMP U–Pb zircon triassic intrusion age of the Finero Mafic Complex (Ivrea–Verbano Zone, Western Alps) and its geodynamic implications. *J. Petrol.* **2013**, *54*, 2235–2265. [CrossRef]

60. Zingg, A.; Handy, M.R.; Hunziker, J.C.; Schmid, S.M. Tectonometamorphic history of the Ivrea Zone and its relationship to the crustal evolution of the Southern Alps. *Tectonophysics* **1990**, *182*, 169–192. [CrossRef]

61. Pistone, M.; Müntener, O.; Ziberna, L.; Hetényi, G.; Zanetti, A. Report on the ICDP workshop DIVE (Drilling the Ivrea–Verbano zonE). *Sci. Dril.* **2017**, *23*, 47–56. [CrossRef]

62. Boriani, A.; Caironi, V.; Colombo, A.; Giobbi Origoni, E. The Cenerigneiss: A controversial metamorphic rock (Southern Alps, Italy and Ticino, CH). *Proc. EUG* **1997**, *9*, 677.

63. Caironi, V.; Colombo, A.; Tunesi, A. Geochemical approach to characterization and source identification of the protoliths of metasedimentary rocks: An example from the southern Alps. *Period. Mineral.* **2004**, *73*, 109–118.

64. Giobbi Origoni, E.; Zappone, A.; Boriani, A.; Bocchio, R.; Morten, L. Relics of pre-Alpine ophiolites in the Serie dei Laghi (western Southern Alps). *Schweiz. Mineral. Petrogr. Mitt.* **1997**, *77*, 187–207. [CrossRef]

65. Boriani, A.; Mancini, E.G.; Villa, I. Pre-Alpine ophiolites in the basement of Southern Alps: The presence of a bimodal association (LAG-Leptyno-Amphibolitic Group) in the Serie del Laghi (N.-Italy, Ticino-CH). *Rend. Fis. Acc. Lincei* **2003**, *14*, 79. [CrossRef]

66. Caironi, V. The zircon typology method in the study of metamorphic rocks: The orthogneisses of the Eastern Serie dei Laghi (Southern Alps). *Rend. Lincei* **1994**, *5*, 37–58. [CrossRef]

67. Boriani, A.; Giobbi Origoni, E.; Pinarelli, L. Paleozoic evolution of southern Alpine crust (northern Italy) as indicated by contrasting granitoid suites. *Lithos* **1995**, *35*, 47–63. [CrossRef]

68. Boriani, A.; Giobbi Origoni, E.; Del Moro, A. Composition, level of intrusion and age of the "Serie dei Laghi" orthogneisses (Northern Italy—Ticino, Switzerland). *Rend. Soc. Ital. Mineral. Petrol.* **1982**, *38*, 191–205. Available online: https://rruff.info/rdsmi/V38/RDSMI38_191.pdf (accessed on 28 April 2020).

69. Boriani, A.; Sacchi, R. Geology of the junction between the Ivrea-Verbano and Strona-Ceneri zones (Southern Alps). *Mem. Ist. Geol. Mineral.* **1973**, *28*, 1–35.

70. Köppel, V.; Grünenfelder, M. Monazite and zircon U–Pb ages from the Ivrea and Ceneri Zones (southern Alps, Italy). *Mem. Sci. Geol.* **1978**, *33*, 55–70.

71. Wilson, M.; Neumann, E.-R.; Davies, G.R.; Timmerman, M.J.; Heeremans, M.; Larsen, B.T. Permo-Carboniferous magmatism and rifting in Europe: Introduction. In *Permo-Carboniferous Magmatism and Rifting in Europe*; Wilson, M., Neumann, E.R., Davies, G.R., Timmerman, M.J., Heeremans, M., Larsen, B.T., Eds.; Geological Society, Special Publications: London, UK, 2004; Volume 223, pp. 1–10. [CrossRef]

72. Rivalenti, G.; Garuti, G.; Rossi, A.; Siena, F.; Sinigoi, S. Existence of different peridotite types and of a layered igneous complex in the Ivrea Zone of the Western Alps. *J. Petrol.* **1981**, *22*, 127–153. [CrossRef]

73. Rivalenti, G.; Rossi, A.; Siena, F.; Sinigoi, S. The layered series of the Ivrea-Verbano igneous complex, western Alps, Italy. *Tscher. Miner. Petrog.* **1984**, *33*, 77–99. [CrossRef]

74. Sinigoi, S.; Quick, J.E.; Demarchi, G.; Peressini, G. The Sesia Magmatic System. *J. Virtual Explor.* **2010**, *36*, 1–33. [CrossRef]

75. Peressini, G.; Quick, J.E.; Sinigoi, S.; Hofmann, A.W.; Fanning, M. Duration of a large mafic intrusion and heat transfer in the lower crust: A SHRIMP U-Pb zircon study in the Ivrea-Verbano Zone (Western Alps, Italy). *J. Petrol.* **2007**, *48*, 1185–1218. [CrossRef]

76. Snoke, A.W.; Kalakay, T.J.; Quick, J.E.; Sinigoi, S. Development of a deep-crustal shear zone in response to syntectonic intrusion of mafic magma into the lower crust, Ivrea–Verbano zone, Italy. *Earth Planet. Sci. Lett.* **1999**, *166*, 31–45. [CrossRef]

77. Zezza, V.; Meloni, S.; Oddone, M. Rare-earth and large-ion-lithophile element fractionation in late-Hercynian granite massif of the Biellese area (southern Alps, Italy). *Rend. Soc. Ital. Mineral. Petrol.* **1984**, *39*, 509–521. Available online: https://rruff.info/rdsmi/V39/RDSMI39_509.pdf (accessed on 28 April 2020).

78. Quick, J.E.; Sinigoi, S.; Peressini, G.; Demarchi, G.; Wooden, J.L.; Sbisà, A. Magmatic plumbing of a large Permian caldera exposed to a depth of 25 km. *Geology* **2009**, *37*, 603–606. [CrossRef]

79. Panizza, M. The geomorphodiversity of the Dolomites (Italy): A key of geoheritage assessment. *Geoheritage* **2009**, *1*, 33–42. [CrossRef]

80. Lombardo, V.; Piana, F.; Fioraso, G.; Irace, A.; Mimmo, D.; Mosca, P.; Tallone, S.; Barale, L.; Morelli, M.; Giardino, M. The classification scheme of the Piemonte geological map and the OntoGeonous initiative. *Rend. Online Soc. Geol. Ital.* **2016**, *39*, 117–120. [CrossRef]

81. Piana, F.; Fioraso, G.; Irace, A.; Mosca, P.; d'Atri, A.; Barale, L.; Falletti, P.; Monegato, G.; Morelli, M.; Tallone, S.; et al. Geology of Piemonte region (NW Italy, Alps–Apennines interference zone). *J. Maps* **2017**, *13*, 395–405. [CrossRef]

82. Bini, A.; Cita, M.B.; Gaetani, M. Southern Alpine lakes — Hypothesis of an erosional origin related to the Messinian entrenchment. *Mar. Geol.* **1978**, *27*, 271–288. [CrossRef]

83. Staub, R. Grundzüge und probleme alpiner morphologie. *Denkschr. Schweiz. Naturforsch. Ges.* **1934**, *69*, 1–183.

84. Carraro, F. Revisione del Villafranchiano nell'area-tipo di Villafranca d'Asti. *Ital. J. Quat. Sci.* **1996**, *9*, 5–120.

85. Carraro, F.; Giardino, M. Quaternary glaciations in the western Italian Alps—A review. *Dev. Quat. Sci.* **2004**, *2*, 201–208. [CrossRef]

86. Braakhekke, J.; Ivy-Ochs, S.; Monegato, G.; Gianotti, F.; Martin, S.; Casale, S.; Christl, M. Timing and flow pattern of the Orta Glacier (European Alps) during the Last Glacial Maximum. *Boreas* **2020**, *49*, 315–332. [CrossRef]

87. Bini, A.; Buoncristiani, J.-F.; Couterrand, S.; Ellwanger, D.; Felber, M.; Florineth, D.; Graf, H.R.; Keller, O.; Kelly, M.; Schlüchter, C.; et al. *Die Schweiz Während Des Letzteiszeitlichen Maximums (LGM)*; Bundesamt für Landestopografie Swisstopo: Wabern, Switzerland, 2009.

88. Ehlers, J.; Gibbard, P.L. Quaternary glaciations-extent and chronology: Part I: Europe. *Dev. Quat. Sci* **2004**, *2*, 488.

89. Scapozza, C.; Castelletti, C.; Soma, L.; Dall'Agnolo, S.; Ambrosi, C. Timing of LGM and deglaciation in the Southern Swiss Alps. *Géomorphologie* **2014**, *20*, 307–322. [CrossRef]

90. Salvatore, M.C.; Zanoner, T.; Baroni, C.; Carton, A.; Banchieri, F.A.; Viani, C.; Giardino, M.; Perotti, L. The state of Italian glaciers: A snapshot of the 2006-2007 hydrological period. *Geogr. Fis. Din. Quat.* **2015**, *38*, 175–198. [CrossRef]

91. Viani, C.; Leonoris, C.; Giardino, M.; Diolaiuti, G.; Smiraglia, C. Il Monte Rosa Valsesiano. Il Sentiero Glaciologico del Parco Naturale Alta Valsesia. Itinerario N. 8. In *Itinerari Glaciologici Sulle Montagne italiane. Volume 2. Dalle Alpi Marittime all'Alpe Veglia. Guide Geologiche Regionali*; Comitato Glaciologico Italiano, Ed.; Società Geologica Italiana: Rome, Italy, 2017; Volume 2, pp. 173–192.

92. Smiraglia, C.; Diolaiuti, G. *Il Nuovo Catasto Dei Ghiacciai Italiani—The New Italian Glacier Inventory*; Ev-K2-CNR: Bergamo, Italy, 2015; Volume 400.

93. Pelfini, M.; Bollati, I. Landforms and geomorphosites ongoing changes: Concepts and implications for geoheritage promotion. *Quaest. Geog.* **2014**, *33*, 131–143. [CrossRef]

94. Bollati, I.; Pellegrini, M.; Reynard, E.; Pelfini, M. Water driven processes and landforms evolution rates in mountain geomorphosites: Examples from Swiss Alps. *Catena* **2017**, *158*, 321–339. [CrossRef]

95. Cavallo, A.; Bigioggero, B.; Colombo, A.; Tunesi, A. The Verbano Cusio Ossola province: A land of quarries in northern Italy (Piedmont). *Period. Mineral.* **2004**, *73*, 197–210.

96. Dino, G.A.; Cavallo, A. Ornamental stones of the Verbano Cusio Ossola quarry district: Characterization of materials, quarrying techniques and history and relevance to local and national heritage. *Geol. Soc. Lond. Spec. Publ.* **2015**, *407*, 187–200. [CrossRef]

97. Pelfini, M.; Bollati, I.; Giudici, M.; Pedrazzini, T.; Sturani, M.; Zucali, M. Urban geoheritage as a resource for Earth Sciences education: Examples from Milan metropolitan area. *Rend. Online Soc. Geol. Ital.* **2018**, *45*, 83–88. [CrossRef]

98. Cavallo, A.; Dino, G.A. Granites of the Verbano-Cusio-Ossola District (Piedmont, Northern Italy): Possible candidates for the designation of "Global Heritage Stone Province" and a proposal of a geotouristic route. In *Engineering Geology for Society and Territory*; Lollino, G., Manconi, A., Guzzetti, F., Culshaw, M., Bobrowsky, P., Luino, F., Eds.; Springer International Publishing: Cham, Switzerland, 2015; Volume 5, pp. 267–271. [CrossRef]

99. Bollati, I.M.; Crosa Lenz, B.; Caironi, V. A multidisciplinary approach for physical landscape analysis: Scientific value and risk of degradation of outstanding landforms in the glacial plateau of the Loana Valley (Central-Western Italian Alps). *Ital. J. Geosci.* **2020**, *139*, 233–251. [CrossRef]

100. Borghi, A.; Castelli, D.; Corbetta, E.; Dino, G.A. Candoglia Marble and the "Veneranda Fabbrica del Duomo di Milano": A resource for Global Heritage Stone Designation in the Italian Alps. *EGU Geophys. Res. Abstr.* **2015**, *17*, 13002.

101. Cavallo, A. Soapstones from the Ossola Valley (Piemonte, northern Italy). *EGU Geophys. Res. Abstr.* **2017**, *19*, 5548.

102. Arzarello, M.; Daffara, S.; Berruti, G.; Berruto, G.; Berté, D.; Berto, C.; Gambari, F.M.; Peretto, C. The Mousterian Settlement in the Ciota Ciara Cave: The oldest evidence of Homo Neanderthalensis in Piedmont (Northern Italy). *Boll. Soc. Ital. Biol. Sper.* **2012**, *85*, 71–75. [CrossRef]

103. Graue, J. Die Gräberfelder von Ornavasso: Eine Studie zur Chronologie der späten Latene-und frühen Kaiserzeit. In *Hamburger Beiträge zur Archäologie*; Buske, E., Ed.; Buske: Hamburg, Germany, 1974; Volume 1, p. 272.

104. Panizza, M. Geomorphosites: Concepts, methods and examples of geomorphological survey. *Chin. Sci. Bull.* **2001**, *46*, 4–5. [CrossRef]

105. Reynard, E.; Fontana, G.; Kozlik, L.; Scapozza, C. A method for assessing the scientific and additional values of geomorphosites. *Geog. Helv.* **2007**, *62*, 148–158. [CrossRef]

106. Pelfini, M.; Fredi, P.; Bollati, I.; Coratza, P.; Fubelli, G.; Giardino, M.; Liucci, L.; Magagna, A.; Melelli, L.; Padovani, V.; et al. Developing new approaches and strategies for teaching Physical Geography and Geomorphology: The role of the Italian Association of Physical Geography and Geomorphology (AIGeo). *Rend. Online Soc. Geol. Ital.* **2018**, *45*, 119–127. [CrossRef]

107. Bollati, I.; Fossati, M.; Zanoletti, E.; Zucali, M.; Magagna, A.; Pelfini, M. A methodological proposal for the assessment of cliffs equipped for climbing as a component of geoheritage and tools for Earth Science education: The case of the Verbano-Cusio-Ossola (Western Italian Alps). *J. Virtual Explor.* **2016**, *49*, 1–23.

108. Pica, A.; Reynard, E.; Grangier, L.; Kaiser, C.; Ghiraldi, L.; Perotti, L.; Del Monte, M. GeoGuides, urban geotourism offer powered by mobile application technology. *Geoheritage* **2018**, *10*, 311–326. [CrossRef]

109. Pelfini, M.; Bollati, I.; Pellegrini, L.; Zucali, M. Earth Sciences on the field: Educational applications for the comprehension of landscape evolution. *Rend. Online Soc. Geol. Ital.* **2016**, *40*, 56–66. [CrossRef]

110. Pelfini, M.; Parravicini, P.; Fumagalli, P.; Graffi, A.; Grieco, G.; Merlini, M.; Porta, M.; Trombino, L.; Zucali, M. New methodologies and technologies in Earth Sciences education: Opportunities and criticisms for future teachers. *Rend. Online Soc. Geol. Ital.* **2019**, *49*, 4–10. [CrossRef]

111. Maskall, J.; Stokes, A. Designing effective fieldwork for the environmental and natural sciences. Plymouth: The Higher Education Academy Subject Centre for Geography, Earth and Environmental Sciences. *Prog. Phys. Geog.* **2008**, *33*, 133–134. [CrossRef]

112. King, C. Geoscience education: An overview. *Stud. Sci. Educ.* **2008**, *44*, 187–222. [CrossRef]

113. Bollati, I.M.; Crosa Lenz, B.; Golzio, A.; Masseroli, A. Tree rings as ecological indicator of geomorphic activity in geoheritage studies. *Ecol. Indic.* **2018**, *93*, 899–916. [CrossRef]

114. Viani, C.; Machguth, H.; Huggel, C.; Godio, A.; Franco, D.; Perotti, L.; Giardino, M. Potential future lakes from continued glacier shrinkage in the Aosta Valley Region (Western Alps, Italy). *Geomorphology* **2020**, *355*, 107068. [CrossRef]

115. Dal Piaz, G.V.; Bistacchi, A.; Massironi, M. Geological outline of the Alps. *Episodes* **2003**, *26*, 175–180. [CrossRef]

116. Leonoris, C. *La Scienza Oltre Le Nuvole: 100 Anni Di Storia Dell'istituto Scientifico Angelo Mosso al Col d'Olen Sul Monte Rosa*; Zeisciu Centro Studi: Magenta, Italy, 2007; p. 168.

117. Guindani, N.; Freppaz, M. *Lo Scienziato Alpinista Umberto Monterin: Pioniere Dello Studio Sui Cambiamenti Climatici*; Le Chóteau Edizioni: Aosta, Italy, 2018; Volume 96.

118. Viani, C.; Giardino, M.; Huggel, C.; Perotti, L.; Mortara, G. An overview of glacier lakes in the Western Italian Alps from 1927 to 2014 based on multiple data sources (historical maps, orthophotos and reports of the glaciological surveys). *Geogr. Fis. Din. Quat.* **2016**, *39*, 203–214. [CrossRef]

119. Colombo, N.; Salerno, F.; Martin, M.; Malandrino, M.; Giardino, M.; Serra, E.; Godone, D.; Said-Pullicino, D.; Fratianni, S.; Paro, L.; et al. Influence of permafrost, rock and ice glaciers on chemistry of high-elevation ponds (NW Italian Alps). *Sci. Total Environ.* **2019**, *685*, 886–901. [CrossRef]

120. Leonoris, C.; Freppaz, M.; Caimi, A.; Filippa, G.; Dal Piaz, G.V.; Dal Piaz, G.; Schiavo, A. *Parco Naturale Alta Valsesia, Percorso Geologico-Pedologico di Cimalegna*; Parco Naturale Alta Valsesia: Varallo Sesia, Italy, 2008; p. 36. Available online: https://www.areeprotettevallesesia.it/it/filedownload?T=5&I=348 (accessed on 28 April 2020).

121. Freppaz, M.; Viglietti, D.; Balestrini, R.; Lonati, M.; Colombo, N. Climatic and pedoclimatic factors driving C and N dynamics in soil and surface water in the alpine tundra (NW-Italian Alps). *Nat. Conserv.* **2019**, *34*, 67–90. [CrossRef]

122. Angelucci, D.E.; Zambaldi, M.; Tessari, U.; Vaccaro, C.; Arnaud, J.; Berruti, G.L.F.; Daffara, S.; Arzarello, M. New insights on the Monte Fenera Palaeolithic, Italy: Geoarchaeology of the Ciota Ciara cave. *Geoarchaeology* **2019**, *34*, 413–429. [CrossRef]

123. Canavese, G.; Gianotti, F.; de Varine, H. Ecomuseums and geosites community and project building. *Int. J. Geoheritage Parks* **2018**, *6*, 43–62. [CrossRef]

124. Surace, I.R.; Clauer, N.; Thélin, P.; Pfeifer, H.-R. Structural analysis, clay mineralogy and K–Ar dating of fault gouges from Centovalli Line (Central Alps) for reconstruction of their recent activity. *Tectonophysics* **2011**, *510*, 80–93. [CrossRef]

125. Graeser, S.; Schwander, H.; Hänni, H.; Mattioli, V. Vigezzite, (Ca, Ce) (Nb, Ta, Ti)$_2$ O$_6$, a new aeschynite-type mineral from the Alps. *Mineral. Mag.* **1979**, *43*, 459–462. [CrossRef]

126. Garavaglia, V.; Diolaiuti, G.; Smiraglia, C.; Pasquale, V.; Pelfini, M. Evaluating tourist perception of environmental changes as a contribution to managing natural resources in glacierized areas: A case study of the Forni Glacier (Stelvio National Park, Italian Alps). *Environ. Manag.* **2012**, *50*, 1125–1138. [CrossRef]

127. D'Agata, C.; Diolaiuti, G.; Maragno, D.; Smiraglia, C.; Pelfini, M. Climate change effects on landscape and environment in glacierized alpine areas: Retreating glaciers and enlarging forelands in the Bernina group (Italy) in the period 1954–2007. *Geol. Ecol. Landsc.* **2020**, *4*, 71–86. [CrossRef]

Article

GEOTOURISM as a Tool for Learning: A Geoitinerary in the Cilento, Vallo di Diano and Alburni Geopark (Southern Italy)

Nicoletta Santangelo [1], Vincenzo Amato [2,3], Alessandra Ascione [1], Elda Russo Ermolli [1] and Ettore Valente [1,*]

[1] Department of Earth, Environmental and Resources Sciences, University of Naples Federico II, via Cinthia 21, 80126 Naples, Italy; nicsanta@unina.it (N.S.); alessandra.ascione@unina.it (A.A.); ermolli@unina.it (E.R.E.)

[2] Department of Bioscience and Territory, University of Molise, C.da F. Lappone, 86090 Pesche (IS), Italy; vamato@unisa.it

[3] Department of Science of the Cultural Heritage, University of Salerno, Campus of Fisciano, Via Giovanni Paolo II, 132-84084 Fisciano (SA), Italy

* Correspondence: ettore.valente@unina.it

Received: 13 May 2020; Accepted: 2 June 2020; Published: 4 June 2020

Abstract: "Geotourism" is a particular type of "sustainable tourism" that is still in an embryonic stage, especially in Italy. The main goal is the transmission of geological knowledge to increase the awareness about geoheritage, geo-resources and geo-hazards. The geoparks represent ideal sites, with a strong educational significance for students, teachers, geo-tourists, and guides interested in geological and environmental sciences, though at different levels. With this in mind, we propose a geoitinerary through some of the most geologically interesting coastal areas in the Cilento, Vallo di Diano, and Alburni Geopark. The aim of the geoitinerary is to provide a good example of how geosites could be promoted through geotourism and used as means of divulgation of geological and environmental knowledge. The selected sites are the San Marco coast, the Licosa Cape and the *Elea-Velia* archaeological area. They are included in the official list of geosites and geomorphosites of the Geopark and have a relevant stratigraphic and geoarcheological value. The San Marco coast and the Licosa Cape are the "best sites" in the Geopark where Quaternary coastal deposits and morphologies are represented. The *Elea-Velia* site is one of the most famous archeological sites in the Geopark, which is also representative of complex human environment interactions. Despite their high scientific significance, the sites that we have selected are not included in a specific promoting program. We have so tried to fill this gap by providing the scientific background for their geotouristic promotion that could also serve as an instrument for the increase of the local economy.

Keywords: geoparks; geosites; geotourism; geological knowledge; geoarcheology

1. Introduction

The Cilento and Vallo di Diano National Park was founded in 1991 under the law 394/91 and it was included in the United Nations Educational, Scientific and Cultural Organization (UNESCO) World Heritage list in 1998. It gained the title of Geopark in 2010 and became a UNESCO Global Geopark in 2015. Global Geoparks have been defined as "Single, unified geographical areas where sites and landscapes of international geological significance are managed with a holistic concept of protection, education and sustainable development" [1,2]. Several authors [3–6] suggested, among the main purposes of a geopark, the preservation of geodiversity by means of geoeducation and geotourism. Geological education, that is, how geoscience, particularly the geological processes, and society are linked, is a basic resource for the social and economic development of any community because it increases the sensitivity and awareness of citizens in the respect of the environmental estate and natural disasters. Unfortunately, geological education is still a missing topic [7] as, for instance, in the context of global change and overexploitation of resources [8]. Reference [9] suggests that, by using geological themes significant to the public through interpretive media and educational packages, it is possible to create a "dialogue between the public and Earth's history", leading people towards an understanding of the important processes that control our planet. In this meaning, Geoparks and Geotourism may be very useful. According to several authors ([10–12] and references therein) "Geotourism" is an emerging type of sustainable tourism, which focuses on geosites and furnishes visitor knowledge, environmental education, and also amusement.

In this paper, we propose a geoitinerary along the Cilento, Vallo di Diano and Alburni Geopark coastal area, with the aim to provide a good example of how some geosites and geomorphosites could be promoted through geotourism and used as means of geological and environmental knowledge divulgation. Geologists always use field activity as a tool to teach their students; our goal is to demonstrate that the experience of coming directly in touch with a geological subject may be the most effective approach also for non-expert people. The coastal area of the Cilento Vallo di Diano and Alburni Geopark includes several geomorphosites, which represent good examples of both coastal features and evidence of past sea level fluctations that occurred in response to global climate change. Several papers have already stressed the importance of increased awareness about climate change in coastal environments, also suggesting geotourism as a key action to reach this goal [13–17]. Coastal zones are typical examples of dynamic and sensitive environments, which evolve through different phenomena that act at different temporal and spatial scales [18]. They represent the interface where the land meets the sea and include river deltas, coastal plains, wetlands, beaches and dunes, reefs, lagoons, cliffs and other coastal features. All these areas are very sensitive to changes in response to sea-level rise or extreme weather events, which may cause or accelerate coastal erosion and retreat [19–21]. Approximately 60% of the world's population lives along coastal zones [22,23], which often represent a major environmental and economic resource, though under risk in the light of the present climatic trend [23] that is enhancing the naturally induced sea level rise. We planned a geoitinerary focused on the geological and geomorphological evidence (namely, sediments and landforms) of both the active dynamics and past sea level fluctations along various coastal environments of the Cilento, Vallo di Diano and Alburni Geopark. Our aim was to give an example of how geomorphosites may be useful for educational purposes and providing teachers and touristic guides with didactic/explanatory material.

We selected some sites among those listed in the Geopark's geosites and geomorphosites inventory, choosing the ones with the following requirements: representativeness, availability, and potential educational approach. For each site, we chose one or more geological and geomorphological topics that are represented at the "best" and explained them in the simplest way, with the aim to transfer their scientific significance to an audience as wide as possible. To reach our educational goal, we followed the suggestions of Macadam [24], i.e.,: (i) choosing the message we want the tourists bring home, (ii) avoiding geological jargon, (iii) using more pictures and figures than words. In particular, we focused on the following messages:

(i)   We want to make the tourists aware that natural environments are not static in space and time, but they may evolve and thus interact with human activities. The selected sites are excellent places to spread this message that could have a strong influence on people worldwide because coastal hazard phenomena are spread all over the world.

(ii)   We want the tourists to become aware that the climate on Earth changed periodically, causing sea level fluctations and, consequently, significant modifications to coastal environments. Understanding past climate changes and their effects on the environment may help to understand the present and future climate changes. It should be the occasion to reflect on natural induced climate changes and to become aware that in the postglacial period we are now living in, the human-induced climate changes are contributing to quicken the natural ones.

2. Study Area

2.1. Geological Setting

The Cilento, Vallo di Diano, and Alburni Geopark is bounded by the Sele River plain to the north, the Vallo di Diano and the Tanagro River valley to the north-east, the Gulf of Policastro to the south, and the Tyrrhenian Sea to the west (Figure 1). The Cilento region is part of the Southern Apennines, a NE-directed fold and thrust belt developed in Neogene to Quaternary times [25–27]. At the surface, the mountain belt consists of both open marine Mesozoic stratigraphic units and continental margin successions composed of platform carbonates and pelagic deposits, covered by Neogene and late Miocene wedge-top basin clastics [28–30]. Due to the heterogeneous nature of outcropping rocks and its rugged topography, the Geopark is characterized by a very high degree of geodiversity [31,32]. Carbonate massifs with summit karst landscapes, bounded by steep structural slopes, are typical of the inner area of the park (Figure 1). In such ridges, karst cave systems associated with the most important underground water reservoirs of the region are preserved [33–35]. The carbonate massifs alternate with a hilly landscape, with gentle slopes and dendritic drainage pattern, where pelagic and clastic successions dominate [36–40]. They are represented by the Cilento Group deposits (age between 17.7 and 10.8 million years), with the Pollica and S. Mauro Formations, mainly made up of alternating sandstones, marls and puddingstones [41–43]. The basinal successions (age between 65 and 22 million years) are well exposed in the southern part of the study area (Ascea-Velia) and are made up of the Crete Nere formation, dark clayey successions more than 500 m thick [36,44]. All these units have been involved in the complex tectonic history leading to the formation of the Southern Apennine chain, and for this reason they appear strongly deformed by fault and fold systems [37,45]. During Quaternary times (last 2.6 million years), the study area emerged from the sea, and its landscape was shaped by the actions of different geomorphic processes (fluvial, coastal, aeolian) under the control of strong climate variations. Continental and marine Quaternary succession, mainly made up by conglomerates and sandstones, accumulated along the valleys of the main rivers (Alento, Mingardo and Bussento rivers) and along the coast in the areas of Santa Maria di Castellabate and San Marco di Castellabate (hereinafter and in the map of Figure 1 labelled Santa Maria and San Marco, respectively), Palinuro and Marina di Camerota [46–54]. Different kinds of geosites (structural, stratigraphic, paleontological, geomorphological) [55,56] are witnesses of this significant geological heritage and have been listed in the geosites inventory of the Geopark [57]. The latter includes 160 geosites, 36% of which with main stratigraphical, paleontological, and geological value, and 54% of which are considered as geomorphosites. Nevertheless, at ~30 years from the National Park foundation and 10 years from the Geopark recognition, the actions aiming at the promotion and divulgation of the geological estate of the area are still very few [34,35,40,58–63].

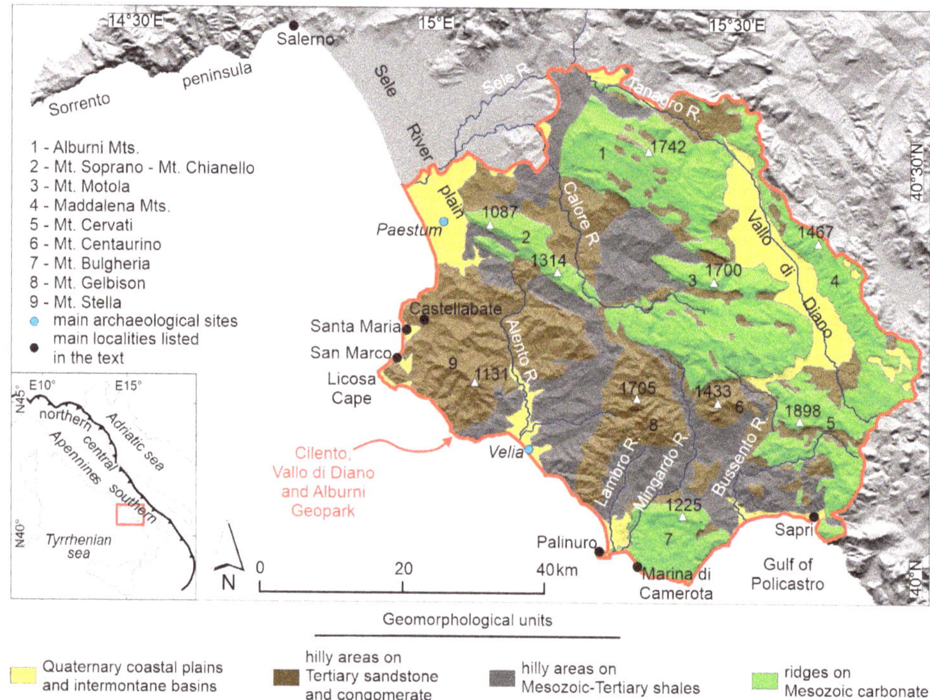

Geomorphological units

Quaternary coastal plains and intermontane basins	hilly areas on Tertiary sandstone and congomerate	hilly areas on Mesozoic-Tertiary shales	ridges on Mesozoic carbonate	

Figure 1. Location of the Cilento, Vallo di Diano, and Alburni Geopark, with indication of the geomorphological units and locations of the main archeological sites.

2.2. Tourism in the Cilento, Vallo di Diano and Alburni Geopark

It is important to bear in mind that Geotourism needs tourists. Unfortunately, the number of tourists that every year visit the Geopark is not monitored by both the Geopark management and the local administrations, e.g., municipalities, Provincia di Salerno and Regione Campania. Such a condition hampers the reconstruction of the tourist flux and its yearly/decadal variations, as already noted by other authors [64]. On the other hand, some information on the number of tourists visiting and spending nights in specific municipalities of the Geopark may be extracted from reports and analyses of the Italian Institute of Statistics (www.istat.it) and the Ente Provinciale del Turismo of Salerno [65]. These data indicate that more than two and up to four million people have visited the Geopark in the last four to five years. Information useful to outline the impact of the tourist traffic in the Geopark area is available from the Italian Ministry of Environment (https://www.minambiente.it/, [66]). Such information indicates that in the Geopark area there are 2931 active tourism companies, with 53,765 beds and 8762 people working in tourism-related economic activities. Within the entire Geopark, the coastal belt gets the largest number of both accommodation services and people working in touristic facilities and attractions [65]. Tourism in the Geopark is mainly concentrated in the summertime, with the beaches and coastal areas representing the main touristic attraction. For instance, the coastal areas caught more than 95% of the more than 2.4 million tourists that visited the Geopark in 2017 [66,67]. Worthy to note, among the Italian National Parks, the Cilento, Vallo di Diano and Alburni Geopark gains large visibility on the internet [65] even if an effective communication system through the web has not been developed so far by the Geopark administration.

3. Materials and Methods

For the selection of sites, we followed the suggestions of Brilha [68], who states that sites with high potential educational value (EV) are those that have geological features that can be easily understood by students of different levels of education, with comfortable and quick access and where students may observe the site under good safety conditions. Similarly, sites with potential tourism value (TV) are those presenting visual beauty enjoyable by the majority of the public, with geological features that can be easily observed and understood by non-specialists, under good safety conditions, and with comfortable and quick access. Therefore, we have followed the Brilha [68] method to assess both the educational (Table 1) and the touristic (Table 2) values of coastal geosites in the Cilento, Vallo di Diano and Alburni Geopark. Moreover, to select the geosites that could be included in the geoitinerary, we analyzed the official list of the geosites of the Geopark reported in Aloia et al. [61] and we selected all the geosites placed along the coastal area that could be considered representative of sea level variations. According to Brilha [68], the educational value of a geosite derives from the sum of the relative scores of twelve indicators and their weights, which are the vulnerability (V), accessibility (AC), use limitations (UL), safety (SA), logistics (L), density of population (DE), association with other values (AS), scenery (SC), uniqueness (UN), observation conditions (OC), didactic potential (DP) and geological diversity (GD). Similarly, the touristic value of a geosite derives from the same indicators of the educational value and their relative weights, with the exception of DP and GD that are replaced by the interpretative potential (IP), economic level (EL), and proximity of recreational areas (RA). Using the criteria listed above, we selected three geosites that gained the highest score of both indexes (Section 4), which are the San Marco site, the marine terrace of Punta Licosa, and the *Elea-Velia* site. In particular, the San Marco site can be considered as the best outcrop of Quaternary coastal deposits with well-preserved sedimentary structures along the entire coast of the Geopark, notwithstanding the presence of other important coastal sites and outcrops [46,51,52,54]. Moreover, the San Marco site is also explicative for the dynamics of rocky coasts and in particular for the formation and evolution of the sea cliffs. The Licosa cape is the best example of a marine terrace in the area of the Geopark. The Licosa cape marine terrace covers an area of more than 4 km in length and up to 500 m in width and has been the object of several scientific publications [47,49,69]. The *Elea-Velia* site is the second most important archaeological site in the Geopark, being only preceded by *Paestum* (Figure 1). Besides its archaeological value, it is also exemplary for the human/environment interaction in a flat coastal area characterised by coastline shifting [47,70–72].

Table 1. List of selected coastal geosites and indicators used to assess their educational value. The geosite number is from Aloia et al. [61]. The number below each indicator represents its relative weight.

	Educational Use of the Geosites												
Indicator Geosite	V (10)	AC (10)	UL (5)	SA (10)	L (5)	DE (5)	AS (5)	SC (5)	UN (5)	OC (10)	DP (20)	GD (10)	Total
22–Marine terrace of Punta Licosa	4	1	4	4	4		2	4	4	4	4	4	3, 6
24–Sandstones of San Marco	3	4	4	4	4	2	4	4	4	4	4	4	3, 8
26–Baia Arena	2	4	4	4	4	2	4	3	2	1	1	2	2, 45
27–Ripe Rosse cliff	4	1	4	3	4	2	4	3	1	3	4	3	3, 1
33–Punta Tresino	4	2	4	3	4	2	4	3	1	3	4	3	3, 2
37–Alento River	1	4	2	4	4	2	4	3	1	2	1	4	2, 5
38–Elea-Velia	3	4	4	4	4	2	4	4	4	4	4	4	3, 8
92–grotta di Cala Fetente	3	1	2	2	4	2	4	3	2	3	2	2	2, 35
93–Cliff of Capo Palinuro	3	1	1	2	4	2	4	3	1	3	4	3	2, 75
94–coastal caves of Camerota	3	1	2	2	4	2	4	4	2	3	2	2	2, 4
96–coast from White Cove to Punta Infreschi	3	1	2	2	4	2	4	4	1	3	4	2	2, 75
97–Fossil dunes of Palinuro	2	4	3	4	4	2	4	1	4	3	1	3	2, 7
98–Tyrrhenian deposits of Palinuro	2	4	3	4	4	2	4	1	4	4	1	3	2, 8
102–Canyon of Marcellino river	4	1	1	2	4	2	4	1	2	2	1	1	1, 9
104–coastal caves of Cala Cefalo	3	1	1	2	4	2	4	3	2	3	2	3	2, 4

Table 2. List of selected coastal geosites and indicators used to assess their touristic value. The geosite number is from Aloia et al. [61]. The number below each indicator represents its relative weight.

| Geosite \ Indicator | Educational Use of the Geosites | | | | | | | | | | | | | Total |
	V (10)	AC (10)	UL (5)	SA (10)	L (5)	DE (5)	AS (5)	SC (15)	UN (10)	OC (5)	IP (10)	EL (5)	RA (5)	
22–Marine terrace of Punta Licosa	4	1	4	4	4	2	4	4	4	4	4	1	4	3, 05
24–Sandstones of San Marco	3	4	4	4	4	2	4	4	4	4	3	1	4	3, 15
26–Baia Arena	2	4	4	4	4	2	4	3	2	1	1	1	4	2, 4
27–Ripe Rosse cliff	4	1	4	3	4	2	4	3	1	3	4	1	4	2, 65
33–Punta Tresino	4	2	4	3	4	2	4	3	1	3	4	1	4	2, 75
37–Alento River	1	4	2	4	4	2	4	3	1	2	1	1	4	2, 25
38–Elea-Velia	3	4	4	4	4	2	4	4	4	4	4	1	4	3, 25
92–grotta di Cala Fetente	3	1	2	2	4	2	4	3	2	3	1	1	3	2, 05
93–Cliff of Capo Palinuro	3	1	1	2	4	2	4	3	1	3	4	1	3	2, 25
94–coastal caves of Camerota	3	1	2	2	4	2	4	4	2	3	3	1	3	2, 3
96–coast from White Cove to Punta Infreschi	3	1	2	2	4	2	4	4	1	3	3	1	3	2, 25
97–Fossil dunes of Palinuro	2	4	3	4	4	2	4	1	4	3	2	1	4	2, 65
98–Tyrrhenian deposits of Palinuro	2	4	3	4	4	2	4	1	4	4	2	1	4	2, 75
102–Canyon of Marcellino river	4	1	1	2	4	2	4	1	2	2	2	1	3	2
104–coastal caves of Cala Cefalo	3	1	1	2	4	2	4	3	2	3	2	1	4	2, 15

The high didactic value of these sites makes them suitable for educational activities addressed to the diffusion of environmental knowledge. With this in mind, we consulted all the available literature data about these sites and tried to "translate" the main basic geological concepts to a wider audience, not necessarily having a geological background. Taking into account the suggestions of UNESCO [73] and Macadam [24], we tried to avoid the use of geological jargon, dumbing down the concepts but without losing their scientific integrity. We used photos, sketches, maps, and 3-D reconstructions accompanied by short and simple sentences to explain our messages. In particular, for each discussed topic, we designed explanatory panels that can serve as a guideline [74–76] for both touristic guides and science teachers at different educational levels. We conceived the panels with a design that could be readable on the most common electronic devices (e.g., smartphone, tablets) thus making the panels available for the users during the entire trip.

The 3-D views have been obtained from detail scale digital terrain models (DTM). The latter includes Light Detection and Ranging (Lidar) data obtained by the Italian Minister of the Environment. Lidar data have been imported in a Geographic Information System (GIS) software, Arcgis 10.7© (Redlands, CA, USA), to derive a hillshade map that has been managed by means of the ArcScene module to obtain the 3-D views. The latter have been obtained by applying a 3-times vertical exaggeration and have been rotated, with respect the north, to obtain the point of view that allows the best appreciation of the coastal landforms.

4. Results

4.1. Geosites Selection

The location of the geosites selected among those listed in the Geopark official list is reported in Figure 2. The results of the application of the Brilha method [68] are reported in Table 1 (educational use of the geosites) and Table 2 (touristic use of the geosites). Moreover, regarding the accessibility indicator (AC), the selected geosites are often placed along the coasts (geosites n. 92, 93, 94, 96, 102 and 104) and they can be reached only by boats, so we have considered distances indicated by Brilha [68] as distances from the closest harbour. As a result of the evaluation, three geosites gained the highest scores in both indexes, which are the geosite n. 22–Marine terrace of Punta Licosa, geosite n. 24–Sandstones of San Marco, and geosite n. 38–*Elea-Velia*. Consequently, we have selected these three geosites and planned the educational geoitinerary.

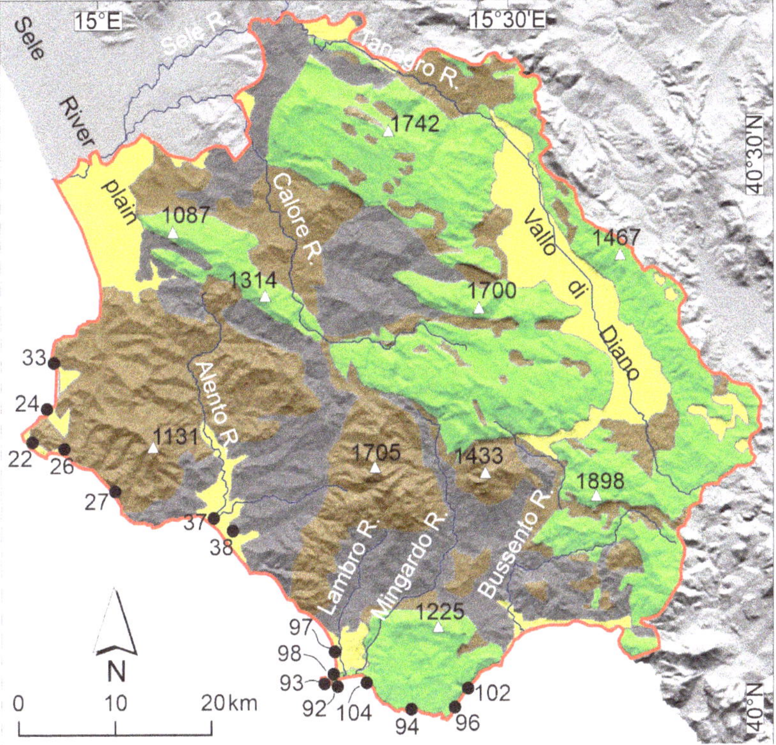

Figure 2. Locations of 15 geosites placed along the Geopark coast and listed in Tables 1 and 2. Numbers refer to geosite numbers reported in both tables.

4.2. Geoitinerary Topics

The sites we have chosen are exemplary to explain the sea level fluctuations that occurred on Earth over the last 250,000 years [77–79] in response to the climate cyclicity (eustatic sea level fluctuations). As shown in Figure 3, during this time interval, the sea level has fluctuated between "High" and "Low" conditions, with variations in the range of about 100 m. The "High" conditions relate to "Interglacial" periods, meaning that climate conditions were similar to the present one or even warmer and more humid. In these periods, ice caps and continental glaciers melted and reduced their extensions, causing the rise of sea level at a global scale. The "Low" conditions are correlated with moments (Glacial periods) when the climate was colder and dryer than the present one. During these periods, the ice masses extended, and the sea level dropped down to 120 m below the present sea level.

Looking at the figure, we can observe that 130,000 years ago (Last Interglacial) the sea level was higher than the present one, at the Last Glacial Maximum (20,000 years ago) it moved down to −120 m and, from then, it started rising up to the present altitude. These relative motions of the sea level obviously caused significant variations in coastal environments, such as the shifting of the coastline position.

We think that these simple concepts on the extreme dynamicity of the coastal environment and on the past climate changes should be fundamental for the education of each citizen, from students to administrators and politicians. We planned the geoitinerary trying to explain these main sea level fluctuations, choosing one site for each peak described in Figure 3.

The San Marco site has preserved the evidence of the Penultimate Interglacial, the Licosa Cape site shows forms and deposits of the Last Interglacial, and finally the *Elea-Velia* site allows us to discuss on the Post-glacial sea level rise.

Climate change and sea level fluctuations

The last 960,000 years of the Earth's history have been characterised by strong climate changes. Several cold (**glacial**) and warm (**interglacial**) periods alternated, with a ciclicity of about 100,000 years. During the repeated glacial and interglacial periods, the cyclic formation and melting of glaciers and sea ice resulted in alternated trapping of freshwater and its releasing to the oceans. Thus, the changing ice volumes were accompanied by major sea level fluctuations in the range of about 100 metres.

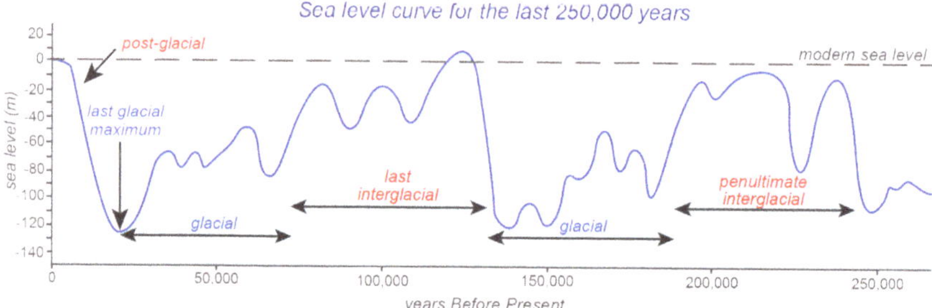

A sea level curve shows the amount of the sea level fluctuations that accompanied the climatic fluctuations. In the curve, the warm, or interglacial, climatic stages correspond to high values of the sea level, while the cold, glacial stages to low sea level values.

In the last 250,000 years, the sea level raised to values close to the modern one (the 0 m value in the curve above) during two main interglacial periods, namely in the *penultimate interglacial*, which occurred around 200,000 years ago, and in the *last interglacial*. During the *last interglacial*, the sea level peaked to values even higher than the modern one around 130,000 years ago. Drops of the sea level down to 120 m were recorded during the glacial stages, including the *last glacial maximum*, which occurred 20,000 years ago. Since then, that is in the *post-glacial* stage, the sea level has been rising to the modern value.

Figure 3. Explanatory panel on eustatic sea level fluctuations for the last 250,000 years. The sea level curve was redrawn and modified after References [78,79].

4.3. The Geoitinerary

The geoitinerary consists of two parts: the first part starts from the center of San Marco village and reaches the Licosa cape. The second part focuses on the archeological site of *Elea-Velia*. It needs at least two days, and the journey from the first stop to the last one implies a transfer by car (32 km). The first part of the geoitinerary includes a nice walk of about 3 km from the San Marco harbourharbour to the Licosa cape, with no significant altitude variations (Figure 4).

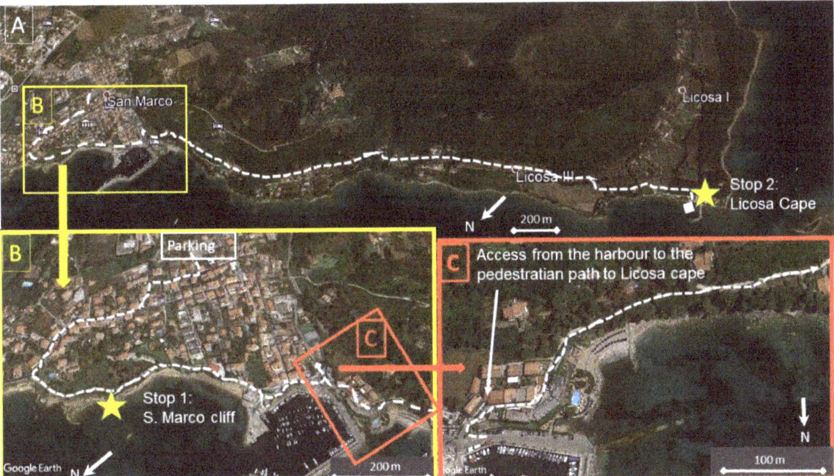

Figure 4. The itinerary of the first day: (**A**) location of stop 1 and 2; (**B**) the walk from the center of San Marco, down to the coastal cliff and the harbour; (**C**) detail of the access to the walk from the San Marco harbour towards Licosa cape.

4.3.1. First Part: the San Marco Sea Cliff and the Licosa Cape Stops

Main topic: evolution of rocky coasts
Contents: sea cliff, marine deposits, marine terrace

Stop 1: San Marco Sea Cliff and Marine Terrace

After parking in the center of the San Marco village, the tourist may direct towards the sea. The pedestrian pathway along the cliff offers a good occasion to explain what a coastal cliff is, how it forms, and how it evolves during time (Figure 5). Following the tracks, it is possible to reach the sea, observing the cliff and the wave-cut platform located at its foot. The latter is well exposed if the sea is flat and in low tide conditions. It is also possible to observe that above the platform, there are several blocks and masses, fallen from the cliff: this site may be useful to understand that a cliff is a dynamic, not static environment, which is moving backward ("retreating") over time. The lower part of Figure 5 (coastal hazard) explains how the normal evolution of a cliff may interact with human activity causing risk conditions for both human properties and lives.

This site is also representative because it preserves the best exposure along the coast of the Geopark of ancient coastal marine sediments related to the Penultimate Interglacial (around 200,000 years ago [55]) with well-preserved sedimentary structures (Figure 6). Looking at the cliff, even a casual tourist may catch the spectacular geometric design drawn by the nature on this rock.

In particular, the cliff is made up of ancient coastal sands, which are now uplifted and form a wide depositional terrace over which the village of San Marco is built (Figure 7). The good exposure of well-preserved sedimentary structures makes this site significant for educational purposes for geological science students. They may come in touch with sandstones and their textures and may observe different kinds of sedimentary structures, testifying for a submerged beach environment. Instead, around 200,000 years ago, the area now occupied by the village of San Marco was invaded by the sea, becoming a gulf, and sands were deposited at the sea bottom. These sands now lie at 15–20 m above the sea level, an altitude higher than the eustatic sea level related to the Penultimate Interglacial (Figure 3). This means that the land moved upward (tectonic uplift) with respect to the sea and the deposited sands emerged, forming the San Marco marine terrace. Geologists name this kind of marine terrace as "raised marine terrace".

The San Marco sea cliff

What is a **sea cliff** (or coastal cliff, or maritime cliff)?
It is defined as a steep rocky slope, often sub-vertical and sometimes overhanging, that rises precipitously from the sea.
A sea cliff is formed by the impact, at its base, of plunging waves and debris picked up by the breaking waves

Day by day, with repeated sea storms, the breaking waves erode the base of the sea cliff. The upper, overhanging part of the sea cliff is made unstable and breaks dowm. The large fallen blocks rest at the base of the sea cliff until they are destroyed by wave action, while the smaller rock fragments are carried away by the sea.
Prolonged erosion makes the sea cliff move back, or **retreat**. As the sea cliff retreats, an **abrasion platform** is carved at the cliff base.

Coastal hazard

The rock falls induced by wave action make the sea cliffs unstable environments. Instability of the sea cliffs makes the man-made structures located at the cliff top, or base, exposed at coastal hazard

Figure 5. Explanatory panel on coastal cliff formation and evolution.

The San Marco sandstones

The rocks that crop out in the San Marco sea cliff are lithified sands, or **sandstones**. It means that the sands contain a cementing material that binds the sand grains after their deposition. The San Marco sandstones, which are rich of fossil shells (mainly bivalve shells), were deposited in a shallow marine environment around 200,000 years ago, during the penultimate interglacial.

During the penultimate interglacial the sea level rose and invaded the land, forming a deep bay in the area where nowadays the San Marco village is located. In that bay the San Marco sandstones were sedimented, while sea cliffs and abrasion platforms were sculpted in the adjacent rocky headlands.

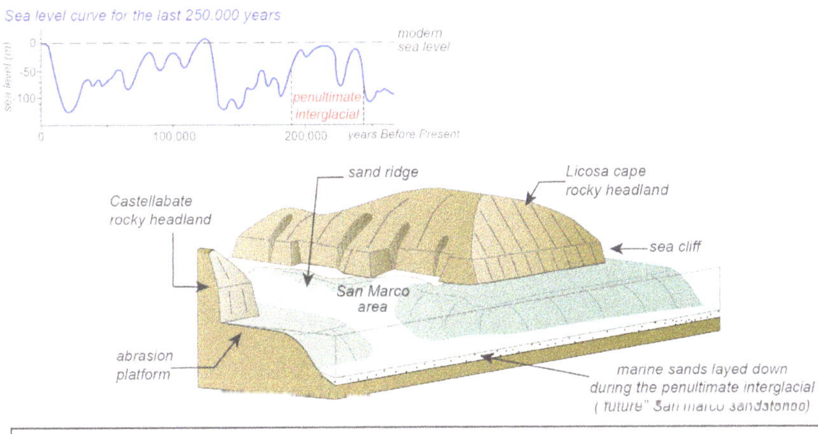

The San Marco sandstones are characterised by a peculiar layering named **cross bedding**, which consists of sets of beds that are inclined to the main horizontal bedding. The cross bedding of the San Marco sandstones developed by the migration of submerged sand ridges (similar to desert dunes) constructed by the action of storm waves on the sea bottom at the time of their deposition.

Figure 6. Explanatory panel on the ancient marine deposits outcropping in the San Marco sea cliff.

The San Marco marine terrace

The San Marco village lies on a plain gently inclined towards the sea, located at an elevation around 15-20 m. That plain is a **marine terrace**.
A marine terrace is a flat area, originally an abrasion platform or a sandy sea bottom, raised above the shoreline. Landwards, a marine terrace is bounded by a **paleo-sea cliff**.

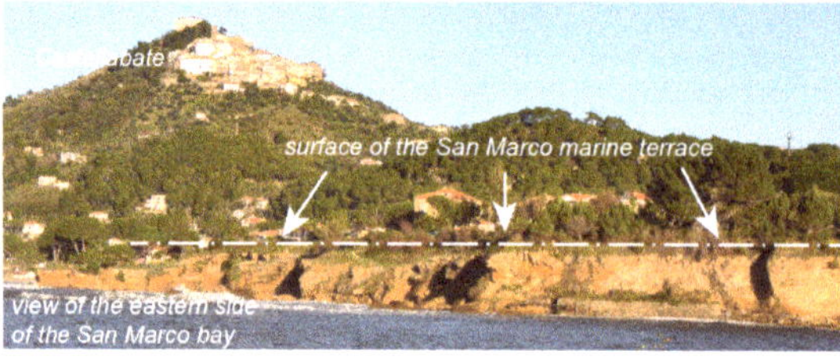

A marine terrace is rasied above the sea level by a relative motion between the land and the sea. The relative land/sea motion may be caused by a sea level fall or uplift of the ground.

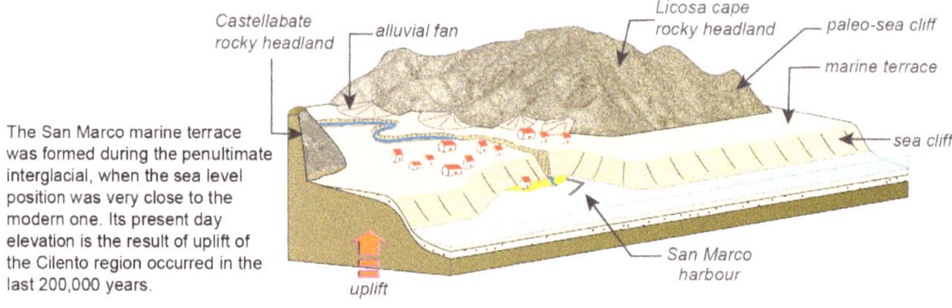

The San Marco marine terrace was formed during the penultimate interglacial, when the sea level position was very close to the modern one. Its present day elevation is the result of uplift of the Cilento region occurred in the last 200,000 years.

Figure 7. Explanatory panel on the San Marco raised marine terrace.

Stop 2: The Licosa Cape

From the harbour of San Marco, the geoitinerary continues towards the Licosa cape. The path starts behind the Approdo Hotel and, after a 1 h walking (Figure 3), it reaches the Licosa cape, where it is possible to have a spectacular view on the little island of Leucosia, so named after one of the mythologic mermaids which distracted Ulysses during its navigation.

The Licosa cape is characterized by the presence of a wide marine terrace (Figure 8) carved by the sea on the Miocene bedrock. It represents an ancient abrasion platform, formed during the Last Interglacial period, when the sea was 6–8 m higher than the present one (Figure 3). Differently from the San Marco terrace, in this case, the altitude of the terrace coincides with that of the sea level (eustatic level; Figure 3) meaning that the land was not uplifted. Geologists refer to this kind of marine terrace as an "eustatic terrace". Along the southern part of the promontory, it is possible to observe that the terrace is locally covered by marine deposits mainly made up of fossil corals and red algae. Geologist were able to define the age of these fossils by means of complex geochemical techniques (in particular, U/Th dating, see Iannace et al. [57] for further details).

The last interglacial marine terrace of the Licosa cape

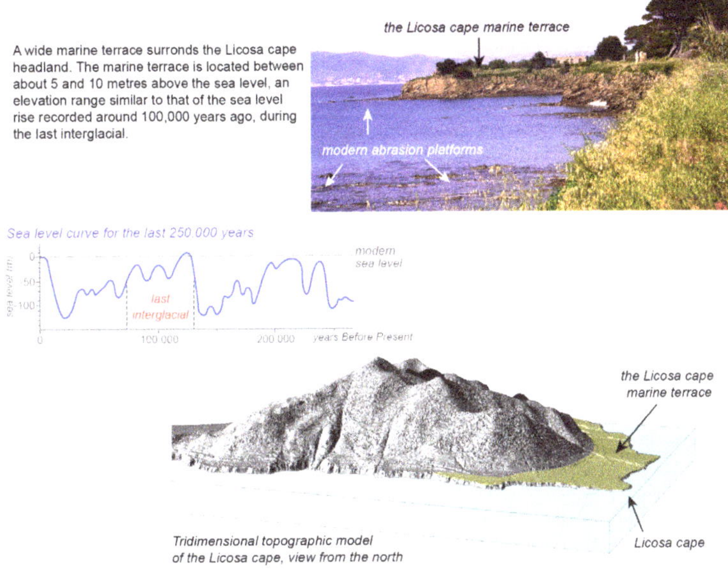

The abrasion platform of the Licosa marine terrace is locally covered by a fossil algal mat composed of algae entrapping sand grains, which provide a nice picture of the Licosa raised sea bottom.

Figure 8. Explanatory panel on the Licosa cape eustatic marine terrace.

4.3.2. The *Elea-Velia* Geo-Archaeological Site

Main Topic: evolution of flat coasts

Contents: coastal progradation; alluvial fan; human-environment interaction

The Archaeological Park of *Elea-Velia* lies in the alluvial-coastal plain of the Fiumarella River and can be reached from San Marco driving towards South the road named SS267 (Figure 9).

Figure 9. Transfer from San Marco to the Archaeological Park of *Elea-Velia*.

The Graeco-Roman town of *Elea-Velia* was founded by the Phoceans in the VI century BC and developed as a Roman town up to the IV century AD when it was abandoned due to both harbour infilling and burial of most of the lower quarters by alluvial deposits [70,71]. For these reasons, this site is explicative of two geomorphological processes: the coastal progradation and the alluvial fan deposition. Both phenomena interacted with human activity from the town foundation up to its abandonment. Within the Archaeological tour of the Park, we suggest two stops which allow for an understanding of the paleogeographical evolution of the area as well as the main environmental changes of the last millennia, including the repeated events of flooding, which caused the burying of the ancient town. Although it is one of the most important touristic attractions of the Geopark, with a number of visitors ranging from 25,000 to 35,000 per year in the period 1996–2018 [80], this peculiar geoarchaeological aspect has not yet been the object of any educational project.

Stop 1: The Acropolis

From the *Acropolis* it is possible to have a wide view on the flat territory surrounding the hill, down to the sea. In this part of the tour, we will focus on what happened during the Postglacial period, when the sea level was rising from −120 m (Last Glacial Maximum) up to its present position. Looking at the eustatic sea level curve of the last 10,000 years (Figure 10), we may observe that there is a variation in the rate of sea level rise at around 7000 years ago. Before that moment, the curve is very steep, suggesting a very rapid sea level rise that resulted in a general submersion of flat coastal areas. For this reason, when the Greeks colonized the territory, the *Acropolis* hill was a promontory bounded by two gulfs, that extended in place of the present Alento (to the NW) and Fiumarella (to the SE) plains [48]. Active sea-cliffs were shaped along the rocky coasts, whereas in front of the flat alluvial-coastal plains, sand ridges isolated lagoon environments that were suitable places for landing.

Why is the Acropolis hill now located hundreds of meters inland with respect to the present coastline? The answer is in the complex interaction between coastal processes and sea level fluctuations [81–83]. From 7000 years ago up to now, the rate of sea level rise decreased and was compensated by the incessant sediment supply from inland (Figure 9). For this reason, the coastline started to move towards the sea, and this phenomenon is called "coastal progradation".

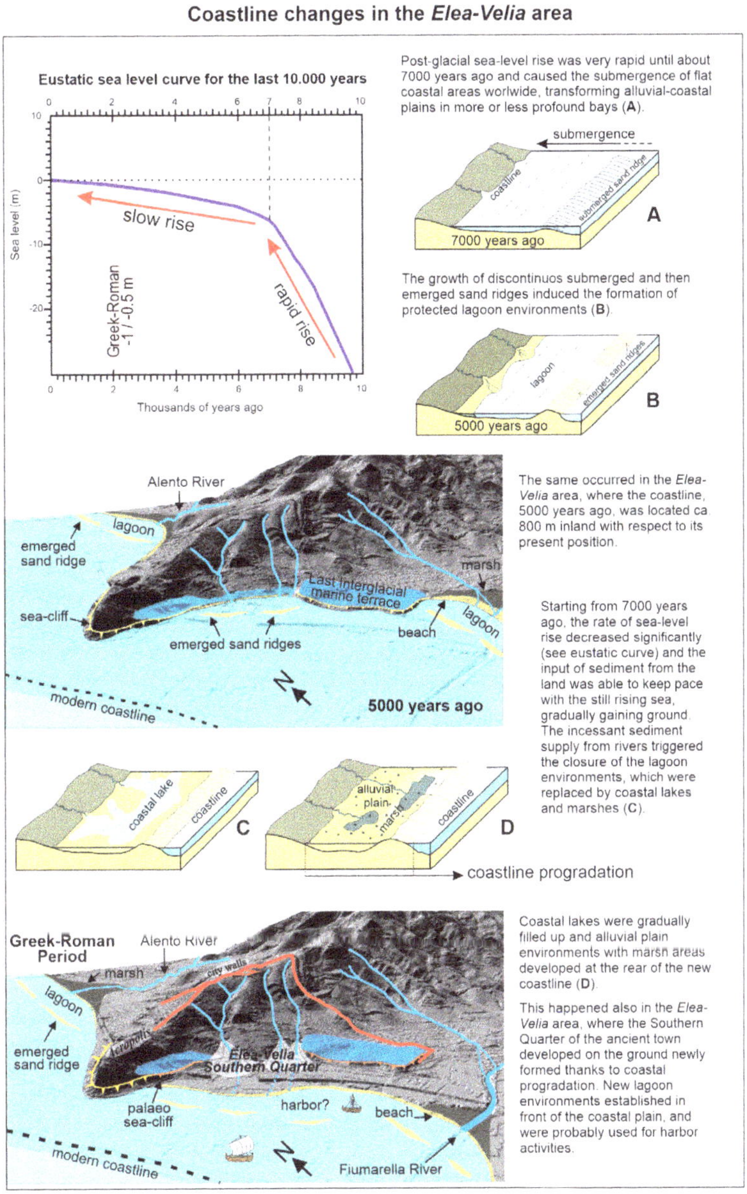

Figure 10. Explanatory panel on paleogeographic condition of the *Elea-Velia* site before and during the Greek colonization.

Stop 2: Via di Porta V

This stop allows the visitors to have a look at the sequence of alluvial fan deposits that damaged the structures of the Southern Quarter all along the life of the Graeco-Roman town. The alluvial deposits came from the erosion of the stream catchments cutting the slopes at the rear of the Southern Quarter, within the town perimeter (Figure 11). The sedimentological characteristic of the alluvial deposits exposed in the section (massive, badly sorted, and poorly stratified pebbles, sands, and silts) clearly indicates that these alluvial events occurred mainly as debris flow, i.e., the water transported high quantity of debris, behaving as a high-density fluid. The intensive fan deposition caused the town abandonment after several phases of restoration. From that moment, a few other floods affected the area up to the V century AD, when flooding was definitively interrupted [72,83].

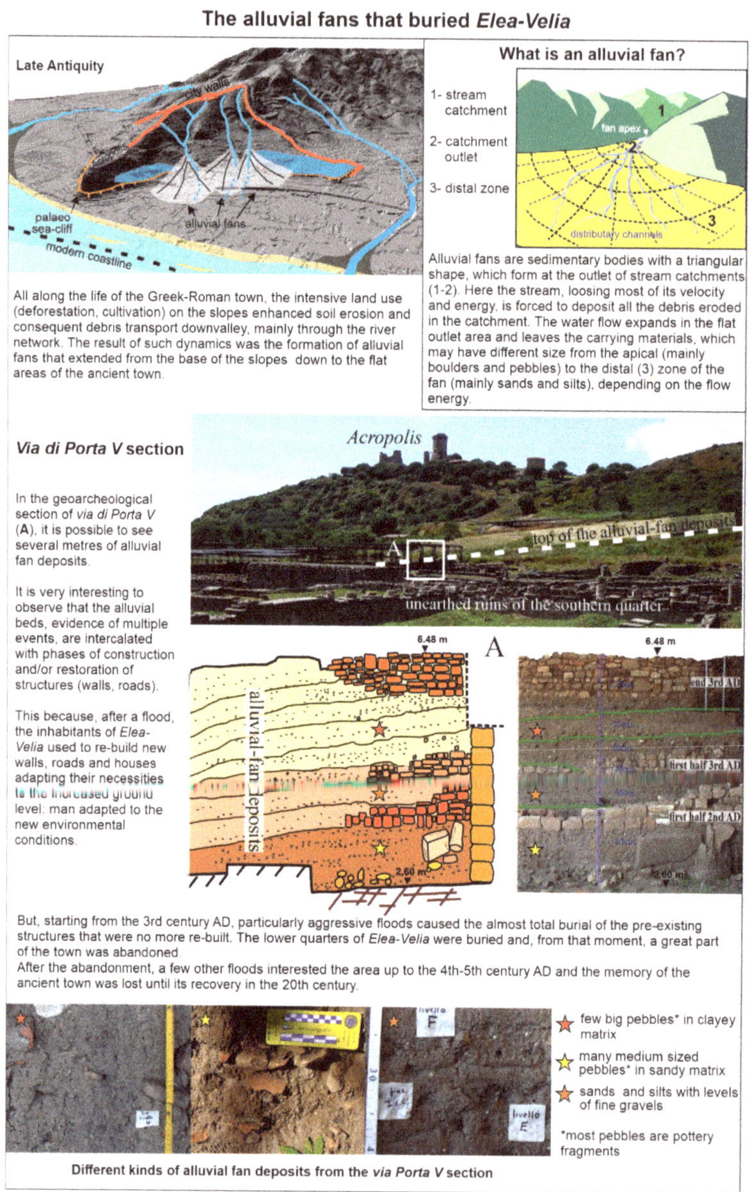

Figure 11. Explanatory panel on the alluvial events that caused the burying of *Elea-Velia*.

5. Discussion and Conclusions

The objective of our work was to provide an example of how geosites, and in particular those with high didactic value, may be used for educational purposes. Bearing in mind our aim, we chose those that are the "best sites" in the coastal belt of the Geopark and built a didactic path. Unfortunately, up to now, the educational itinerary in the area is still a missing topic, and when consulting the Geopark web page it is possible to find only general information about the geology of the Geopark, the wildlife, and the vegetation. The Geopark website, in fact, suggests only one poorly detailed itinerary in the karst area of the Bussento river [84]. In such a context, the paper herein and one recently published [35] address the potential educational value of geosites in the Cilento, Vallo di Diano, and Alburni Geopark, and they may be used as examples of how to use geosites to divulge geological knowledge to a wide audience. By adopting a similar geological approach, didactic routes may be organized within the Geopark, choosing the best sites for various environmental topics.

International tourism is a fundamental economic resource for the coastal areas of the Cilento, Vallo di Diano, and Alburni Geopark [65,66]. In fact, the economy of the coastal areas is mainly based on international tourism, although it is mainly focused during summertime, and the preferred touristic destinations are the beaches and the archeological sites of *Elea-Velia* and Paestum [66,67]. Considering that any development of the Cilento, Vallo di Diano, and Alburni Geopark should be based on the enhancement of local resources through the active cooperation of all local actors [67], our work may be an incitement for the Geopark administrators to develop educational activity as a source for touristic interest. The development of such an activity might also contribute to extending the tourist presence beyond summertime, over most of the year.

To analyze the potential of the geoitinerary and to highlight possible activities addressed to its efficiency and effective use, we also defined the strengths, weaknesses, opportunities, and threats through the SWOT analysis, whose results are reported in Table 3.

We strongly believe that increasing the base geological knowledge of each citizen, from the student to the politician or the administrator, may contribute to increasing the awareness in respect to the evolution of our planet and its environmental systems. Indeed, to achieve integrated management of resources and spaces in a coastal area, it is critical to understand the possible change patterns, which differ substantially in the two environmental scenarios we have focused on, i.e., the sea cliffs and coastal plains. For instance, given the widespread and in some instances dense use of coastal land, it is crucial for people to become acquainted with the risk under which properties, people, and their economic activities are eventually posed by coastal recession and instabilities in both the short and long terms. The scenarios that we selected as sites for visits represent meaningful examples of the instability of both rocky coasts under the current, meteorologically-driven coastal dynamics, and low-lying coastal areas exposed to the impact of future sea level rise. Now more than ever before, it is mandatory to generate public awareness about geology and environmental conditions related to global climate change. Issues such as coastal erosion and submergence of coastlines are becoming popular and, in this respect, evidence of the striking environmental impact of the relatively-subdued sea level rise that occurred in the *Elea-Velia* area appears as a meaningful exemplification of possible future change in comparable scenarios.

Geotourism may be one of the most useful tools to spread knowledge on the dynamicity of the natural environment. However, to reach this goal, individual and/or group visitors should get knowledge in advance through guide boards and educational or informational image data. With our work, we tried to achieve such an objective and our wish is that the explanatory material we designed may constructively support geopark guides and teachers aiming at promoting the Geopark area and spreading geological knowledge, respectively, through geotourism.

Table 3. Results of the SWOT (strengths, weaknesses, opportunities and threats) analysis about the development of Geotourism in the Cilento, Vallo di Diano, and Alburni Geopark.

Strengths	Weaknesses
1. Propose an alternative point of view to increase the touristic attractiveness of the coastal area of the Cilento, Vallo di Diano and Alburni Geopark. This area is normally visited by hundreds of thousands of tourists every year, especially in summer, mostly because of the sea and the beaches.	1. The geoitinerary is a personal initiative by the Authors and lacks, up to now, enough of an intense collaboration with both the Geopark and local administration.
2. The area is easily reachable by public transport.	2. The full development of the geoitinerary needs funding and a management policy.
3. We demonstrated that the three selected geosites have an high educational and touristic values.	3. Local people and local administration have not yet fully understood the high touristic and didactic potential of the area.
4. Landforms are visible and comprehensible to tourists, even if they have no geological background.	4. Tourists reach the study area because of the amazing seas and beaches but they have no cognition of the environmental estate in the surroundings.
5. The geoitinerary aims to increase people awareness about environmental variations due to climate change, which is a challenging topic in this epoch.	5. The tour from San Marco to Licosa cape is flat but needs ~1 h of walking.
6. There are several accommodation facilities that can host high tourist traffic.	

Opportunities	Threats
1. The geoitinerary could promote similar initiatives in other sectors of the Geopark.	
2. The geoitinerary could bring tourists in this area all over the year and not only during the summer period.	1. The San Marco sea-cliff is controlled and managed because it could be affected by rock-fall.
3. The geoitinerary could be used by schools of every level for educational purposes in the field of environmental science.	2. Some fences in the San Marco sea-cliff area needs maintenance
4. The Elea-Velia site give the opportunity to see an archaeological site not only from an historical perspective.	3. Lack of financial resource to develop the geoitinerary
5. The geoitinerary could be split in two days thus providing enough time for the tourists to enjoy the hospitality of local people, the beaches, the sea and the local food.	4. Low interest in local authorities for the development of Geotourism

Finally, we intended for the interdisciplinary perspective we propose for the *Elea-Velia* archaeological site to serve as a contribution to increasing its touristic interest. We selected *Elea-Velia* to suggest that, for an archaeological site, the availability of geological information which integrates information on historical or cultural issues through the explanation of the environmental conditions that existed during the human frequentation may provide the tourist with a more comprehensive view of the interaction between human activities (agriculture, commerce, etc.) and the natural landscape. This may be a new approach for the valorization of archaeological sites. Archaeological heritage is one of the most important economic resources of Italy and, particularly, of the Campania region, where numerous archaeological sites ranging in value from global to national and regional interest are located. Most of those sites, besides telling of history, architecture, and figurative art, bear the legacy of geological and/or geomorphological processes and human—environment interaction. Despite this, there is still no effort, even in the most important sites such as *Pompeii* and *Herculaneum* excavations, to divulge the natural aspect of the archaeological sites [85,86]. As we believe that learning from the past is crucial to understand the present and future environmental crisis, we have put the attention on the "other stories" that an archaeological site may narrate to its visitors.

Author Contributions: Conceptualization, N.S., A.A. and E.R.E.; methodology, N.S. and E.V.; software E.V.; validation, N.S., A.A. and E.R.E.; formal analysis, V.A.; investigation: San Marco Licosa area: N.S., A.A. and E.V.; Elea-Velia: V.A. and E.R.E.; writing—original draft preparation, N.S. and E.V.; writing—review and editing, N.S., A.A., E.V. and E.R.E.; visualization, A.A., E.R.E. and E.V.; supervision, N.S. All authors have read and agreed to the published version of the manuscript.

Funding: This research received no external funding.

Acknowledgments: We wish to thank three anonymous reviewers whose suggestions helped us to improve the manuscript.

Conflicts of Interest: The authors declare no conflict of interest.

References

1. UNESCO Global Geoparks Operational Guidelines. Available online: www.unesco.org (accessed on 29 April 2020).
2. Henriques, M.H.; Brilha, J.B. UNESCO Global Geoparks: A strategy towards global understanding and sustainability. *Episodes* **2017**, *40*, 349–355. [CrossRef]
3. Dowling, R. Geotourism's Global Growth. *Geoheritage* **2010**, *3*, 1–13. [CrossRef]
4. Global Geoparks Network (GGN). Guidelines and Criteria for National Geoparks Seeking UNESCO's Assistance to Join the Global Geoparks Network (GGN) (January 2014). France: European Geoparks Network. Available online: http://www.europeangeoparks.org/wp-content/uploads/2012/03/Geoparks_Guidelines_Jan2014.pdf (accessed on 29 April 2020).
5. Woo, K.S. Qualification and prospect of national and global geoparks in Korea. *J. Geol. Soc. Korea* **2014**, *50*, 3–19. [CrossRef]
6. Koh, Y.-K.; Oh, K.-H.; Youn, S.-T.; Kim, H.-G. Geodiversity and geotourism utilization of islands: Gwanmae Island of South Korea. *J. Mar. Isl. Cult.* **2014**, *3*, 106–112. [CrossRef]
7. Joyce, B.; Brohl, M. Geological and geomorphological features of Australia: How our geosites can be used in geoparks and geotourism to promote better understanding of our geological heritage and as a tool for public education. In Proceedings of the Inaugural Global Geotourism Conference: Discover the Earth Beneath Our Feet, Fremantle, Australia, 17–20 August 2008.
8. Bloise, A.; Catalano, M.; Critelli, T.; Apollaro, C.; Miriello, D. Naturally occurring asbestos: Potential for human exposure, San Severino Lucano (Basilicata, Southern Italy). *Environ. Earth Sci.* **2017**, *76*, 648. [CrossRef]
9. Gozzard, J.R. *WA Coast—Rottnest Island Digital Datasets*; Geological Survey of Western Australia: Perth, Australia, 2011.
10. Dowling, R.; Newsome, D. Chapter 1: Geotourism: Definition, characteristics and international perspectives. In *Handbook of Geotourism*; Edward Elgar Publishing: Cheltenham, UK, 2018; pp. 1–22.
11. Ólafsdóttir, R.; Tverijonaite, E. Geotourism: A Systematic Literature Review. *Geosciences* **2018**, *8*, 234. [CrossRef]
12. Justice, S. UNESCO Global Geoparks, Geotourism and Communication of the Earth Sciences: A Case Study in the Chablais UNESCO Global Geopark, France. *Geosciences* **2018**, *8*, 149. [CrossRef]
13. Zeppel, H. Climate change and tourism in the Great Barrier Reef Marine Park. *Curr. Issues Tour.* **2012**, *15*, 287–292. [CrossRef]
14. Zeppel, H.; Beaumont, N. 2011 Climate change and Australian tourism: A research bibliography. ACSBD Working Paper No. 1. In *Climate Change and Australian Tourism*; Australian Centre for Sustainable Business and Development, University of Southern Queensland: Springfield Central, Australia, 2011.
15. Scott, D.; Hall, C.M.; Stefan, G. *Tourism and Climate Change*; Informa UK Limited: London, UK, 2012.
16. Wang, L.; Tian, M.; Wang, L. Geodiversity, geoconservation and geotourism in Hong Kong Global Geopark of China. *Proc. Geol. Assoc.* **2015**, *126*, 426–437. [CrossRef]
17. Rutherford, J.; Newsome, D.; Kobryn, H.; Jessica, R.; David, N.; Halina, K. Interpretation as a Vital Ingredient of Geotourism in Coastal Environments: The Geology of Sea Level Change, Rottnest Island, Western Australia. *Tour. Mar. Environ.* **2015**, *11*, 55–72. [CrossRef]

18. Castedo, R.; Paredes, C.; De La Vega-Panizo, R.; Santos, A.P.; Santos, R.D.L.V.A.A.P. The Modelling of Coastal Cliffs and Future Trends. In *Hydro-Geomorphology—Models and Trends*; IntechOpen: London, UK, 2017; pp. 53–78.

19. Valiela, I. *Global Coastal Change*; Blackwell Publishing: Oxford, UK, 2006; p. 367.

20. Bird, E. *Coastal Geomorphology: An Introduction*, 2nd ed.; John Wiley & Sons: New York, NY, USA, 2008; p. 436.

21. Trenhaile, A.S. Chapter 2 Climate change and its impact on rock coasts. *Geol. Soc. Lond. Mem.* **2014**, *40*, 7–17. [CrossRef]

22. Nicholls, R.J.; Cazenave, A. Sea-Level Rise and Its Impact on Coastal Zones. *Science* **2010**, *328*, 1517–1520. [CrossRef] [PubMed]

23. Iinternational Panel on Climate Change. Climate Change 2014: Mitigation of Climate Change. In *Contribution of Working Group III to the Fifth Assessment Report of the Intergovernmental Panel on Climate Change*; Cambridge University Press (CUP): Cambridge, UK, 2014.

24. Macadam, J. Geoheritage: Getting the message across. What message and to whom? In *Geoheritage. Assessment, Protection, and Management*; Reynard, E., Brilha, J., Eds.; Elsevier: Amsterdam, The Netherlands, 2018; pp. 267–288.

25. Mazzoli, S.; Helman, M. Neogene patterns of relative plate motion for Africa-Europe: Some implications for recent central Mediterranean tectonics. *Geol. Rundsch.* **1994**, *83*, 464–468. [CrossRef]

26. Turco, E.; Macchiavelli, C.; Mazzoli, S.; Schettino, A.; Pierantoni, P. Kinematic evolution of Alpine Corsica in the framework of Mediterranean mountain belts. *Tectonophysics* **2012**, *579*, 193–206. [CrossRef]

27. Sartori, R. The Tyrrhenian back-arc basin and subduction of the Ionian lithosphere. *Episodes* **2003**, *26*, 217–221. [CrossRef]

28. Ciarcia, S.; Mazzoli, S.; Vitale, S.; Zattin, M. On the tectonic evolution of the Ligurian accretionary complex in southern Italy. *GSA Bull.* **2011**, *124*, 463–483. [CrossRef]

29. Ascione, A.; Ciarcia, S.; Di Donato, V.; Vitale, S.; Mazzoli, S. The Pliocene-Quaternary wedge-top basins of southern Italy: An expression of propagating lateral slab tear beneath the Apennines. *Basin Res.* **2011**, *24*, 456–474. [CrossRef]

30. Mazzoli, S.; Szaniawski, R.; Mittiga, F.; Ascione, A.; Capalbo, A. Tectonic evolution of Pliocene–Pleistocene wedge-top basins of the southern Apennines: New constraints from magnetic fabric analysis. *Can. J. Earth Sci.* **2012**, *49*, 492–509. [CrossRef]

31. Gray, J.M. Geodiversity: Developing the paradigma. *Proc. Geol. Assoc.* **2008**, *119*, 287–298. [CrossRef]

32. Ruban, D.A. Quantification of geodiversity and its loss. *Proc. Geol. Assoc.* **2010**, *121*, 326–333. [CrossRef]

33. Santangelo, N.; Santo, A. Endokarst processes in the Alburni massif (Campania, southern Italy): Evolution of ponors and hydrogeological implications. *Z. Geomorphol.* **1997**, *41*, 229–246.

34. Santangelo, N.; Romano, P.; Santo, A. Geo-itineraries in the Cilento Vallo di Diano Geopark: A Tool for Tourism Development in Southern Italy. *Geoheritage* **2014**, *7*, 319–335. [CrossRef]

35. Valente, E.; Santo, A.; Guida, D.; Santangelo, N. Geotourism in the Cilento, Vallo di Diano and Alburni UNESCO Global Geopark (Southern Italy): The Middle Bussento Karst System. *Resources* **2020**, *9*, 52. [CrossRef]

36. Vitale, S.; Ciarcia, S.; Mazzoli, S.; Zaghloul, M. Tectonic evolution of the 'Liguride' accretionary wedge in the Cilento area, southern Italy: A record of early Apennine geodynamics. *J. Geodyn.* **2011**, *51*, 25–36. [CrossRef]

37. Vitale, S.; Ciarcia, S. Tectono-stratigraphic and kinematic evolution of the southern Apennines/Calabria–Peloritani Terrane system (Italy). *Tectonophysics* **2013**, *583*, 164–182. [CrossRef]

38. Vitale, S.; Ciarcia, S.; Fedele, L.; Tramparulo, F.D.; Vitale, S.; Sabatino, C.; Lorenzo, F.; D'Assisi, T.F. The Ligurian oceanic successions in southern Italy: The key to decrypting the first orogenic stages of the southern Apennines-Calabria chain system. *Tectonophysics* **2019**, *750*, 243–261. [CrossRef]

39. Vitale, S.; Ciarcia, S. Tectono-stratigraphic setting of the Campania region (southern Italy). *J. Maps* **2018**, *14*, 9–21. [CrossRef]

40. Vitale, S.; Amore, O.; Ciarcia, S.; Lo Schiavo, L.; Santangelo, N.; Tramparulo, F. Itinerario 11- Cilento settentrionale. In *Guide Geologiche Regionali—Campania e Molise*; Calcaterra, D., D'Argenio, B., Ferranti, L., Pappone, G., Petrosino, P., Eds.; Società Geologica Italiana: Rome, Italy, 2016; pp. 69–80, ISBN 9788894022728.

41. Geological Map of Italy at Scale 1:50000, Sheet 502—Agropoli. Available online: http://www.isprambiente. gov.it/Media/carg/502_AGROPOLI/Foglio.html (accessed on 30 April 2020).

42. Geological Map of Italy at Scale 1:50000, Sheet 503—Vallo. Available online: http://www.isprambiente.gov.it/Media/carg/503_VALLO/Foglio.html (accessed on 30 April 2020).

43. Geological Map of Italy at Scale 1:50000, Sheet 519—Capo Palinuro. Available online: http://www.isprambiente.gov.it/Media/carg/519_CAPO_PALINURO/Foglio.html (accessed on 30 April 2020).

44. Vitale, S.; Ciarcia, S.; Mazzoli, S.; Iannace, A.; Torre, M. Structural analysis of the 'Internal' Units of Cilento, Italy: New constraints on the Miocene tectonic evolution of the southern Apennine accretionary wedge. *C. R. Geosci.* **2010**, *342*, 475–482. [CrossRef]

45. Cello, G.; Mazzoli, S. Apennine tectonics in southern Italy: A review. *J. Geodyn.* **1998**, *27*, 191–211. [CrossRef]

46. Romano, P. La distribuzione del Pleistocene marino lungo le coste della Campania: Stato delle conoscenze e prospettive di ricerca. *Stud. Geol. Camert.* **1992**, *1*, 265–269.

47. Cinque, A.; Romano, P.; Rosskopf, C.; Santangelo, N.; Santo, A. Morfologie costiere e depositi quaternari tra Agropoli e Ogliastro Marina (Cilento—Italia Meridionale). *Il Quat.* **1994**, *7*, 3–16.

48. Cinque, A.; Rosskopf, C.; Barra, D.; Campajola, L.; Paolillo, G.; Romano, P. Nuovi dati stratigrafici e cronologici sull'evoluzione recente della piana del fiume Alento (Cilento, Campania). *Il Quat.* **1995**, *8*, 323–338.

49. Iannace, A.; Romano, P.; Santangelo, N.; Santo, A.; Tuccimei, P. The OIS 5c along Licosa cape promontory (Campania region, Southern Italy). *Z. Geomorphol.* **2001**, *45*, 307–319.

50. Esposito, C.; Filocamo, F.; Marciano, R.; Romano, P.; Santangelo, N.; Scarciglia, F.; Tuccimei, P. Late Quaternary shorelines in southern Cilento (Mt. Bulgheria): Morphostratigraphy and chronology. *Il Quat.* **2003**, *16*, 3–14.

51. Antonioli, F.; Cinque, A.; Ferranti, L.; Romano, P. Emerged and Submerged Quaternary Marine Terraces of Palinuro Cape (Southern Italy). *Mem. Descr. Carta Geol. d'It* **1994**, *52*, 293–319.

52. Antonioli, F.; Ascione, A.; Cinque, A.; Ferranti, L.; Romano, P. Coastal and Underwater Geomorphology of Capo Palinuro Area, Guidebook to the field—Sea trip. In Proceedings of the Guide for Excursion, International Meeting on Underwater Geology, Rome, Italy, 8–10 June 1994.

53. Brancaccio, L.; Cinque, A.; Romano, P.; Rosskopf, C.; Russo, F.; Santangelo, N.; Santo, A. L'evoluzione delle pianure costiere della Campania: Geomorfologia e neotettonica. *Mem. Soc. Geogr. Ital.* **1995**, *53*, 313–336.

54. Ascione, A.; Romano, P. Vertical movements on the eastern margin of the Tyrrhenian extensional basin. New data from Mt. Bulgheria (Southern Apennines, Italy). *Tectonophysics* **1999**, *315*, 337–356. [CrossRef]

55. Santangelo, N.; Santo, A.; Guida, D.; Lanzara, R.; Siervo, V. The geosites of the Cilento-Vallo di Diano national park (Campania region, southern Italy). *Il Quat.* **2005**, *18*, 101–112.

56. Aloia, A.; Guida, D.; Valente, A. Geodiversity in the Geopark of Cilento and Vallo di Diano as heritage and resource development. *Rend. Online Soc. Geol. Ital.* **2012**, *21*, 688–690.

57. Cilento, Vallo di Diano and Alburni Geopark Geosites Database. Available online: http://www.cilentoediano.it/it/geositi-gli-ambiti-paesaggio (accessed on 5 May 2020).

58. Aloia, A.; Guida, D. *The Geosites: Geopark's Gaia Synphony in the Cilento, Vallo Diano and Alburni Geopark, Geopark Book n. 1*; Cilento, Vallo di Diano and Alburni Geopark: Vallo della Lucania, Italy, 2014.

59. Aloia, A.; Calcaterra, D.; Guida, D.; Valloni, R. Field trip Guide book. In Proceedings of the 12th European Geopark Conference, Ogliastro Cilento, Italy, 4–7 September 2013; pp. 42–52.

60. Aloia, A.; De Vita, A.; Guida, D.; Toni, A.; Valente, A. National Park of Cilento and Vallo di Diano: Geodiversity, geotourism, geoarchaeology and historical tradition. In Proceedings of the 9th European Geopark Conference—European Geopark Network, Mytilene Lesvos, Greece, 1–5 October 2010; p. 41.

61. Aloia, A.; De Vita, A.; Guida, D.; Toni, A.; Valente, A. La geosiversità del Parco Nazionale del Cilento e Vallo di Diano: Verso il Geoparco. In Proceedings of the Atti del Convegno Nazionale "II Patrimonio Geologico: Una risorsa da proteggere e valorizzare", Sasso di Castalda, Italy, 29–30 April 2010; pp. 188–202.

62. Santangelo, N.; Amato, M.; Ascione, A.; Cafaro, S.; Calcatera, D.; Romano, P. Geological field trip n. 4: From Vallo di Diano to Angelo caves. In *Field Trip Guide Book 12th European Geopark Conference*; Aloia, A., Calcaterra, D., Guida, D., Valloni, R., Eds.; National Park of Cilento Vallo di Diano and Alburni: Vallo della Lucania, Italy, 2013; pp. 42–52.

63. Santangelo, N.; Bravi, S.; Santo, A. Itinerario 12 Piana del Sele, Monti Alburni e Vallo di Diano. In *Guide Geologiche Regionali—Campania e Molise*; Calcaterra, D., D'Argenio, B., Ferranti, L., Pappone, G., Petrosino, P., Eds.; Società Geologica Italiana: Rome, Italy, 2016; pp. 227–242, ISBN 9788894022728.

64. Chua, A.; Servillo, L.; Marcheggiani, E.; Moere, A.V. Mapping Cilento: Using geotagged social media data to characterize tourist flows in southern Italy. *Tour. Manag.* **2016**, *57*, 295–310. [CrossRef]

65. Amodio, T. La dimensione territoriale dell'ospitalità turistica in provincia di Salerno. *Boll. dell'Assoc. Ital. Cartogr.* **2016**, *157*, 105–116. [CrossRef]

66. Italan Ministry of the Environment, Technical Analysis 2017. Available online: https://www.minambiente.it/sites/default/files/archivio/allegati/biodiversita/Rapporto_Natura_Cultura.pdf (accessed on 31 May 2020).

67. Di Martino, P.; Petrillo, C.S. An International Project to Develop Networking for Promoting a Specific Destination: Emigration as a Tool to Enhance Tourism in Cilento Area. In *Tourism Local Systems and Networking*; Elsevier BV: Amsterdam, The Netherlands, 2006; pp. 219–233.

68. Brilha, J.B. Inventory and Quantitative Assessment of Geosites and Geodiversity Sites: A Review. *Geoheritage* **2015**, *8*, 119–134. [CrossRef]

69. Marciano, R.; Munno, R.; Petrosino, P.; Santangelo, N.; Santo, A.; Villa, I. Late quaternary tephra layers along the Cilento coastline (southern Italy). *J. Volcanol. Geotherm. Res.* **2008**, *177*, 227–243. [CrossRef]

70. Greco, G.; Krinzinger, F. *Velia. Studi e Ricerche*; Panini Franco Cosimo: Modena, Italy, 1994; ISBN 8876863087.

71. Cicala, L. Il Quartiere occidentale di Elea-Velia. *Mélanges l'Éc. Fr. Rome Antiq.* **2013**, *125*. [CrossRef]

72. Ermolli, E.R.; Romano, P.; Ruello, M.R. Human-Environment Interactions in the Southern Tyrrhenian Coastal Area: Hypotheses from Neapolis and Elea-Velia. In *The Ancient Mediterranean Environment between Science and History*; Brill: Leyden, The Netherlands, 2013; pp. 213–231.

73. UNESCO. *UNESCO Global Geoparks. Celebrating Earth Heritage, Sustaining local Communities*; UNESCO: Paris, France, 2016.

74. Moreira, J.C. Interpretative Panels About the Geological Heritage—A Case Study at the Iguassu Falls National Park (Brazil). *Geoheritage* **2012**, *4*, 127–137. [CrossRef]

75. Migoń, P.; Pijet-Migoń, E. Interpreting Geoheritage at New Zealand's Geothermal Tourist Sites—Systematic Explanation Versus Storytelling. *Geoheritage* **2016**, *9*, 83–95. [CrossRef]

76. Crawford, K.R.; Black, R. Visitor Understanding of the Geodiversity and the Geoconservation Value of the Giant's Causeway World Heritage Site, Northern Ireland. *Geoheritage* **2011**, *4*, 115–126. [CrossRef]

77. Vacchi, M.; Marriner, N.; Morhange, C.; Spada, G.; Fontana, A.; Rovere, A. Multiproxy assessment of Holocene relative sea-level changes in the western Mediterranean: Sea-level variability and improvements in the definition of the isostatic signal. *Earth Sci. Rev.* **2016**, *155*, 172–197. [CrossRef]

78. Chappell, J.; Omura, A.; Esat, S.; McCulloch, M.; Pandolfi, J.M.; Ota, Y.; Pillans, B. Reconciliaion of late Quaternary sea levels derived from coral terraces at Huon Peninsula with deep sea oxygen isotope records. *Earth Planet. Sci. Lett.* **1996**, *141*, 227–236. [CrossRef]

79. Waelbroeck, C.; Labeyrie, L.; Michel, E.; Duplessy, J.; McManus, J.; Lambeck, K.; Balbon, E.; Labracherie, M. Sea-level and deep water temperature changes derived from benthic foraminifera isotopic records. *Quat. Sci. Rev.* **2002**, *21*, 295–305. [CrossRef]

80. Visitors to Italian Museum and Archaeological Areas. Available online: http://www.statistica.beniculturali.it/Visitatori_e_introiti_musei_16.htm (accessed on 6 May 2020).

81. Amorosi, A.; Colalongo, M.; Pasini, G.; Preti, D. Sedimentary response to Late Quaternary sea-level changes in the Romagna coastal plain (northern Italy). *Sedimentology* **1999**, *46*, 99–121. [CrossRef]

82. Kayan, I. Holocene stratigraphy and geomorphological evolution of the Aegean coastal plains of Anatolia. *Quat. Sci. Rev.* **1999**, *18*, 541–548. [CrossRef]

83. Amato, V.; Filocamo, F. Geoarchaeotourism along the coast of the Campania region (southern Italy). In Proceedings of the Geomorphic Processes and Geoarchaeology: From Landscape Archaeology to Archaeotourism, International Conference, Moscow, Russia, 20–24 August 2012; pp. 25–28.

84. Itinerary Proposed by the Cilento, Vallo di Diano and Alburni Geopark. Available online: http://www.cilentoediano.it/en/escursioni (accessed on 27 May 2020).

85. Migoń, P.; Pijet-Migoń, E. Natural Disasters, Geotourism, and Geo-interpretation. *Geoheritage* **2018**, *11*, 629–640. [CrossRef]

86. Coratza, P.; De Waele, J. Geomorphosites and Natural Hazards: Teaching the Importance of Geomorphology in Society. *Geoheritage* **2012**, *4*, 195–203. [CrossRef]

Article

The "Fan of the Terre Peligne": Integrated Enhancement and Valorization of the Archeological and Geological Heritage of an Inner-Mountain Area (Abruzzo, Central Apennines, Italy)

Tommaso Piacentini [1,2,*], Maria Carla Somma [3], Sonia Antonelli [3], Marcello Buccolini [1], Gianluca Esposito [1], Vania Mancinelli [1] and Enrico Miccadei [1,2]

[1] Department of Engineering and Geology, Università degli Studi "G. d'Annunzio" Chieti-Pescara, Via dei Vestini 31, 66100 Chieti Scalo, Italy; marcello.buccolini@unich.it (M.B.); gianesp11@gmail.com (G.E.); vaniamancinelli@gmail.com (V.M.); enrico.miccadei@unich.it (E.M.)

[2] Istituto Nazionale di Geofisica e Vulcanologia, Sezione Roma 1, Via di Vigna Murata 605, 00143 Rome, Italy

[3] DISPUTer, Università degli Studi "G. d'Annunzio" Chieti-Pescara, Via dei Vestini 31, 66100 Chieti Scalo, Italy; mariacarla.somma@unich.it (M.C.S.); sonia.antonelli@unich.it (S.A.)

* Correspondence: tommaso.piacentini@unich.it

Received: 27 May 2019; Accepted: 20 June 2019; Published: 24 June 2019

Abstract: The outstanding cultural heritage of Italy is intimately related to the landscape and its long-lasting history. Besides major cities, famous localities, and park areas, several minor places and areas hide important features that allow the enhancing of inner-mountain and hilly areas as well as local natural reserves. This enhancement is supported by combining different types of cultural tourism, such as the archeological and geological ones. In this paper, an integrated geological–archeological itinerary is presented, which aims to valorize both these aspects in the inner-mountain areas of the central Apennines. The itinerary, called the "Fan of the Terre Peligne", is focused on the Terre Peligne area located in the Sulmona basin, in the central-eastern part of the Apennines chain (Abruzzo region, central Italy). It is composed of five sectors (one for each of the municipalities included) and incorporates traditional physical tools and digital ones. Here, the evidence of the Apennines formation is preserved from the origin of marine carbonate rocks to their deformation and the landscape shaping. The Terre Peligne intermontane basin became—and still is—one of the main transit areas for crossing the Italian peninsula since before Roman times and here many stages of Italian history are preserved. This allows outlining of the presence of man since prehistoric times, and here the name "Italia" was defined for the first time, in Corfinio, and to testify the connection between human and landscape history. A SWOT (strengths–weaknesses–opportunities–threats) analysis highlighted the main strengths, weaknesses, opportunities, and threats. Combining geological and archeological elements, which are intimately connected in this area, this itinerary intends to be an instrument for the enhancement and awareness of the natural and cultural heritage of a poorly known area that features outstanding geological, landscape, and human elements of the history of the inner Apennines.

Keywords: geotourism resources; cultural tourism; archeology; touristic itinerary; valorization; inner-mountain areas; Apennines; central Italy

1. Introduction

The cultural heritage of Italy is one of the most noticeable in the world and is intimately related to the landscape and its long-lasting history. This is valued mainly in the major cities, popular localities, and national parks [1–8], while it is poorly developed in many minor villages, country areas, local natural reserves, and less-known areas, such as in the inner-mountain areas of the Apennines. However, there are many places, archeological sites, and landscapes that preserve the geological history of our country, as well as important stages of human history [9–14]. Tourism, in particular cultural tourism and geotourism combining natural and cultural features with geological and geomorphological ones, is constantly expanding in the management, use, and enhancement of geological and cultural heritage of poorly known areas [15–18]. The valorization of natural and cultural features combined with geological and geomorphological features is the base of this kind of tourism, which can become a crucial promotion tool for inner-mountain areas and for connecting them to the most known touristic places and park areas. It can also become a contribution to resilience, turning geological hazards (e.g., earthquakes, landslides, flooding) into opportunities to raise the people's awareness, to set up a culture of preventing natural hazards and also to create new job opportunities in inner-mountain areas [19]. Many different types of activities have been realized in recent decades to enhance these valuable elements with varying kinds of approach. To well-known major and less-known minor archeological sites, abiotic elements of geodiversity were added (in the meaning of Brilha [20]). These include geosites—elements, areas, or places of geological interest of significant value and important witnesses of Earth's history [21]—as well as geomorphosites—as areas with geological features and landforms that have acquired a scientific, cultural/historical, aesthetic and/or social/economic value due to human perception or exploitation [7,22–24]. The realization of maps, itineraries, and illustrative materials in this area has aided not only in disseminating information about geological features and heritage but also in increasing awareness of geo-environmental and historical issues through sustainable tourism [10–14,25–29]. Moreover, in recent years, integrated proposals were defined based on tried and tested itineraries incorporating geological elements (e.g., geosites) combined to more common topics, such as flora and fauna, architecture, archeology, etc. [28,30–35].

This paper presents an integrated geological–archeological itinerary in its overall structure and main features. It aims to enhance the natural (geological) and cultural (archeological) heritage of an inner-mountain area of the Apennines, precisely of the Terre Peligne in the Sulmona basin (Abruzzo, central Italy; Figure 1, Figure 2). The itinerary is located between the main national parks of Abruzzo. It connects the Majella National Park and the Velino-Sirente Regional Park, crossing the Natural Reserve of the San Venanzio Gorges. It is also placed south of the Gran Sasso Laga mountains National Park and north of the Abruzzo, Lazio, and Molise National Park. To contribute to the enhancement of the precious elements of the "Terre Peligne", various activities have been carried out so far. Concerning the archeological aspects, several sites have been investigated as places keeping evidence of human activities in different historical periods. For geological and geomorphological aspects, geosites and geomorphosites have been identified, as areas with specific characteristics that have acquired a scientific value [4,36].

The integrated itinerary of the "Terre Peligne" combines geological–archeological sites trying to enhance them in an overall integrated view under the historical, cultural, aesthetic, social, and economic profiles [20,22,28]. The itinerary was realized, in a collaboration among geologists and archeologists of the "G. d'Annunzio" University of Chieti-Pescara, local authorities (the municipalities of Roccacasale, Prezza, Raiano, Vittorito, Corfinio, gathered in the Terre Peligne Association), professionals, and technicians, working in the area. It stems from the awareness of the poor sensitivity of the population on: (i) the long-lasting geological and geomorphological history that has created the current mountain landscape from an ancient tropical sea (and that today determine natural hazards) and (ii) the poor dissemination and popularization of the archeological features that bear witness to the human history of the inner mountainous areas of Abruzzo.

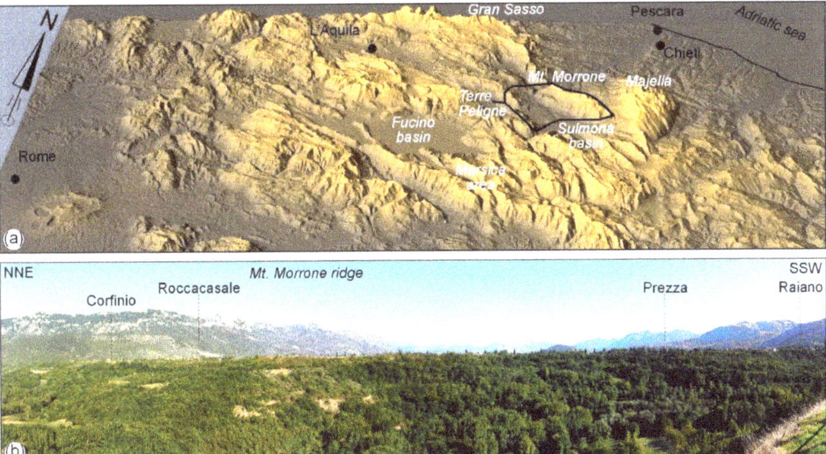

Figure 1. (**a**) Location Map of the Terre Peligne area (black line) within the central Apennines (three-dimensional view from 90 m DEM, SRTM). (**b**) Panoramic view of the Terre Peligne area within the Sulmona basin (view from western sector, in the municipality of Vittorito).

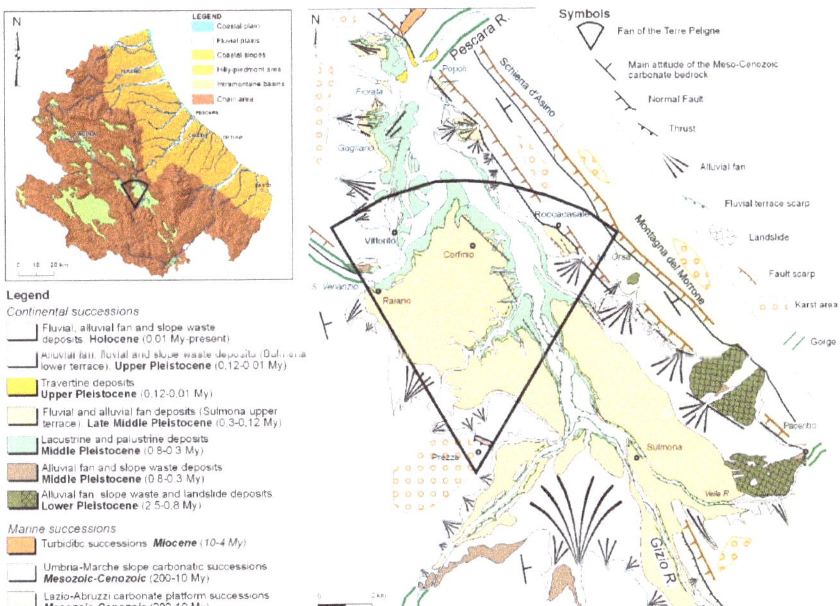

Figure 2. Location of the Terre Peligne (black line) in the Abruzzo Region (top left) and Geological scheme of the Sulmona basin showing the location of the "Fan of the Terre Peligne" (black line) (modified from [37,38]).

The "Fan of the Terre Peligne" itinerary focuses on how the landscape and geological features can be intimately related and connected to human history, settlements, and activities. It incorporates in five sectors dedicated to different themes, one for each of the Terre Peligne Association municipalities. These sectors define a fan-shaped area that named the itinerary as "Fan of the Terre Peligne". The purpose is the integration of archeological elements with geological-geomorphological and other attractive aspects of the territory, including all forms of cultural heritage (both tangible and intangible), to increase the tourist attractiveness of this area. Here, the evidence of the geological history of the Apennines is preserved from the origin of marine carbonate rocks to their deformation and from the formation of the major mountain ridges to the landscape shaping. From an archeological point of view, the territory has maintained significant evidence, which allows reconstruction of the settlement evolution as well as the development of the relationship between man, rocks, and landscape in these internal areas of Abruzzo. The economy of this region in ancient times was based on the wise exploitation of available landscape resources in an integrated economic system through agriculture, forestry/herding, and sheep farming [39]. Moreover, in this area, the name "Italia" was defined for the first time, in Corfinio, the capital city of the ancient Italic League.

2. Materials and Methods

The "Fan of the Terre Peligne" itinerary runs across the Sulmona basin and is defined as a multicolor fan. The selection of the path and the main stop sites was qualitatively based on three main criteria similar for archeology and geology: (i) location and spatial connection along the Terre Peligne area; (ii) site type and origin, and (iii) temporal distribution in the geological/archeological time scale [28,40–42]. Even if at this stage we did not run a quantitative assessment of the sites [20], this approach is in agreement with what has been done in many cases for geosites evaluation (e.g., within Environmental Impact Assessment and territorial planning [11,12,43]; inventories of natural heritage sites, [31,35,40,44]; tourist promotion [14,45]; management of nature parks [4,6,22,46,47]).

The "Fan of the Terre Peligne" results from the fan-shape of the itinerary and is composed of five geological and archeological thematic sectors, one for each of the municipalities joined the Association of Terre Peligne (Figure 3). The sectors are identified by a geographic orientation: north-east, south, west, north-west, north. Each of them is associated with a color that refers to the geological-geomorphological theme, according to the international coded conventions (Geological Survey of Italy, ISPRA, and International Association of Geomorphology, IAG) (Figure 3): red for tectonics, which created the ridges; orange for karst landscape and dark green for fluvial and water-related processes shaping the main ridges; light green for lacustrine environments and blue for hydrography and rivers-related processes in the Sulmona basin. Each sector is also connected to a specific archeological theme: protect and dominate, the invisible history, water and stone, man and highlands, from the capital of the italics to Pentima. This arrangement is realized to enhance the intrinsic value of the villages, the variable views of the landscape and the geological features of the area, as well as the various archeological elements. It is defined to shed a new light on the values of this territory rich in cultural heritage, in which natural and human history are strictly combined.

The geological-geomorphological features presented in the different sectors and sites along the itinerary were selected and defined through research activity and studies in the Sulmona area and the central-eastern Apennines in recent decades. The studies were mostly based on field geological-geomorphological investigations and mapping, as well as morphometric and morphotectonic investigations, focused on the geological features and geomorphological landforms of this area [37,38,48–54]. From these studies, the primary and most exciting findings were selected and underwent specific field investigations in order to be presented to a broad public, mainly focusing on the evidence of the different stages of the geological and landscape evolution as well as the hazard connected to the recent tectonic and geomorphological processes. The latest field investigations provided the best sites and outcrops to show and explain the formation of marine and continental sedimentary rocks and to outline their structure. The best exposures of fault plains were selected to explain their role in the landscape shaping, as well as their connection to earthquakes and seismic hazard. The geomorphological analysis outlined the most interesting and well-exposed landforms shaped by water-related processes (i.e., fluvial and lacustrine processes, gorges/valley incisions, karst processes), which explained of the landscape evolution.

The archeological findings presented in the selected sites are also the results of several activities and studies carried out in the Terre Peligne area during recent decades [36].

The stages of development and transformation of this territory—and their extensive diachrony—were reconstructed, being so crucial for the understanding of the settlement history in this part of the central Apennines area. From the late 90s of the last century extensive archeological excavations have been carried out in Corfinio, thanks to an urban archeological project from the Abruzzo Archeological Department, the University of Chieti and the University of Rome "La Sapienza" as well as the municipality of Corfinio [55–60]. Excluding these large projects, the Terre Peligne area has not been studied systematically from an archeological point of view. Previously, the findings have been mostly occasional, except for some broader research carried out between the end of the 19th and the beginning of the 20th centuries by Antonio De Nino, gathered mainly in the volume of *Forma Italiae* published in 1984 by van Wonterghem [61].

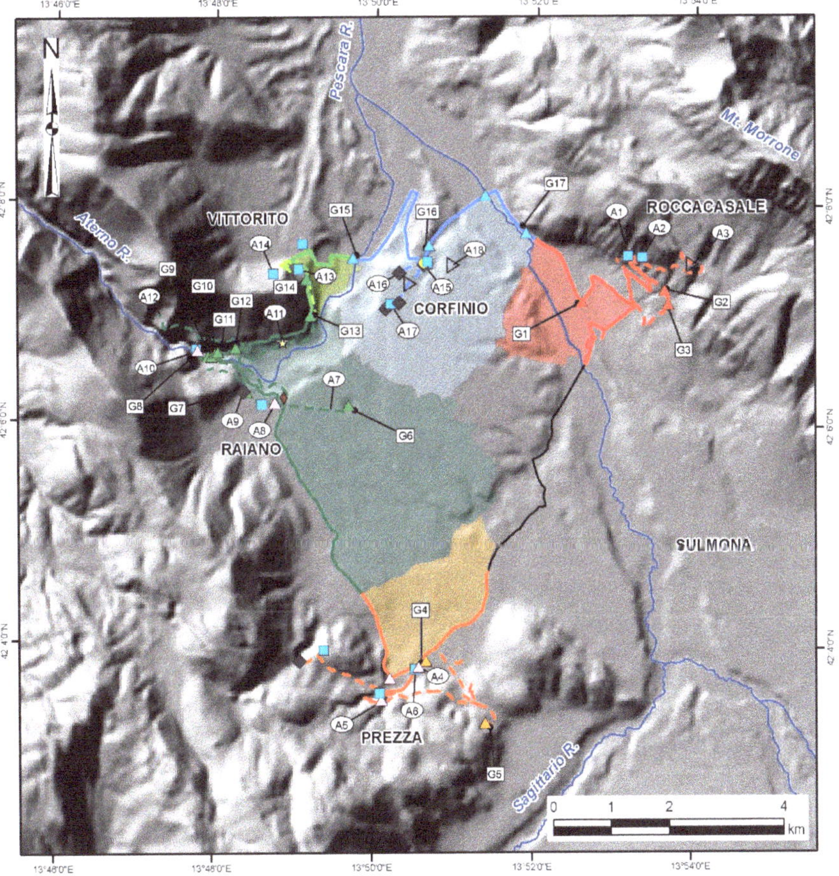

Figure 3. Scheme of the "Fan of the Terre Peligne" path (see Table 1 for details). The colors are those officially coded for the geological and geomorphological features (Geological Survey of Italy, ISPRA, and International Association of Geomorphology, IAG) that characterize each sector: red for tectonics, orange for karst landscape, dark green for fluvial processes, light green for lacustrine environments, and blue for rivers and hydrography.

These overall studies highlighted the high value of the archeological heritage, which however in most cases is hardly usable by the general public and is disseminated throughout the territory, often in areas of high landscape value. In this study, the main sites were selected across the area defining main groups or themes, which combines historical perspective and landscape connections. These thematic groups and the single sites were then organized and explained with lay and easy to understand language to make them suitable for the general public and for enhancing these sites in a touristic and educational perspective.

The geological and archeological element of the "Fan of the Terre Peligne" are therefore arranged in five sectors. Each sector of the itinerary features a triangular "totem" with the introduction to the geological and archeological setting of the specific area and municipality. In the most significant sites, geological and archeological stops are defined, marked by plates and logos, which through digital devices (tablet, smartphone) allow access to the information about the site. For the main sites (1–2 for each sector), larger and more descriptive panels are defined. The entire itinerary features five triangular totems, ten panels in the main stops and 26 plates or logos, as summarized in Figure 3 and Table 1. In each sector, stops of geological (G1, G2, etc.) and archeological (A1, A2, etc.) interest are identified. Secondary routes are connected (by walk or bicycle, dashed in Figure 3) are connected to the main ones.

3. Geological and Archeological Values of the Terre Peligne Area

The study area is located in the Sulmona intermontane basin (also known as Peligna valley or Terre Peligne area) on the eastern side of the central Apennines chain. This area is located in a key node in terms of both geological (junction of marine paleogeographies Mesozoic-Cenozoic era, tectonic features, and Quaternary terrestrial landscape) and archeological framework (transit zone from the mountain areas to the Adriatic and Tyrrhenian coast for the entire central Apennines). The itinerary runs all along the northern part of the Sulmona intermontane basin and incorporates the main elements of the geological and human history of this area (Figure 1, Figure 2, Figure 3).

3.1. Geological Setting

The landscape of the Abruzzo Apennines, as well as of the entire chain, is the result of the geological evolution in the Neogene–Quaternary period (from ~15 My ago to present) of an east-verging orogenic system composed of several Neogene thrust sheet displaced by Pliocene (5–2.5 My) strike-slip and Quaternary (2.5 My to present) extensional tectonics and regional uplift [52,62]. These tectonic processes led to the emersion of the Apennines orogen and the progressive development of the main ridges with the beginning of the geomorphological evolution forming an initial landscape (at least from the Late Miocene-Early Pliocene, ~5 My). From the chain area to the piedmont and coastal area, the landscape evolution was closely connected with a complex combination of endogenous (morphotectonics) and exogenous processes (slope, fluvial, karst and glacial processes) [38,45,63–66]. During the Quaternary, regional uplift combined with extensional tectonics defined a system dominated by calcareous ridges, by valleys incised in Neogene sandstone and claystone rocks, and by intermontane basins filled by Quaternary continental (mainly slope and fluvial) deposits.

The Sulmona basin (Figure 1, Figure 3) is one of the main and more eastern intermontane basins of the Apennines surrounded by main NW-SE elongated ridges (Mt. Morrone, Maiella Mts., Peligne Mts. [37,49–51,67]). These ridges are composed of calcareous rocks formed in different paleogeographic marine environment from Jurassic to Miocene (200–10 My): a thick sequence of shallow sea carbonate platform rocks (southern and central sector of Mt. Morrone) and a sequence of slope to basin limestone-marl-chert rocks (northern sector of Mt. Morrone and Peligne mountains in the western side of the basin) (Figure 2) [53,54,67]. They are affected by N-S thrust faults and systems of NW-SE normal faults. Specifically, the Sulmona basin is bordered by a primary NW-SE normal fault system along the Mt. Morrone SW slope, affected by Quaternary and recent (~1 My to present) tectonic activity responsible for the formation of the basin [49–51,53,54,68,69]. The basin was occupied by a large lake during the Middle Pleistocene (~0.8–0.3 My) and is partially filled by slope, alluvial fan deposits, and by a thick sequence of lacustrine deposits related to an ancient Peligno lake [50,70] covered by fluvial deposits during the late Middle Pleistocene (~0.3–0.12 My) and incised by river valleys (Late Pleistocene-Holocene, 0.12 My to present). The combination of extensional tectonics and lacustrine-fluvial processes defined the formation of the gentle landscape of the intermontane basins in which the Terre Peligne are placed. The combination of fluvial and karst processes, also controlled by tectonic features, is well documented along the slopes and in the valleys. This combination defined the shaping of the rugged landscape of the ridges and the formation of deep valleys, gorges, and incisions, parallel or transversal to the main ridges (e.g., Aterno Gorges, Sagittario Gorges, Popoli Gorges; [38]), which define the main transit routes across the Apennines since historic and protohistoric times.

In the "Fan of the Terre Peligne" itinerary, the most significant geological and geomorphological themes are therefore related to (i) the origin and formation of the calcareous rocks in marine environment, (ii) the junction between marine and continental environment testifying the emersion of the ridges, (iii) the tectonic processes affecting the landscape through major fault systems; (iv) the formation and evolution of the ancient Peligno lake, and finally (v) the progressive development of the main ridges and the shaping of the landscape due to fluvial, slope, and karst processes.

3.2. Archeological Framework

The area of the Terre Peligne is a significant node of the Abruzzo territory also for the human history testified by archeological findings. This area has been an obligatory transit point since prehistoric times for those who moved from the internal mountain areas of the central Apennines to the hilly and coastal Adriatic sector. An articulated network of paths closely connected to the morphology of the territory has allowed and supported the link of various settlements that tells a story thousands of years long marked by the interaction of man with the environment. Significant findings dating back to the prehistoric age testify the presence of man in this area. Thereafter, the Italic population settled in the Peligna valley before the Roman conquest. In addition, here, in Corfinio, the capital city of the ancient Italic League, the name "Italia" was defined for the first time [71]. The Roman conquest led to a widespread occupation focused on high-ground sites, where massive defensive walls are still visible, and around sanctuaries located in places with geological and landscape features strictly related to road links [55,72,73].

Corfinio, even after the Social War, continued to be the primary center in this area, a Roman municipality which, especially between the first century B.C. and the first century A.D., would show the presence of significant monuments [55]. His role as a political and religious center in the territory would continue throughout the Early Middle Ages. The extensive reconstruction of the Episcopal complex in Valva carried out by the bishop Trasmondo at the end of the XI century may represent the last large-scale work in this area [74]. Since then, the settlement layout has been increasingly centered on the close network of castles, some of them built on previous pre-Roman settlements, which in many cases still exists in the old centers of present towns. These events gave rise to the arrangement of Corfinio's archeological areas which are now preserved also in the local Municipal Museum "Antonio De Nino" [75].

In the itinerary, the main archeological elements are therefore related to (i) the history form the capital of the Italics to Pentima, (ii) the protection and domination of the Terre Peligne area from the surrounding ridges, (iii) the man occupation in the highlands, (iv) the role of water and stones in the transitways from the inner Apennines to the Adriatic area, and finally (v) the invisible history hidden in the landscape of the Terre Peligne.

4. The "Fan of the Terre Peligne": Five Villages, Five Colors, Five Themes

The "Fan of the Terre Peligne" connects five different municipalities (Roccacasale, Prezza, Raiano, Vittorito, and Corfinio) across the Terre Peligne area. As mentioned above, each village corresponds to a sector and a specific geological-geomorphological theme as well to an archeologic one and includes two to seven stops for geological and archeological observations (Figure 3, Table 1).

Table 1. Main sectors of "Fan of Terre Peligne" itinerary, geological and archeological themes and local stops along the path (see Figure 3 for location). The colors are those officially coded for the geological and geomorphological features (Geological Survey of Italy, ISPRA, and International Association of Geomorphology, IAG) that characterize each sector: red for tectonics, orange for karst landscape, dark green for fluvial processes, light green for lacustrine environments, and blue for rivers and hydrography.

EASTERN SECTOR (Roccacasale)	Geology the born of the basin and the slope	Stop G1: Panoramic view on the Mt. Morrone fault escarpment (plate) Stop G2: The Roccacasale fault and the fault rocks (panel) Stop G3: The alluvial fans of the Mt. Morrone fault escarpment (plate)
	Archeology protect and dominate	Stop A1: The Roccacasale Castle (plate) Stop A2: The San Michele cave (plate) Stop A3: The hill of Fairies (plate)
SOUTHERN SECTOR (Prezza)	Geology The karst landscape of the slopes: the water dissolves the rocks:	Stop G4: Small lakes in the plain, sinkholes (plate) Stop G5: Dolines and the karst landscape of Prezza (panel)
	Archeology The invisible history	Stop A4: The fortification and the village (plate) Stop A5: The Roman inscription and the Santa Lucia Church (plate) Stop A6: Colle village: the Roman inscription and the Santa Maria Church (panel)
WESTERN SECTOR (Raiano)	Geology From ancient rocks to new mountains	Stop G6: The Quaglia Lake (plate) Stop G7: The San Venanzio Gorges (panel) Stop G8: A hermitage in the Apennines, the San Venanzio Hermitage (panel—connected to A9) Stop G9: The mill turns like the rocks (plate) Stop G10: The join of two complex and ancient geological "world" join: the marine and the continental (plate) Stop G11: The climate changed, and the rocks from red turn white (plate) Stop G12: The history of the ancient Peligno Lake and human presence in this area started here (plate)
	Archeology The water and the stone	Stop A7: The Transhumance Route (plate) Stop A8: Santa Maria Maggiore Church (plate) Stop A9: The Vuccole aqueduct (plate—connected to G8) Stop A10: A hermitage in the Apennines, the San Venanzio Hermitage (panel) (connected to G8) Stop A11: Santa Maria di Contra Church (panel) Stop A12: Cave paintings (plate)
NORTH-WESTERN SECTOR (Vittorito)	Geology Appearance and disappearance of a lake	Stop G13: A panoramic view on the ancient Peligno Lake (panel) Stop G14: The sediments of the ancient lake (plate)
	Archeology Men and highlands	Stop A13: San Michele Arcangelo Church (panel) Stop A14: the Castle (plate)
NORTHERN SECTOR (Corfinio)	Geology From ancient to new rivers	Stop G15: The ancient rivers (Upper and Lower Terrace) and the present-day rivers (panel) Stop G16: The Aterno River (plate) Stop G17: The Sagittario River (plate)
	Archeology From the Italics' capital to Pentima	Stop A15: The theatre (plate) Stop A16: San Giacomo plain (plate) Stop A17: Valvensis complex, San Pelino Church (panel) Stop A18: Sant'Ippolito Church (panel)

4.1. Roccacasale: Eastern Sector (Red)

The east part of the itinerary is in the municipality of Roccacasale and starts the "Fan of Terre Peligne" itinerary (Figure 2) and run at the boundary of the Majella National Park. From the Roccacasale castle, it shows a glance on the overall landscape of the Terre Peligne area explaining the formation of the Sulmona basin and the steep south-west slope of the Mt. Morrone (north-east side of the basin). In this sector, it is also well documented how this landscape was exploited for human activities along historical time, specifically for defense and protection as well as for domination of the settlements on the plain.

4.1.1. Geology—The Born of the Basin and The Slope

This sector is arranged to explain the processes that led to the formation of the slope of Mt. Morrone (Figure 4) and of the Sulmona intermontane basin, as well as how these processes are connected to one of the main faults of the central Apennines (the Mt. Morrone fault) forming the structure of the Terre Peligne landscape. The red color is coded in geology for tectonics and faults, the main topic of the sector: what are they and how can be observed, when and where have been formed, how are connected to earthquake and seismicity and what have been their effects on the landscape? Faults are explained as a well-known hazard element connected to earthquakes but also as natural events that contributed to building the variety of the landscape of the Terre Peligne area and the entire Apennines, featuring sharp ridges, wide basins, and deep valleys.

Figure 4. (**a**) Panoramic view of the SW escarpment of the Mt. Morrone ridge; the landscape is marked by several fault scarp related to the Mt. Morrone fault system (red lines). (**b**) Fault scarp located SE of the Roccacasale village; it is a well-polished rock fault plane, over 3 m high, NW-SE oriented and SW-dipping of about 50°; in the upper part it weathered and eroded due to the rock jointing; it is a normal fault with the SW side (hanging wall) downthrown and the NE side uplifted (foot wall), as explained in the inset cartoon. (**c**) Alluvial fan sediments consisting of blocks, gravel, and sand; they result from the fault-related escarpment weathering and erosion and the accumula-tion at the base of the slope; they are arranged in layers declining from the fault scarp to the plain.

After a panoramic view on the Mt. Morrone fault-related escarpment (Figure 4a), in the Roccacasale area, it is possible to see and touch a large fault plane (Figure 4b). It is part of the Mt. Morrone fault system to which strong historical earthquakes were connected as well as large ancient landslides and recent slope mass movements [51,69]. The fault plane is visible as a flat and smoothed surface on calcareous rocks NW-striking and SW-dipping around 50° forming a steep rock scarp (known as fault scarp; Figure 4b). It is a normal fault, meaning that the SW side (Sulmona plain) has been downthrown while the NE side (Mt. Morrone ridge) has been uplifted forming the steep slope separating ridge and plain. This process induced strong jointing in the calcareous rocks, especially along the fault, and their intense weathering. The last part of this sector outlines how the progressive weathering of the escarpment and the incision of the jointed rocks has produced a significant accumulation of debris deposits resulting in large alluvial fans forming the junction between slope and plain (Figure 4c).

4.1.2. Archeology—Protect and Dominate

The steepness and straightness, strictly connected to the tectonic origin as a fault-related slope, made the escarpment of the Mt. Morrone and the Roccacasale area a key sector for dominating and protecting the whole Terre Peligne area during the human history since pre-Roman times (Figure 5).

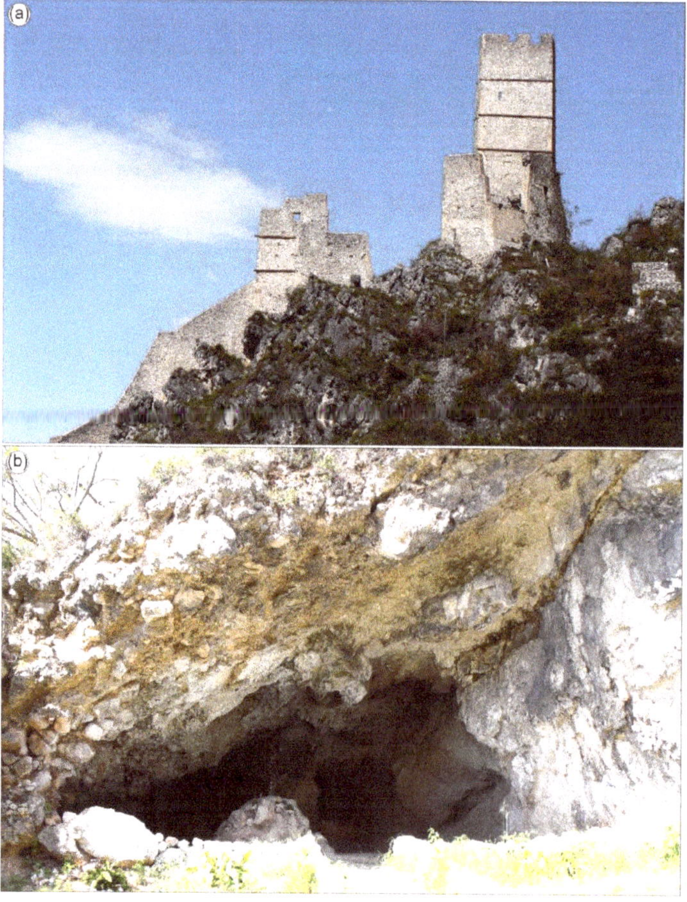

Figure 5. Roccacasale hillside: (**a**) remains of the Roccacasale castle, (**b**) S. Michele cave located above the village.

The Morrone slope was ideal for placing some natural terraces and settlements to control the valley routes, just close to Roccacasale. Here, within the Terre Peligne area, the principal north–south transit axes parallel to the Apennines chain (from L'Aquila to Sulmona toward Isernia), known as the "via degli Abruzzi" in the Middle Ages, crossed the east–west transit axis, represented by the Tiburtina-Valeria roads perpendicular to the Apennines chain [36]. This situation is already evident in the protohistoric site of Colle delle Fate and in the triangular fortification of Roccacasale. Colle delle Fate (770 m a.s.l.) preserves the remains of a typical high-ground inhabited area, probably built in the late Bronze Age. It bears evidence of three defensive walls, the last and higher one in polygonal bonding, and two cisterns for rainwater [55].

The Roccacasale castle, which dominates the eponymous village expanding like a fan from the base of the triangular enclosure, represents one of the best-preserved examples in the region of this type of fortification. Its specific shape was determined by the need to adapt the defensive requirements to its position on the slope and the slope shape. Specifically, it is located on the steep fault-related escarpment between two minor valleys perpendicular to the slope. The valleys' incision left a remnant of the fault-related escarpment with triangular shape (also called triangular facet in geomorphology) to which the castle shape is adapted. The upper end of the triangle was reinforced by a high tower meant to protect it from any attacks from higher ground. The original settlement was traced back to between the 10th and the 11th centuries A.D., but in the following centuries, it went through several restorations to be suitable for residential uses, as shown by the structure of the building that sits against the basis of the enclosure [76].

4.2. Prezza: Southern Sector (Orange)

Moving clockwise, the southern part of the itinerary lies in the municipality of Prezza (Figure 3). It outlines an undulated landscape with several small circular depression and lakes, which have controlled the development of the historical settlements of this area. This arrangement let to explain how the slopes surrounding the Sulmona plain and the Terre Peligne area were deeply affected and shaped by karst processes related to water infiltration in the calcareous rocks.

4.2.1. Geology—The Karst Landscape of the Slopes: The Water Dissolves the Rocks

After observing how ridges and slopes are formed leading to the formation of the Terre Peligne landscape in the eastern sector, this sector begins the explanation of the different processes that have shaped the landscape of the plain and the surrounding ridges. This is the "realm" of karst, the water-related process of chemical dissolution of the calcareous rocks, which has led to the slow and progressive weathering of rocks and the shaping of the ridges (Figure 6a). The karst process has created spectacular landforms, both superficial (e.g., dolines) and underground (e.g., caves) (Figure 6a,b), which have often hosted human settlements in prehistoric times. The orange color is the one coded in geomorphology to represent the karst processes and landforms.

The itinerary moves from the plain toward the surrounding Mt. Prezza slope. In the plain, filled by alluvial gravel sediments, the karst landforms consist of small circular-shaped ponds (also known as sinkholes), large up to about 100 m (Figure 6b,c). They are mostly connected to the collapse of underground caves due to the dissolution of buried calcareous rocks. They were useful in ancient and recent times as water resource and led to the growth of stories and myths around these places.

On the slopes, on calcareous rocks that surround Prezza, the karst landforms consist of large dolines, funnel-shaped depressions, up to over 100 m wide and several tens of meters deep (Figure 6b).

This sector of the itinerary shows the karst landforms affecting the landscape, connected to one of the main processes that, combined with slope gravity-induced and fluvial processes, are responsible for the shaping of the landscape as shown in the following sectors.

Figure 6. Prezza areas: (**a**) Didactic scheme of all forms related to karst processes; (**b**) Aerial photo of the Prezza area (from "Portale Cartografico Nazionale"); in orange the main superficial karst forms are highlighted: dolines (continuous line) e sinkholes (line hatched); (**c**) detail of a sinkhole (in the locality "il Colle").

4.2.2. Archeology—The Invisible History

The territory of Prezza has plenty of archeological findings, mostly from the protohistoric and Roman eras, which unfortunately are almost or entirely invisible today, recalling in some way the karst process as a kind of slow "invisible" process affecting the rocks. Anyway, these findings still testify the presence of a diffuse settlement over time, perfectly integrated with the forms and resources of the territory. Near the villages of Castiglione and Castellone, there were probably two protohistoric high-ground sites, and a necropolis with circular tombs (recalling the circular shape of the dolines) traced back to this period (Figure 7).

Figure 7. Prezza area: (**a**) the Prezza main village incorporating the old castle; (**b**) rural house in the Colle village.

The Roman Age findings are very numerous, and it is possible to identify at least three medium-small inhabited areas, which were mostly dependent on the exploitation of agricultural resources: one in the district named Colle, one in the hamlet named Campo di Fano, and one at Colle San Giovanni [55]. These were possibly connected to the water resources available in the karst ponds. Here, the presence of a Roman inscription mentioning the names of Magistri Laverneis has prompted the hypothesis that the Roman Vicus of Lavernae, known from the sources, was located in this territory. Prezza is also related to the oldest mention of a castle in the Peligna valley: its fortification was first mentioned in the second half of the IX century [77]. The castle is no longer visible, as it has been "incorporated" and modified by the development of the current residential area. The castle stood on a rocky spur at 580 m a.s.l. (left above the karst landforms), dominating the entire Peligna valley, and it was in sight of other fortifications over the territory (e.g., Roccacasale castle). As the landscape underwent a progressive change due to surface processes (karst one in this area), the castle is the proof of the transformations also occurred in the human settlement layout in this area, determined by the process of fortification during the Middle Ages.

4.3. Raiano: Western Sector (Green)

The western sector of the itinerary runs from the village of Raiano to the San Venanzio Gorges passing through the spectacular homonymous hermitage (Figure 8a) and the Natural Regional Reserve of the San Venanzio Gorges. This sector is a key area not only in the landscape of the Terre Peligne area but also in the whole Apennines geological and archeological history. This has been a primary area of connection through time as a junction through the geological and landscape history of the chain and as one on the main transit routes across the Apennines through human history. For all these reasons, in this area, a specific geosite was defined and included in the ISPRA Geosite Inventory, and a geological touristic map was previously realized [35].

Figure 8. (**a**) Panoramic of the San Venanzio Gorges. (**b**) Contact between continental sedimentary rocks, ancient slope deposits (up), and marine sedimentary rocks, limestones (down). (**c**) Detail of the erosive contact between white calcareous debris (older and coarser) and red limestone debris (more recent and finer). (**d**) Lake deposits in the San Venanzio Gorges and throughout the northern part of the Sulmona Basin.

4.3.1. Geology—From Ancient Rocks to New Mountains

In the rocks and landforms of this sector, the main stages of the geological history of this part of the Apennines, over 200 million years long, can be recognized: from the formation of the rocks to their deformation; from the creation of the Sulmona basin and the surrounding ridges to their progressive shaping by the water of the Aterno River. The dark green color is coded in geomorphology for the water-related landforms.

The calcareous rocks surrounding the gorges bear witness of a very ancient marine environment along scarps surrounding atolls and carbonate platform environments during the Jurassic and

Cretaceous. In this paleogeographic environment, a thick sequence of calcareous rocks, which now constitute the backbone of the ridges, were formed over several tens millions of years (from 200 to 10 million years). The calcareous strata show evidence of faults and folds that explain the tectonic deformation of the rocks during the formation of the Apennines chain.

In the lower part of the gorges, toward the Sulmona basin, the contact between different types of rocks is well exposed (Figure 8b). It testifies a fundamental change from marine environment to present mountain landscape: marine calcareous rocks are overlain by breccias, gravels, and sands pertaining to slope and alluvial fan continental deposits, as well as siltstone of lacustrine deposits (developed in the last million year). This contact shows the separation of the ancient Jurassic-Cretaceous marine "world", which led to the formation of the rocks, and the more recent Quaternary continental "world", which led to the deformation and shaping of the mountain landscape and of the Peligno basin.

Through the Quaternary rocks (from breccias to gravel and siltstones), it is possible to understand the landscape changes from the border of the basin, with slope and alluvial fan deposits, to its center with the presence of an ancient lake (Figure 8c,d). Finally, the last step of the geological history is preserved in the San Venanzio Gorges. A primary fault runs along the valley, and the Aterno River has incised a deep incision, a fluvial gorge, in the calcareous rock, while toward the basins has shaped a wide valley. Once again, as in Roccacasale sector, the combination of tectonics and geomorphological processes control the landscape. In this case, it created the San Venanzio Gorges, which, since Roman times and earlier, have been an essential transit way for crossing the Apennines from the Tyrrhenian coast to the Adriatic coast (one of the main Roman consular roads, the Tiburtina-Valeria road, still runs along the gorges).

Moreover, in this sector of the itinerary, the evidence of a further connection between the geological and human history is presented. The ancient Roman Vuccole aqueduct is dug in the calcareous rocks along the slope of the gorges; the San Venanzio Hermitage overlooks the gorges incision; thanks to San Venanzio, there are ancient testimonies of the "lithotherapy", according to which some rocks would have healing power.

4.3.2. Archeology—The Water and the Stone

The Aterno River crosses this part of the Peligna valley and, as seen for the landscape, also its archeological evidence has been deeply characterized by the presence of water, and its relationship with human beings, as it was essential for the settlements. On one side the narrow gorges attracted forms of hermit's lives, which in the Middle Ages reached a peak with the building of the impressive S. Venanzio Hermitage (Figure 9a); the same morphological structure allowed the construction of several mills, which before the industrial revolution represented the most essential "factories", as they were required for the transformation of produce [78]. The presence of the Aterno River provided the catchment for an important aqueduct that during the Roman Age served the Corfinio municipality, through extensive works of hydraulic engineering; that is the Vuccole aqueduct, which was entirely excavated through the limestone [55,79]. Finally, the river has been the major center of the residential area (Vicus/Pagus?) whose memory is preserved today by the church of S. Maria di Contra (Figure 9b). The settlement located along the ancient road connecting the area of Raiano to Vittorito probably dates back to the Roman Age, as confirmed by numerous materials from that age which were reused in the church, as is the case of the roof tiles still covering the oldest part of the building. This settlement was still significant during the Early Middle Ages when the S. Maria church went through substantial building works from the Longobards, who had one of their major power centers (gastald) over the territory in the nearby Corfinio. The splendidly decorated slab that was part of the church ornaments dates back to that period and today is preserved as the altar mensa in the parish church of Raiano [36]. All this evidence shows the close connection between water and landscape evolution from one side and human history from the other.

Figure 9. Raiano area: (**a**) San Venanzio Hermitage; (**b**) Santa Maria di Contra Church.

4.4. Vittorito: North-Western Sector (Light Green)

Moving on, the north western part of the itinerary runs from the San Venanzio Gorges to the Vittorito village. It provides a scenic view on the present landscape of the Terre Peligne and on the ancient landscape of the Peligno Lake, already outlined in the western sector, which was part of a system of lakes, characterizing the landscape of the central Apennines during Middle Pleistocene (~0.8–0.3 million years ago; Figure 10a) [38].

4.4.1. Geology—Appearance and Disappearance of a Lake

The panoramic view from this part of the itinerary is particularly meaningful. It covers the entire Terre Peligne area from Roccacasale to Prezza, Raiano, and Corfinio, and from the Mt. Morrone Ridge to the Sulmona basin (Figure 10b). However, the landscape that can be observed today allows viewing of the evidence of an ancient landscape resulting from a long history dominated first by an ancient lake (the light green color characterizes the lake environments) that occupied the Sulmona intermontane basin, and then by rivers that first filled in the lake and then incised its sediments. In the landforms and the rocks of this sector of the itinerary, the remains of the ancient lake that occupied the entire basin during the Middle Pleistocene (~0.8–0.3 My) can be observed (Figure 10a,b). The valley sides of the Aterno River hide lacustrine deposits as those seen in the Raiano area (Figure 8d). At the top of the valley sides, a sharp scarp borders a vast plain known as Sulmona "Upper Terrace" [50], hanging 90–100 m above the Aterno fluvial plain, developed on ancient (late Middle Pleistocene) fluvial conglomerate deposits (Figure 10a,b). These deposits can be directly observed along the itinerary

(Figure 10c) and outline the disappearance of the lake, filled in by a large fluvial and alluvial plain, which covered the entire basin about 100–300 thousand years ago (late Middle Pleistocene). In the last part of the itinerary is shown how, from this time on, the entire landscape of the Terre Peligne has been dominated by the rivers that have incised the valleys and shaped the landscape that is observed today. This evolution created the present landscape composed of flat surfaces or "terraces" and broad water-rich valleys that made this area very suitable for human settlements since prehistorical times.

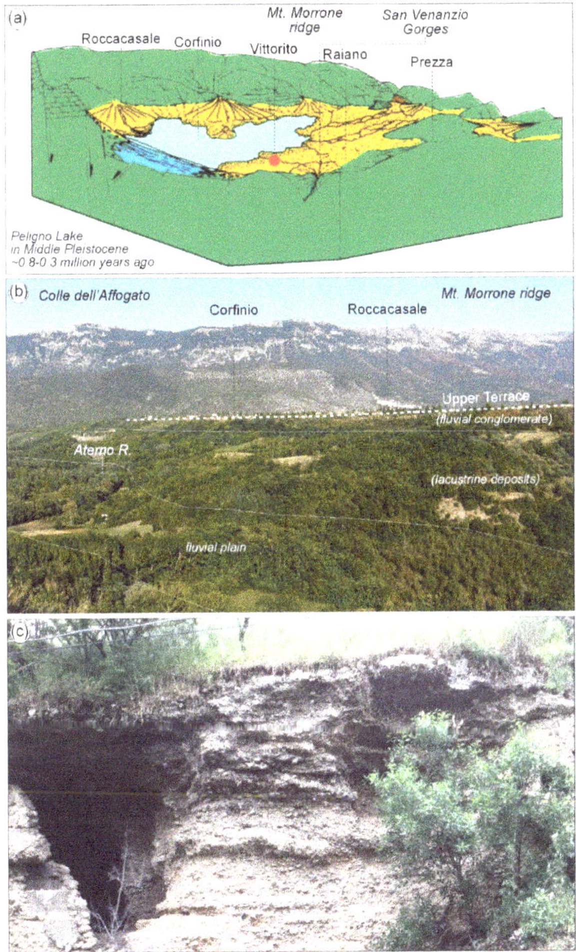

Figure 10. (**a**) Terre Peligne area landscape during Middle Pleistocene (~0.8–0.3 million years ago), while filled in by the Peligno Lake. Calcareous rocks (green) formed the backbone of the ridges and were cut by faults, and covered by slope deposits and alluvial fans, flowing into the lake and the lacustrine deposits (light blue). (**b**) Panoramic view of the Terre Peligne area outlining the "Upper Terrace" on the fluvial deposits covering the lacustrine deposits. (**c**) Vittorito (375 m a.s.l.), conglomerate rocks on the "Upper Surface" referable to the fluvial deposits filling the Peligno Lake during late Middle Pleistocene (anthropic caves are dug into them).

4.4.2. Archeology—Men and Highlands

The Vittorito area has returned archeological findings attesting the presence of human already in the protohistoric age, with a marked propensity to occupy the "terraces" delimiting this area of the valley and originated from the Pleistocene landscape evolution. Some remaining walls in polygonal stonework (currently no longer visible) testify a protohistoric settlement on the Castellano mount. The area where today stands the San Michele Arcangelo church (Figure 11a), which has retained significant evidence from the Early Middle Ages, was the location of a famous sanctuary dedicated to Hercules, occupying a dominant position over a settlement stretching along the Aterno River [36]. Still in the Middle Ages, on the northern side of the Castellano mount, the castle of Vittorito was erected, which later originated the present village (Figure 11b). Today, only a quadrangular tower, dominating the village, remains of the medieval castle. The fortification was shaped like a triangular enclosure, and the village expanded "fanwise" along the sides of the mountain. The fortification was wide enough and divided into residential and functional buildings [76]. The castle tower, *Turris Bectorrita*, was first mentioned in a document from the 1098 A.D. *Chronicon Casauriense*, which indicated, among other things, that it was the property of the Bishop of Valva (present Corfinio). On the edges of the Vittorito village stands the S. Michele Arcangelo church, built in the Middle Ages just over an old pagan temple. Several architectural fragments and funerary inscriptions have been reused in its masonry works. One find of particular interest is an Early Middle Ages slab still holding the signature of the stonecutter URSUS [36].

This intense occupation through time, mainly developed on the upper "terrace", again confirms and explains, in an easy to understand way, how strong the connection between the human activities and the ancient landscape evolution has been.

Figure 11. Vittorito area: (**a**) the San Michele Church; (**b**) the medieval castle above the present-day village.

4.5. Corfinio: Northern Sector (Blue)

The "Terre Peligne fan" ends in the northern sector. It started in Roccacasale, explaining how the landscape of the Terre Peligne was first formed by the tectonics along the Mt. Morrone normal faults. It ends in Corfinio explaining how, particularly in the last stage, the landscape has been shaped by rivers and fluvial processes after being occupied by a large lake for a long time and how this has controlled the occupation of the area in historical times.

4.5.1. Geology—From Ancient to New Rivers

In this area, the itinerary runs across the Aterno River valley and explains the last stages of the landscape evolution (meaning the last 100 thousand years), dominated by river-related processes (blue is the coded color for rivers and hydrography in geomorphological conventions). A large fluvial-alluvial plain was formed in the late Middle Pleistocene (300–100 thousand years ago), as seen in Vittorito sector. In the Late Pleistocene (last 100 thousand years), this plane was carved by the incision of the main rivers incising the Aterno and Sagittario gorges forming large valleys in the Sulmona basin and leaving the "Upper Terrace" hanging (Figure 12a [50]). Further stages of fluvial sedimentation and river incision formed a second smaller hanging surface, known as "Lower Terrace", preserved on the valley sides, and the present-day fluvial plain (Figure 12b). This arrangement explains how alternating incision and sedimentation have formed a typical staircase landscape in the main valleys of the Terre Peligne area (Figure 12b) and how the fluvial "terraces" are imprinted in the landscape [80].

The landscape shape, defined by planar surfaces hanging over the present valleys and rich of ancient and present water, was very favorable and largely contributed to the development of human settlement since prehistorical times. This arrangement outlines a further connection between landscape evolution and human history, which is explained in the archeological sites.

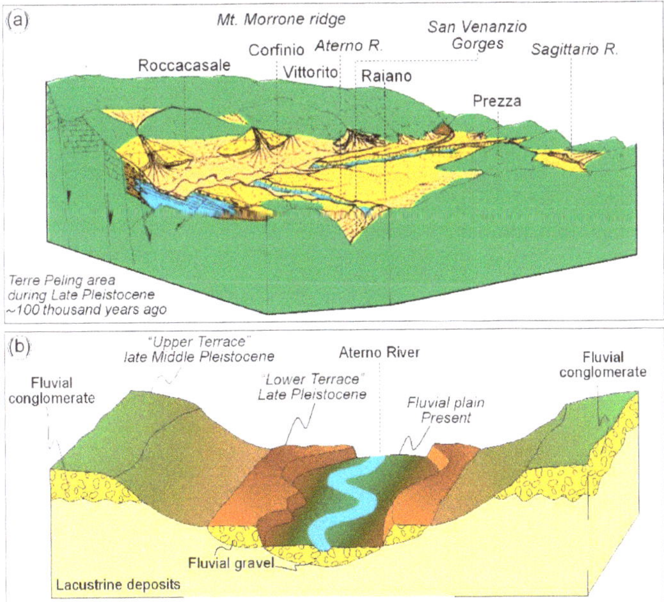

Figure 12. (**a**) Terre Peligne area landscape during Late Pleistocene (~100 thousand years ago), during the incision of the river valleys. Fluvial deposits (pale orange) filled the river valleys incised in the lacustrine deposits. (**b**) Geomorphological scheme explaining the fluvial terraces (Upper and Lower Terrace) along the Aterno River.

4.5.2. Archeology—From the Italics' Capital to Pentima

The wide fluvial "terrace" dominating the center of the Terre Peligne area has been occupied since protohistoric times by the principal settlement in the area, Corfinio. This settlement initially occupied the northernmost edge of the fluvial terrace, for its evident defensive qualities, which in the Middle Ages prompted the construction of the Pentima castle (Figure 13a). During the Roman Age and especially after the Social War the city expanded to the south. At this time, Corfinio was the capital of the Italic League and contributed to the first formulation of the name "ITALIA". The terrace was progressively occupied with monumental buildings, known from epigraphic evidence, and a few archeological remains, punctuating the present village (the temple, mosaic Domus, Morroni) [55]. From late antiquity to the XI century the southern area of the city, along the Tiburtina road, acquired a unique role as the seat of both secular and religious powers and took the name of Valva, which is still connected to the magnificent Romanesque complex of S. Pelino (Figure 13b) [36,81,82]. In the area currently occupied by the cathedral, significant archeological evidence exists concerning a Paleochristian funerary area probably related to a venerated burial site; a fortification incorporating the funerary and worship space and connected to the ancient Roman campus, possibly used between the end of Late Antiquity and the Early Middle Ages as "urban" settlement; and finally a *palatium* from the Longobard period. This latter was coexisting with the church built in honor of the martyred Saint Pelino, already in the Early Middle Ages, to define the institution in the southern suburb of Corfinio during the Roman Age, a pole of civil powers (the gastald and later as a county) and religious ones (diocese). This controlled a wide internal area of Abruzzo, from the Popoli Gorges to the high Sangro River valley, from the Morrone to the Sirente mountains.

Figure 13. Corfinio area: (**a**) a *castrum Pentime* drone view; (**b**) the valvensis complex of San Pelino.

5. SWOT Analysis

To evaluate the real potential of geological heritage and archeological heritage development in the Terre Peligne area, an analysis was performed summarizing and comparing strengths, weaknesses, opportunities, and threats (SWOT analysis) of the integrated itinerary (see among many others [83,84]).

5.1. Strengths

The itinerary integrates geological and human history in a single framework. Moreover, in geological, archeological, and landscape terms, it is an itinerary of connections through time from hundreds million years to present (i.e., between rocks, landscapes, ancient human histories and present municipalities and park reserve areas). The Terre Peligne area intersects two of the main national and regional parks of central Italy (Maiella National Park and Velino-Sirente Regional Park) and are very close (10–30 km) to two other main national parks (Abruzzo, Lazio, Molise National Park, and Gran Sasso Laga Mountains National Park). Moreover, it intersects the Gole di San Venanzio Natural Reserve. This might provide the support of already existing infrastructures and dissemination policies. One of the main highways crossing central Italy from Rome to the Adriatic coast passes through the Terre Peligne area and easily connect this area to main cities and main airports (>60 km local airport; 180 km international airport) and railway stations. A national road (a former Roman consular road) and two interregional railways pass through the area too. This provides an easy accessibility to the area. Several archeological and geological studies and projects have been carried out in the area in collaboration between local authorities, universities (e.g., University of Chieti-Pescara, University La Sapienza of Rome, University of RomaTre and many others) and research centers (e.g., ISPRA, INGV). These provided valuable scientific information, which already provides a high-level knowledge of this area and are converted into the itinerary for scientific dissemination. The results of scientific and dissemination studies and activities were already presented at national and international scientific congresses. The itinerary combines "on-site" tools (totems and panels) and "digital" tools (explanatory material readable through smartphones and tablets) and might attract people of a wide range of age and digital alphabetization. There are pre-existing cultural, geological, archeological, and landscape attractions and many tools and features already exist and focusing on these valuable elements, such as already known archeological sites (e.g., the Corfinio area and San Pelino), museums (e.g., the Corfinio Archeological Museum [75]) and geosites (e.g., the Gole di San Venanzio geosite [35]), as well as books and geotourist maps [34]. This provides a pre-existing tourist development of this area, which results in at least several thousands of visitors per year in these specific sites. Moreover, local schools have started performing field trip and visits in the Terre Peligne area.

5.2. Weaknesses

The realization of the itinerary and the emplacements of all the structures is not completed. The itinerary is supported by a rather small group of people in the local municipalities and in the universities involved. It is mostly based on regional funding. An actual marketing strategy has not been activated so far. A small reach of the road included in the itinerary is closed due to safety reason related to landslide risk waiting to be fixed; however, an alternative path is possible. There is a poor connection between the itinerary and the other local values, specifically in terms of quality food and wine. More in general, a management system for the identification, assessment, and divulgation of the itinerary and its values in connection with the surrounding areas is still lacking. Community residents have not yet realized the high value and the potential opportunities of geological and archeological heritage and still have poor cognition and consciousness concerning its protection.

5.3. Opportunities

The highway, roads and railways, provide good public accessibility opportunities. The numbers of visitors already coming in the Terre Peligne area could be surely improved by creating a network connecting the "Fan of the Terre Peligne" itinerary with the park and reserve areas in the area. The itinerary focused on a large territorial diffusion networking the resources of the entire Peligna valley, including connections to other internal areas of the Abruzzo and the Apennines area. This might contribute to induce a tourist flow from the main parks of central Italy through the Terre Peligne area. This flow might support an expansion of the tourism offer and an increase of economic opportunities and possible investments, in terms of hospitality (hotels, bed and breakfast, agritourisms, restaurants, etc.), existing and new cultural events, etc. This can also create new jobs as local geological–archeological touristic guides. The economic-touristic opportunities can be specifically supported and increased by integrated management strategy of the archeological and geological itinerary in connection with the surrounding national, regional parks, and natural reserves, with the local landscape features, and the local quality food and wine values.

5.4. Threats

The final realization of the itinerary and the emplacements of all the structures depends on the Terre Peligne Association and on the local municipalities and is mostly supported by regional funding. This can result in management problems (many bodies involved) and in irregular funding (according to the variable funding opportunities and to the variable local-regional political conditions). If an integrated management strategy is not arranged, overlapping management of the geological and archeological sites (e.g., from local municipalities, reserve areas and parks) can result in a poor enhancement of the "Fan of the Terre Peligne" itinerary as a connection through time between rocks, landscapes, ancient human histories, and present municipalities and park reserve areas.

6. Concluding Remarks

The "Fan of the Terre Peligne" is an archeological-geological integrated touristic itinerary along the northern part of the Sulmona intermontane basin in the central-eastern Apennines (Abruzzo). It is organized in five thematic sectors focused on the main elements of the geological history and landscape evolution of this area and outlining the joining of archeological and geological-geomorphological features of the Terre Peligne. Through a combined arrangement of "on site" and "digital" tools, it contributes to the valorization of the geological and archeological sites and to make them understandable for the general public, through lay language and reliable information. The main stages of the history of the landscape are described, from the formation of the rocks of the mountain ridges (in marine environment in Jurassic-Miocene times) to their deformation, along thrust and folds and normal faults (Miocene-Quaternary); from the formation of the mountain ridges to the shaping of the landscape mostly due to the competition of gravitational, fluvial and karst processes and of recent-active tectonics and seismicity (Quaternary). The tectonics and seismicity are explained as processes inducing seismic hazard but also as positive elements that formed the landscape of the Apennines. For each of the sectors and in the entire "fan" the close connection between man and landscape is revealed in different perspectives outlining that the presence of man since prehistoric times has given rise to a careful interaction with the territory. The landscape is dominated by high standing rocky ridges surrounding the Sulmona intermontane basin, which through history have been used for defensive purposes, castles, and hermitages. Conversely, the low standing hills and plains have been advantageously exploited for farming, as shown by the terracing and the remains from the Roman centuriation, while the valleys and gorges, rich of water resource, have been used as transit ways across the Apennines. This latter element also explains the close connection between the historical economic development and the landscape shape. The archeological heritage, which enriches the entire area, is mostly centered on

Corfinio resulting from historical events in which the city was the protagonist as well as from its placement at the cross of north–south and east–west transit ways across the Apennines.

Therefore, the itinerary features connections among different elements: ancient marine geological environments and recent continental landscapes; between Tyrrhenian and Adriatic coasts through main transit ways across the Apennines from historical to present times; among mountain rugged and plain gentle landscapes; and among the main park areas and natural reserves of the Abruzzo Apennines. It outlines specifically the connection, through time, of water and rocks with the landscape and human history.

In summary, the itinerary mostly focuses on the territory with the intent to network the specificities and resources of the Peligna valley, without neglecting the ramifications and connections to the nearby areas (territorial diffusion). From this point of view, the geological and landscape elements joined to the archeological heritage are the aspects that best showcase the specificities of this territory taking into account a very wide diachrony, so as to give greater prominence to the conditions and times that have shaped this territory in its present forms. This "account", as well as the connection between landscape and human history and development, have been rendered to be understandable to the general public. This approach, based on scientifically correct information, is an instrument to involve the local population, schools, and tourists and to sensitize them about the "history" of their territory and the related hazard, as well as a useful instrument for enhancing tourism inducing direct and indirect economic return. The purpose of the study is in line with the stream of similar initiatives aimed at the knowledge, valorization, and fruition of the internal mountain areas. Just to remain within the Abruzzo region, among many others see the cases within the Parks of Majella and Velino-Sirente [6,85], or the case of the APSAT (Ambiente e Paesaggi dei Siti d'Altura Trentini) Project for the Trentino region [86] or the transhumance route in southern Italy [87] at international level (e.g., Egypt, Russia, Malta [7,88,89]).

The "Fan of the Terre Peligne" features a strong territorial nature and diffusion, connecting the Peligna valley with other internal areas of the Abruzzo and the whole Apennines area. The SWOT analysis revealed the great strength of the itinerary, in terms of integration of different themes and features, high accessibility, connection with already existing reserve and park areas and tourist attractions. This can lead to wide opportunities for networking the itinerary to the surrounding parks and reserves and for increasing the tourist flow between them, also creating new development and jobs not conflicting with the existing one but enhancing them. This can largely support local communities and economic development [3]. However, the analysis summarized also the weaknesses (e.g., itinerary structures not completed, lacking of an integrated management system) and the threats to be faced (e.g., completion of the itinerary depending on irregular funding and results connected to an integrated management, which should overcome overlapping management from local bodies), On this basis, the itinerary wants to be a resource and instrument to contribute to the popularization and enhancement of the cultural heritage of the human and geological history of one of the key areas of the central Apennines. Finally, improving sustainable tourism developed on valuable and less-known sites, the itinerary aims to strengthen the awareness toward the themes of natural hazards and risks of the territory increasing the development and resilience of the inner areas of the Apennines.

Author Contributions: Conceptualization, E.M. and M.C.S.; methodology, E.M., M.C.S., S.A., T.P. and M.B.; investigation, S.A., T.P. and V.M.; data curation, T.P. and V.M.; writing—original draft preparation, S.A., T.P. and G.E.; writing—review and editing, T.P., M.C.S., E.M., G.E. and M.B.; supervision, E.M. and M.C.S.; funding acquisition, E.M. and M.C.S.

Funding: This research was funded by Regione Abruzzo, "Progetto di valorizzazione e sviluppo dei beni storico-culturali, archeologici, ambientali delle tradizioni delle Terre dei Peligni", Reg. Abruzzo PAR FSC 2007–2013. L.A. 1.2.4.a. "Definizione ed attuazione di un programma di sviluppo della Valle Peligna". The APC was funded by University "G. d'Annunzio" of Chieti-Pescara (Miccadei fund, and Piacentini fund).

Acknowledgments: The authors wish to thank the municipalities of Raiano, Roccacasale, Prezza, Vittorito, Corfinio and the Terre Peligne Association for the support in the project and the study for the realization of the itinerary. The authors also wish to thank the Cartographic Office of Abruzzo Region using the Open Geodata Portal (http://opendata.regione.abruzzo.it/), for providing the topographic data and orthophotos used for this work.

Conflicts of Interest: The authors declare no conflict of interest.

References

1. Hose, T.A. European geotourism—Geological interpretation and geoconservation promotion for tourists. In *Geological Heritage: Its Conservation and Management*; Barretino, D., Wimbledon, W.P., Gallego, E., Eds.; Instituto Tecnologico Geominero de España: Madrid, Spain, 2000; pp. 127–146.

2. Joyce, B. Geotourism, geosites and geoparks: Working together in Australia. *Aust. Geol.* **2007**, *144*, 26–29.

3. Dowling, R.K.; Newsome, D. *Global Geotourism Perspectives*; Goodfellow Publishers: Oxford, UK, 2010.

4. Miccadei, E.; Piacentini, T.; Esposito, G. Geomorphosites and geotourism in the parks of the Abruzzo region (central Italy). *Geoheritage* **2011**, *3*, 233–251. [CrossRef]

5. Miccadei, E.; Sammarone, L.; Piacentini, T.; D'Amico, D.; Mancinelli, V. Geotourism in the Abruzzo, Lazio and Molise national park (central Italy): The example of Mount Greco and Chiarano Valley. *GeoJ. Tour. Geosites* **2014**, *13*, 38–51.

6. Liberatoscioli, E.; Boscaino, G.; Agostini, S.; Garzarella, A.; Scandone, E.P. The Majella national parck: An Aspiring UNESCO geopark. *Geosciences* **2018**, *8*, 256. [CrossRef]

7. Cappadonia, C.; Coratza, P.; Agnesi, V.; Soldati, M. Malta and Sicily joined by geoheritage enhancement and geotourism within the framework of land management and development. *Geosciences* **2018**, *8*, 253. [CrossRef]

8. Del Monte, M.; D'Orefice, M.; Luberti, G.M.; Marini, R.; Pica, A.; Vergari, F. Geomorphological classification of urban landscapes: The case study of Rome (Italy). *J. Maps* **2016**, *12*, 178–189. [CrossRef]

9. Reynard, E.; Coratza, P. The importance of mountain geomorphosites for environmental education. Examples from the Italian Dolomites and the Swiss Alps. *Acta Geogr.* **2016**, *56*, 291–303. [CrossRef]

10. Piacentini, T.; Miccadei, E.; Berardini, G.; Aratari, L.; De Ioris, A.; Calista, M.; Carabella, C.; d'Arielli, R.; Mancinelli, V.; Paglia, G.; et al. Geological tourist mapping of the Mount Serrone fault geosite (Gioia dei Marsi, Central Apennines, Italy). *J. Maps* **2019**, *15*, 298–309. [CrossRef]

11. Castaldini, D.; Valdati, J.; Ilies, D.C.; Barozzini, E.; Bartoli, L.; Dallai, D.; Del Prete, C.; Sala, L. *Carta turistico ambientale dell'alta valle delle tagliole, parco del frignano. Parco del Frignano 2005*; Eliofototecnica Barbieri: Parma, Italy, 2005.

12. Castaldini, D.; Valdati, J.; Ilies, D.C.; Chiriac, C.; Bertogna, I. Geo-tourist map of the natural reserve of Salse di Nirano (Modena apennines, northern Italy). *Alp. Mediterr. Quat.* **2005**, *18*, 245–255.

13. Castaldini, D.; Coratza, P.; Bartoli, L.; Dallai, D.; Del Prete, C.; Dobre, R.; Panizza, M.; Piacentini, D.; Sala, L.; Zucchi, E. *Carta turistico ambientale del Monte Cimone, Parco del Frignano*; Eliofototecnica Barbieri: Parma, Italy, 2008.

14. Miccadei, E.; Esposito, G.; Innaurato, A.; Manzi, A.; Basile, P.; Berti, C. *Carta Geologico-Turistica e relative "Note" della Comunità Montana Aventino-Medio Sangro—Zona Q*; Comunità Montana Aventino-Medio Sangro: Provincia di Chieti, Abruzzo, Italy, 2008.

15. Coratza, P.; Reynard, E.; Zwolinski, Z. Geodiversity and geoheritage: Crossing disciplines and approaches. *Geoheritage* **2018**, *10*, 525–526. [CrossRef]

16. Gordon, J.E. Geoheritage, geotourism and the cultural landscape: Enhancing the visitor experience and promoting geoconservation. *Geosciences* **2018**, *8*, 136. [CrossRef]

17. Ólafsdóttir, R.; Tverijonaite, E. Geoturism: A systematic licterature review. *Geosciences* **2018**, *8*, 234. [CrossRef]

18. Ólafsdóttir, R. Geoturism. *Geosciences* **2019**, *9*, 48. [CrossRef]

19. Migoń, P.; Pijet-Migoń, E. Natural disasters, geotourism, and geo-interpretation. *Geoheritage* **2019**, *11*, 629–640. [CrossRef]

20. Brilha, J. Geoheritage: Inventories and evaluation. In *Geoheritage: Assessment, Protection and Management*; Reynard, E., Brilha, J., Eds.; Elsevier: Amsterdam, The Netherlands, 2018; pp. 69–86.

21. ISPRA. Geological Survey of Italy. National Inventory of Geosites. Available online: www.isprambiente.gov.it/en/projects/soil-and-territory/protection-of-geological-heritage-parks-geo mining-geoparks-and-geosites/national-inventory-of-geosites (accessed on 31 January 2019).

22. Panizza, M. Geomorphosites: Concepts, methods and examples of geomorphological survey. *Chin. Sci. Bull.* **2001**, *46*, 4–6. [CrossRef]

23. Coratza, P.; Panizza, M. Geomorphology and cultural heritage. In *Memorie Descrittive della Carta Geologica d'Italia, 87/2009*; ISPRA: Rome, Italy, 2010; pp. 1–190.

24. Carton, A.; Coratza, P.; Marchetti, M. Guidelines for geomorphological sites mapping: Examples from Italy. *Géomorphol. Relief Process. Environ.* **2005**, *11*, 209–218. [CrossRef]

25. Bertacchini, M.; Benito, A.; Castaldini, D. Carta geo-archeo-turistica del territorio di Otricoli (Terni, Umbria). In Proceedings of the Geologia e Turismo—Beni Geologici e Geodiversità. Atti 3° Congresso Nazionale Associazione Italiana Geologia e Turismo, Bologna, Italy, 1–3 March 2007; pp. 213–220.

26. Bissig, G. Mapping geomorphosites: An analysis of geotourist maps. *Geoturystika* **2008**, *3*, 3–12. [CrossRef]

27. Regione Emilia-Romagna. *Il Paesaggio Geologico dell'Emilia Romagna. Carta a scala 1: 250.000*; SELCA: Florence, Italy, 2009.

28. Panizza, M. *Un'escursione Nello Spazio e nel Tempo. Via GeoAlpina Itinerari Italiani. APAT-ISPRA, G&T Associazione Italiana di Geologia e Turismo, EuroGeoSurveyes*; CSR Publisher: Rome, Italy, 2010; pp. 1–192.

29. Piacentini, T.; Castaldini, D.; Coratza, P.; Farabollini, P.; Miccadei, E. Geotourism: Some examples in northern-central Italy. *GeoJ. Tour. Geosites* **2011**, *2*, 240–262.

30. Piacente, S.; Coratza, P.; L'associazione Nazionale Geologia & Turismo. Per una Nuova Prospettiva di Sviluppo Culturale e Sociale del Turismo in Italia. In *Contributi alle Giornate del Turismo 2003-2004*; Patron Editore: Bologna, Italy, 2005; pp. 383–390.

31. Coratza, P.; Ghinoi, A.; Piacentini, D.; Valdati, J. Management of geomorphosites in high tourist vocation area: An example of geo-hiking maps in the Alpe di Fanes (Natural Park of Fanes-Senes-Braies, Italian Dolomites). *GeoJ. Tour. Geosites* **2008**, *2*, 106–117.

32. Brandolini, P.; Faccini, F.; Robbiano, A.; Bulgarelli, F. Geomorphology and cultural heritage of the Ponci Valley (Finalese karstic area, Ligurian Alps). *Geogr. Fis. Din. Quat.* **2011**, *34*, 65–74.

33. Sacchini, A.; Imbrogio Ponaro, M.; Paliaga, G.; Piana, P.; Faccini, F.; Coratza, P. Geological landscape and stone heritage of the genoa walls urban park (NW Mediterranean, Italy). *J. Maps* **2018**, *14*, 528–541. [CrossRef]

34. Miccadei, E.; Mancinelli, V.; Piacentini, T.; Paglia, G.; Carabella, C. Cartografia per geologia e turismo in abruzzo. In *Atti del convegno "Il patrimonio geologico: Una risorsa scientifica, paesaggistica, culturale e turistica"*; Associazione Italiana di Geologia e Turismo: Bologna, Italy, 2018.

35. Faccini, F.; Gabellieri, N.; Paliaga, G.; Piana, P.; Angelini, S.; Coratza, P. Geoheritage map of the Portofino natural park (Italy). *J. Maps* **2018**, *14*, 87–96. [CrossRef]

36. Miccadei, E.; Piacentini, T.; Mancinelli, V. Il Geosito delle Gole di San Venanzio a Raiano (AQ). Scheda e Descrizione Tecnico-Scientifica. Inventario Nazionale dei Geositi. ISPRA. 2016. Available online: http://sgi.isprambiente.it/geositiweb/scheda_geosito.aspx?id_geosito_x=2814 (accessed on 20 May 2019).

37. Miccadei, E.; Barberi, R.; Cavinato, G.P. La geologia quaternaria della Conca di Sulmona (Abruzzo, Italia centrale). *Geol. Romana* **1999**, *34*, 58–86.

38. Piacentini, T.; Miccadei, E. The role of drainage systems and intermontane basins in the quaternary landscape of the Central Apennines chain (Italy). *Rend. Lincei* **2014**, *25*, 139–150. [CrossRef]

39. Somma, M.C.; Antonelli, S.; La Salvia, V. Paesaggi ed insediamenti in area montana: Il caso del territorio valvense tra persistenze e trasformazioni. In *Storia e Archeologia Globale dei Paesaggi Rurali D'Italia fra Tardoantico e Medioevo*; Volpe, G., Ed.; Edipuglia: Bari, Italy, 2018; pp. 463–504.

40. Hose, T.A. *Geoheritage and Geotourism: A European Perspective (Heritage Matters)*; Boydell Press, Newcastle University: Newcastle, UK, 2016.

41. Reynard, E.; Perret, A.; Bussard, J.; Grangier, L.; Martin, S. Integrated approach for the inventory and management of geomorphological heritage at the regional scale. *Geoheritage* **2016**, *8*, 43–60. [CrossRef]

42. Kubalíková, L. Geomorphosite assessment for geotourism purposes. *Czech. J. Tour.* **2013**, *2*, 80–104. [CrossRef]

43. Coratza, P.; de Waele, J. Geomorphosites and natural hazards: Teaching the importance of geomorphology in society. *Geoheritage* **2012**, *4*, 195–203. [CrossRef]

44. Coratza, P.; Regolini-Bissig, G. Methods for mapping geomorphosites. In *Geomorphosites*; Reynard, E., Coratza, P., Regolini-Bissig, G., Eds.; Pfeil: München, Germany, 2009; pp. 89–103.

45. Miccadei, E.; Mascioli, F.; Piacentini, T. Quaternary geomorphological evolution of the Tremiti islands (Puglia, Italy). *Quat. Int.* **2011**, *233*, 3–15. [CrossRef]

46. Panizza, M.; Piacente, S. Geomorphological asset evaluation. *Z. Geomorph.* **1993**, *87*, 13–18.

47. Panizza, M.; Piacente, S. Geomorphosites: A bridge between scientific research, cultural integration and artistic suggestion. *Il Quat.* **2005**, *18*, 3–10.

48. Miccadei, E. Geologia dell'area alto sagittario-alto sangro (Abruzzo, Appennino centrale). *Geol. Romana* **1993**, *29*, 463–481.

49. Ciccacci, S.; D'Alessandro, L.; Dramis, F.; Miccadei, E. Geomorphologic evolution and neotectonics of the Sulmona Intramontane Basin (Abruzzi, Apennine, Central Italy). *Zeirschrift Geomorph. Suppl. Bd.* **1999**, *118*, 27–40.

50. Miccadei, E.; Piacentini, T.; Barberi, R. Uplift and local tectonic subsidence in the evolution of intramontane basins: The example of the Sulmona Basin (Central Appennines, Italy). In Proceedings of the International INQUA Workshop Large-Scale Vertical Movements and Related Gravitational Processes, Camerino/Rome, Italy, 21–26 June 1999; Dramis, F., Farabollini, P., Molin, P., Eds.; Studi Geologici Camerti: Camerino/Rome, Italy, 2002; pp. 119–133.

51. Miccadei, E.; Paron, P.; Piacentini, T. The SW escarpment of the Montagna del Morrone (Abruzzi, Central Italy): Geomorphology of a fault-generated mountain front. *Geogr. Fis. Din. Quat.* **2004**, *27*, 55–87.

52. Miccadei, E.; D'Alessandro, L.; Parotto, M.; Piacentini, T.; Praturlon, A. *Note Illustrative della Carta Geologica d'Italia alla scala 1:50,000, foglio 378 SCANNO*; Servizio Geologico d'Italia: Ispra, Italy, 2014.

53. APAT—Agenzia per la Protezione dell'Ambiente e per i servizi Tecnici. *Carta Geologica d'Italia alla Scala 1:50,000, foglio 369*; Servizio Geologico d'Italia, Dipartimento Difesa del Suolo: Sulmona, Italy, 2006. Available online: http://www.isprambiente.gov.it/Media/carg/abruzzo.html (accessed on 15 January 2019).

54. ISPRA—Istituto Superiore per la Protezione e la Ricerca Ambientale. *Carta Geologica d'Italia alla scala 1:50,000, foglio 378 SCANNO*; Servizio Geologico d'Italia: Ispra, Italy, 2014. Available online: http://www.isprambiente.gov.it/Media/carg/abruzzo.html (accessed on 15 January 2019).

55. Antonelli, S. La ceramica dallo scavo di S. Pelino a Corfinio (AQ): Problematiche e spunti di riflessione. In Proceedings of the VII Congresso Nazionale di Archeologia Medievale, Lecce, Italy, 9–12 September 2015; Arthur, P., Imperiale, M.L., Eds.; All'Insegna del Giglio: Sesto Fiorentino, Italy, 2015; pp. 223–227.

56. Antonelli, S. Corfinio (AQ), loc. Fonte, Sant'Ippolito, 2014–2015. *Archeologia Medievale* **2015**, *XLII*, 247.

57. Antonelli, S.C.S. Urbanistica e cristianizzazione a Corfinio: Alcuni spunti per una ricerca. In Proceedings of the Ancient Modern Towns. I centri urbani a continuità di vita: Archeologia e valorizzazione, Convegno internazionale di studi a dieci anni dalla scomparsa di Anna Maria Giuntella, Chieti-Corfinio, Italy, 30 September–2 October 2015.

58. Antonelli, S.; Somma, M.C. Il palatium altomedievale di Valva (Corfinio, AQ): Forme e funzioni. In Proceedings of the VII Congresso Nazionale di Archeologia Medievale, Lecce, Italy, 9–12 September 2015; Arthur, P., Imperiale, M.L., Eds.; All'Insegna del Giglio: Sesto Fiorentino, Italy, 2015; pp. 122–125.

59. Somma, M.C.; Antonelli, S. La problematica cultuale e funeraria del complesso di S. Pelino a Corfinio (Aq). In Proceedings of the VII Congresso Nazionale di Archeologia Medievale, Lecce, Italy, 9–12 September 2015; Arthur, P., Imperiale, M.L., Eds.; All'Insegna del Giglio: Sesto Fiorentino, Italy, 2015; pp. 192–196.

60. Somma, M.C.; Antonelli, S.; Casolino, C. Culto delle acque tra continuità e trasformazioni: Il caso di Fonte, S. Ippolito a Corfinio. In Proceedings of the VII Congresso Nazionale di Archeologia Medievale, Lecce, Italy, 9–12 September 2015; Arthur, P., Imperiale, M.L., Eds.; All'Insegna del Giglio: Sesto Fiorentino, Italy, 2015; pp. 197–200.

61. Van Wonterghem, F. *Superaequum, Corfinium, Sulmo*; Olschki: Firenze, Italy, 1984.

62. Cosentino, D.; Cipollari, P.; Marsili, P.; Scrocca, D. Geology of the central Apennines: A regional review. *Geol. Italy J. Virtual Explor.* **2010**, *36*, 1–37. [CrossRef]

63. Miccadei, E.; Piacentini, T.; Gerbasi, F.; Daverio, F. Morphotectonic map of the Osento River basin (Abruzzo, Italy), scale 1:30,000. *J. Maps* **2012**, *8*, 62–73. [CrossRef]

64. Miccadei, E.; Piacentini, T.; Dal Pozzo, A.; La Corte, M.; Sciarra, M. Morphotectonic map of the Aventino-Lower Sangro valley (Abruzzo, Italy), scale 1:50,000. *J.Maps* **2013**, *9*, 390–409. [CrossRef]

65. Miccadei, E.; Piacentini, T.; Buccolini, M. Long-term geomorphological evolution in the Abruzzo area, Central Italy: Twenty years of research. *Geologica Carpathica* **2017**, *68*, 19–28. [CrossRef]

66. Carabella, C.; Miccadei, E.; Paglia, G.; Sciarra, N. Post-wildfire landslide hazard assessment: The case of the 2017 Montagna del Morrone fire (Central Apennines, Italy). *Geosciences* **2019**, *9*, 175. [CrossRef]

67. Miccadei, E.; Parotto, M. Assetto geologico-strutturale delle dorsali di M. Rotella—M. Pizzalto—M. Porrara (Abruzzo, Appennino centrale). *Geologica Romana* **1999**, *34*, 87–114.

68. Vannoli, P.; Burrato, P.; Fracassi, U.; Valensise, G. A fresh look at the seismotectonics of the Abruzzi (Central Apennines) following the 6 April 2009 L'Aquila earthquake (Mw 6.3). *Ital. J. Geosci.* **2012**, *131*, 309–329.

69. Gori, S.; Falcucci, E.; Ladina, C.; Marzorati, S.; Galadini, F. Active faulting, 3-D geological architecture and plio-quaternary structural evolution of extensional basins in the central Apennine chain, Italy. *Solid Earth* **2017**, *8*, 319–337. [CrossRef]

70. Sagnotti, L.; Giaccio, B.; Liddiocoat, J.C.; Nomade, S.; Renne, P.R.; Scardia, G.; Sprain, C.J. How fast was the Matuyama–Brunhes geomagnetic reversal? A new subcentennial record from the Sulmona Basin, central Italy. *Geophys. J. Int.* **2016**, *204*, 798–812. [CrossRef]

71. Somma, M.C.C.S. Corfinio al tempo della guerra sociale: Il quadro archeologico. Il contributo di Sara Santoro. In Proceedings of the L'Italie entre Déchirements et Réconciliations: Revisiter la Guerre Sociale (91–88) et ses Lendemains, Colloque International D'histoire et D'histoire du droit Organisé par l'Université Paris 1, Paris, France, 13–15 October 2016.

72. Mattiocco, E. *Centri Fortificati Preromani Nella Conca di Sulmona*; Sopr. Archeologica per l'Abruzzo: Chieti, Italy, 1981.

73. Mattiocco, E. *Centri Fortificati Preromani nel Territorio dei Peligni, Mostra documentale*; Sopr. Archeologica per l'Abruzzo: Sulmona, Italy, 1983; p. 81.

74. Somma, M.C. Fondare castelli e monasteri: La politica di Trasmondo vescovo di Valva (Abruzzo). In *'Fondare' Tra Antichità e Medioevo*; Galetti, P., Ed.; Fondazione Centro Italiano di Studi sull'Alto Medioevo: Spoleto, Italy, 2016; pp. 187–199.

75. Campanelli, A.; Finarelli, C. Città antiche a continuità di vita dell'Abruzzo romano. In Proceedings of the I siti archeologici: Un problema di musealizzazione all'aperto, II Seminario di Studi, Roma gennaio 1994, Roma, Italy, 1995; pp. 145–160.

76. Chiarizia, G.; Latini, M.L.; Properzi, P. *Atlante dei Castelli d'Abruzzo, Repertorio Sistematico delle Fortificazioni*; Carsa: Pescara, Italy, 2002.

77. Wickham, C. *Studi sulla società degli appennini nell'Alto Medioevo. Contadini, signori e insediamento nel territorio di Valva (Sulmona)*; Clueb: Bologna, Italy, 1982.

78. Micati, E.; Boesch Gajano, S. *Eremi d'Abruzzo e Luoghi di Culto Rupestri*; Carsa Edizioni: Pescara, Italy, 1996.

79. Burri, E. L'antico Acquedotto delle Vuccole nelle Gole di S. Venanzio (Raiano L'Aquila Italia centrale) contributo preliminare. *Rivista di Topografia Antica* **2004**, *14*, 57–60.

80. Summerfield, M.A. Tectonic geomorphology. *Progr. Phys. Geogr. Earth Environ.* **1991**, *15*, 193–205. [CrossRef]

81. Somma, M.C. Da corfinio a valva: Lo sviluppo urbano di un municipio romano tra tardantichità e alto medioevo. In Proceedings of the VII Congresso Nazionale di Archeologia Medievale, Lecce, Italy, 9–12 September 2015; Arthur, P., Imperiale, M.L., Eds.; All'Insegna del Giglio: Sesto Fiorentino, Italy, 2015; pp. 282–286.

82. Somma, M.C. Corfinio (AQ), Basilica di San Pelino, Valva, 2013–2015. *Archeologia Medievale* **2015**, *XLII*, 246–247.

83. Kubalíková, L.; Kirchner, K. Geosite and geomorphosite assessment as a tool for geoconservation and geotourism purposes: A case study from vizovická vrchovina highland (eastern part of the Czech Republic). *Geoheritage* **2016**, *8*, 5. [CrossRef]

84. Carrión Mero, P.; Herrera Franco, G.; Briones, J.; Caldevilla, P.; Domínguez-Cuesta, M.J.; Berrezueta, E. Geotourism and local development based on geological and mining sites utilization, Zaruma-Portovelo, Ecuador. *Geosciences* **2018**, *8*, 205. [CrossRef]

85. Agostini, S.; Colecchia, A. Ecomusei e geoturismo nell'Abruzzo montano: Dalle esperienze locali a una progettazione allargata. *Scienze del Territorio, Riabitare la Montagna* **2016**, *4*, 88–93.

86. Angelucci, D.E.; Casagrande, L.; Colecchia, A.; Rottoli, M. *APSAT 2. Paesaggi d'altura del Trentino. Evoluzione Naturale e Aspetti Culturali*; SAP Società Archeologica: Mantova, Italy, 2013.

87. Meini, M.; Di Felice, G.; Petrella, M. Geotourism perspectives for transhumance routes. Analysis, requalification and virtual tools for the geoconservation management of the drove roads in Southern Italy. *Geosciences* **2018**, *8*, 368. [CrossRef]
88. Taha, M.M.N.; El-Asmar, H.M. Geo-archeoheritage sites are at risk, the Manzala Lagoon, NE Nile Delta Coast, Egypt. *Geoheritage* **2019**, *11*, 441–457. [CrossRef]
89. Moroni, A.; Gnezdilova, V.; Ruban, D.A. Geological heritage in archaeological sites: Case examples from Italy and Russia. *Proc. Geol. Assoc.* **2015**, *126*, 244–251. [CrossRef]

Article

Devil Landforms as Resources for Geotourism Development: An Example from Southern Apulia (Italy)

Paolo Sansò

Dipartimento di Scienze e Tecnologie Biologiche e Ambientali, Università del Salento, 73100 Lecce, Italy; paolo.sanso@unisalento.it

Received: 20 June 2019; Accepted: 21 July 2019; Published: 26 July 2019

Abstract: The landscape of *Murge Tarantine* limestone ridge (southern Apulia, Italy) is marked by the presence of an isolated relief showing a singular shape and name, the *Monte del Diavolo* (i.e., the Devil's Mount). The *Monte del Diavolo* is located in a very interesting area from a geological point of view since it shows an E–W trending high-fault scarp, the morphological effect of the right-lateral transtensive North Salento Fault Zone. The *Monte del Diavolo* is a small isolated conical relief reaching at its top 115 m above m.s.l.; it elevates about 20 m from the surrounding plain surface, stretching at about 95 m altitude. Its evolution has been influenced by the occurrence of strongly cemented *breccia* deposits, most likely due to cave roof collapse and calcite precipitation, which are more resistant to the karst denudation process than surrounding limestones. This paper would be the first step towards the cultural promotion of the *Monte del Diavolo* area, which is marked by geological and geomorphological peculiar features and by a relevant archaeological and natural heritage as well.

Keywords: isolated relief; geological heritage; southern Apulia; Italy

---

1. Introduction

Geotourism can be defined as "the provision of interpretative and service facilities for geosites and geomorphosites and their encompassing topography, together with their associated in situ and ex situ artefacts, to constituency-build for their conservation by generating appreciation, learning and research by and for current and future generations" [1] (p. 11). It employs an easily and globally accurate translatable vocabulary for the nature, focus and location of modern geology-based geotourism with a geoconservation purpose.

Geotourists can be divided in two groups, "casual" and "dedicated" [2]. The former occasionally visit geosites and geomorphosites mainly for recreation and pleasure; populist guides, trails and visitor centres have recently been provided for them. The latter intentionally visit geosites for the purpose of personal educational or intellectual improvement and enjoyment; field guides and journal papers are a long-standing provision for them.

The geotourism offer is usually based on geoheritage sites of preeminent scientific value which directly translate into educational opportunities. However, at least from the perspective of tourism industry, the scenic component of geosites is equally important since modern geotourism provision meets geotourists' needs by attracting them to particular localities with spectacular or readily-appreciated geomorphological features [3]. This refers particularly to landforms marked by intriguing, unusual or even bizarre shapes that justifies the "devil" name attributed to them by local communities or first explorers. For example, the Times World Atlas reports numerous "devil" landforms such as mountains, hills, gorges, lakes and deserts. Some examples are the Devil's City, a Serbian area marked out by about 200 earth pillars or the Devil's Balls, a number of spheroidal boulders occurring at Northern Territories (Australia). However, the most famous devil's landform in the world is most likely the Devil's Tower placed in the Wyoming (United States), whose top surface has been the theatre of the first close meeting of mankind with aliens in the Spielberg's movie "Close Encounters of the Third Kind" of 1997. This spectacular isolated relief of cylindrical shape shows its flat top surface at 1558 m altitude and elevates about 400 m from the surrounding plain. The sub-vertical slopes of this laccolith are marked by a well-developed columnar jointing. This relief is placed inside a natural park which receives about 400,000 visitors each year.

Thanks to their peculiar morphology, "devil" landforms can attract tourists whose interest in Earth history is minimal or non-existent. Then, an opportunity arises to provide more in depth explanation and interpretation. Thus, it is assumed that telling the story about rocks may be easier at natural rock outcrops rather than in quarries, particularly since in the former case the story would be more comprehensive, involving near-surface processes and landform evolution too. In addition, strange landforms usually give rise to strange stories so that Serbian earth pillars are seen by local people as petrified men or aliens, Australian spheroidal blocks are the Rainbow Snake eggs, and columnar jointing at Devil's Tower are the scratches produced by the paws of a huge grizzly which tried to catch seven little girls saved by the Great Spirit on the relief flat top surface. For this, devil landforms are generally useful links with the cultural heritage of the area, covering in this way a leading role in the geotourism development of the area. Geotourism, in fact, allows tourists to appreciate local geology but also to better understand its relationship to other assets of the territory, such as biodiversity, archaeological and cultural values [4]. The geotourism offer should thus include abiotic, biotic and cultural components. These last ones include cultural, spiritual and historic meanings (e.g., folklore, sacred sites and sense of place) [5].

This paper reports the results of the geomorphological analysis carried out in a coastal area of southern Apulia area which is marked by a singular isolated small relief, the *Monte del Diavolo* (i.e., the "Devil's Mount") (Figure 1). Research was carried out by means of field survey integrated with interpretation of aerial photos; an accurate bibliographic analysis allowed data collection about geological features and archaeological and cultural heritage of the area as well.

Notwithstanding its morphological peculiarity, *Monte del Diavolo* has received little attention from researchers so far [6] and it has not been enclosed in the sites of the relevant geological interest inventory realized by Regione Puglia Administration [7]. Nevertheless, the *Monte del Diavolo* relief belongs to the history of the geological knowledge development of Apulia region since Earl Michele Milano (1820), in his pioneering paper about the geology of this region [8], wrongly considered *Monte del Diavolo* relief a geological proof of past volcanic activity in southern Apulia. He reported smoke coming from surficial fractures of the small relief as well as some strange lights; moreover, he compared the singular relief near *Manduria* village to the homonymous relict volcano placed in the *Verona* area and to the *Fossa del Diavolo* (i.e., the Devil's Graben) at *Lipari* island, where, according to the local traditions, flames came out from a pothole. A further evidence of *Monte del Diavolo* volcanic origin would have been the presence of a sulphurous water spring at the nearby *Li Cuturi* wood which dried up in the 1778.

A detailed geomorphological study of the area was carried out aiming to fill up this gap. Collated data are the first step towards the construction of a geotourism offer in this area, which is already a well-known tourism locality.

In addition, the *Monte del Diavolo* isolated relief is placed in an area of high archaeological interest (*Li Castelli* locality), marked by the remains of a Messapic settlement referred to the period spanning between VIII century to III century B.C. [9] and includes some regional natural reserves (*Riserva naturale regionale orientata del Litorale Tarantino Orientale*) of high ecological value. Finally, it is crossed by several hiking trails and biking routes, which allow its natural heritage to be exploited in a sustainable way.

Figure 1. A view of Monte del Diavolo from SE. The relief elevates about 20 m from the surrounding plain surface.

2. Geotourism in Apulia Region

The Apulia region is among the most visited Italian regions, with more than 15 million visitors recorded in 2018. The local tourism industry is rapidly growing with a tourism presence increase of about 12.2% in the period spanning from 2015 to 2018. Nowadays it represents about 5% of added regional value [10].

Data collated by Regione Puglia Administration show that domestic tourists prevail (about 80%); they concentrate during summer because of the attractiveness of local beaches, the high quality of coastal waters, and the eno-gastronomic tradition as well. On the other hand, international tourists are more interested in the cultural and natural heritage of the region and their presence is recorded on a wider period of the year spanning from March to October.

The Apulia region shows a rich geological heritage and geodiversity since it comprises foreland–foredeep–chain domains. It has been the focus of the Puglia Regional Lawn, 33/2009 *"Tutela e valorizzazione del patrimonio geologico e speleologico"* (Conservation and Promotion of Geological and Speleological Heritage), which promoted the compilation of a regional geosites inventory comprising 440 records [7] and economic support to a number of measures for their valorisation and protection.

However, the geotourism potential of this region is largely unexploited since an organized geotourism provision is presently restricted to very few spots, like the Castellana Caves (more than 320,000 visitors in the 2018) or Zinzulusa Cave (more than 100,000 visitors per year). Guided field trips for the public, often carried out in the context of national and international events dedicated to geology themes as well as those organized by cultural associations, do not improve significantly the current situation.

3. Geological and Geomorphological Outline of the Area

The *Monte del Diavolo* is placed between the *Manduria* village (Province of *Taranto*) and the coastline (Figure 2), at the easternmost part of the *Murge Tarantine*, one of the five morphological districts recognized in southern Apulia [11] (Figure 3), a narrow and flat peninsula stretching between the Ionian and the Adriatic Seas.

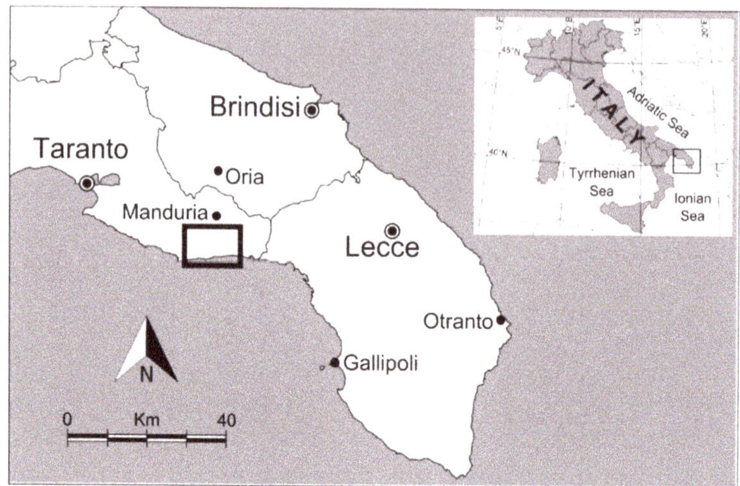

Figure 2. Geographical position of Monte del Diavolo area.

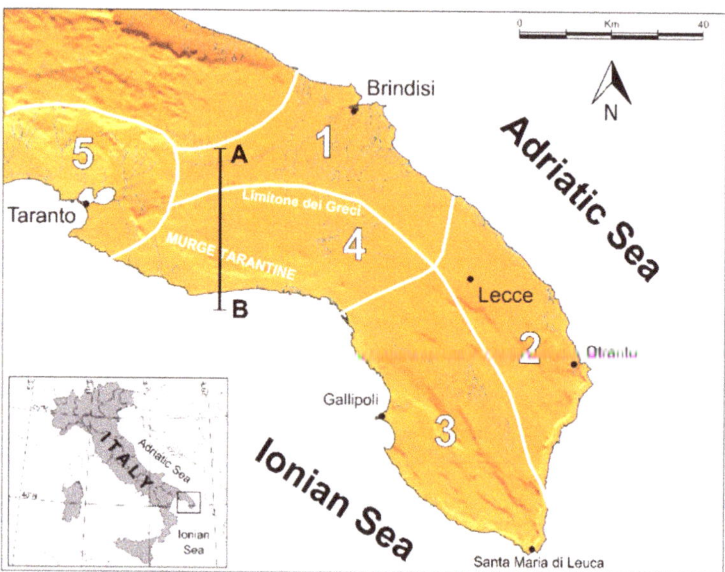

Figure 3. The landscape of southern Apulia can be subdivided in five morphological districts. In particular, district 1 is characterized by a low-elevated Middle Pleistocene sedimentary plain gently sloping from west to east; it is drained by a relict hydrographic network flowing toward the Adriatic coast. District 2 emerged most likely at the beginning of the Pleistocene period and is mostly shaped on pre-Quaternary carbonatic rocks. District 3 is marked by wide sedimentary plains emerged definitively during the Middle Pleistocene, interposed among NW–SE trending morphostructural carbonatic ridges, the *Serre*. District 4 comprises the *Murge Tarantine* area, a low-elevated ridge stretching in the W–E direction, placed between the *Limitone dei Greci* scarp and the Ionian coastline. Finally, district 5 shows a well-known sequence of Middle–Upper Pleistocene marine terraces. The A–B line is the track of the topographic section reported in Figure 4.

The *Murge Tarantine* area (district 4 in Figure 3) is made of a low-elevated and strongly asymmetric ridge, stretching in the W–E direction, interposed between the *Murge* highplain, to the north, which is considered a horst bordered by NW–SE fault scarps [12], and the *Salento* peninsula, to the southeast, marked by narrow horsts and grabens elongated in NNW–SSE direction. These two different areas are separated by the *Brindisi-Taranto* plain which is strongly affected by an E–W tectonic alignment [13,14], the North Salento Fault Zone, a dextral strike-slip fault of regional importance (Figure 5).

From a geomorphological point of view, the *Murge Tarantine* area is a relief elongated in the E–W direction interposed between the *Brindisi-Taranto* plain to the north and the coastline to the south. Its western border is the *S. Giorgio Jonico* NNW–SSE horst whereas its altitude gradually lowers eastward. Its cross profile is highly asymmetric since the northern slope very gently dips toward the *Limitone dei Greci* scarp, marked by the *Oria* relict high dune belt, whereas southward it shows a steep slope bordered by three low elevated and narrow marine terraces (Figure 4).

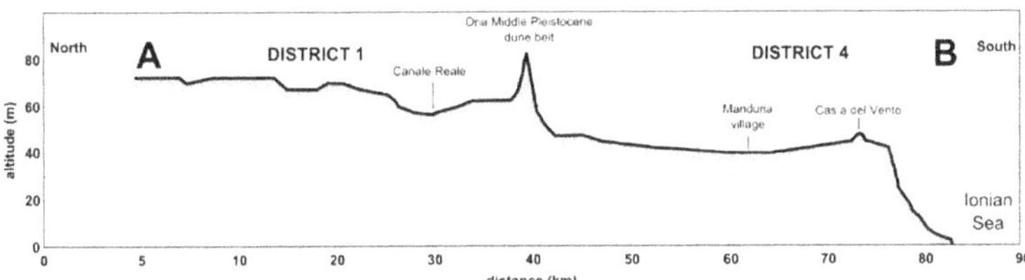

Figure 4. Topographic section of *Murge Tarantine* area. The *Limitone dei Greci* scarp is the northern border of this area and is marked by the *Oria* Middle Pleistocene dune belt. The top surface of *Murge Tarantine* is joined to the Ionian coastline by a steep fault scarp.

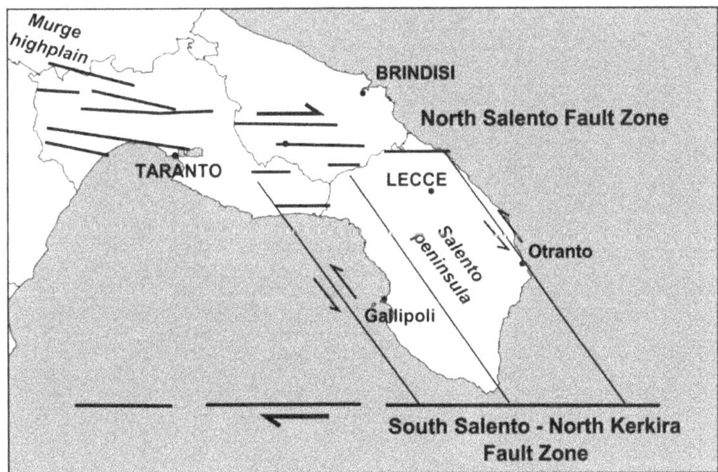

Figure 5. Tectonic scheme of Salento Peninsula (from [14], mod.).

The local relief is in detail articulated by a sequence of NNW–SSE elongated graben and horst, the first ones filled up by Plio-pleistocene sediments and the latter made of Cretaceous limestones.

4. The *Monte del Diavolo* Local Geology and Landscape

Different rock units crop out in the *Monte del Diavolo* area (Figure 6). The oldest one is constituted by the *Calcari di Altamura*, a Late Cretaceous limestone unit characterized by shallowing upward cycles with rudists facies and developed in inner carbonate-platform environments [15]. It is covered by three younger formations: the *Calcarenite di Gravina* formation, the Marine Terraced Deposits and the

Aeolianites. The *Calcarenite di Gravina* formation is a very fossiliferous biodetritical calcareous sediment with *Artica islandica* Linneo deposited during the Lower Pleistocene. Marine Terraced Deposits formed during the Middle–Upper Pleistocene due to the superimposition of glacioeustatic sea level change and regional uplift [16,17]; they are made of bioclastic grainstones with high fossil content. Aeolianites are often associated to Marine Terraced Deposits.

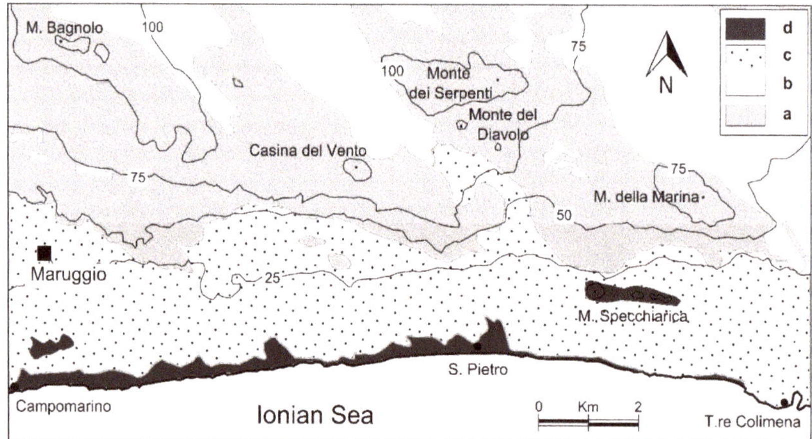

Figure 6. Geological map of *Monte del Diavolo* area. Legend: **a**—*Calcare di Altamura* unit (Late Cretaceous); **b**—*Calcareniti di Gravina* unit (Plio-Pleistocene); **c**— Marine Terraced Deposits (Middle—Upper Pleistocene); **d**—Aeolian deposits (Middle Pleistocene—Holocene).

The local landscape of *Monte del Diavolo* area is dominated by degraded fault scarps; a relict top palaeosurface and a sequence of marine terraces have been also recognized.

The field survey along with aerial photo analysis point out that fault scarps can be clustered into three main systems (Figure 7). The first one is about E–W oriented and comprises two main scarps which constitute the southern steep border of *Murge Tarantine* ridge and strongly influence shoreline position along *Campomarino—S. Pietro in Bevagna* coastal tract.

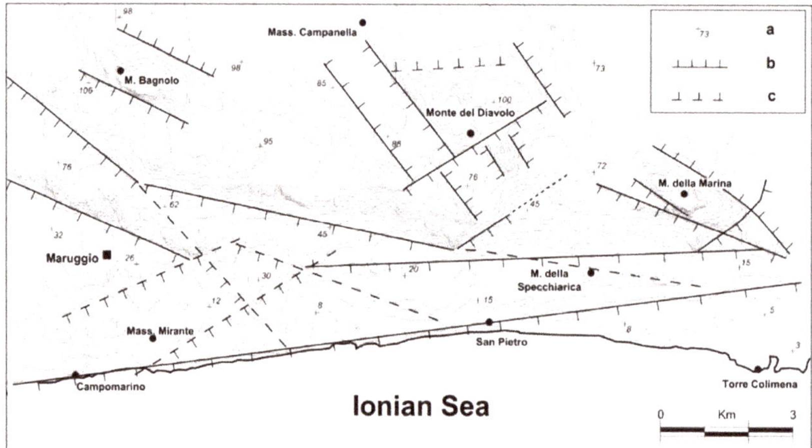

Figure 7. Main fault scarps detected in the *Monte del Diavolo* area. Legend: **a**—elevation point; **b**—main fault scarp; **c**—scarp probably due to fault activity.

The other two systems are easily recognizable on the *Murge Tarantine* ridge, whereas they are less evident on the low elevated coastal area. The second system comprises NW–SE and NE–SW fault scarps which produce horst and graben structures in the area closest to the *Monte del Diavolo* relief. The last system, showing WNW–ESE- and WSW–ENE-oriented faults, is responsible for the development of horst structures (*Monte Bagnolo* and *Monte della Marina*) and indents the main complex tectonic scarp which locally divides the coastal plain from the *Murge Tarantine* top surface. Fault scarps belonging to the E–W tectonic alignment clearly cuts the other two fault scarp systems producing the main features of the landscape. They are most likely the most evident morphological effects of North Salento Fault Zone occurring in the region.

The landscape of *Monte del Diavolo* area is marked at the highest parts of *Murge Tarantine* ridge by the remains of an undulating palaeosurface showing a number of low dome-shaped reliefs whose elevation is comprised between 124 m, to the west, and 100 m to the east (Figure 8): *Monte Bagnolo* (124 m), *Casina del Vento* (108 m), *Monte dei Castelli* (112 m), *Monte dei Serpenti* (109 m), *Monte della Marina* (100 m) (Figure 8).

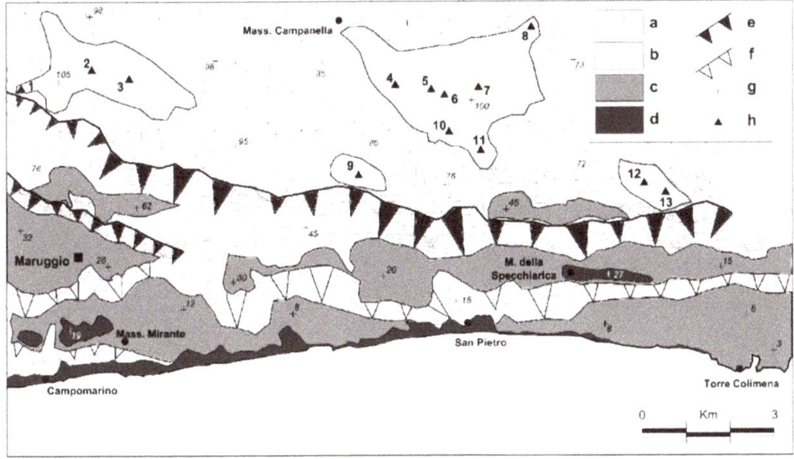

Figure 8. Geomorphological map of *Monte del Diavolo* area. Legend. **a**—denudation surface; **b**—undulating palaeosurface; **c**—marine terrace; **d**—dune belt; **e**—fault scarp; **f**—degraded fault scarp; **g**—elevation point; **h**—main dome-shaped reliefs: 1 *Monte Furlano* (101 m), 2 *Monte Bagnolo* (124 m), 3 Unnamed (126 m), 4 *Monte dei Castelli* (112 m), 5 Unnamed (117 m), 6 Unnamed (117 m), 7 *Monte dei Serpenti* (109 m), 8 *Monti d'Arena* (91 m), 9 *Casina del Vento* (107 m), 10 *Monte del Diavolo* (115 m), 11 Unnamed (103 m), 12 *Cannelli* (99 m), *Monte della Marina* (100 m).

Three narrow eastward sloping marine terraces mark the area stretching between the *Murge Tarantine* ridge and the shoreline. The first and the highest one is mainly a narrow abrasion platform cut on the limestone basement between 65 and 45 m above m.s.l. The second and the third marine terraces shows a thin sedimentary body covering Mesozoic limestone as well as Lower Quaternary deposits. The higher of two stretches between 32 m above m.s.l. to the west and 15 m above m.s.l. to the east whereas the lower one can be recognized between 12 and 3 m of elevation.

Relict dune belts are associated to marine terraces. They have been found at *Specchiarica* locality, where the maximum altitude of 27 m is reached, and near *Mass. Mirante* locality (19 m).

A continuous well-developed mid-Holocene dune belt, up to 14 m high and 150 m wide, borders the present shoreline [18].

5. The *Monte del Diavolo* Isolated Relief Morphology and Evolution

The *Monte del Diavolo* isolated relief is placed on the top surface of *Monti Castelli* limestone horst, close to two main fault scarps. It elevates about 20 m from a plain surface, stretching around 95 m altitude, reaching, at its top, 115 m above m.s.l.

The *Monte del Diavolo* shows an oblique conical shape; its ellipsoidal basal area has a major axis 1700 m long and a minor one 900 m long. The steeper southern slope has a mean inclination of 5.7%, whereas the northern gentler slope is about 2.8% (Figures 9 and 10).

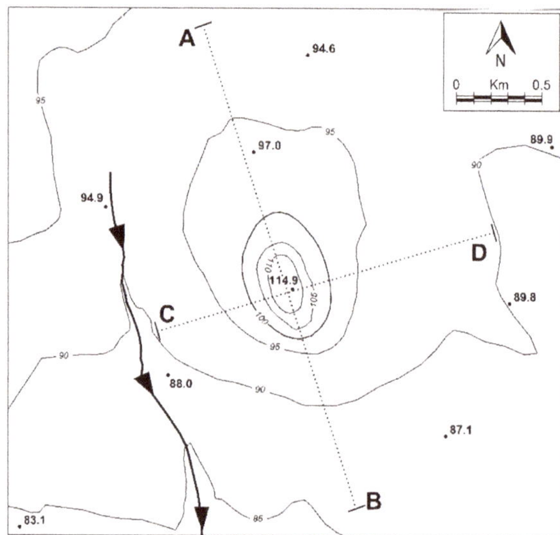

Figure 9. Contour lines at *Monte del Diavolo* locality. Dotted lines are the tracks of topographical profiles reported in Figure 10. Oriented bold line is a low incised stream.

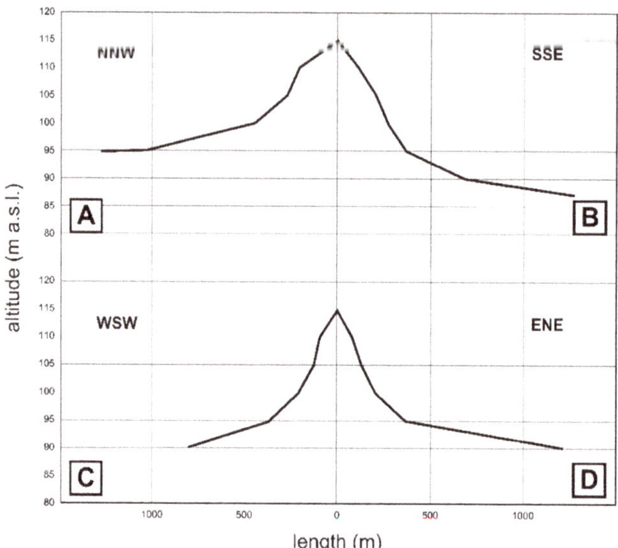

Figure 10. Topographical profiles of *Monte del Diavolo*.

In the area, Mesozoic limestones (*Calcare di Altamura* formation) widely crop out. However, at *Monte del Diavolo*, a massive and well-cemented grain-supported *breccia* made of hazel limestone clasts of decametric size was detected; voids are filled with a matrix-supported finer *breccia* marked by iron oxides and centimetric limestone clasts (Figure 11).

Figure 11. A view of the very well cemented breccia deposits cropping out at the top of *Monte del Diavolo*.

According to the classification of palaeo cave deposits proposed by [19] the detected clastic deposits can be referred to the coarse-clast chaotic *breccia* facies produced by cave ceiling and wall collapse. This facies is characterized by a mass of very poorly sorted, granule- to boulder-sized chaotic breccia clasts (i.e., clasts approximately 0.3 to 3 m long), that form a ribbon to tabular shaped body as much as 15 m across and hundreds of meters long.

The deposit is affected by a sub-vertical system of joints dipping northward (Figure 12); joints widening due to present karst solution is responsible for the detachment of *breccia* blocks by gravity (Figure 13).

Figure 12. A set of high angle joints dipping northward affects breccia deposits so that *Monte del Diavolo* developed an asymmetric transverse profile.

Figure 13. *Monte del Diavolo* slopes are covered by sparse breccia blocks due to rockfalls produced by the widening of joints by karst solution.

Unfortunately, the exposure conditions do not allow the geometry of *Monte del Diavolo breccia* deposits to be reconstructed. However, a sharp change in the slope angle would suggest that breccia deposits crop out in the upper half of the studied relief.

The lithological difference between the layered limestone bedrock cropping out widely around *Monte del Diavolo* area and the *breccia* deposits detected at its top could explain the genesis of this isolated relief (Figure 14). *Breccia* deposits are most likely the filling of karst underground voids and are more resistant to denudation processes than the limestone bedrock. The progressive lowering of ground surface, mainly due to surface karst processes, has been faster in the bedrock than areas where *breccia* deposits crop out so that an isolated relief gradually has been shaped. The asymmetry of the relief can be attributed to the high-angle joint system that affects *breccia* deposits miming stratification, so that a dip and a scarp slope developed.

Small isolated hills due to this particular evolution have been not recorded yet in classical karst areas. However, small hills with a conical or truncated conical shape, or similar to a dome or a tower, have been surveyed in the *Santa Ninfa* karst area (*Trapani* province, Sicily) [20]. Their origin is due to the evolution of jointed evaporitic limestone caprock which is deformed along its edge because of chemical erosion of underlying gypsum deposits. Chemical erosion is linked to water infiltrating through the carbonate caprock, which is higher in the area close to its border, whereas it is lower in its centre.

In particular, [21] defines karst features like *Monte del Diavolo* as "*breccia* pipe hills", which are areas of relatively high relief in gypsum karst areas that result from the selective denudation of *breccias* composed of different rock types. *Breccias* originate due to the collapse of gypsiferous beds and other rocks into cavities formed within deep-seated gypsum beds. Such *breccias* commonly have a pipe-like form and locally they can offer more resistance to erosion than do the surrounding rocks. In conclusion, the *Monte del Diavolo* isolated relief can be defined as a "*breccia* hills" due to differential karstic processes in a limestone area.

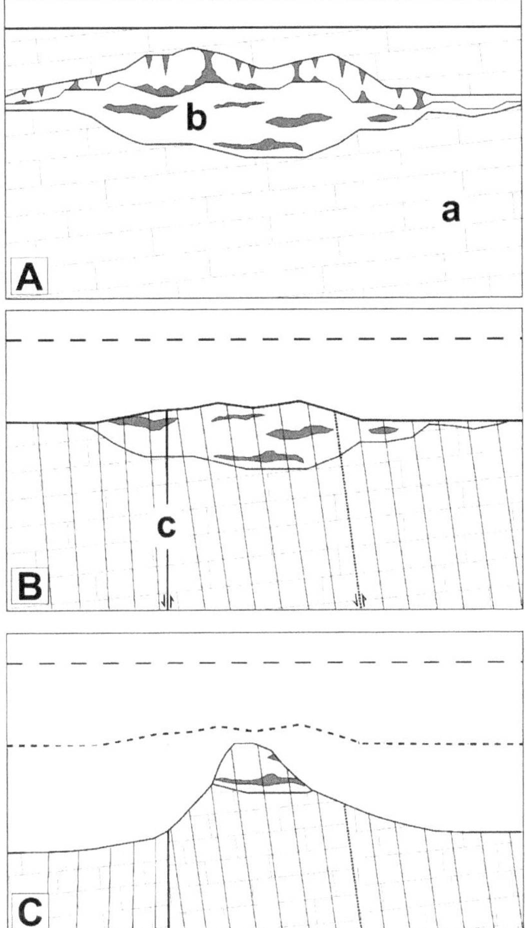

Figure 14. Model of the geomorphological evolution of *Monte del Diavolo.* **Stage A**: A cave develops in the vadose zone into the *Calcari di Altamura* limestone (**a**). The cave is partly filled with coarse *breccia* deposits and speleothems (**b**). **Stage B:** The lowering of the ground surface is accompanied by joints development due to fault activity (**c**). **Stage C**: Cave *breccia* deposits offer greater resistance to the karst dissolution process than surrounding limestones so that ground surface lowering led to the development of a small isolated relief, the *Monte del Diavolo.*

6. Developing Geotourism in the *Monte del Diavolo* Area

The *Monte del Diavolo* isolated relief could be a valuable source for the development of a local geotourism offer since it shows a number of favourable features. In particular:

(1)   Its name is particularly attractive;
(2)   it is part of the history of the local geological knowledge development;
(3)   it shows peculiar geological and geomorphological features;
(4)   it displays a unique natural scenery not affected by urbanization;
(5)   it is inside an archaeological area of great interest;
(6)   it comprises some natural reserve areas;

(7)   it is crossed by hiking trails and bike routes;

(8)   it is close to a coastal area already very popular to tourists.

The promotion of geotourism in the *Monte del Diavolo* area can be realized following different strategies dedicated to the two main types of involved tourists.

Maps and field guide books illustrating the area stretching from *Oria* relic belt to the coastline crossing the *Monte del Diavolo* area could attract "specialized" tourists which visit geosites for the purpose of personal educational or intellectual improvement and enjoyment. The availability of GPS tracks and waypoints as well as remote resources to deepen main topics would facilitate the self-guided visit of the area.

Guided group field trips along bike routes and footpaths should be the main strategy to promote geotourism in the *Monte del Diavolo*, involving casual tourists which visit geosites and geomorphosites mainly for recreation and pleasure. A geotourism guide can lead a group safely through existing trails, introducing, with a popular language, the main geological, geomorphological, cultural and environmental aspects of the area. In this way the recreation activity is enriched without further effort by a cultural experience. A visitors' centre at *S. Pietro in Bevagna*, the most frequented locality occurring along the coast, is needed to promote guided field trips and as a meeting point for interested people. Popular guides can be realized both to advertise field trips and to keep the memory of the lived experience.

7. Conclusions

A small isolated relief, the *Monte del Diavolo*, marks the morphological district of *Murge Tarantine*, an area of great geological interest since it shows the morphological effects of the North Salento Fault Zone, a dextral strike-slip fault of regional importance.

The geomorphological survey of *Monte del Diavolo* revealed that its peculiar morphology can be referred to the presence of well cemented karst breccia deposits produced by cave ceiling and wall collapse. The major resistance to the karst solution of *breccia* deposits compared to surrounding limestones would be responsible for the shaping of the unusual isolated relief.

Monte del Diavolo is placed in an area marked by a valuable archaeological, cultural and natural heritage that has been unexploited so far notwithstanding the remarkable tourism flow along the nearby coastline. Thanks to its peculiar morphology and name, the *Monte del Diavolo* isolated relief could represent the focal point of bike routes and footpaths crossing the area between the *Oria* relic dune belt and the coastline. Maps and field guide books illustrating the main geological and geomorphological features of the area could be produced for "specialized" tourists, whereas "casual" tourist can take advantage of guided field trips, thus joining recreation activity with geotourism.

Funding: This research received no external funding.

Conflicts of Interest: The author declares no conflict of interest.

References

1.   Hose, T.A. 3G's for modern geotourism. *Geoheritage* **2012**, *4*, 7–24. [CrossRef]

2.   Hose, T.A. Three centuries (1670–1970) of appreciating physical landscapes. In *Appreciating Physical Landscapes: Three Hundred Years of Geotourism*; Special Publications 417; Hose, T.A., Ed.; The Geological Society: London, UK, 2016; pp. 1–22.

3.   Michniewicz, A.; Różycka, M.; Migoń, P. Granite tors of Waldviertel (Lower Austria) as sites of geotourist interest. *Geotourism* **2015**, *1–2*, 19–36. [CrossRef]

4.   Martini, G.; Alcalá, L.; Brilha, J.; Iantria, L.; Sá, A.; Tourtellot, J. Reflections about the geotourism concept. In Proceedings of the 11 European Geoparks Conference, AGA—Associação Geoparque Arouca, Arouca, Portugal, 19–21 September 2012; pp. 187–188.

5.   Gray, M. *Geodiversity: Valuing and Conserving Abiotic Nature*, 2nd ed.; Wiley Blackwell: Hoboken, NJ, USA, 2013; p. 495.

6. Fiore, A.; Valletta, S. 2010: Il patrimonio geologico della Puglia. Territorio e geositi. *Suppl. Geol. dell'Ambiente* **2010**, *4*, 1–160.
7. Mastronuzzi, G.; Valletta, S.; Damiani, A.; Fiore, A.; Francescangeli, R.; Giandonato, P.B.; Iurilli, V.; Sabato, L. *Geositi Della Puglia*; Regione Puglia: Bari, Italy, 2015.
8. Milano, M. *Cenni Geologici Sulla Provincia di Terra d'Otranto*; Mausi, G., Ed.; Masi: Livorno, Italy, 1820; p. 56.
9. Alessio, A. Li Castelli. In *Archeologia dei Messapi*; Edipuglia: Bari, Italy, 1990; pp. 1–350.
10. Regione Puglia. *Osservatorio Turistico Regionale*; Regione Puglia: Bari, Italy, 2019.
11. Mastronuzzi, G.; Sansò, P. The Salento Peninsula (Apulia, Southern Italy): A Water-Shaped Landscape Without Rivers. In *Landscapes and Landforms of Italy*; Soldati, M., Marchetti, M., Eds.; Springer: Berlin/Heidelberg, Germany, 2017; pp. 421–430.
12. Ricchetti, G.; Ciaranfi, N.; Luperto Sinni, E.; Mongelli, F.; Pieri, P. Geodinamica ed evoluzione sedimentaria e tettonica dell'Avampaese Apulo. *Mem. Soc. Geol. It.* **1988**, *41*, 57–82.
13. Tozzi, M. Assetto tettonico dell'avampaese apulo meridionale (Murge meridionali-Salento) sulla base dei dati strutturali. *Geol. Romana* **1993**, *29*, 95–111.
14. Gambini, R.; Tozzi, M. Tertiary geodynamic evolution of the Southern Adria microplate. *Terranova* **1996**, *8*, 593–602. [CrossRef]
15. Ciaranfi, N.; Pieri, P.; Ricchetti, G. Note alla Carta Geologica delle Murge e del Salento (Puglia Centro-Meridionale). *Mem. Soc. Geol. It.* **1992**, *41*, 449–460.
16. Hearty, P.J.; Dai Pra, G. The age and stratigraphy of Middle Pleistocene and Younger deposits along the Gulf of Taranto (Southeast Italy). *J. Coast. Res.* **1992**, *8*, 82–105.
17. Belluomini, G.; Caldara, M.; Casini, C.; Cerasoli, M.; Manfra, L.; Mastronuzzi, G.; Palmentola, G.; Sanso', P.; Tuccimei, P.; Vesica, P.L. Age of Late Pleistocene shorelines, morphological evolution and tectonic history of Taranto area, Southern Italy. *Quat. Sci. Rev.* **2002**, *21*, 427–454. [CrossRef]
18. Mastronuzzi, G.; Sansò, P. Holocene coastal dune development and environmental changes in Apulia (southern Italy). *Sediment. Geol.* **2002**, *150*, 139–152. [CrossRef]
19. Loucks, R.G.; Mescher, P. Paleocave facies classification and associated pore types—American Association of Petroleum Geologists. In Proceedings of the Southwest Section, Annual Meeting, Dallas, TX, USA, 11–13 March 2001; p. 18.
20. Agnesi, V.; Macaluso, T.; Meneghel, M.; Sauro, U. Geomorfologia dell'area carsica di S. *Ninfa. Mem. Ist. Speleol.* **1989**, *3*, 23–48.
21. Sauro, U. Geomorphological aspects of gypsum karst areas with special emphasis on exposed karst. *Int. J. Speleol.* **1996**, *25*, 105–114. [CrossRef]

Article

Geoheritage as a Tool for Environmental Management: A Case Study in Northern Malta (Central Mediterranean Sea)

Lidia Selmi [1], Paola Coratza [1,*], Ritienne Gauci [2] and Mauro Soldati [1]

1   Department of Chemical and Geological Sciences, University of Modena and Reggio Emilia, Via Campi 103, 41125 Modena, Italy; lidia.selmi@unimore.it (L.S.); soldati@unimore.it (M.S.)
2   Department of Geography, Faculty of Arts, University of Malta, MSD 2080 Msida, Malta; ritienne.gauci@um.edu.mt
*   Correspondence: paola.coratza@unimore.it; Tel.: +39-059-2058448

Received: 1 October 2019; Accepted: 22 October 2019; Published: 26 October 2019

Abstract: The recognition, selection and quantitative assessment of sites of geological and geomorphological interest are fundamental steps in any environmental management focused on geoconservation and geotourism promotion. The island of Malta, in the central Mediterranean Sea, despite having a steadily increasing growth in population and tourism, still conserves geological and geomorphological features of great relevance and interest, both for their contribution to the understanding of the geological processes acting through time on landscape and for their aesthetic importance. The present work proposes an inventory for northern Malta, through three main stages, with the outcome of a final list of geosites that have the potential to be recognized as both natural heritage and tourist resources with potential economic benefits. In particular, the assessment methodology applied combines scientific value and additional and use-values, showing the links existing between geoheritage and other aspects of nature and culture of the sites. The results provide useful knowledge for the definition of strategies aimed at the development of a sustainable and responsible tourism.

Keywords: geoheritage; geosites; quantitative assessment; Malta

---

1. Introduction

Recent global trends have shown heightened appreciation of the variety of abiotic natural resources, known as geodiversity. This variety of non-living natural resources is defined by Gray [1] as the natural range (diversity) of geological, geomorphological and soil features. It describes the diversity of physical processes operating on Earth and the resultant rocks, minerals, fossils, sediments, soils, landforms, landscapes and habitats found on the world's surface today [1–3]. Geodiversity, a resource still little known and which can create potential economic growth that has been largely untapped, allows for the definition of geosites, that together form the geological heritage. In this regard, geoheritage is considered as a natural resource and can be used in local and regional development, especially for promoting a territory for geotouristic purposes [4,5].

The Maltese archipelago, which lies at the center of the Mediterranean Sea, is a European country with a rich cultural heritage endowed with a great variety of natural features of international significance. Indeed, the small geographic scale of the islands is inversely proportional to the richness and frequency of places and artefacts of major importance, and it encompasses, as well, a large number of sites of geoscientific interest, showing a considerable geodiversity. This applies in particular to northern Malta, a sector of the island moderately populated, but which still conserves landscapes of great relevance and interest from a scenic and scientific point of view. These sites are mainly located along coastal areas, and have to co-exist with the island's main economic activity of tourism. This industry has in fact been capitalizing on some of the most impressive coastal sceneries of the Maltese archipelago for over half a century. However, there is still remarkable potential on how the rich natural and cultural heritage of the archipelago is valued and promoted especially with regard to its geological and geomorphological heritage.

It is a widely shared opinion that any action aiming to promote or protect geoheritage implies a good knowledge of the resource in terms of its location and characteristics. For this reason, an inventory, based on the analysis and assessment of the most valuable elements that define the geoheritage of a territory, represents the first necessary step towards its effective management. A number of European countries have already carried out a similar national inventory, such as Czech Republic, Denmark, Estonia, Finland, France, Iceland, Ireland, Italy, Lithuania, Netherlands, Poland, Portugal, Slovakia, Spain, Switzerland and United Kingdom [6]. More work is, however, required on a global scale.

Recently in Malta, considerable geological and geomorphological research, especially in the north of the archipelago, has been undertaken by scientists in order to showcase the international geological and geomorphological significance of Maltese landscapes [7–13]. Nevertheless, the Maltese Islands still lack an official inventory of sites of geological interest and the government has not yet assigned geological heritage as a specific (or separate) legal provision related to the conservation and management of natural sites. Though the Maltese natural landscapes are governed by a comprehensive legal framework, such instrument mainly (but not only) sustains the importance of biodiversity and ecological conservation at local and international levels. Recently, efforts to recognize elements of geological heritage of the Maltese Islands were primarily channeled to urban landscapes, through the historical and cultural use of the Maltese Lower Globigerina Limestone over the centuries for heritage buildings. These efforts resulted in this limestone unit receiving the status of Global Heritage Stone Resource (GHSR) by the International Union of Geological Sciences in 2019.

In this context, a study for the inventory and assessment of sites of geological interest, highlighting their location and characteristics (e.g., integrity, state of activity, attractiveness and accessibility) in the northern part of the island of Malta has been conducted and the results are here presented. This work aims at providing a better understanding of the geological and geomorphological characteristics of the study area and facilitating the recognition of the opportunities and threats, in order to strengthen the argument for the setting-up an effective environmental management plan, which would directly include both geoconservation and geotourism actions.

2. Maltese Context

Despite the small geographic size of the archipelago, the protection of the natural heritage of the Maltese Islands is governed by a fair number of main legislative acts, related legal chapters and subsidiary legislation (Table 1). These legal instruments are regularly updated in order to transpose European and international laws, mainly from the United Nations (including the Mediterranean Action Plan), the Council of Europe and the European Union [14]. A number of subsidiary legislations are also in force (Table 1), a few of which have replaced earlier legal notices, in order to also transpose international legal obligations into national law.

The Environment Protection Act is the main legal instrument that safeguards the protection of the 'landscape and its features' under the relatively broad umbrella term of 'environment'. A number of natural landscape features are classified as areas of high landscape value (AHLV) under the Development and Planning Act, mainly coastal cliffs, valley systems, karstic plateaus, escarpments, woodland and agricultural settings. Most of these natural features intrinsically incorporate geological and geomorphological properties; however, the value of these features is primarily recognized for its support function to biodiversity and ecological systems, rather than specifically (or exclusively) for their geological properties in their own right. Under the Cultural Heritage Act, the definition of cultural heritage also includes 'geological sites and deposits' and 'landscapes'; however, the act has no specific provisions related to their geoheritage value. The Fertile Soil (Preservation) Act primarily addresses the maintenance of terraced landscapes, so typical in Malta's rural setting, by offering direct protection to soil as a resource. A number of islets around the Maltese Islands, such as Filfla and St. Paul's Islands have been legally established as nature reserves and limiting human access only for scientific purposes. In addition to that, 13.1% of terrestrial areas of the Maltese Islands and 35% of their territorial waters form part of the EU Natura 2000 Network as protected areas under various designations (Table 2, [15]).

Table 1. Maltese legal instruments related to the natural landscape management and protection.

Type of Legal Instruments	Designations
Acts	Environment Protection Act (Chapter 549) Development Planning Act (Chapter 552) Cultural Heritage Act (Chapter 445) Fertile Soil (Preservation) Act (Chapter 236) Filfla Nature Reserve Act (Chapter 323)
Subsidiary Legislations	Flora, Fauna and Natural Habitats Protection Regulations (SL 549.44) Trees and Woodland Protection Regulations (SL 549.64) Selmunett Islands (St. Paul's Islands) Nature Reserve Regulations (SL 549.03) Fungus Rock (il-Ġebla tal-Ġeneral) Nature Reserve Regulations (SL 549.01) Motor Vehicles Off-roading Regulations (SL 552.01) Rubble Walls and Rural Structures (Conservation and Maintenance) Regulations (SL 552.02). Conservation of Wild Birds Regulations (SL 549.42) Establishment of the Majjistral, Nature and History Park Regulations (SL 549.48) Establishment of the Park Nazzjonali tal-Inwadar Regulations (SL 549.109) Protected Beaches (SL 549.42) Tree Protection Areas (SL 549.123)

Table 2. The number of protected sites according to designation type (Source: Compiled from the Environment and Resource Authority (ERA) [15]).

Designation Type	Number of Sites
Tree Protection Areas	60
Area of Ecological Importance and Site of Scientific Importance	41
Special Areas of Conservation - International Importance	35
Bird Sanctuary	26
Area of Ecological Importance	22
Special Protection Areas	21
Area of High Landscape Value	13
Protected Beaches	11
Site of Scientific Importance	10
Special Areas of Conservation—National Importance	7
List of Historical Trees Having an Antiquarian Importance	6
Nature Reserve	3

The legal framework of natural heritage protection of the Maltese Islands is thus a mosaic of different provisions, with a number of sites protected by more than one designation (Table 2). Within this legal context, the importance of geoheritage as a conservation rationale remains, however, diluted, when compared with that for ecological and biodiversity protection. Despite this, the interest of the scientific community in geoheritage and geotourism has been growing over a number of years. With respect to the Maltese archipelago, the importance of developing studies to investigate the linkage between environment and cultural heritage and the relationship between geoheritage and tourism was initially explored in April 2007 during the International Workshop on the 'Integration of the geomorphological environment and cultural heritage for tourism promotion and hazard prevention' held in Malta [16,17]. The papers presented dealt with different aspects of the integration of the physical environment and cultural heritage through case studies from different parts of the world including Malta (e.g., [18]). More recently, Gauci et al. [11] and Gauci and Inkpen [19] have highlighted the geoheritage value of shore platforms in Malta by examining the close relationship between the physical landscape of the foreshore and human cultural development. The significance of Maltese coastal landforms for societal wellbeing was also investigated by Satariano and Gauci [20] who examined the intense reactions experienced by both the Maltese and international community following the sudden loss of an iconic sea arch at Dwejra (Gozo) in March 2017. This latter work forms part of a collection of contributions recently edited by Gauci and Schembri [13] and which illustrate the rich diversity of the Maltese physical landscapes under the World Geomorphological Landscapes series (Springer). Specific studies on geoheritage and geosites inventory and assessment have been carried out on the north-west coast of Malta, especially in the area of Il-Majjistral Nature and History Park and environs [7,12]. With respect to the island of Gozo, this theme was explored by Coratza et al. [8] who examined spectacular sinkholes having highly scientific, ecological, aesthetic, cultural and use-values as geomorphosites. Specific research on Dwejra area, on the western coast of Gozo [9,21], has highlighted how the integration of environmental and cultural heritage aspects makes this area a site of remarkable value to be promoted for a more holistic and varied tourism.

3. Study Area

The study area is located in the north of Malta, the largest island of the Maltese archipelago (Figure 1). It is sparsely inhabited and characterized by a high tourism vocation. According to the National Tourism Policy 2015–2020, northern Malta is defined as a 'tourism zone' due to its tourism infrastructures, hosting a further 42% of tourist accommodation [22].

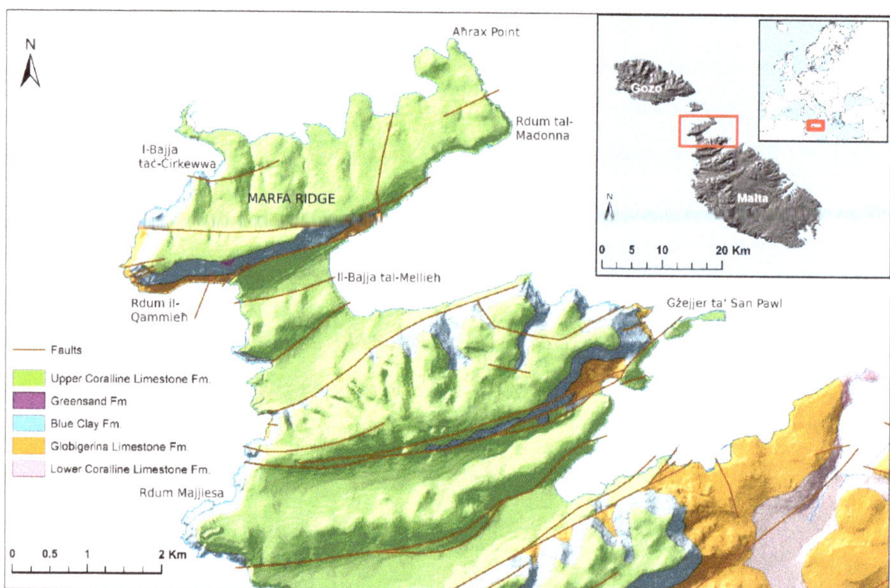

Figure 1. Location and geological setting of the study area.

The island attracts many tourists, also thanks to its mild Mediterranean climate characterized by an average rainfall of 530 mm per year and mean temperatures ranging from 12 to 27 °C.

The rocks exposed in the island comprise a marine sedimentary succession, mostly composed of limestones and marls and deposited in a period between Upper Oligocene and Miocene [23,24]. In the study area, all five geological formations constituting the Maltese archipelago outcrop (Figure 1). From the oldest to the youngest the formations are Lower Coralline Limestone Fm., Globigerina Limestone Fm., Blue Clay Fm., Greensand Fm. and Upper Coralline Limestone Fm. (Figure 2).

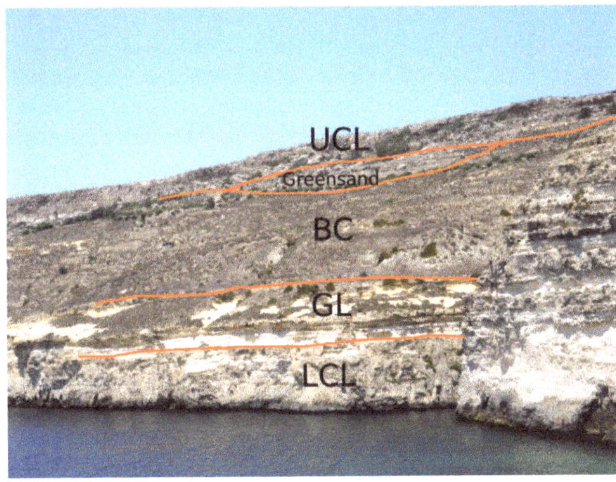

Figure 2. View of Il-Qammieħ, showing the entire geological/stratigraphic sequence. From the bottom: Lower Coralline Limestone Fm. (LCL), Globigerina Limestone Fm. (GL), Blue Clay Fm. (BC), Greensand and Upper Coralline Limestone Fm. (UCL).

The Lower Coralline Limestone Fm., composed of pale grey, hard, shallow marine biomicrites and biospartites [23,25], outcrops in a restricted coastal stretch between Rdum il-Qammieħ and Iċ-Ċumnija, in the eastern part of the study area. The sequence continues with the soft and yellowish Globigerina Limestone Fm., named on account of the high percentage of planktonic foraminifera present in the unit (Figure 3a). The usual color of the formation is pale-yellow, although a pale-grey subdivision bounded both above and below by phosphorite conglomerate horizons, occurs in the middle of the sequence [25,26]. It outcrops on the Ras il-Qammieħ coast and in Selmun Bay, in proximity of St. Paul's Islands.

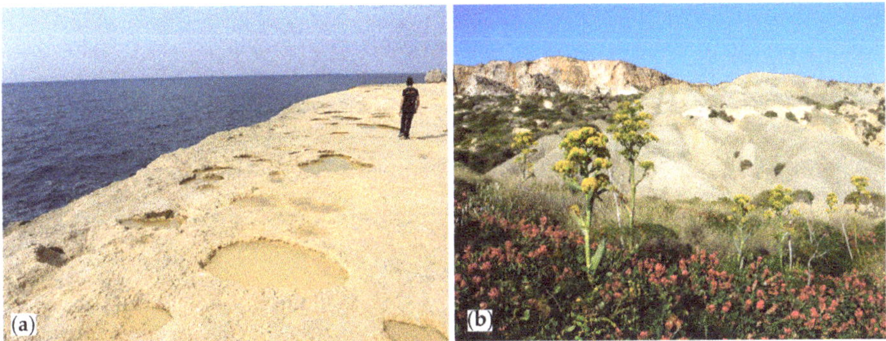

Figure 3. Landscape features of the study area: (**a**) Terrace in Lower Globigerina Limestone at Il-Qammieħ with typical honey pots dissolution structures; (**b**) badland topography in Blue Clay slopes overlain by Upper Coralline Limestone cliffs at Il-Qammieħ.

It is followed by the Blue Clay Fm., formed in a deep-sea depositional setting and is made up of fine-grained sediments with a large component of organic material derived from planktonic organisms.

It consists of sequences of alternating pale-grey and dark-grey banded marls (Figure 3b), with lighter bands containing a higher proportion of carbonate [27]. The uppermost part of the Blue Clay Fm. Shows an increase in brown phosphatic sand grains and green grains of glauconite, together with abundant fossil fragments, often separated by an erosional surface. This level is known as Greensand Fm. and underlines the passage to the overlying Upper Coralline Limestone Fm. The fossiliferous content is mostly represented by mollusks, gastropods, brachiopods, echinoids, bryozoans, algae, shark teeth, and remains of marine mammals [23,24,28]. It shows its maximum thickness of 11 m in Gozo, but the formation is rarely thicker than one meter in the area under study at Il-Qammieh point. The upper part of the sequence is made up of the Upper Coralline Limestone Fm., a hard, pale grey limestone unit, very similar to the Lower Coralline Limestone especially in color and coralline algal content, of shallow water environment. It usually makes up plateaus and steep cliffs affected by weathering and mass movements [26]. It is often affected by a dense network of tectonic discontinuities which provide the rock masses with a brittle behavior (Figure 3b) [29,30]. This formation largely covers the study area, with a thickness even higher than 100 m.

The geological formations lie almost horizontally across the islands, although they are displaced by tectonic structures [25,31,32]. From a tectonic viewpoint, the archipelago is crossed by two fault systems, the NW-SE trending Pantelleria Rift and the WSW-ENE graben system [23]. The latter is the most ancient and is responsible for a horst and graben structure that characterizes the northern sector of the island of Malta [33,34]. Indeed, the study area is part of the North Malta Graben, one of the three main structural regions of the Maltese Islands. The North Malta Graben is characterized by typical ridge-trough morphology and bounded by the Great Fault to the south [32].

The geomorphological landscape is largely controlled by the different physical and mechanical properties of the lithostratigraphic units and by tectonic features.

The coastal landscape is mainly shaped by marine processes, that produce inlets and bays with small pocket beaches [35–37]. Due to the presence of resistant conglomerate beds and hardgrounds within the stratigraphy of Globigerina Limestone, a number of shore platforms have developed at sea level as a result of differential erosion [19]. On the contrary, plunging cliffs are the dominating features in Upper Coralline Limestone, at times shaped in sea caves. Mass movements are widespread all along the northwestern part of the study area, due to the fragile behavior of limestones, which cap Blue Clay Fm. characterized by visco-plastic properties. Rock falls and topples are abundant along the coastline and mainly affect the Upper Coralline Limestone plateaus which are characterized by persistent fissures and cracks of tectonic origin [29,37–41]. Evidence of rock spreading and block sliding phenomena characterize the stretch of coast at Rdum il-Qammieħand Rdum il-Qawwi, in the northwestern part of the Marfa Ridge Peninsula, and at Rdum il-Majjiesa, located inside the Il-Majjistral Park boundaries. The lateral extension of rock masses tends to evolve into block sliding whose onset is extensively witnessed by scattered blocks of variable size lying on the Blue Clay slopes which gently slide toward the sea and protect the shoreline from the marine erosion (Figure 4a) [29,42–44].

Karstic features are well developed on the surface topography of plateaus, characterized by highly irregular and rugged surface morphology, resulting from solution processes. Karst pavements, solution holes and solution pans are also particularly relevant. Sinkholes have been found in the area, usually caused by the collapse of cave roofs (Figure 4b). They are characterized by a flat bottom and may reach a few hundreds of meters in diameter and stratigraphic throw [45,46].

The area under study is relatively less urbanized compared with the rest of the island, but has been significantly influenced over time by human activity for agricultural and tourism purposes [47]. Coastal and inland slopes have been remodeled into terraced fields retained by dry stone walls and utilized as terraced agricultural land [48,49]. The terraced fields and agricultural land are usually installed on V-shaped dry valleys, relict of former pluvial conditions and extensive groundwater sapping. The presence of archaeological features and British military architecture can also be encountered.

Figure 4. Landscape views of the study area: (**a**) Aerial photo of Rdum il-Qammieħ, showing the impressive rock fall and block slides, typical of the area; (**b**) remarkable example of karstic feature in Upper Coralline Limestone Fm. at Aħrax point.

4. Materials and Methods

During the last 30 years, the increasing interest in geoheritage has led to the development of methodologies for its inventory and assessment [50,51] and references therein. In fact, the scientific literature is rich in examples of geosite inventories both at national (e.g., [52–56], regional (e.g., [51,57–59]) and local scale (e.g., [60–62]). Numerous methods are described in literature for the qualitative and quantitative assessment of geoheritage and geosites in various contexts (cf. [63,64]): Environmental Impact Assessment and territorial planning (e.g., [65–68]); inventory of natural heritage sites (e.g., [53,58,69–71]); tourist promotion (e.g., [72–76]); management of nature parks and geoheritage (e.g., [77–81]). A complete review of methods for the assessment of geosites has been recently published by Brilha [82]. In general, it should be emphasized that all methods inevitably imply a degree of subjectivity since their intrinsic value cannot be measured. In order to reduce subjectivity and properly evaluate the various components of a geosite, it is necessary to define clear and transparent criteria, which can vary according to the aim, working scale and subject of the assessment. Even though there is no generally accepted method for the numerical assessment of geosites, recurrent criteria are used in literature, such as rarity, representativeness and integrity, ecological value, paleogeographic importance, educational value etc. [64].

Based on published literature, as well as on knowledge achieved in previous research on geoheritage in various morphoclimatic contexts, the methodological approach adopted for the identification of geosites in northern Malta comprises the following operational phases (Figure 5): (i) Recognition and selection of sites of geological and geomorphological interest (i.e., potential geosites), based on their representativeness in terms of geohistory and geo(morpho)diversity [51,79]; (ii) analysis and characterization of potential geosites; (iii) quantitative assessment of potential geosites and final selection of geosites.

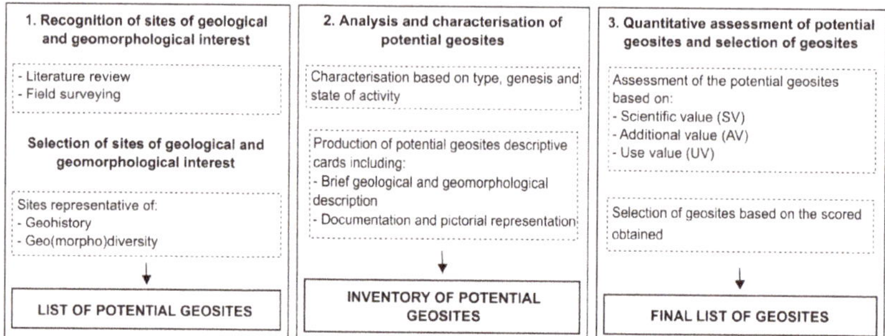

Figure 5. The three stages of the methodological approach.

4.1. Recognition and Selection of Sites of Geological and Geomorphological Interest

In order to recognize sites of geological and geomorphological interest, the first phase consists of a literature review of papers and maps of the area under study and field surveys. A number of papers dealing with the geological and geomorphological features of the Maltese archipelago compiled in the last decades are available, some of which specifically devoted to the geoheritage of the northwestern sector of the island.

Literature review and field survey are fundamental for the recognition of sites of geological and geomorphological interest to be qualitatively assessed, considering the different morphoclimatic conditions, geomorphological processes and lithological and structural constraints that controlled their development. This enables us to account for a variety of features that can finally be considered as geosites. Two main criteria have been taken into account in the assessment procedure (cf. [51]):

- The sites have to be representative of the geo-history and geomorphological evolution of the study area at a regional scale. Both active and inherited geological and geomorphological features can be considered as potential geosites.
- The sites have to represent the regional geo(morpho)diversity, i.e., a complete set of geomorphological processes that acted over time in the study area. Unique or rare landforms, as well as more common and abundant ones, can be useful to provide an overview of the landforms visible in the area (cf. [12]).

4.2. Analysis and Characterisation of Potential Geosites

The second phase foresees the analysis and characterization of potential geosites to be selected among the sites of geological and geomorphological interest previously identified. The analysis provides for the identification of a series of parameters characterizing each potential geosites. These parameters are collected in a descriptive card including elements of textual description and pictorial data. In particular, each descriptive card collects the following headings:

(1) Feature: name of the potential geosite;
(2) Location: as precise as possible;
(3) Coordinates: international system;
(4) Type (according to [51,58,60,83,84], distinguished on its geometrical characters in: (i) punctiform, small-size isolated single form or object (e.g., a sinkhole or a spring); (ii) linear, one or more simple forms developed preferentially in a single direction (e.g., a canyon, or a paleo riverbed) and/or stratigraphical sequences; (iii) areal: a set of large simple landforms related to just one type of genetic process (e.g., a karren field);
(5) Lithology;

(6) Genesis/main interest: e.g., tectonic, geomorphological, stratigraphic; regarding the geomorphological interest, a morphogenetic division related to a group of processes (coastal, fluvial, karstic, gravity-induced etc.) can be applied;

(7) State of activity (e.g., [85–93]): active sites, those that allow the visualization of geological and geomorphological processes in action (e.g., fluvial systems); inherited sites defined as inherited landforms, which testify to past processes and have a particular heritage value since they are symbols of Earth's history and evolution (e.g., stack);

(8) Brief geological and geomorphological description based on field observations and literature survey;

(9) Documents, archive material and pictorial representations: e.g., photographs, sketches.

4.3. Quantitative Assessment of Potential Geosites and Selection of Geosites

The employment of a quantitative assessment is considered necessary in order to decrease the subjectivity associated with any evaluation. The methodology adopted by Coratza et al. [8], already applied with positive results on the northwestern coast of the island of Malta, in a similar geological and geomorphological context [12], has been considered as the most suitable for the assessment of potential geosites. This methodology is inspired by methods previously proposed by Serrano and Gonzàlez Trueba [69], Bruschi and Cendrero [68], Pererira et al. [77] and Reynard et al. [70]. The geosite value assessment is based on 16 criteria divided into three main groups of value, i.e., scientific value (SV), additional value (AV) and use-value (UV), each one producing a final score for its category (Table 3). The scientific value aims to reveal the value of the site for the geosciences and it is assessed according to four criteria (paleogeomorphological model, rareness, representativeness and integrity) scored on a scale from 0 to 1. The additional value is linked to the importance that a geosite assumes owing to non-geological aspects which increase its overall value and is made up of three independent sub-values: ecological, aesthetic and cultural. The use-value refers to the possible utilisation of geosites by society. The scores given for each criterion are reported in Table 3.

Table 3. Values and criteria of geosite assessment methodology and related scores.

Value		Criteria	Score
Scientific value (SV)		Paleogeomorphological model	0–1
		Rareness	0–1
		Representativeness	0–1
		Integrity	0–1
Additional value (AV)	Ecological value	Ecological role support	0–1
	Aesthetical value	Panoramic quality	0–0.25
		Color diversity	0–0.25
		Vertical development	0–0.25
		Naturalness	0–0.25
	Cultural value	Religious importance	0–0.33
		Historical importance	0–0.33
		Artistic importance	0–0.33
Use value (UV)		Accessibility	0–0.75
		Visibility	0–0.75
		Services	0–0.75
		Importance for education	0–0.75

The value of a geosite results from the total of the scores obtained from all criteria, with 10 being the highest score possible. Once completed, the assessment will provide a set of total scores for each of the three observed values (scientific, additional and use-value). On the basis of both the range and the total of these scores, a series of score-defined thresholds were established in order to allow also the inclusion of sites, which though they may have limited scientific value, they nonetheless may hold potential for geotourism and educational activities. The score thresholds were established on both the basis of the highest scores and the scope of the study. The sites that reach such thresholds can be finally considered as geosites.

5. Results

5.1. Recognition and Selection of Sites of Geological and Geomorphological Interest

As a first stage, a literature review has been carried out referred to more than 50 scientific references comprising 13 theses, ca. 40 national and international papers, 5 geological and geomorphological maps and several reports of Maltese environmental agencies (Planning Authority, Environmental and Resources Authority, Malta Environment and Planning Authority). In particular, the scientific papers analyzed (Figure 6) deal with various geological aspects including geomorphology (33%), structural geology (19%), stratigraphy (10%), paleontology (7%), geoheritage (6%) and miscellaneous geological topics (25%).

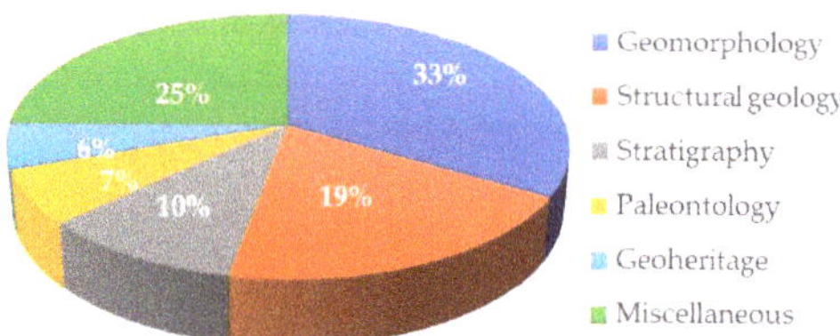

Figure 6. Distribution of geological literature according to the main topics of the scientific papers analyzed. The item miscellaneous comprises papers on geology l.s.

This detailed literature review combined with several field surveys led to the identification of sites in the study area with geological and geomorphological interest. The field surveys were essential to integrate the list of sites previously identified with new sites not mentioned in literature. In addition, field surveys were also fundamental to collect site-specific updated information—i.e., state of conservation, state of activity, accessibility, visibility and presence of services—relevant to the completion of the descriptive cards and the quantitative assessment of potential geosites.

Through literature review and field surveys, sites with geological and geomorphological interest were recognized and 31 were selected as potential geosites considering the two criteria mentioned in paragraph 4.1, i.e., geohistory and geo(morpho)diversity. The sites selected are representative evidence of the main geological and geomorphological processes acting through time in the study area (Figure 7).

5.2. Analysis and Characterization of Potential Geosites

The 31 potential geosites selected were analyzed and for each site a descriptive card has been compiled including the information reported in paragraph 4.2 and Figure 8. The data collected in this phase were stored in a GIS database.

Figure 7. Location of the 31 sites selected within the study area. The numbers correspond to the ID of the sites.

Figure 8. Example of a descriptive card of a potential geosite.

Regarding the type of sites (Figure 9a), 19 sites were classified as areal (61%), 10 sites as punctiform (31%) and 2 sites as linear (7%). It can be stated that the selected sites refer to the three main lithological formations of the area under study (Upper Coralline Limestone Fm., Blue Clay Fm. and Globigerina Limestone Fm.) and most of the sites consist of two or more different lithologies (Figure 9b). Regarding the main scientific interest, 26 of the selected sites have mainly geomorphological interest (84%), 3 sites display evidence of anthropogenic activity (10%) and the last 2 sites have tectonic origin (6%). As reported in Figure 9c, almost half of the geomorphological sites (45%, 14 sites) feature gravitational movements, followed by 7 sites (23%) shaped by sea action and 3 sites (16%) by karstic processes. Most of them are located along the coast (Figure 9d), where impressive lateral spreading phenomena dominate the landscape and where wave action and litho-structural processes shape cliffs and bays. For their representativeness, karst morphologies have also been selected, such as the surface topography on limestone plateaus that present small irregular rock pools colonized by typical Mediterranean vegetation and a large number of endemic communities [94]. Other selected sites in the area are two sinkholes at the eastern and western ends of the Marfa Ridge peninsula.

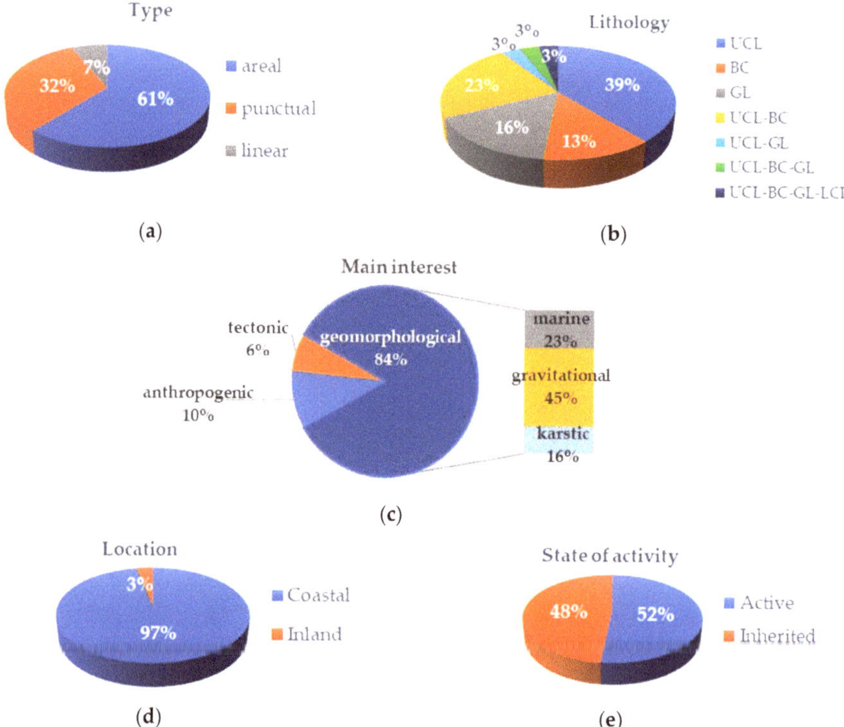

Figure 9. Distribution of potential geosites according to (**a**) type; (**b**) lithology; (**c**) main interest; (**d**) location; (**e**) state of activity.

Besides the sites with entirely natural origin, sites of geological and geomorphological interest strictly linked with the anthropogenic activity were selected. In fact, the geology has greatly influenced the location of settlement and activity of human civilization. The term 'anthropogenic site' was then used to differentiate these types of sites from the pristine ones. The best example is the large area of industrial salinas (ID26, ID27), not in use anymore, that covers approximately 1 km along the shore platforms of Blata l-Bajda in Selmun [95]. The rocky shore platforms in soft Globigerina Limestone developed the ideal coastal landscape for the formation of natural pools filled with seawater. This natural feature was extended and built from humans in order to collect seawater for the production of salt [11,19]. This site shows how geological features influence traditional practices and how sites with geological and geomorphological interest can be considered as part of the cultural heritage. Another example of anthropogenic site is the presence of cart ruts (ID12), on Wied Musa battery. This site is evidence of ancient agricultural civilizations that hewn the rock below the field, using a slide-car or wheeled cart.

Two sites have been chosen mainly for the geological/geotectonic interest such as St. Paul's Islands that are crossed by one of the major SW-NE faults in the island and which affected the horizontal transition between Upper Coralline Limestone and Upper Globigerina Limestone. The 52% of the sites (16 sites) are active landforms which provide clear evidence of geological and geomorphological processes in action. The remaining 48% (15 sites) consist of inherited landforms, that testify to inactive processes which are evidence of past geological and geomorphological processes (Figure 9e).

5.3. Quantitative Assessment and Selection of Geosites

The 31 potential geosites have been assessed through the methodology described in paragraph 4.3 in order to establish the final selection of geosites.

Once the potential geosites have been evaluated, the total scores for each value (scientific, additional and use-value) were plotted on a graph plane according to the cartesian coordinate system. The total scores of the scientific value plus additional and use values (combined) were plotted on the cartesian plane as x-axis and y-axis respectively (Figure 10). A score value of ≥4.5 was established for the total value, along which to define the potential geosites as (final) geosites, provided that the scientific value was ≥2.0.

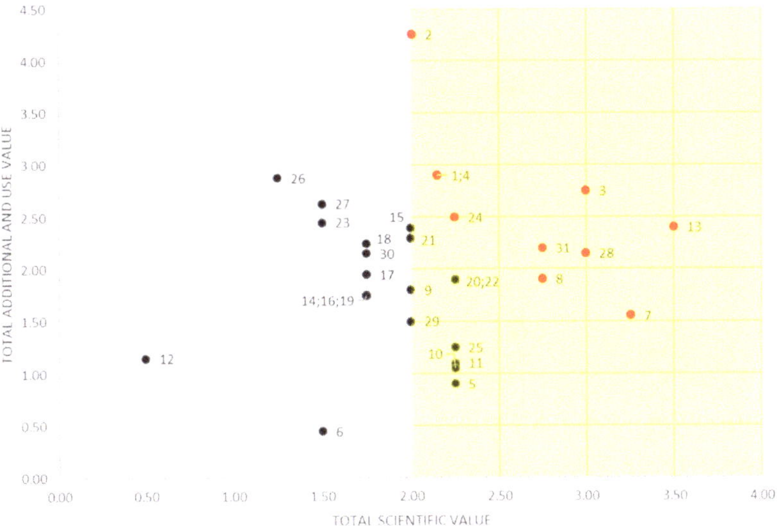

Figure 10. Total scientific value vs total additional and use-value of potential geosites. Sites finally selected as geosites are those displaying a scientific value of ≥2.0 and a total value of ≥4.5 (red dots).

The results are presented in Table 4 where the values of each geosite are shown. 10 sites have been selected as geosites for the high score in scientific interest and total additional and use-value. Not only sites with high scientific value were selected, but also sites with potential as geotourist destination and ideal for educational activities, according to the aim of the present research. As shown in Figure 10, two salinas at Blata l-Bajda (ID26 and ID27), despite the high potential as tourist attractions, were not selected as geosite due to the lack of relevant scientific importance. Instead, areas affected by rock spreading (ID1 and ID4) and rock topple (ID2), even though the low relevance as scientific site, are selected as geosite due to their important score in additional and use-value. All the identified geosites are examples that well represent geohistory and geo(morpho)diversity of the study area and are capable of being exploited as geotourist resources. The sinkhole at Il-Ponta tal-Ahrax (ID 13) is considered the only occurrence of this type in the island of Malta and the fault at Il-Qammieh (ID31) is the only spot where all the Maltese geological formations outcrop in the study area. Considering the use-value, the rock topple at Il-Bajja tać-Ċirkewwa (ID2) is the only site presenting a complete range of services and facilities, thus having the possibility to host geotourism activities. Almost all the sites are accessible without obstacles, except ID24 that is located on St. Paul's Islands, a protected nature reserve with limited access. All the sites have educational potential at different levels. All the geosites show a high total aesthetic value, making them attractive also to a public of non-specialists.

Table 4. Final quantitative assessment of potential geosites (sites finally selected as geosites are highlighted in yellow).

ID	Feature	Location	Geosite Values				
			SV	AV	UV	AV+UV	SV+AV+UV
1	Area affected by rock spreading	Ta' Qassisu	2.15	1.15	1.75	2.90	5.05
2	Rock topple	Il-Bajja tać-Ċirkewwa	2.00	1.25	3.00	4.25	6.25
3	Sinkhole	Ċirkewwa	3.00	1.10	1.65	2.75	5.75
4	Area affected by rock spreading	Rdum il-Qawwi	2.15	1.40	1.50	2.90	5.05
5	Rock window	Ta' Qassisu	2.25	0.65	0.25	0.90	3.15
6	Marine cave	Ġebel Imbark	1.50	0.30	0.15	0.45	1.95
7	Lower Globigerina Limestone terrace	Rdum il-Qammieħ	3.25	1.15	0.40	1.55	4.80
8	Badland topography in Blue Clay slopes	Rdum il-Qammieħ	2.75	1.00	0.90	1.90	4.65
9	Dissolution structure (Globigerina pavement)	Rdum il-Qammieħ	2.00	1.30	0.50	1.80	3.80
10	Shore platform	Ras il-Qammieħ	2.25	1.10	0.00	1.10	3.35
11	Badland topography in Blue Clay slopes	Ras il-Qammieħ	2.25	0.80	0.25	1.05	3.30
12	Cart ruts	Il-Palazz tal-Marfa	0.50	0.25	0.90	1.15	1.65
13	Sinkhole	Il-Ponta tal-Aħrax	3.50	1.00	1.50	2.40	6.00
14	Marine Cave	Rdum l-Aħmar	1.75	0.60	1.15	1.75	3.50
15	Karst landform (limestone pavement)	Aħrax Point	2.00	1.15	1.25	2.40	4.40
16	Rock topple	Rdum l-Aħmar	1.75	0.25	1.50	1.75	3.50
17	Rock topple	Rdum tal-Madonna	1.75	0.45	1.50	1.95	3.70

Table 4. *Cont.*

18	Area affected by rock spreading	Il-Marbat	1.75	0.75	1.50	2.25	4.00
19	Rock topple	Rdum il-Ħmar	1.75	0.50	1.25	1.75	3.50
20	Badland topography in Blue Clay slopes	Rdum il-Ħmar	2.25	0.75	1.15	1.90	4.15
21	Area affected by rock spreading	Il-Parsott	2.00	0.80	1.50	2.30	4.30
22	Sinkhole	Ta' L-Imgharrqa	2.25	0.65	1.25	1.90	4.15
23	Area affected by rock spreading	Rdum il-Bies	1.50	0.95	1.50	2.45	3.95
24	Fault	Gżejjer ta' San Pawl	2.25	1.00	1.50	2.50	4.65
25	Marine cave	Gżejjer ta' San Pawl	2.25	0.25	1.00	1.25	3.50
26	Salinas	Blata l-Bajda	1.25	0.88	2.00	2.88	4.13
27	Salinas	Blata l-Bajda	1.50	0.88	1.75	2.63	4.13
28	Badland topography in Blue Clay slopes	Tal-Blata	3.00	0.90	1.25	2.15	5.15
29	Tsunami deposit	Il-Ponta tal-Aħrax	2.00	0.25	1.25	1.50	3.50
30	Area affected by rock spreading	Ghajn Ħadid	1.75	1.15	1.00	2.15	3.90
31	Fault	Il-Qammieh	2.75	2.20	1.25	2.20	6.20

Description of the Geosites

The geosites finally selected are reported in descending order, considering the total score value achieved.

ID31: Il-Qammieh Fault

Il-Qammieh, on the south side of the Marfa Ridge, is one of the most striking geological features which exposes the entire Oligo-Miocene Maltese lithological sequence, including the Greensand Fm. The place is already designated under the Flora, Fauna and Natural Habitats Protection Regulations (SL 549.44) in view of its diverse and endemic ecology. Upper beds of the Lower Coralline Limestone are well exposed along the base of an elevated platform. The top of this formation is marked by the abundance of the echinoid *Scutella subrotunda* and constitutes the important marker Scutella Bed. The succession continues with the exposure of Globigerina Limestone Fm., all the three members, and passes transitionally up into the banded Blue Clay deposit. Overlying the Blue Clay Fm. there is approximately 1 m of the Greensand Fm., occurring as a friable, green and brown colored glauconitic micrite. On the top of this stratigraphic section, the Upper Coralline Limestone Fm. outcrops, typically cream colored by fossiliferous algal limestones (Mtarfa Member) containing abundant spherical rhodoliths [23,96]. The site is one of the most accessible and clear spots showing the intact transition of the five formations, suitable for educational activities (Figure 2).

ID13: Id-Dragonara Sinkhole

The site is a subsidence structure found at Il-Ponta tal-Aħrax. This structure is created as a result of the corrosive action of rainwater with limestone which enlarges a cave to an extent where the cave's roof becomes unstable and collapses. This unique site is connected with the sea and it is a place of interest for diving and kayaking. It lies 10 m above sea level; for this reason, it is regarded as a panoramic lookout point from where it is possible to view all Marfa peninsula, Gozo and Comino. The site has already considered as a site with aesthetical value frequented by recreational activities, but it has also scientific relevance being a unique sinkhole in Malta connected with the sea (Figure 4b). It is known by the locals as Id-Dragonara.

ID3: Ċirkewwa Sinkhole

A semi-circular sinkhole is found in Ċirkewwa, northwest of Malta, known from the locals as Latnija, or as Għajn Tuta, the latter being the name of the local area in which it is situated. It is probably Quaternary in age [46,97] and represents the collapse of a limestone roof of a small cave. It shows a semi-circular shape and has a diameter of 35 m, on the ground level and it is surrounded by a rocky pavement with soil infills on its karstic surface. Partly obscured by typical Mediterranean scrubland, the site is highly affected by human activities such as rock climbing, camping and recreation such as barbeque. The geodiversity content of the site can be linked with other subjects as ecology and biology, due to the presence of Mediterranean vegetation. The site is a unique example of inland sinkhole in the area under study and the second in all Malta (consequent only by Il-Maqluba in the south of Malta). It is a perfect spot to appreciate the karst processes that acted and act nowadays on the archipelago (Figure 11a).

Figure 11. Views of the selected geosites: (**a**) Latinija sinkhole (ID3); (**b**) Il-Bajja taċ-Ċirkewwa where mass movements affect Upper Coralline Limestone overlaying Blue Clays (ID2); (**c**,**d**) badland topography in Blue Clay slopes (ID28 and ID8); (**e**) Lower Globigerina terrace (ID7); (**f**) lateral spreading affecting Upper Coralline Limestone overlying Blue Clays (ID1).

ID2: Il-Bajja Taċ-Ċirkewwa Rock Topple

A spectacular site with lateral spreading and rock topple in Upper Coralline Limestone Fm., this embayment represents a highly-sought-after bay on the island with a pocket sandy beach. The site is called Il-Bajja taċ-Ċirkewwa, better known as Paradise Bay due to the clear sea waters that fringe the white sandy beach. The high score is assigned not only from a scientific point of view for its landslide features, but also for the presence of recreational facilities. Services as bars, hotels, car parks and a bus station are found within the site. The bay is also popular for shore diving. The whole area is easily accessible via public services and directly connected with the national road that could favor educational activities (Figure 11b).

ID28 and ID8: Badland Topography on Blue Clay Slopes

Both sites have a high visual impact and make up exemplary cases that help to understand the geomorphological evolution of coastal areas. ID28 (Figure 11c) is located at Blata l–Bajda, between the salinas in Globigerina Limestone and the fragmented plateau of Upper Coralline Limestone. The site is easily accessible and it could be the destination of a number of activities related to other subjects, such as history, ecology and biology due to the presence of salinas, military fortifications and green areas. Despite the fragility of the environment, this site is widely used by locals for recreational activities such as hiking, cycling, motorcycling and hunting. These recreational activities are a potential threat to the exposed Blue Clay slopes. The second site, ID8 (Figure 11d), has high scenic impact and remarkable educational value. The Blue Clay slopes outcrop, gently corrugated, between the Upper Coralline Limestone plateau and a unique terrace (ID7) in Globigerina Limestone. In this site it is possible to understand how detached blocks of Upper Coralline Limestone move on the underling Blue Clays slopes and how badland topography develops.

ID7: Rdum Il-Qammieh Terrace

A 1 km terrace in Lower Globigerina Limestone Fm. extends along the coast of Rdum il-Qammieh, featuring with typical examples of karst terrain. Chemical weathering is the main process shaping the surface of the platform and forming small solution pools, also known as honeycomb structures. High scientific value and high aesthetic value are assigned to this unique terrace in northern Malta, which is spectacularly flat and yellow-colored. In addition, it conserves a substantial number of fossils. The intensive network of fossilized burrowing channels over the surface of the Lower Globigerina Limestone scallop shells especially within the Lower Conglomerate bed, and the fossils of Echinoids species exposed at the surface. It can be considered an unspoiled outbound site, without services nearby. The Upper Coralline Limestone forms a plateau at the top of the slope profile and is the source of numerous boulders that are deposited on the Blue Clay slopes and the terrace. These boulders, different in size, are used by climbers for boulder activity (Figure 11e).

ID1 and ID4: Areas Affected by Lateral Spreading

Located on the west coast, respectively in Ta' Qassisu (Figure 11f) and Rdum il-Qawwi (Figure 12a), these two sites are representative of gravity-induced processes active on the coast. In particular, it is possible to appreciate deep fissures on the carbonatic plateau, block sliding and lateral spreading, constantly expanding towards the sea. The geodiversity content of the area can be combined with two other subjects: biology and history. Indeed, as additional value, the plateau hosts a variety of endemic flowers and plants and offers a spectacular view of Gozo. Rich also from the cultural-historical point of view, both sites host remains of old villages and pillboxes of the Second World War. ID4 presents a higher number of blocks located on the coast and a small rock window shaped by sea action.

Figure 12. Views of the selected geosites: (**a**) Lateral spreading affecting Upper Coralline Limestone overlying Blue Clays (ID4); (**b**) St. Paul's Islands fault (ID24).

ID24: St. Paul's Islands Fault

The islets of St. Paul, protected as Nature Reserve, lie 800 m from Selmunett Bay. A direct fault across the island has brought the Upper Coralline Limestone in juxtaposition with Upper Globigerina Limestone [95]. The Upper Coralline Limestone is predominant on the surface and represents the entire surface morphology of the islets; the Upper Globigerina Limestone outcrops as a small cliff with a narrow shore platform at the base. The coast of islets also features a number of marine caves. The islets are a Level 2 Site of Scientific Importance (SSI) for its geomorphology (GN 827 of 2002). Access to the islets is only permissible between sunrise and sunset and then only against an entry permit obtainable from the Environment and Resource Authority (ERA) (Figure 12b).

6. Conclusions

This work aims to increase the knowledge of the rich geological heritage of northern Malta, providing a better understanding of the geological and geomorphological characteristics of the study area and facilitating the recognition of the opportunities, in order to strengthen the argument for the setting-up an effective environmental management plan, taking into full account the geological component as well. The present research shows that, considering the small geographic scale of the island, there is a high level of geodiversity of features primarily controlled by the interaction between geomorphological processes, structure and stratified geology. An assessment of geosites has been carried out based on a set of criteria that links geological and geomorphological importance with additional values of the sites, as aesthetic, cultural, ecological and economic. The accurate description and characterization of potential geosites and their inventory aim to help the government administration become more aware of the sites of geological interest in the area, giving useful information for their effective management which includes both geoconservation and geotourism actions. As found in the result (Section 5.1), we classified the sites in active or inherited, not only to note their state of activity, but to take into consideration their vulnerability and fragility. Active geosites, in fact, are fragile and may necessitate management and protection measures. Similar to most geosites, they are exposed to natural and man-made processes that threaten their integrity and may compromise their value. Therefore, their conservation is a complex issue since it should address the problem of both possible destruction by natural active processes and man-induced damage. In addition, very often dynamic sites are highly sensitive features, susceptible to modifications due to processes' changes in time, frequency and intensity. Many coastal environments are very sensitive areas, particularly vulnerable to disturbance and prone to change, where climate change impacts are very acute. Changes are visible at very short time scales and may generate active processes, very evident to observe. The same consideration can be done to the size of the sites. The limited study area comprises small isolated features that are usually more vulnerable due to their dimension and can stand a lower tourist pressure compared to extensive areal geosites [58]. Geoheritage inventory and assessment are therefore the first steps in the process of effective conservation and promotion. Some degree of legal protection already exists in a few sites. A wide part of the study area falls under Natura 2000 management as Special Areas of Conservation (SAC) or Special Protected Areas (SPA). In addition, some sites, such as Gżejjer ta' San Pawl, are scheduled as Nature Reserves and so protected under the Nature Reserve legislation (Table 1) or established as nature parks, such as the Majjistral Nature and History Park. The integration of the geoheritage character of the area would mean both strengthening the landscape value for its geological and geomorphological component and unifying the whole study area under geoheritage conservation rather than leaving it as an area with single components of conservation.

Geoheritage, combined with the rich cultural heritage, could be considered as the heart of tourism and educational activities, with Malta's tourism direct contribution to GDP being among the highest in the EU. Data from the World Travel and Tourism Council (WTTC) show that the travel and tourism industry's total contribution to Malta's GDP stood at 27.1% in 2017. This was the highest share recorded within the Mediterranean region by a notable margin and was also well above the Mediterranean, European Union and World averages, which ranged between 10% and 12%. The total contribution of travel and tourism industries to employment including indirect and induced impacts was estimated to reach 55,000 jobs in 2017 (28.3% of total employment) [98,99]. Concerning the kind of tourism, leisure tourism remains the main purpose of visit for the vast majority of tourist arrivals to the Maltese Islands, with a share of 85.3% of total inbound tourists in 2017. The number of visitors for business purposes stood at 7.9% (2017), whilst "other" tourist segments, such as for educational, religious and health-related purposes, stood at 6.8%. Most importantly, there has also been some evidence of diversification within the Maltese holiday product itself, which departs from the stereotypical image of the islands as a 'sun and sea' destination. The Market Profile Survey (for 2017) undertaken by the Malta Tourism Authority's [100] has in fact shown that only 15.7% of inbound tourists chose Malta based on traditional 'sun and sea' destinations. The largest share of tourists (42.9%) chose Malta for its culture and heritage. Moreover, the tourism industry in Malta has gradually also shifted from package to non-package holidaymakers. This reflects the emergence of a more independent type of tourist who wishes to experience the Maltese Islands in a more autonomous and dynamic way.

Given the increasing number of tourists (currently standing at 2.6 million tourists in 2018), geotourism, as a form of sustainable tourism, is the best solution that sustains and enhances the identity of the territory, especially rural areas, taking in consideration its geology, environment, culture, aesthetics, heritage and the well-being of its residents [101]. Geotourism will ensure benefits for traveler that will discover the geoheritage, cultural heritage and traditions of the archipelago in an innovative and green way, respecting the environment and ensuring a sustainable economic growth. At the same time, geotourism may offers to locals a high-quality standard of life, helping to build a local identity and promote the unique and authentic heritage in their territory, being involved and architects of geotourism activities. In addition, the need to establish geoheritage recognition of these sites is also paramount to provide long-term sustainable measures [102], especially in view of the recent trends of construction boom on the islands to meet the demands of a growing population. The latter is primarily driven by the influx economic migrant workers (EU and non-EU) to support the current growing economy of the Maltese Islands, with 14.1% of the population in 2017 being foreign citizens.

The establishment of a geopark could align well with the recent vision announced by the Maltese government to improve not only the quality of the tourists' experiences but also increase high expenditure and demand-oriented tourists [103,104], over and above the already high annual number of tourists reaching the islands (2.6 million in 2018). Geoparks have the strong potential to maximize the quality of these experiences expected by such higher-expenditure tourists and it would directly inject further policy actions in both geoconservation and geotourism strategies for the islands. In this framework, the recognition of viewpoint geosites, intended as "a specific locality which allows for unobstructed observation of the surrounding landscape and comprehension of Earth history recorded in rocks, structures and landforms visible from this locality" [105], would be crucial for geo-education and outreach activities, and future research will be addressed to this.

Author Contributions: Conceptualization, P.C. and L.S.; methodology, P.C. and L.S.; investigation, R.G. and L.S.; writing—original draft preparation, P.C., R.G. and L.S.; writing—review and editing, P.C., R.G. and L.S.; visualization, L.S.; supervision, M.S.

Funding: The present work was carried out in the framework of the Project 'Hazard and Vulnerability Assessment—The Path to Identifying Risk' funded by the EUR-OPA Major Hazards Agreement of the Council of Europe (Ref No.: GA/2019/11; Scientific Responsible: Mauro Soldati).

Acknowledgments: The authors gratefully acknowledge Darren Saliba, manager at Il-Majjistral Nature and History Park, Malta.

Conflicts of Interest: The authors declare no conflict of interest.

References

1. Gray, M. *Geodiversity—Valuing and Conserving Abiotic Nature*, 1st ed.; Wiley, J., Ed.; The Atrium, Southern Gate: Chichester, UK, 2004.

2. Sharples, C. *Concepts and Principles of Geoconservation. Tasmanian Parks & Wildlife Service*; PDF Document; Tasmanian Parks and Wildlife Service: Hobart, Australia, 2002. Available online: https://dpipwe.tas.gov.au/Documents/geoconservation.pdf (accessed on 17 October 2019).

3. Gray, M. Geodiversity: The backbone of Geoheritage and Geoconservation. In *Geoheritage: Assessment, Protection, and Management*; Reynard, E., Brilha, J., Eds.; Elsevier: Amsterdam, The Netherland, 2018; pp. 13–25.

4. Newsome, D.; Dowling, R. Geoheritage and Geotourism. In *Geoheritage: Assessment, Protection, and Management*; Reynard, E., Brilha, J., Eds.; Elsevier: Amsterdam, The Netherland, 2018; pp. 305–322.

5. Coratza, P.; Hobléa, F. The Specificities of Geomorphological Heritage. In *Geoheritage: Assessment, Protection, and Management*; Reynard, E., Brilha, J., Eds.; Elsevier: Amsterdam, The Netherland, 2018; pp. 87–106.

6. IUGS. Heritage Sites and Collections Subcommission, National Heoheritage Inventories. Available online: https://geoheritage-iugs.mnhn.fr/index.php?catid=19&blogid=1 (accessed on 15 May 2019).

7. Coratza, P.; Bruschi, V.M.; Piacentini, D.; Saliba, D.; Soldati, M. Recognition and Assessment of Geomorphosites in Malta at the Il-Majjistral Nature and History Park. *Geoheritage* **2011**, *3*, 175–185. [CrossRef]

8. Coratza, P.; Galve, J.P.; Soldati, M.; Tonelli, C. Recognition and assessment of sinkholes as geosites: Lessons from the Island of Gozo (Malta). *Qua. Geor.* **2012**, *31*, 22–35. [CrossRef]

9. Coratza, P.; Gauci, R.; Schembri, J.A.; Soldati, M.; Tonelli, C. Bridging Natural and Cultural Values of Sites with Outstanding Scenery: Evidence from Gozo, Maltese Islands. *Geoheritage* **2016**, *8*, 91–103. [CrossRef]

10. Agnesi, V.; Angileri, S.E.; Cappadonia, C.; Coratza, P.; Costanzo, D.; Soldati, M.; Tonelli, C. Geositi nel paesaggio mediterraneo: Confronto tra aree costiere maltesi e siciliane. *Culture Territori Linguaggi* **2014**, *4*, 1–12.

11. Gauci, R.; Schembri, J.A.; Inkpen, R. Traditional Use of Shore Platforms: Mapping the Artisanal Management of Coastal Salt Pans on the Maltese Islands (Central Mediterranean). In *Traditional Wisdom: Before It's Too Late*; York, K., Ed.; SAGE Open Special Issue; Sage Publications: Thousand Oaks, CA, USA, 2017; Volume 7, pp. 1–16.

12. Cappadonia, C.; Coratza, P.; Agnesi, V.; Soldati, M. Malta and Sicily Joined by Geoheritage Enhancement and Geotourism within the Framework of Land Management and Development. *Geosci. J.* **2018**, *8*, 253. [CrossRef]

13. Gauci, R.; Schembri, J.A. (Eds.) *Landscapes and Landforms of the Maltese Islands*; Springer: Cham, Switzerland, 2019.

14. Spiteri, L.; Stevens, D. Landscape Diversity and Protection in Malta. In *Landscapes and Landforms of the Maltese Islands*; Gauci, R., Schembri, J.A., Eds.; Springer: Cham, Switzerland, 2019; pp. 359–372.

15. Environment and Resource Authority (ERA). Database on Designated Areas in National Law. 2018. Available online: https://era.org.mt/en/Pages/Database-on-Designated-Areas-in-National-Law.aspx (accessed on 3 June 2019).

16. Soldati, M.; Buhagiar, S.; Coratza, P.; Pasuto, A.; Schembri, J.A. Integration of the geomorphological environment and cultural heritage: A key issue for present and future times. *Geogr. Fis. Dinam. Quat.* **2008**, *31*, 95–96.

17. Soldati, M.; Buhagiar, S.; Coratza, P.; Pasuto, A.; Schembri, J.A. (Eds.) Integration of the geomorphological environment and cultural heritage for tourist promotion and hazard prevention. *Geogr. Fis. Dinam. Quat.* **2008**, *31*, 93–247.

18. May, V. Integrating the geomorphological environmental, cultural heritage, tourism and coastal hazards in practice. *Geogr. Fis. Dinam. Quat.* **2008**, *31*, 187–194.

19. Gauci, R.; Inkpen, R. The physical Characteristics of Limestone Shore Platforms on the Maltese Islands and Their Neglected Contribution to Coastal Land Use Development. In *Landscapes and Landforms of the Maltese Islands*; Gauci, R., Schembri, J.A., Eds.; Springer: Cham, Switzerland, 2019; Volume 27, pp. 343–356.

20. Satariano, B.; Gauci, R. Landform Loss and Its Effect on Health and Wellbeing: The collapse of the Azure Window (Gozo) and the Resultant Reactions of the Media and the Maltese Community. In *Landscapes and Landforms of the Maltese Islands*; Gauci, R., Schembri, J.A., Eds.; Springer: Cham, Switzerland, 2019; Volume 23, pp. 289–303.

21. Calleja, I.; Tonelli, C. Dwejra and Maqluba: Emblematic sinkholes in the Maltese Islands. In *Landscapes and Landforms of the Maltese Islands*; Gauci, R., Schembri, J.A., Eds.; Springer: Cham, Switzerland, 2019; Volume 11, pp. 129–139.

22. Ministry of Tourism. National Tourism Policy 2015–2020. Available online: https://tourism.gov.mt/en/Documents/FINALBOOKLETexport9.pdf (accessed on 17 October 2019).

23. Pedley, H.M.; Clarke, M.H.; Galea, P. *Limestone Isles in a Crystal Sea. The Geology of the Maltese Islands*; PEG Ltd.: San Gwann, Malta, 2002.

24. Baldassini, N.; Di Stefano, A. Stratigraphic features of the Maltese Archipelago: A synthesis. *Nat. Hazards* **2017**, *86*, 203–231. [CrossRef]

25. Pedley, H.M.; Waugh, B. Easter Field Meeting to the Maltese Islands. *Proc. Geol. Assoc.* **1976**, *87*, 343–358. [CrossRef]

26. Scerri, S. Sedimentary Evolution and Resultant Geological Landscapes. In *Landscapes and Landforms of the Maltese Islands*; Gauci, R., Schembri, J.A., Eds.; Springer: Cham, Switzerland, 2019; pp. 31–47.

27. Pedley, H.M.; House, M.R.; Waugh, B. The Geology of the Pelagian Block: The Maltese Islands. In *The Ocean Basins and Margins, The Western Mediterranean*; Nairn, A.E.M., Kanes, W.H., Stehli, F.G., Eds.; Plenum Press: London, UK, 1978; Volume 4B, pp. 417–433.

28. Bianucci, G.; Gatt, M.; Catanzariti, R.; Sorbi, S.; Bonavia, C.G.; Curmi, R.; Varola, A. Systematics, biostratigraphy and evolutionary pattern of the Oligo-Miocene marine mammals from the Maltese Islands. *Geobios* **2011**, *44*, 549–585. [CrossRef]

29. Devoto, S.; Biolchi, S.; Bruschi, V.M.; González, D.A.; Mantovani, M.; Pasuto, A.; Soldati, M. Landslides Along the North-West Coast of the Island of Malta. In *Landslide Science and Practice: Landslide Inventory and Susceptibility and Hazard Zoning*; Margottini, C., Canuti, P., Sassa, K., Eds.; Springer: Berlin, Germany, 2013; Volume 1, pp. 57–63.

30. Prampolini, M.; Gauci, C.; Micallef, A.S.; Selmi, L.; Vandelli, V.; Soldati, M. Geomorphology of the north-eastern coast of Gozo (Malta, Mediterranean Sea). *J. Maps* **2018**, *14*, 402–410. [CrossRef]

31. House, M.R.; Dunham, K.C.; Wigglesworth, J.C. Geology and Structure of the Maltese Islands. In *Malta: Background for Development*; Bowen Jones, H., Dewdney, J.C., Fisher, W.B., Eds.; University of Durham: Durham, UK, 1962; pp. 24–33.

32. Gauci, R.; Scerri, S. A Synthesis of Different Geomorphological Landscapes on the Maltese Islands. In *Landscapes and Landforms of the Maltese Islands*; Gauci, R., Schembri, J.A., Eds.; Springer: Cham, Switzerland, 2019; pp. 49–65.

33. Alexander, D. A review of the physical geography of Malta and its significance for tectonic geomorphology. *Quat. Sci. Rev.* **1988**, *7*, 41–53. [CrossRef]

34. Galea, P. Central Mediterranean Tectonics—A key Player in the Geomorphology of the Maltese Islands. In *Landscapes and Landforms of the Maltese Islands*; Gauci, R., Schembri, J.A., Eds.; Springer: Cham, Switzerland, 2019; pp. 19–30.

35. Biolchi, S.; Furlani, S.; Devoto, S.; Gauci, R.; Castaldini, D.; Soldati, M. Geomorphological identification, classification and spatial distribution of coastal landforms of Malta (Mediterranean Sea). *J. Maps* **2016**, *12*, 87–99. [CrossRef]

36. Foglini, F.; Prampolini, M.; Micallef, A.; Angeletti, L.; Vandelli, V.; Deidun, A.; Soldati, M.; Taviani, M. Late Quaternary Coastal Landscape Morphology and Evolution of the Maltese Islands (Mediterranean Sea) Reconstructed from High-Resolution Seafloor Data. In *Geology and Archaeology: Submerged Landscapes of the Continental Shelf*; Harff, J., Bailey, G., Lüth, L., Eds.; Geological Society, Special Publication: London, UK, 2016; Volume 411, pp. 77–95.

37. Prampolini, M.; Foglini, F.; Biolchi, S.; Devoto, S.; Angelini, S.; Soldati, M. Geomorphological mapping of terrestrial and marine areas, Northern Malta and Comino (central Mediterranean Sea). *J. Maps* **2017**, *13*, 457–469. [CrossRef]

38. Magri, O.; Mantovani, M.; Pasuto, A.; Soldati, M. Geomorphological investigation and monitoring of lateral spreading along the north-west coast of Malta. *Geogr. Fis. Dinam. Quat.* **2008**, *31*, 171–180.

39. Devoto, S.; Biolchi, S.; Bruschi, V.M.; Furlani, S.; Mantovani, M.; Piacentini, D.; Soldati, M. Geomorphological map of the NW coast of the Island of Malta (Mediterranean Sea). *J. Maps* **2012**, *8*, 33–40. [CrossRef]

40. Devoto, S.; Forte, E.; Mantovani, M.; Mocnik, A.; Pasuto, A.; Piacentini, D.; Soldati, M. Integrated Monitoring of Lateral Spreading Phenomena Along the North-West Coast of the Island of Malta. In *Landslide Science and Practice: Early Warning, Instrumentation and Monitoring*; Margottini, C., Canuti, P., Sassa, K., Eds.; Springer: Berlin, Germany, 2013; Volume 2, pp. 235–241.

41. Mantovani, M.; Devoto, S.; Forte, E.; Mocnik, A.; Pasuto, A.; Piacentini, D.; Soldati, M. A multidisciplinary approach for rock spreading and block sliding investigation in the north-western coast of Malta. *Landslides* **2013**, *10*, 611–622. [CrossRef]

42. Mantovani, M.; Devoto, S.; Piacentini, D.; Prampolini, M.; Soldati, M.; Pasuto, A. Advanced SAR interferometric analysis to support geomorphological interpretation of slow-moving coastal landslides (Malta Mediterranean Sea). *Remote Sens.* **2016**, *8*, 443. [CrossRef]

43. Soldati, M.; Barrows, T.T.; Prampolini, M.; Fifield, K.L. Cosmogenic exposure dating constraints for coastal landslide evolution on the Island of Malta (Mediterranean Sea). *J. Coast. Conserv.* **2018**, *22*, 831–844. [CrossRef]

44. Soldati, M.; Devoto, S.; Prampolini, M.; Pasuto, A. The spectacular landslide-controlled landscape of the northwestern coast of Malta. In *Landscapes and Landforms of the Maltese Islands*; Gauci, R., Schembri, J.A., Eds.; Springer: Cham, Switzerland, 2019; Volume 14, pp. 167–178.

45. Saliba, D. Cave Formations and Processes in Malta. Bachelor' Thesis, Geography, University of Malta, Msida, Malta, 2008.

46. Soldati, M.; Tonelli, C.; Galve, J.P. Geomorphological evolution of palaeosinkhole features in the Maltese archipelago (Mediterranean Sea). *Geogr. Fis. Dinam. Quat.* **2013**, *36*, 189–198.

47. Schembri, J.A. The Geographical Context of the Maltese Islands. In *Landscapes and Landforms of the Maltese Islands*; Gauci, R., Schembri, J.A., Eds.; Springer: Cham, Switzerland, 2019; pp. 9–17.

48. Cyffka, B.; Bock, M. Degradation of field terraces in the Maltese Islands-reasons, processes, and effects. *Geog. Fis. Dinam. Quat.* **2008**, *31*, 119–128.

49. Micallef, S. The Terraced Character of the Maltese Rural Landscape: A Case Study of Buskett Area. In *Landscapes and Landforms of the Maltese Islands*; Gauci, R., Schembri, J.A., Eds.; Springer: Cham, Switzerland, 2019; pp. 153–165.

50. Carton, A.; Cavallin, A.; Francavilla, F.; Mantovani, F.; Panizza, M.; Pellegrini, G.; Tellini, C. with the coll. of Bini, A.; Castaldini, D.; Giorgi, G.; Floris, B.; Marchetti, M.; Soldati, M.; Surian, N. Ricerche ambientali per l'individuazione e la valutazione dei beni geomorfologici. Metodi ed esempi. *Il Quat.* **1994**, *7*, 365–372.

51. Reynard, E.; Perret, A.; Bussard, J.; Grangier, L.; Martin, S. Integrated Approach for the Inventory and Management of Geomorphological Heritage at the Regional Scale. *Geoheritage* **2016**, *8*, 43–60. [CrossRef]

52. Lapo, A.V.; Davydov, V.I.; Pashkevich, N.G.; Petrov, V.V.; Vdovets, M.S. Methodic principles of study of geological monuments of nature in Russia. *Stratigr Geol. Correlat.* **1993**, *1*, 636–644.

53. Lima, F.F.; Brilha, J.B.; Salamuni, E. Inventorying geological heritage in large territories: A methodological proposal applied to Brazil. *Geoheritage* **2010**, *2*, 91–99. [CrossRef]

54. García-Cortés, A.; Carcavilla Urquí, L.; Díaz-Martínez, E.; Vegas Salamanca, J. *Inventario de lugares de interés geológico de la Cordillera Ibérica*; Instituto Geológico y Minero de España: Madrid, Spain, 2012; Available online: http://www.igme.es/patrimonio/Informe%20Ib%C3%A9rica%20Final.pdf. (accessed on 24 October 2019).

55. Joyce, E.B. Australia's geoheritage: History of study, a new inventory of geosites and applications to geoturism and geoparks. *Geoheritage* **2010**, *2*, 39–56. [CrossRef]

56. De Wever, P.; Alterio, I.; Egoroff, G.; Cornée, A.; Bobrowsky, P.; Collin, G.; Duranthon, F.; Hill, W.; Lalanne, A.; Page, K. Geoheritage, a National Inventory in France. *Geoheritage* **2015**, *7*, 205–247. [CrossRef]

57. White, S.; Mitchell, M. *Geological Heritage Sites: A Procedure and Protocol for Documentation and Assessment*; AESC: Melbourne, Australia, 2006.

58. Fuertes-Gutiérrez, I.; Fernández-Martínez, E. Geosites inventory in the Leon Province (Northwestern Spain): A tool to introduce geoheritage into regional environmental management. *Geoheritage* **2010**, *2*, 57–75. [CrossRef]

59. Mastronuzzi, G.; Simone, O. L'individuazione Dei Siti, il Rilevamento e la Redazione Della Scheda. In *Ricognizione e Verifica dei Geositi e Delle Emergenze Geologiche della Regione Puglia*; Regione Puglia: Bari, Italy, 2017; pp. 28–32.

60. Fuertes-Gutiérrez, I.; Fernández-Martínez, E. Mapping geosites for geoheritage management: A methodological proposal for the Regional Park of Picos de Europa (León, Spain). *Environ. Manag.* **2012**, *50*, 789–806. [CrossRef]

61. Poiraud, A.; Chevalier, M.; Claeyssen, B.; Biron, P.E.; Joly, B. From geoheritage inventory to territorial planning tool in the Vercors massif (French Apls): Contribution of statistical and expert cross approaches. *Appl. Geogr.* **2016**, *71*, 69–82. [CrossRef]

62. Mauerhofer, L.; Reynard, E.; Asrat, A.; Hurni, H. Contribution of a geomorphosite inventory to the geoheritage knowledge in developing countries: The case of the Simien Mountains National Park, Ethiopia. *Geoheritage* **2018**, *10*, 559–574. [CrossRef]

63. Reynard, E. The Assessment of Geomorphosites. In *Geomorphosites*; Reynard, E., Coratza, P., Regolini-Bissig, G., Eds.; Pfeil: Munchen, Germany, 2009; pp. 63–71.

64. Reynard, E.; Coratza, P. Scientific research on geomorphosites. A review of the activities of the IAG working group on geomorphosites over the last twelve years. *Geogr. Fis. Dinam. Quat.* **2013**, *36*, 159–168.

65. Barba, F.J.; Remondo, J.; Rivas, V. Propuesta de un procedimento para armonizar la valoraciòn de elementos del patrimonio geològico. *ZUBIA* **1997**, *15*, 11–20.

66. Rivas, V.; Rix, K.; Frances, A.; Cendrero, A.; Brunsden, D. Geomorphological indicators for environmental impact assessment: Consumable and non-consumable geomorphological resources. *Geomorphology* **1997**, *18*, 169–182. [CrossRef]

67. Coratza, P.; Giusti, C. Methodological proposal for the assessment of the scientific quality of geomorphosites. *Ital. J. Quat. Sci.* **2005**, *18*, 303–313.

68. Bruschi, V.M.; Cendrero, A. Geosite evaluation; can we measure intangible values? *Il Quat.* **2005**, *18*, 293–306.

69. Serrano, E.; Gonzàlez-Trueba, J.J. Assessment of geomorphosites in natural protected areas: The Pico de Europa National Park (Spain). *Géomorphologie* **2005**, *3*, 197–208. [CrossRef]

70. Reynard, E.; Fontana, G.; Kozlik, L.; Scapozza, C. A method for assessing «scientific» and «additional values» of geomorphosites. *Geog. Helv.* **2007**, *62*, 148–158. [CrossRef]

71. García-Cortés, A.; Carcavilla Urquí, L.; Díaz-Martínez, E.; Vegas, J. *Documento Metodológico Para la Elaboración del Inventario Español de Lugares de Interés Geológico (IELIG)*; Version 12; Instituto Geológico y Minero de España: Madrid, Spain, 2014; Available online: http://www.igme.es/patrimonio/descargas/METODOLOGIA%20IELIG%20V16%20actualizaci%C3%B3n%202018.pdf. (accessed on 24 October 2019).

72. Pralog, J.P. A method for assessing the tourism potential and use of geomorphological sites. *Geomorphologie* **2005**, *3*, 189–196. [CrossRef]

73. Vujicic, M.D.; Vasiljevic, D.E.; Markovic, S.B.; Hose, T.A.; Lukic, T.; Hadzic, O.; Janicevic, S. Preliminary geosite assessment model (GAM) and its application on Fruska Gora Mountain, potential geotourism destination of Serbia. *Acta Geogr. Slov.* **2011**, *51*, 361–377. [CrossRef]

74. Pereira, P.; Pereira, I.I. Assessment of Geosites Tourism Value in Geoparks: The example of Arouca Geopark (Portugal). In Proceedings of the 11th European Geoparks Conference, Arouca, Portugal, 19–21 September 2012; pp. 231–232.

75. Kubalíková, L. Geomorphosite assessment for geotourism purposes. *Czech. J. Tour.* **2013**, *2*, 80–104. [CrossRef]

76. Corbí, H.; Fierro, I.; Aberasturi, A.; Ferris, E.J.S. Potential use of a significant scientific geosite: The Messinian coral reef of Santa Pola (SE Spain). *Geoheritage* **2018**, *10*, 427–441. [CrossRef]

77. Pereira, P.; Pereira, D.; Caetano, M.I. Geomorphosite assessment in Monteshino Natural Park (Portugal). *Geogr. Helv.* **2007**, *62*, 159–168. [CrossRef]

78. Zouros, N. Geomorphosite assessment and management in protected areas of Greece. The case of the Lesvos island coastal geomorphosites. *Geogr. Helv.* **2007**, *62*, 169–180. [CrossRef]

79. Feuillet, T.; Sourp, E. Geomorphological heritage of the Pyrenees National Park (France): Assessment, clustering, and promotion of geomorphosites. *Geoheritage* **2011**, *3*, 151–162. [CrossRef]

80. Fassoulas, C.; Mouriki, D.; Dimitriou-Nikolakis, P.; Iliopoulos, G. Quantitative assessment of geotopes as an effective tool for geoheritage management. *Geoheritage* **2012**, *4*, 177–193. [CrossRef]

81. Coratza, P.; Vandelli, V.; Fiorentini, L.; Paliaga, G.; Faccini, F. Bridging Terrestrial and Marine Geoheritage: Assessing Geosites in Portofino Natural Park (Italy). *Water* **2019**, *11*, 2112. [CrossRef]

82. Brilha, J. Geoheritage: Inventories and Evaluation. In *Geoheritage: Assessment, Protection, and Management*; Reynard, E., Brilha, J., Eds.; Elsevier: Amsterdam, The Netherland, 2018; pp. 69–86.

83. Carton, A.; Coratza, P.; Marchetti, M. Guidelines for geomorphological sites mapping: Examples from Italy. *Géomorphologie* **2005**, *3*, 209–218. [CrossRef]

84. Bollati, I.; Pellegrini, M.; Reynard, E.; Pelfini, M. Water driven processes and landforms evolution rates in mountain geomorphosites: Examples from Swiss Alps. *Catena* **2017**, *158*, 321–339. [CrossRef]

85. Hooke, J.M. Strategies for Conserving and Sustaining Dynamic Geomorphological Sites. In *Geological and Landscape Conservation*; O'Halloran, D., Green, C., Harley, M., Knill, J., Eds.; The Geological Society: London, UK, 1994; pp. 191–195.

86. Strasser, A.; Heitzmann, P.; Jordan, P.; Stapfer, A.; Stürm, B.; Vogel, A. Geotope und der Schutz erdwissenschaftlicher. *GeoInfo* **1995**, *8*, 3–30.

87. Reynard, E. Protecting Stones: Conservation of Erratic Blocks in Switzerland. In *Dimension Stone. New Perspectives for a Traditional Building Material*; Prykril, R., Ed.; Balkema: Leiden, The Netherland, 2004; pp. 3–7.

88. Reynard, E.; Panizza, M. Geomorphosites: Definition, assessment and mapping. An introduction. *Géomorphologie* **2005**, *3*, 177–180.

89. Koster, E.A. The European aeolian sand belt: Geoconservation of drift sand landscapes. *Geoheritage* **2009**, *1*, 93–110. [CrossRef]

90. Bollati, I.; Pelfini, M.; Pellegrini, L. A geomorphosites selection method for educational purposes: A case study in Trebbia Valley (Emilia Romagna, Italy). *Geogr. Fis. Dinam. Quat.* **2012**, *35*, 23–35.

91. Pelfini, M.; Bollati, I. Landforms and geomorphosites ongoing changes: Concepts and implications for geoheritage promotion. *Qual. Geor.* **2014**, *33*, 131–143. [CrossRef]

92. Reynard, E.; Coratza, P. The importance of mountain geomorphosites for environmental education. Examples from the Italian Dolomites and the Swiss Alps. *Acta Geogr. Slov.* **2016**, *2016 56*, 291–303. [CrossRef]

93. Zgłobicki, W.; Poesen, J.; Cohen, M.; Del Monte, M.; García-Ruiz, J.M.; Ionita, I.; Niacsu, L.; Machová, Z.; Martín-Duque, J.F.; Nadal-Romero, E.; et al. The Potential of Permanent Gullies in Europe as Geomorphosites. *Geoheritage* **2019**, *11*, 217–239. [CrossRef]

94. Lanfranco, S.; Briffa, K. Limestone Dissolution and Temporary Freshwater Rockpools of the Maltese Islands. In *Landscapes and Landforms of the Maltese Islands*; Gauci, R., Schembri, J.A., Eds.; Springer: Cham, Switzerland, 2019; Volume 15, pp. 179–191.

95. Sammut, S.; Gauci, R.; Inkpen, R.; Lewis, J.J.; Gibson, A. Selmun: A Coastal Limestone Landscape Enriched by Scenic Landforms, Conservation Status and Religious Significance. In *Landscapes and Landforms of the Maltese Islands*; Gauci, R., Schembri, J.A., Eds.; Springer: Cham, Switzerland, 2019; Volume 26, pp. 325–341.

96. Pedley, H.M. Miocene Sea-floor Subsidence and Later Subaerial Solution Subsidence Structures in the Maltese Islands. *Proc. Geol. Assoc.* **1975**, *85*, 533–547. [CrossRef]

97. Tonelli, C.; Galve, J.P.; Soldati, M.; Gutiérrez, F. Erosional Morphostructures Related to Miocene Paleosinkholes in the Island of Gozo, Malta. In Proceedings of the XII Reunión Nacional de Geomorfologia, Santander, Spain, 17–20 September 2012; pp. 409–412.

98. Central Bank of Malta. The Evolution of Malta's Tourism Product over Recent Years. *Q. Rev.* **2018**, *4*, 41–55. Available online: https://www.centralbankmalta.org/file.aspx?f=72256 (accessed on 17 October 2019).

99. World Travel and Tourism Council. *Travel & Tourism Economic Impact 2018 Malta*; World Travel and Tourism Council: London, UK, 2018.

100. Malta Tourism Authority. Tourism in Malta—Facts and Figures 2017. 2008. Available online: https://www.mta.com.mt/en/file.aspx?f=32328 (accessed on 16 October 2019).

101. Arouca Declaration. In Proceedings of the International Congress AROUCA 2011—Geotourism in Action, Arouca Geopark, Portugal, 9–13 November 2011; Available online: https://www.dropbox.com/s/q41gbd0cp2nt73o/Declaration_Arouca_%5BEN%5D.pdf?dl=0 (accessed on 17 October 2019).

102. Attard, M. The Sustainability of Landforms and Landscapes. In *Landscapes and Landforms of the Maltese Islands*; Gauci, R., Schembri, J.A., Eds.; Springer: Cham, Switzerland, 2019; pp. 373–380.

103. Pace, F. Country is Faced with a Choice of Making a Leap Forward—PM. *TVM News*. 27 January 2019. Available online: https://www.tvm.com.mt/en/news/country-is-faced-with-choice-of-making-a-leap-forward-pm (accessed on 18 August 2019).

104. Costa, M. Malta Needs to Reach Higher Economic Dimension, Muscat Says, as He Defends Corinthia Project. *Malta Today*. 27 January 2019. Available online: https://www.maltatoday.com.mt/news/national/92535/malta_needs_to_reach_higher_economic_dimension_muscat_says_as_he_defends_corinthia_project#.XWaF8ugzbDc (accessed on 18 August 2019).

105. Migoń, P.; Pijet-Migoń, E. Viewpoint geosites—Values, conservation and management issues. *PGA* **2017**, *128*, 511–522. [CrossRef]

Article

Assessing Geotourism Resources on a Local Level: A Case Study from Southern Moravia (Czech Republic)

Lucie Kubalíková

Department of Geology and Pedology, Faculty of Forestry and Wood Technology, Mendel University in Brno, Zemědělská 3, 61300 Brno, Czech Republic; Lucie.Kubalikova@ugn.cas.cz

Received: 11 July 2019; Accepted: 20 August 2019; Published: 22 August 2019

Abstract: In the last decades, the geotourism has shown a considerable growth all over the world and it is appreciated and accepted as a useful tool for promoting natural and cultural heritage and for fostering local and regional economic development, especially within rural areas. Geotourism focus especially on the geological and geomorphological aspects of the landscape; however, according to the current holistic approach, it also builds on the close relations between geodiversity and other assets of the territory, such as biodiversity, archaeological and cultural values, gastronomy or architecture. Currently, geotourism activities are promoted mainly within geoparks, but other regions also possess an important geotourism potential. A complex assessment of the geotourism resources of a particular area is crucial for geotourism-development. The paper presents two case studies from Southern Moravia (Czech Republic) where the assessment of geotourism's potential was made by using the geomorphosite concept and extended SWOT analysis. Results show that these areas (situated outside the geoparks or large-scale protected areas and not far from a big city) have considerable potential for geotourism development, and geodiversity can be considered an important resource for local and regional development. Based on this, conclusions about the possibilities of geotourism development outside the geoparks are outlined.

Keywords: geodiversity; SWOT analysis; rural regions; geomorphosites

---

1. Introduction

In the last few decades, the geotourism has shown considerable growth all over the world [1–4]. Originally, and in a strict sense, the geotourism was defined in the 1990s as "the provision of interpretive and service facilities to enable tourists to acquire knowledge and understanding of the geology and geomorphology of a site (including its contribution to the development of the Earth sciences) beyond the level of mere aesthetic appreciation" [5]. A similar approach was presented by Słomka and Kicińska-Świderska [6], Joyce [7], and Dowling and Newsome [8].

In the broader sense, the geotourism is understood as a form of nature-tourism that focuses on landscape and geology, but also on the biotic and cultural features that are linked to the abiotic nature [9]. It is a so-called ABC (abiotic—biotic—culture) approach. This approach is also reflected in Arouca declaration [10] where geotourism is defined as "tourism which sustains and enhances the identity of a territory, taking into consideration its geology, environment, culture, aesthetics, heritage and the well-being of its residents." The economic and environmental aspects of geotourism are emphasized as well: Dowling [9] defines geotourism as "sustainable tourism with a primary focus on experiencing the Earth's geologic features in a way that fosters environmental and cultural understanding, appreciation and conservation, and is locally beneficial. Geotourism product protects, communicates and promotes geoheritage, helps build communities and works with a wide range of different people." Martini et al. [11] present a more literal and comprehensible definition: "Geotourism allows tourists to know the local geology but also to better understand that this geology is closely related to all the other assets of the territory, such as biodiversity, archaeological and cultural values, gastronomy, etc." The growing interest in geotourism and the interdisciplinary approach adopted for geotourism studies is reflected in the increasing number of papers and the wide scope of particular topics; e.g., geocultural heritage, geotourism's role in regional development or geotourist perceptions [4].

Currently, this holistic approach is widely respected [2], but geological, geomorphological, pedological and hydrological aspects (the components of geodiversity as defined by Gray [12]) stay in the center of attention and represent the basic resource for geotourist activities. Nevertheless, it has to be remembered that setting the links between geodiversity, biodiversity, culture, and history can help to appreciate the geodiversity as a full value resource for tourist activities, and thus, as an important resource for local and regional development. This approach is widely applied, especially within geoparks, which are defined as areas with particular geological heritage and a sustainable territorial development strategy [13,14]. This is also the case of the Central European countries, including the Czech Republic: Geotourist activities are developed in geoparks [15,16] and in some cases, in large-scale protected areas such as National Parks or Protected Landscape Areas [17]. However, outside the geoparks, the geodiversity represents an important resource for geotourism development too (see case studies in [2,8,18]).

Two study areas in the South Moravian Region (shortly Southern Moravia) in the southeastern part of the Czech Republic were assessed by using the selected criteria within the geomorphosite concept [19,20] and extended SWOT analysis [21]. These areas are not a part of any geoparks or large-scale protected areas in the sense of the Law 114/1992 Coll [22]. The areas of interest (Deblínská vrchovina Highlands and Sýkořská hornatina Mountains) were already a subject of scientific research, including the description of the potential sites of geoconservation and geotourist interest [23–27]; however, a complex assessment of geotourism resources was not elaborated—only the pilot assessment of geotourism resources of these areas and several sites was a subject of conference papers [23,24,28]. In these terms, the article brings a more complex view on the geotourism resources and their potential in these areas.

2. Materials and Methods

2.1. Assessment of the Geotourism Potential

To recognize the potential of an area for geotourism, it is necessary to undertake the detailed literature and map review and detailed fieldwork which takes into account both abiotic resources (geodiversity) and other types of resources that are related to geodiversity (biotic, cultural aspects) [18]. Numerous methods for assessing the geotourism potential have been developed (for an overview the works of [20,29–34] are relevant), but they have been limited to an assessment of particular geological or geomorphological sites. Larger areas were assessed by the methods using the GIS-based analysis, e.g., [35–39], but those procedures were usually focused on geoheritage management or implications for geoconservation and did not include the cultural or economical aspects that are essential for geotourism development.

As the geomorphosites are defined as landforms that have acquired a scientific, cultural/historical, aesthetic and/or social/economic value due to human perception or exploitation, and these landforms can be represented both by single geomorphological objects and wider landscapes [19,40], it is supposed that the criteria used within this concept for the assessment of single geomorphological objects can be applied for the qualitative assessment of larger areas ("wider landscapes") as well.

Within the geomorphosite assessment, the assessment criteria are usually divided into several groups (e.g., [19,20,41–44]): Scientific value, added value, economic value, and conservation value. For the qualitative and semi-quantitative assessment of the areas and specific sites of geotourist interest, the method proposed by Reynard et al. [20] is used (Table 1). This method was already applied in several cases; e.g., [45]. It includes all the groups of the values which correspond with a holistic approach to geotourism and respect five pillars of geotourism defined by [18]. The criteria for the site assessment are applied without changes and numerical scoring is used. In the case of the assessment of "wider landscapes," some criteria were excluded, adapted and assessed qualitatively. The qualitative assessment is based on the detailed literature review, fieldwork and partly on the discussions with local people and it takes into account the assessment of particular sites.

Table 1. Criteria used for the qualitative and semi-quantitative assessment of the geotourism potential of the study areas and selected sites of geotourism interest (based on [20,42,45]).

	Criterion	Brief Explication
scientific value	integrity	current status of the site or area, degree of degradation of Earth-science features; assessed on the scale from 0 (null) to 1 (excellent); for the areas, the overall landscape quality is assessed
	representativeness	the site's or area's exemplarity; assessed on the scale from 0 (null) to 1 (excellent)
	rareness	the existence of features that are unique on the national level; assessed on the scale from 0 (null) to 1 (excellent)
	paleogeographical interest	importance of the site for the Earth or climate history; the sites assessed on the scale from 0 (null) to 1 (excellent); this criterion was not applied for the assessment of the areas
	synthesis	the average of the values (applicable for site assessment)
added value	ecological	specific or rare species, important ecosystems; assessed on the scale from 0 (no ecological value) to 1 (high ecological value)
	aesthetical	viewpoints, contrasts, space structuration; assessed on the scale from 0 (no aesthetical value) to 1 (high aesthetical value)
	cultural	archaeological, historical, artistic aspects of the area related to geodiversity, anthropogenic landforms; assessed on the scale from 0 (no cultural value) to 1 (high cultural value)
	synthesis	the average of the values (applicable for site assessment)
use characteristics	protection status	legal protection and conservation of the Earth-science features, the sites assessed on the scale from 0 (no protection) to 1 (Earth-science feature as a subject of protection)
	threats	risks and hazards: threats to geodiversity—both anthropogenic and natural, assessed on the scale from 0 (existing threats) to 1 (no considerable threats)
	accessibility	both by public and individual transport, location of the transport facilities in the proximity (in the case of sites); assessed on the scale from 0 (site with a limited accessibility) to 1 (site with a very good accessibility); for the areas, the "permeability of the landscape" is taken into account
	security	safety and limitations on specific sites, assessed on the scale from 0 (problems with safety) to 1 (no considerable limitations); this criterion was not applied to the area assessment
	site context	applicable only on the site assessment
	tourist infrastructure	catering, accommodation, shelters, tourist paths leading to the sites, proximity of these features to the specific sites; assessed on the scale from 0 (missing infrastructure) to 1 (present and diverse infrastructure)
	interpretive facilities	existing interpretive facilities, promotion of the sites/area, supporting products, the common knowledge of the area, assessed on the scale from 0 (missing interpretive facilities) to 1 (present and diverse interpretive facilities)
	educational interest	the potential for interpretation, comprehensibility for the lay public; assessed on the scale from 0 (low potential for interpretation) to 1 (high potential for interpretation)
	synthesis	the average of the values (applicable for site assessment)

The assessment is accompanied by extended SWOT analysis (Table 2) which is widely used as a common tool for local development strategies. Basic SWOT analysis has been already employed for the assessment of geotourist resources, e.g., [23,44,46,47], but extended SWOT analysis (or so-called "TOWS matrix") offers a more complex view on the geotourist resources as it provides important information about the applicability and feasibility of geotourism-development [21].

Table 2. Extended SWOT analysis ("TOWS matrix").

	Strengths	Weaknesses
Opportunities	Strengths—Opportunities (S-O) strategy (maxi-maxi): use strengths to take advantage of opportunities	Weaknesses—Opportunities (W-O) strategy (mini-maxi): overcome weaknesses by taking advantages of opportunities
Threats	Strengths—Threats (S-T) strategy (maxi-mini): use strengths to avoid the threats	Weaknesses—Threats (W-T) strategy (mini-mini): minimize weaknesses and avoid threats

Coming out of this complex assessment, the possibilities of geotourism-development are presented.

2.2. Study Areas

The geotourist potential was analyzed in the two areas of interest: The Sýkořská hornatina Mountains (the southern part) and Deblínská vrchovina Highland (Figure 1) which both belong to the geological unit of Svratka Dome ([27,48], Figure 2). Between these areas, Tišnovská kotlina and Šerkovická kotlina basins are situated and form a natural connection between the two areas. These areas are not a part of any geoparks or large-scale protected area (Protected Landscape Area or National Park according to the Czech legislative, [22]); they are of rural character and they already partly serve as a recreation base for people from the Moravian metropolis Brno and nearby towns.

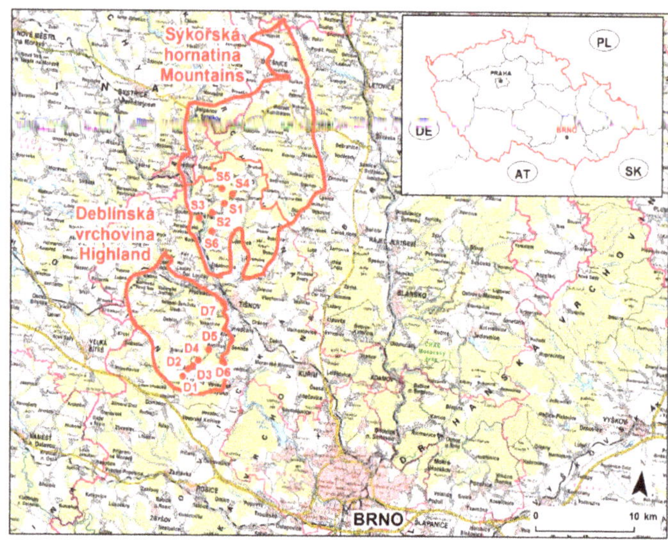

Figure 1. Position of the study areas within the southeastern part of the Czech Republic and selected sites of geotourist-interest: S1—Dobrá studně, S2—Hrušín, S3—pod Sokolí skálou, S4—Synalovské kopaniny, S5—Míchovec, S6—Veselský chlum, D1—Skalky, D2—kaolin pit, D3—abandoned limestone quarry, D4—karst spring, D5—Marškovský and Pejškovský potok streams, D6—Svratka valley, and D7—Vokoun's viewpoint.

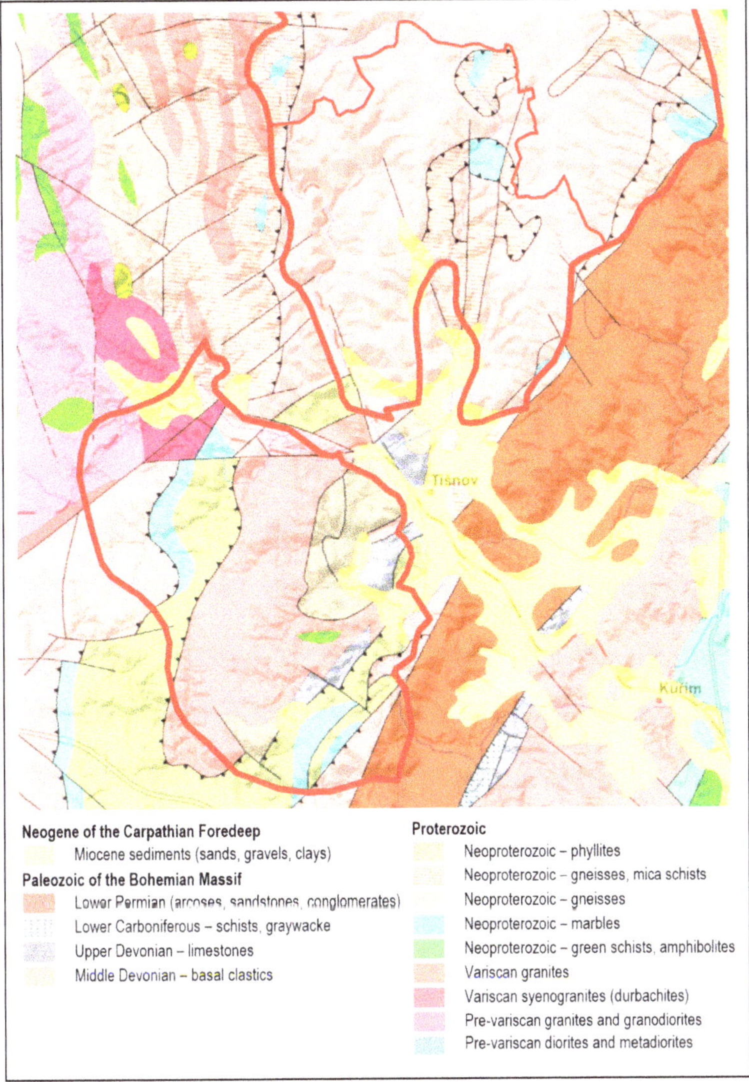

Figure 2. Geological settings of the Deblínská vrchovina Highland and the southern part of the Sýkořská hornatina Mountains (source data: [48,49]).

The detailed inventory and description of these areas were already undertaken [23,24,26,28]. The description of the specific sites which are important from the Earth-science point of view are included in the Database of Geological Localities [25], so only the brief characteristics of the study areas and selected sites of geotourist-interest are presented. The position of the sites can be seen in Figure 1.

2.2.1. Sýkořská Hornatina Mountains

The Sýkořská hornatina Mountains are situated 30 km north from Brno city, which is the second largest city in the Czech Republic (approximately 380,000 inhabitants, but the real number of people living here is higher). The harmonious landscape with well-conserved natural resources proves the sustainable use of them and represents a good example of how the people exploited the landscape in the past (Figure 3a). The part of the area is legally protected within the Svratecká Hornatina Natural Park (which represents the lowest category of general nature conservation) and there are 12 sites protected within the category of National Reserve or National Monument [50]. Geologically, the area belongs to the northern part of the Svratka Dome [48]. The basement is rather monotonous and it is formed by biotite-muscovitic, sericite-muscovitic gneisses of the Bíteš group (Figure 3b) with limited occurrences of limestone and schists covered by quaternary sediments. In specific places, there are remnants of the marine sediments of the Ottnang age [49]. Despite the relatively monotonous geological composition, the morphological diversity of the area is very high. The landscape has been affected by several geomorphological processes, but the most significant landforms were created mainly by periglacial and cryogenic processes: Tors, castle-koppies, structural ridges, block accumulations and flows, nivation depressions, cryoplanation terraces, frost-riven cliffs, isolated boulders or congelifluction scree talus cones (Figure 3e). The anthropogenic landforms are present as well, especially those of agricultural origin (heaps, terraces, ramparts, and small walls). Due to the unique combination of geology and geomorphological landforms, the Sýkořská hornatina Mountains belong to the best-preserved areas with periglacial and cryogenic landforms in the Czech Republic [24].

Figure 3. Sýkořská hornatina Mountains: (**a**) A view on the Synalov village (harmonic landscape with scattered settlements). (**b**) Bíteš orthogneiss—the main rock that builds up the area. (**c**) Jewish cemetery in Lomnice—an important part of cultural heritage with links to geodiversity (use of local stone for tombs); (**d**) small sacral monuments are common within the area and contribute to the typical character of the landscape; (**e**) congelifluction scree talus cones, isolated boulders and block accumulations on the slopes of the Sýkoř Hill (the highest peak of the study area).

The study area is rich in cultural features. The historically and architectonically valuable objects in the Lomnice Township on the southern part of the area (e.g., the Jewish cemetery (Figure 3c), synagogue, plaque column, castle, and church) and in the Lysice Township in the eastern part of the area (e.g., the chateau and church) are the most important. In the villages, sacral buildings, traditional agricultural buildings, and other objects of folk architecture can be found. In the open landscape, the small sacral objects, e.g., crosses or small chapels, are common (Figure 3d). Usually, the local building stone was used there [24].

Based on the literature review and fieldwork, the sites of geotourist interest were identified. Within the selected six sites (S1–S6, displayed in Figure 1), the above mentioned geological and geomorphological features and their relationships to the biodiversity and cultural heritage of the area can be observed.

Site 1 (S1)—Dobrá studně represents a complex of cryogenic landforms, especially the solifluction ones: Solifluction streams of several generations, terraces and occasional wet depressions can be found here. Locally, the massive gneiss boulders can be observed here. The largest solifluction stream is over 100 m long and 50 m wide, and together with others, indicates the existence of permafrost in the Pleistocene. The landforms have a crucial role for the differentiation of the vegetation cover; some of the endangered species live there.

As mentioned in the description of the geological settings of the area, the Bíteš gneiss is the main rock that builds the southern part of the Sýkořská hornatina Mountains. It has a porfyroblastic structure and it is clearly stratified with the clearly distinctive layers and direction of the metamorphosis [49]. The typical example of this rock can be observed on Hrušín (S2). The site is rich in cryogenic landforms (frost cliffs, boulder fields with plate boulders, debris accumulations, and cryoplanation terraces) and mezoforms of polygenetic origin: Small fissure caves, abris, mushroom rock, or bedding cavities. The block streams and debris accumulations are important from the ecological point of view: Thanks to the specific geomorphological and pedological settings, a natural debris forest with a high diversity of plants and the occurrence of protected species is conserved here.

The Sokolí skála Rock (S3) is a massive outcrop built of marginal facies of the Bíteš orthogneiss. The gneiss layers alternate with amphibolite beds there, which is important from the petrographic point of view. Besides this, the site possesses a significant geomorphological aspect: The outcrop was formed thanks to the erosional activity of the Svratka River which formed the deeply incised valley there. This valley is of epigenetic origin: During the tertiary uplift of the eastern margin of the Bohemian Massive, the Svratka River eroded the Miocene sediments, and then it continued to erode the gneiss bedrock. This is especially important from the paleogeographic point of view.

Synalovské kopaniny (S4) represents an example of congelifluction scree talus cones on the slopes of the Sýkoř Hill which are the result of Pleistocene cryogenic processes. Within the locality, the traces of recent slope movements can be observed. In the past, the site was used mainly as pasture land. Thanks to this, the typical mosaic of meadows, pastures, forests, and boulders has been conserved here until now.

Míchovec (S5) represents typical cryogenic landforms of the area: Tors, nivation depressions, and block streams. The cryogenic landforms are similar to those in other localities, but thanks to specific microclimatic conditions, the nivation processes were relatively intensive and strong here—there are several nivation depressions with abri with a height over 4 m. Besides this, numerous fissure caves can be found there, and recumbent folds are observable on the walls of frost cliffs. The site is also important from the ecological point of view: The occurrence of well-conserved debris forests with a massive population of endangered species *Lunaria rediviva*.

Veselský chlum (S6) displays specific aspects of the study area's history and shows evidence of how the people in the past used the land and natural resources. Numerous anthropogenic landforms (especially agrarian terraces, ramparts, and unpaved walls made of flat gneiss stones collected from the surrounding fields and pastures) can be found here. The site is protected by law and the reason for protection is the well-conserved segment of the harmonious cultural landscape with a unique mosaic of pasture land, orchards, scattered greenery, and anthropogenic landforms with high aesthetic value. Moreover, the site is an important viewpoint geosite (as defined by [51]): It offers a view on the Svratka River valley and its surroundings, so the geomorphological context of the study area can be studied and observed here.

2.2.2. Deblínská Vrchovina Highland

Deblínská vrchovina Highland lies about 25 km northwest from the Brno city. The area has a very varied geology, thanks to its position on the eastern margin of the Bohemian Massif. High lithological diversity implies a high diversity of landforms and processes. The area represents the harmonic landscape characterized by a mosaic of fields, forests, meadows, and ancient orchards. The southern margin of the area is a part of Bílý potok Natural Park [50]. The only Nature Reserve situated in the study area is represented by beech forests at Slunná; however, numerous geological and geomorphological sites (rock outcrops and abandoned quarries) are included in the Database of Geological Localities [25]. Currently, the area represents similar recreational and touristic background for the Brno City as Sýkořská hornatina Mountains (described in Section 2.2.1 Sýkořská Hornatina Mountains); however, they both remain in the shade of popular, and geologically and geomorphologically spectacular Moravian Karst [52] which is visited more frequently.

The area is situated in the southern part of the Svratka Dome, a structure which includes Svratka massif (composed of the oldest rocks of the area: Prepalaeozoic intrusive and metamorphic rocks, Devonian basal clastics and limestone, and Carboniferous siliciclastic), and the Moravicum nappe which is made up of a weak metamorphosed volcano-sedimentary complex with prevailing phyllites and orthogneiss (metamorphosed Cadomian granite) [48,49]. Neogene is represented by Miocene and Pliocene freshwater sediments that fill older valleys and depressions between the Maršov and Lažánky villages. Here, the lower Miocene sediments with abundant fauna are overburden with clays, sands, and gravels [53]. Pleistocene is represented by fluvial sandy gravel, which often forms terraces at different heights above the present valley bottom. Loess sediments are also common and reach the thicknesses of up to 5 m. Holocene flood sediments are not very thick (maximum 2 m). The Holocene also includes anthropogenic sediments (heaps and dumps of the quarries) [53].

The fluvial, karst and anthropogenic landforms, together with polygenetic rock formations modeled by slope and cryogenic processes, represent the most significant features of the study area [23,26]. The origin of the remarkable landforms is often linked to the lithology; e.g., the resistant rocks (limestone, basal clastics, quartzite, and gneiss) formed significant outcrops and elevations (Figure 4a). The most important fluvial landforms are represented by the Svratka Valley (Figure 4b); typical fluvial mezoforms can be observed in Svratka's tributaries. Anthropogenic landforms are represented by the abandoned kaolin pit (Figure 4c), limestone quarries (Figure 4d), and remains from the medieval mining of ores (adits and heaps). The use of limestone can be traced back to the Middle Ages and until the present; the remains of old lime kilns are preserved (Figure 4e) and represent an important part of local cultural heritage [23]. Water management landforms are related to the streams and allow tracing the use of natural resources in the past (Figure 4f). Other cultural features of the area are represented by historical buildings in the Tišnov city (situated on the border of the area) and Předklášteří village (especially Cistercian convent Porta Coeli) where the local building stone and material from nearby quarries were used.

Based on the literature review and fieldwork, the sites of geotourist-interest were identified. The selected sites (D1 D7, displayed in Figure 1) allow observing and studying specific geological and geomorphological features, and their relationships to the cultural heritage and ecological aspects.

Site 1 (D1)— the rock outcrop Skalky, is built of resistant quartzite. Geomorphologically, it can be described as a monadnock. Similar outcrops are situated approximately 700 m southwest of the site, in the valley of Salašský potok Stream. There, they form natural steps and during the wet seasons, there are small waterfalls. The position of this lithological member of the parautochtonal Svratka Dome sediments is not clear yet, so the site is important as a study locality. Generally, these outcrops represent a typical example of selective erosion and on the surface; numerous meso- and microrelief phenomena (especially small caverns filled with calcite and baryte) can be seen.

The kaolin deposit in the old kaolin pit (D2) is situated on the contact zone of granodiorite and phyllites. Kaolin was exploited here at the beginning of the 20th century, but it was stopped in 1939 because of the bad quality of the material. During the active exploitation, several prospection shafts were dug there, and several lignite seams were discovered. Adjacent sediments (clays) are paleontologically rich and accompanied by gravels. They represent a relic of ancient valley fill. The site is interesting from the geomorphological point of view: Small abrasion cliffs and landslides can be observed on the pit slopes.

Figure 4. Deblínská vrchovina Highland: (**a**) Quartzite outcrops of Skalky—illustration of the role of the resistance of the rock. (**b**) Svratka River valley with gravel banks. (**c**) The flooded kaolin pit near Maršov; (**d**) an abandoned limestone quarry—one of the sites where the limestone for lime burning was extracted; numerous karst features are present there. (**e**) Remains of the Havílkova lime kiln near Lažánky, an important part of industrial heritage; (**f**) water management anthropogenic landforms—channels and water races were used by mills.

The evidence of limestone quarrying is represented by abandoned limestone quarry (D3). Within the area, there are several old quarries that are currently very well incorporated into the landscape, and increase landscape diversity. Slight karstification can be observed here, including karren, small caves, and cavities filled with calcite. The limestone extracted here was suitable for lime burning; near the site, an old lime kiln is situated. The site is thus important from the historical point of view as it brings forth the evidence of using the natural resources in the past.

Under the active limestone quarry (currently closed for public), on the right slope of Maršovský potok Stream, the karst spring is situated (site D4). It is probably connected with cave systems situated in the active quarry because during dry periods, the cavemen found the continuation towards the active quarry. In this quarry, (situated just several tens of meters north of the spring), several caves with were found and documented in the 1980s, but due to the progressive quarrying, these caves were destroyed. However, the existence of the spring, its hydrological aspects, and its continuation into the limestone massive suggest that there are uncovered cave systems situated beneath the current level of the lowest quarry bench.

The site D5 (Maršovský potok and Pejškovský potok streams) represents the complex of fluvial landforms. Both valleys are rich in meanders, empty oxbow lakes, cutoffs, alluvial ramparts, gravel banks, and other fluvial landforms. At Maršovský potok Stream, there is an observable alteration of floodplains and deeply incised segments of the valley, which follow the alteration of bedrock. Moreover, the traces of anthropogenic use of the watercourses can be seen here (old water races and small dams)

The Svratka Valley (D6) is an epigenetic valley where the relics of fluvial terraces in different heights above the present valley bottom can be found. Thus, the site has high paleogeographic importance. The site is also interesting from the geomorphological point of view: Numerous cryogenic landforms (frost cliffs, and boulder and debris accumulations) are situated here. Moreover, specific vegetation communities with the occurrence of rare and endangered species can be found here. Several thermophilic species have the northernmost border of their areal here thanks to the specific geomorphologic and climatic conditions (dry and steep southwestern slopes without forest).

The Vokoun's viewpoint (D7) represents a viewpoint geosite which allows for the observing of the Tišnovská kotlina Basin (with the Svratka floodplain, Květnice Hill and Dřínová quarry which are important from an Earth-science point of view, but situated outside the study areas) and a southern part of the Sýkořská hornatina Mountains. The viewpoint is situated on the steep slope on the southern end of the village of Předklášteří, not far from an old gneiss quarry. The terrain was badly accessible and the view was obstructed by trees; however, thanks to the activity of local enthusiasts, the tourist facilities (steps, shelter, and information panel) were constructed and a newly marked tourist path leads there.

Both areas (Sýkořská hornatina Mountains and Deblínská vrchovina Highland) were recently the subject of several large-scale paintings of Adam Kašpar who introduced them at a temporary exhibition in Tišnov. They have also been the subject of other painters in the past (e.g., J. Jambor). An important social event related to geodiversity is represented by traditional mineral exhibitions which are held two times per year in Tišnov.

3. Results

The study areas and the sites of geotourist-interest were assessed by using the methods described in chapter 2.1. The assessment of the sites of geotourist interest is presented in Table 3; the assessments of the areas were elaborated separately (Table 4 for Sýkořská Hornatina Mountains and Table 5 for Deblínská vrchovina Highland). The SWOT analysis was elaborated for the two territories together, as they are situated close to each other and most of the characteristics of Sýkořská hornatina Mountains and Deblínská vrchovina Highland are in common or very similar. Table 6 thus presents the basic SWOT analysis for both areas; Table 7 shows the extended SWOT analysis. Where the differences between the particular areas occur, they are marked by indexes (S for Sýkořská hornatina Mountains, and D for Deblínská vrchovina Highland).

Table 3. Assessment of the sites of geotourism interest in the Sýkořská hornatina Mountains (S1—S6) and Deblínská vrchovina Highland (D1—D7).

Criterion/Site	S1	S2	S3	S4	S5	S6	D1	D2	D3	D4	D5	D6	D7
integrity	1.00	1.00	1.00	0.75	1.00	0.75	1.00	0.50	0.50	0.75	0.75	1.00	1.00
representativeness	1.00	1.00	1.00	0.75	1.00	1.00	1.00	1.00	1.00	1.00	1.00	1.00	1.00
[illegible]	0.75	0.25	0.75	0.50	0.50	0.50	0.75	1.00	0.25	0.50	0.25	0.75	0.50
paleogeographical interest	0.75	0.50	1.00	0.50	0.75	0.50	1.00	0.75	0.50	1.00	0.50	1.00	0.50
scientific value (synthesis)	**0.88**	**0.69**	**0.94**	**0.63**	**0.81**	**0.69**	**0.94**	**0.81**	**0.56**	**0.81**	**0.63**	**0.94**	**0.75**
ecological	1.00	1.00	0.50	1.00	1.00	1.00	0.25	0.50	0.50	0.50	0.50	1.00	0.50
aesthetical	0.50	0.50	0.75	0.75	0.50	1.00	0.50	0.75	0.50	0.25	0.50	0.75	1.00
cultural	0.00	0.00	0.00	0.75	0.00	1.00	0.00	1.00	1.00	0.25	1.00	0.50	1.00
added value (synthesis)	**0.50**	**0.50**	**0.42**	**0.83**	**0.50**	**1.00**	**0.25**	**0.75**	**0.67**	**0.33**	**0.67**	**0.75**	**0.83**
protection status	1.00	1.00	1.00	1.00	1.00	1.00	0.50	0.00	0.00	0.00	0.00	0.00	0.00
threats	0.50	0.75	0.75	0.50	0.75	0.50	0.75	0.25	0.50	0.50	0.50	0.25	0.50
accessibility	0.50	0.50	0.25	0.75	0.50	0.75	0.75	0.50	0.50	0.50	1.00	0.50	0.75
security	0.75	0.75	0.50	0.75	0.75	1.00	1.00	0.50	0.50	0.25	1.00	0.25	0.75
site context	0.50	0.50	0.75	0.75	0.50	1.00	0.75	0.75	0.75	0.25	0.50	0.50	1.00
tourist infrastructure	0.25	0.25	0.25	0.50	0.25	0.50	0.75	0.25	0.25	0.50	0.75	0.75	1.00
interpretive facilities	0.25	0.50	0.25	0.50	0.50	0.50	0.00	0.00	0.25	0.00	0.50	0.25	0.75
educational interest	0.50	0.75	0.50	0.50	0.50	0.50	0.50	0.75	0.75	0.50	0.75	0.75	0.75
use characteristics (synthesis)	**0.53**	**0.63**	**0.53**	**0.66**	**0.59**	**0.72**	**0.63**	**0.38**	**0.44**	**0.31**	**0.63**	**0.41**	**0.69**

Table 4. Assessment of the Sýkořská hornatina Mountains.

	Criterion	Qualitative Assessment
scientific value	integrity	The current status of the study area is good and it represents a typical example of the sustainable and regardful use of the natural resources (both in past and present), the current status of Earth-science phenomena in general is good as well.
	representativeness	The area represents a well-conserved landscape with traces of past sustainable use of natural resources. Particular sites represent typical examples of cryogenic, fluvial and anthropogenic landforms and processes.
	rareness	Similar harmonic landscapes can be found in different areas within Moravia, so the degree of rarity is not high. However, the geomorphological diversity is high as numerous landforms are concentrated at relatively small area.
	paleogeographical interest	n/a
added value	ecological	Most of the landscape segments which are legally protected are home to the specific and rare species, so the ecological value of the study area is quite high.
	aesthetical	Within the study area, there are numerous viewpoints to the open landscape. The landscape pattern is quite diverse (small pieces of fields, forests, villages, meadows, alleys), so the study area is aesthetically attractive. Moreover, there are not any large constructions which would disturb the landscape character.
	cultural	Cultural features are concentrated in the settlements, there are numerous small sacral objects both in the villages (chapels) and in the open landscape (wayside crosses). Also, there are old agricultural buildings and other objects of folk architecture. The landforms of anthropogenic origin (especially agricultural landforms) are important from the historical point of view as they serve evidence of the use of the landscape in the past. A series of paintings of young artist Adam Kašpar and some paintings of Josef Jambor (Moravian landscape painter) reflect the geodiversity of the area. Existence of several legends about Sýkoř Hill.
use characteristics	protection status	The conservation of the specific geological and geomorphological phenomena and adjacent ecosystems is adequate—most of these landforms are protected within the category of Natural Reserve or Natural Monument.
	threats	Concerning the environmental, respectively geological and geomorphological hazards, the area is not at risk. There may be anthropogenic pressure connected to the construction activity (new houses, communications).
	accessibility	The public transport is sufficient as the area is partly included in the Integrated transport system of the South-Moravian region. The permeability of the landscape is quite good thanks to the presence of the network of paths and local communications (both marked and not marked).
	security	n/a
	site context	n/a
	tourist infrastructure	Some of the marked paths lead through the most attractive segments of the area accompanied by shelters. The limited accommodation capacities in Lomnice or Lysice. As the area is rather used for one-day trips, the current tourist infrastructure is relatively sufficient. There are local restaurants even in the smaller villages.
	interpretive facilities	The area is promoted especially via web pages of the local communities and web pages devoted to the touristic attractions of the South-Moravian region. The knowledge and popularity of the area are rather local/regional (it is not well-known on the national level).
	educational interest	The cryogenic landforms are well visible (especially during the season without vegetation) and if the short explanation is given (e.g., via information panels), they are also comprehensive for the public. Anthropogenic landforms and processes are also easy to understand as they are related to the common activities of humans (e.g., picking the stones from the fields and accumulating them on the agrarian heaps or ramparts).

Table 5. Assessment of the Deblínská vrchovina Highland.

	Criterion	Qualitative Assessment
scientific value	integrity	The current status of the landscape is relatively good, however, particular sites can suffer from human activities (e.g., active quarrying, transport, agriculture, expansion of buildings into the open landscape).
	representativeness	The area represents a relatively well-conserved landscape with traces of past sustainable use of natural resources (limestone, kaolin, water resources). Thanks to the high lithological diversity, the morphological diversity is also high (at a relatively small area, there are landforms of different origin).
	rareness	Similar type of landscape can be found in different areas within Moravia, so the degree of rarity is not high. However, at specific sites, the rareness of the Earth-science phenomena can be considered high at regional level.
	paleogeographical interest	n/a
added value	ecological	Numerous geosites are accompanied by important ecosystems and protected species. The abandoned quarries also play a specific role regarding the biodiversity and ecosystems. Karst caves and old mining landforms (adits) are home to the protected species (bats).
	aesthetical	The mosaic of fields, meadows, and forests is aesthetically valuable, abandoned quarries increase the overall diversity of the landscape. The landscape character is disturbed by extensive built-up areas (inadequate development of living, land occupation) and partly by quarrying.
	cultural	There is a lot of buildings that use local stone. Generally, the cultural heritage is concentrated in the Tišnov city and Předklášteří village (convent Porta Coeli) and small sacral buildings in the villages within the study area. Thanks to the historical exploitation of natural resources (limestone, ores) and partly conserved mining landforms, the cultural value is also very high. A series of paintings of young artist Adam Kašpar reflect the geodiversity of the area.
use characteristics	protection status	The southern part of the area is protected in the category "Natural park"—the category of general protection of nature. The protection of the geological and geomorphological phenomena is not sufficient (particular sites are included in the Database of CGS, but they have no legal protection with the exception of karst caves which are generally protected by law).
	threats	Abandoned quarries are often used as dumps and suffer from vandalism. These undesirable activities can affect or damage natural karst features. Spreading the area of the fields can disturb the harmonic landscape as well as the spreading of the family houses and intensifying the transport.
	accessibility	The public transport is sufficient as the area is included in the Integrated transport system of the South-Moravian region. The accessibility to the particular sites is in most cases easy, the terrain is not difficult. The permeability of the landscape is quite good thanks to the presence of the network of paths and local communications (both marked and not marked).
	security	n/a
	site context	n/a
	tourist infrastructure	Tourist paths lead through the area, however, some geosites remain out of the reach of these paths. Accommodation and catering are accessible especially in Tišnov and Veverská Bítýška, but in small villages too. The area is usually visited within one-day trips, so currently, the tourist infrastructure is sufficient.
	interpretive facilities	The area is promoted especially via web pages of the local communities and web pages devoted to the touristic attractions of the South-Moravian region. The knowledge and popularity of the area are rather local/regional (it is not well-known on the national level). Specific sites are well promoted on local guides and websites of the municipalities, but some sites with high scientific and added values remain "unexplored."
	educational interest	Karst, fluvial and other features that are present here, are not important in size, but they can provide a solid basis for explanation and educational activities for local schools. Abandoned quarries are a good example of using natural resources in the past and together with cultural aspects can be an important resource for education.

Table 6. The basic SWOT analysis for both study areas.

Strengths	Weaknesses
1. harmonic landscape with well-conserved nature, high lithological (D) and morphological diversity 2. landforms and processes are well visible and comprehensible for the public 3. high added values (cultural, ecological) 4. marked paths leading to the most attractive natural features, good permeability of the landscape 5. the areas do not suffer from excessive attendance 6. good accessibility by public transport 7. sufficient legal conservation of specific sites (S)	1. the tourist infrastructure is not sufficient if the visitors want to spend here more time 2. the educational, recreational and tourist potential is not still fully recognized by locals 3. the geodiversity is not promoted as a resource for tourism and education 4. landscape and landforms have been affected by anthropogenic activity, e.g., active quarrying, urban spreading, transport or agriculture (D) 5. lack of legal protection of the geological and geomorphological phenomena (D) 6. lack of interest from the municipalities for the geotourism-development (the geotourism-development is not the priority of the local stakeholders)
Opportunities	Threats
a. study areas as a good option for one-day trips from the Brno city b. both areas can be seen as an alternative to overcrowded Moravian Karst situated nearby c. promotion of close links between geodiversity and culture/history can raise the awareness of geodiversity and foster the local identity d. geotourist and geoeducational potential can be used both for the lay public (visitors) and organized groups of students of local/regional schools e. reasonable developing of the geotourism as a driving force for the local economic development f. possibility to cooperate with local communities, schools, voluntary associations of the municipalities or subjects within Local Action Groups etc.	a. the fast and inadequate development of the tourist infrastructure can cause the disturbances and damages to the landscape and particular geological and geomorphological phenomena b. the continuing anthropogenic activity (inadequate land use) can negatively affect the character of villages or generally, the harmonic character of the landscape and it can change the aesthetic quality of the landscape c. further preference of construction activity before nature conservation and sustainable development

Table 7. The extended SWOT analysis for both study areas.

S-O Strategy (maxi-maxi)	W-O Strategy (mini-maxi)
- promotion of the natural and cultural heritage related to geodiversity can attract visitors from nearby towns and metropolis (1, 2, 3, a, b) - both areas can be presented as an alternative to overcrowded Moravian Karst (5, b) - good accessibility and good permeability of the landscape can be presented as an advantage and interesting investment opportunity for the potential "developers" of tourist infrastructure (4, 6, e) - developing a new geotourist product or including the geodiversity aspects into the current tourist offer can be used both by visitors/tourists and by students and pupils of local schools and schools (1, 2, d, f)	- focus on the short-term recreation and tourism (1, a, b) - promotion of the links between abiotic-biotic-cultural components of the area can help to raise the awareness of the geotourist and geoeducational resources (2, 3, 6, c, d) - promotion of the geotourist resources and their good accessibility as an advantage for tourism and economic development (3, 4, e, f) - involving local communities and subjects (e.g., schools) to raise the awareness of the value of geodiversity (3, 4, d, f) - emphasizing the close links between abiotic—biotic—cultural components of the landscape can help to justify the need for conservation (4, 5, c)
S-T Strategy (maxi-mini)	W-T Strategy (mini-mini)
- present the geodiversity as an important resource for tourism and as an entity that has to be conserved for future generations (1, 2, a, b) - maintaining and fostering sufficient legal conservation of the specific sites can help to avoid the disturbing activities that can negatively affect the landscape and particular geodiversity components (7, b) - promotion of geotourist resources (incl. examples of good practice from different regions) on the meetings of municipalities or local stakeholders (e.g., Local Action Groups) can overcome the lack of interest from the local stakeholders (1, 2, c)	- to avoid the anthropogenic pressure and uncontrolled development of tourist infrastructure, especially via landscape planning, development strategies and conservation measures (1, 4, a, b) - to promote geotourism concept as an alternative to traditional (demanding) tourism, to stress the sustainability of this form of tourism, - to cooperate with successful subjects and regions that use the concept of geotourism (2, 3, 6, c) - maintain the development of tourist infrastructure in accordance with geotourism and nature conservation principles (4, 5, a, b) - to avoid the future damage of geological and geomorphological sites by using the landscape planning and management strategies and public discussion with stakeholders (4, 5, 6, b, c)

4. Discussion and Concluding Remarks

The criteria used within geomorphosite concept proved to be a simple and comprehensive tool for qualitative and semi-quantitative assessment of geotourist resources within larger areas. The assessment of larger (wider) areas within a geomorphosite concept has some specifics—this assessment is rather qualitative and was based on expert knowledge, numerical assessment of particular sites, detailed fieldwork, or discussions with residents. A degree of subjectivity exists there; however, the qualitative assessment is probably more comprehensible for the local authorities or stakeholders than the numerical one. In the future, the assessment of geotourist resources can be accompanied, e.g., by an approach presented by Martins and Pereira [33], which is based on the perception of local people. The numerical assessment of the sites of geotourist interest which served as one of the bases for the qualitative assessment are more objective; however, the assessment of specific criteria within the Reynard's method remains relatively subjective (e.g., aesthetic value).

In comparison with the geomorphosite concept, the SWOT analysis represents an even more comprehensible tool for assessing geotourist resources. It is easily understandable for authorities, members of Local Action Groups, and other subjects that aim to participate in geotourism development, and thus can serve a simple way for assessing geotourist resources and setting the directions and possibilities of geotourism development as proved by numerous studies; e.g., [21,44,46,54,55].

Qualitative and semi-quantitative assessment, basic SWOT and extended SWOT analysis thus allowed us to identify the directions of geotourism development, and to propose particular activities to use the geotourist resources in a sustainable way. Based on this, specific strategies for geotourism development can be proposed:

1) The geodiversity of the assessed areas is not unique on the national level, but the educational value is high: Landforms and processes are illustrative, visible, and relatively simple to understand (e.g., thq role of the rock resistance in the shaping of significant outcrops at Skalky in the Deblínská vrchovina Highland, or typical cryogenic landforms in the Sýkořská vrchovina Mountains) which is supported by numerical assessment of specific localities. Integrity and conditions of landforms are relatively high thanks to the position of the areas outside the main tourist destinations. The landscape is well-preserved and it shows a good example of the co-existing of man and nature. Moreover, specific sites are very important from the paleogeographic point of view (especially epigenetic valleys of Svratka). These issues were assessed as the main resource with high potential for developing sustainable geotourism and educational activities (both for local people and visitors). It has to be emphasized that geotourism provides economic, cultural, and social benefits for both visitors and hosting communities [56].

2) Added value is closely linked to the geodiversity. Both areas can present numerous examples of the mutual relationships between abiotic, biotic, and cultural components of the landscape (historical values, geomythological aspects, the traces of the landscape memory, and local materials used for local buildings and constructions). This is supported by the high value of several sites—especially in the Sýkořská hornatina Mountains, where numerous sites have an important biotic element which is legally protected there (together with geo-elements). This holistic approach has to be taken into account when planning the management of the landscape and conservation measures: The existing links can help improve acceptance of conservation measures (in Deblínská vrchovina Highland) and can increase the overall attractiveness of the area in the terms of interpretation of the heritage. The mutual links between abiotic, biotic, and cultural components can be used for environmental education as well, and can help to raise awareness about geodiversity in the study areas.

3) Accessibility of the areas is relatively good; the tourist facilities are average or below the average. It is subject to further discussions about whether the adjacent tourist infrastructure has to be developed. If decided to support the geotourism in these areas, some additional tourist infrastructure should be built; however, this has to be balanced with geoconservation

principles. According to Dowling and Newsome [18], the geotourism should be sustainable and environmentally friendly, so this has to be respected while developing the tourist infrastructure, improving the access to the particular sites or building accommodation capacities.

4) The number of visitors and knowledge/popularity of the areas is not high. The promotion is very irregular. In order to develop the geotourism, the promotion should be assured and should take into account two aspects: The promotion of specific sites of geotourist interest and the promotion of the area as a whole with its cultural heritage related to geodiversity, with its history or with its specific characteristics. Due to the fact that the geotourist resources of these areas cannot compete with the Moravian Karst Protected Landscape Area with its caves, springs, and spectacular outcrops and valleys, these areas will probably never reach high popularity. Nevertheless, they can be promoted as a calm alternative to the overcrowded Moravian Karst or an accessible and pleasant area for short-term recreation and tourism. New geotourism products (e.g., educational path connecting significant sites of geological and geomorphological interest, local products related to the geodiversity resources, and information panels on websites) can attract both visitors and local people.

5) If it is decided to support the development of geotourist activities, close communication with local communities and initiatives is needed in order to develop the effective management of geotourist resources. Cooperation with research institutions is important, as academic research provides the background for further activities supporting the promotion of geoheritage [57]; however, they have to be implemented by local communities themselves. Thus, a bottom-up approach has to be respected. Moreover, the volunteer activities can increase the local awareness and appreciation of geoheritage [58], and can foster the local identity in general. As there are active NGOs, volunteer associations of municipalities, or Local Action Groups in these areas (as indicated in SWOT analysis), it can be supposed that the bottom-up approach can have success. There have already been several specific cases recorded where the local NGOs made the sites accessible or visible.

6) The geotourist activities have to accept the intrinsic value of geodiversity and respect the principles of nature conservation (respectively geoconservation as defined by Prosser et al. [59]). Legal protection of specific sites is already set up (in Sýkořská vrchovina Mountains); however, other sites are not protected, thus can be endangered by human activities. The involvement of local subjects, and informing them about these geotourist resources can improve acceptance of the conservation measures. As the geoethical practice is an essential part of geotourism [56], this aspect of using the geotourist resources should be also taken into account.

As geodiversity represents a basis for the geotourism, it can be considered an important resource for the local and regional development. In order to use this resource in a responsible and sustainable way, the inventory and assessment of the geodiversity and geoheritage are the initial steps which have to be reflected in the plans for geotourism-development; e.g., [18,60,61]. In these terms, cooperation with universities and research institutions is more than desirable.

The particular outcomes from the assessment and basic/extended SWOT analysis can be used in strategic development document or planning. The plans have to follow the geoconservation rules and principles of sustainable development, and in the future, they can become a respected part of a local and regional planning and development conceptions and strategies.

Funding: The research was supported by Internal Grant Agency MENDELU, project number VP_2018024.

Acknowledgments: The Author thanks the three anonymous reviewers for their helpful comments.

Conflicts of Interest: The author declares no conflict of interest.

References

1. Hose, T.A. 3G's for Modern Geotour. *Geoheritage* **2012**, *4*, 7–24. [CrossRef]

2. Dowling, R.K.; Newsome, D. *Handbook of Geotourism*; Edward Elgar Publishing: Cheltenham, UK, 2018; ISBN 978-1-78536-885-1.

3. Ólafsdóttir, R. Geotourism. *Geosciences* **2019**, *9*, 48. [CrossRef]

4. Olafsdóttir, R.; Tverijonaite, E. Geotourism: A Systematic Literature Review. *Geosciences* **2018**, *8*, 234. [CrossRef]

5. Hose, T.A. Selling the Story of Britain's Stone. *Environ. Interpret.* **1995**, *10*, 16–17.

6. Słomka, T.; Kicińska-Świderska, A. Geoturystyka—Podstawowe pojęcia. *Geoturystyka* **2004**, *1*, 5–7.

7. Joyce, E.B. Geomorphological sites and the new geotourism in Australia. *Geol. Soc. Aust. Melb.* **2006**, *2006*, 1–4.

8. Dowling, R.K.; Newsome, D. (Eds.) *Geotourism*; Elsevier Butterworth-Heinemann: Oxford, UK; Burlington, MA, USA, 2006; ISBN 978-0-7506-6215-4.

9. Dowling, R.K. Global geotourism—An emerging form of sustainable tourism. *Czech J. Tour.* **2013**, *2*, 59–79. [CrossRef]

10. Arouca Declaration on Geotourism November 12, 2011 Portugal. Available online: http://www.europeangeoparks.org/?p=223 (accessed on 25 September 2018).

11. Martini, G.; Alcalá, L.; Brilha, J.; Iantria, L.; Sá, A.A.; Tourtellot, J. Reflections about the geotourism concept. In Proceedings of the 11th European Geoparks Conference, Arouca Geopark, Portugal, 19–21 September 2012.

12. Gray, M. *Geodiversity: Valuing and Conserving Abiotic Nature. Second Edition*; Wiley-Blackwell: Chichester, UK, 2013; ISBN 978-0-470-74215-0.

13. McKeever, P.; Zouros, N. Geoparks: Celebrating Earth heritage, sustaining local communities. *Episodes* **2005**, *28*, 274–278.

14. What Is a Geopark. Available online: http://www.europeangeoparks.org/?page_id=165 (accessed on 2 July 2019).

15. Štrba, L.; Kršák, B.; Sidor, C. Some Comments to Geosite Assessment, Visitors, and Geotourism Sustainability. *Sustainability-Basel* **2018**, *10*, 2578. [CrossRef]

16. The Network of National Geoparks. Available online: http://www.geology.cz/narodnigeoparky/o-siti (accessed on 1 July 2019).

17. Kubalíková, L. Promoting geomorphological heritage: Bringing geomorphology to people. In *Landscapes and Landforms of the Czech Republic*; Pánek, T., Hradecký, J., Eds.; Springer International Publishing: Basel, Switzerland, 2016; pp. 399–410. ISBN 978-3-319-27537-6.

18. Newsome, D.; Dowling, R. *Geotourism: The Tourism of Geology and Landscape*; Goodfellow Pub Ltd.: Woodeaton, Oxford, UK, 2010; ISBN 978-1-906884-09-3.

19. Panizza, M. Geomorphosites. Concepts, methods and examples of geomorphological survey. *Chin. Sci. Bull.* **2001**, *46*, 4–6. [CrossRef]

20. Reynard, E.; Perret, A.; Bussard, J.; Grangier, L.; Martin, S. Integrated Approach for the Inventory and Management of Geomorphological Heritage at the Regional Scale. *Geoheritage* **2015**, *8*, 43–60. [CrossRef]

21. Carrión Mero, P.; Herrera Franco, G.; Briones, J.; Caldevilla, P.; Domínguez-Cuesta, M.J.; Berrezueta, E. Geotourism and Local Development Based on Geological and Mining Sites Utilization, Zaruma-Portovelo, Ecuador. *Geosciences* **2018**, *8*, 205. [CrossRef]

22. Law 114/1992 Coll. on Nature Conservation and Landscape Protection. Czech Republic, Prague. 1992. Available online: https://www.zakonyprolidi.cz/cs/1992-114 (accessed on 22 August 2019).

23. Kubalíková, L.; Kirchner, K. Relief assessment methodology with respect to geoheritage based on example of the Deblinska vrchovina highland. In *Public Recreation and Landscape Protection—With Man Hand in Hand. Conference Proceeding*; Fialová, J., Kubíčková, H., Eds.; Mendel University: Brno, Czech Republic, 2013; pp. 131–141.

24. Kirchner, K.; Kubalíková, L.; Bajer, A. Local geoheritage: Its importance and potential for geotourist and recreational activities (a case study from Lomnicko area). In *Public Recreation and Landscape Protection—With Nature Hand in Hand*; Fialová, J., Pernicová, D., Eds.; Mendel University: Brno, Czech Republic, 2017; pp. 202–211.

25. Significant Geological Localities of the Czech Republic—Czech Geological Survey. Available online: http://lokality.geology.cz/d.pl?item=1&l=e (accessed on 26 June 2019).

26. Kubalíková, L. Hodnocení Geomorfologických Lokalit V Kontextu Ochrany Neživé Přírody. Ph.D. Thesis, Masaryk University, Brno, Czech Republic, 2011.

27. Demek, J.; Mackovčin, P. *Hory a Nížiny—Zeměpisný Lexikon ČR*; Nature Conservation Agency of the Czech Republic: Brno, Czech Republic, 2006; ISBN 978-80-86064-99-9.
28. Kubalíková, L. Geodiversity as a hidden resource for local development: Possibilities of geotourism growth outside the geoparks. In *Proceedings of the Region in the Development of Society 2018, Proceedings of the International Scientific Conference*; Mendel University: Brno, Czech Republic, 2018; pp. 333–343.
29. Pralong, J.-P. A method for assessing tourist potential and use of geomorphological sites. *Géomorphologie Relief Process. Environ.* **2005**, *11*, 189–196. [CrossRef]
30. Kubalíková, L. Geomorphosite assessment for geotourism purposes. *Czech J. Tour.* **2013**, *2*, 80–104. [CrossRef]
31. Štrba, L.; Rybár, P.; Baláž, B.; Molokáč, M.; Hvizdák, L.; Kršák, B.; Lukáč, M.; Muchová, L.; Tometzová, D.; Ferenčíková, J. Geosite assessments: Comparison of methods and results. *Curr. Issues Tour* **2015**, *18*, 496–510. [CrossRef]
32. Brilha, J. Inventory and Quantitative Assessment of Geosites and Geodiversity Sites: A Review. *Geoheritage* **2016**, *8*, 119–134. [CrossRef]
33. Martins, B.; Pereira, A. Residents' Perception and Assessment of Geomorphosites of the Alvão—Chaves Region. *Geosciences* **2018**, *8*, 381. [CrossRef]
34. Ovreiu, A.B.; Comănescu, L.; Bărsoianu, I.A. Evaluating Geomorphosites and the Geomorphological Hazards that Impact them: Case Study—Cozia Massif (Southern Carpathians, Romania). *Geoheritage* **2019**. [CrossRef]
35. Serrano, E.; Ruiz-Flaño, P. Geodiversity: A theoretical and applied concept. *Geogr. Helv.* **2007**, *62*, 140–147. [CrossRef]
36. Pellitero, R.; González-Amuchastegui, M.; Ruiz-Flaño, P.; Serrano, E. Geodiversity and Geomorphosite Assessment Applied to a Natural Protected Area: The Ebro and Rudron Gorges Natural Park (Spain). *Geoheritage* **2011**, *3*, 163–174. [CrossRef]
37. Zwoliński, Z.; Stachowiak, J. Geodiversity map of the Tatra National Park for geotourism. *Quaest. Geogr.* **2012**, *31*, 99–107. [CrossRef]
38. Pereira, D.; Pereira, P.; Brilha, J.; Santos, L. Geodiversity Assessment of Parana State (Brazil): An Innovative Approach. *Environ. Manag.* **2013**, *52*, 541–552. [CrossRef] [PubMed]
39. Zwoliński, Z.; Najwer, A.; Giardino, M. Methods for assessing geodiversity. In *Geoheritage: Assessment, Protection, and Management*; Reynard, E., Brilha, J., Eds.; Elsevier: Amsterodam, The Netherlands, 2018; pp. 27–52. ISBN 978-0-12-809531-7.
40. Reynard, E.; Panizza, M. Geomorphosites: Definition, assessment and mapping. An introduction. *Géomorphologie: Relief Process. Environ.* **2005**, *11*, 177–180.
41. Pereira, P.; Pereira, D.; Alves, M. Geomorphosite assessment in Montesinho Natural Park (Portugal). *Geogr Helv* **2007**, *62*, 159–168. [CrossRef]
42. Reynard, E.; Fontana, G.; Kozlik, L.; Scapozza, C. A method for assessing "scientific" and "additional values" of geomorphosites. *Geogr. Helv.* **2007**, *62*, 148–158. [CrossRef]
43. Pereira, P.; Pereira, D. Methodological guidelines for geomorphosite assessment. *Géomorphologie* **2010**, *16*, 215–222. [CrossRef]
44. Kubalíková, L.; Kirchner, K. Geosite and Geomorphosite Assessment as a Tool for Geoconservation and Geotourism Purposes: A case Study from Vizovická vrchovina Highland (Eastern Part of the Czech Republic). *Geoheritage* **2016**, *8*, 5–14. [CrossRef]
45. Boukhchim, N.; Fraj, T.B.; Reynard, E. Lateral and "Vertico-Lateral" Cave Dwellings in Haddej and Guermessa: Characteristic Geocultural Heritage of Southeast Tunisia". *Geoheritage* **2018**, *10*, 575–590.
46. Nazaruddin, D.A. Systematic Studies of Geoheritage in Jeli District, Kelantan, Malaysia. *Geoheritage* **2017**, *9*, 19–33. [CrossRef]
47. Ateş, H.Ç.; Ateş, Y. Geotourism and Rural Tourism Synergy for Sustainable Development—Marçik Valley Case—Tunceli, Turkey. *Geoheritage* **2019**, *11*, 207–215. [CrossRef]
48. Jaroš, J.; Mísař, Z. Nomenclature of the tectonic and lithostratigraphic units in the Moravian Svratka Dome (Czechoslovakia). *Věstník Ústředního Ústavu Geol.* **1976**, *51*, 113–122.
49. Maps—Czech Geological Survey. Available online: http://www.geology.cz/extranet-eng/maps (accessed on 26 June 2019).
50. Central Nature Conservation List. Available online: https://drusop.nature.cz/portal/ (accessed on 2 July 2019).
51. Migoń, P.; Pijet-Migoń, E. Viewpoint geosites—Values, conservation and management issues. *Proc. Geol. Assoc.* **2017**, *128*, 511–522. [CrossRef]

52. Mackovčin, P. *Chráněná území ČR. IX., Brněnsko*; Nature Conservation Agency of the Czech Republic: Brno, Czech Republic, 2007; ISBN 978-80-86064-66-6.

53. Hanžl, P.; Buriánková, K.; Čtyroká, J.; Čurda, J.; Gilíková, H.; Gürtlerová, P.; Kabátník, P.; Kratochvílová, H.; Manová, M.; Maštera, L.; et al. *Vysvětlivky k Základní geologické mapě České republiky 1: 25 000, list 24-321 Tišnov*; Czech Geological Survey: Brno, Czech Republic, 2001.

54. Farsani, N.T.; Coelho, C.; Costa, C. Geotourism and Geoparks as Gateways to Socio-cultural Sustainability in Qeshm Rural Areas, Iran. *Asia Pac. J. Tour Res.* **2012**, *17*, 30–48. [CrossRef]

55. Cai, Y.; Wu, F.; Han, J.; Chu, H. Geoheritage and Sustainable Development in Yimengshan Geopark. *Geoheritage* **2019**, *11*, 1–13. [CrossRef]

56. Gordon, J.E. Geoheritage, Geotourism and the Cultural Landscape: Enhancing the Visitor Experience and Promoting Geoconservation. *Geosciences* **2018**, *8*, 136–160. [CrossRef]

57. Pijet-Migoń, E.; Migoń, P. Promoting and Interpreting Geoheritage at the Local Level—Bottom-up Approach in the Land of Extinct Volcanoes, Sudetes, SW Poland. *Geoheritage* **2019**, 1–10. [CrossRef]

58. Miles, E. Involving local communities and volunteers in geoconservation across Herefordshire and Worcestershire, UK—The Community Earth Heritage Champions Project. *Proc. Geol. Assoc* **2013**, *124*, 691–698. [CrossRef]

59. Prosser, C.; Bridgland, D.R.; Brown, E.; Larwood, J. Geoconservation for science and society: Challenges and opportunities. *Proc. Geol. Assoc* **2011**, *122*, 337–342. [CrossRef]

60. Bouzekraoui, H.; Barakat, A.; Mouaddine, A.; El Youssi, M.; Touhami, F.; Hafid, A. Mapping geoheritage for geotourism management, a case study of Aït Bou Oulli Valley in Central High-Atlas (Morocco). *Environ. Earth Sci.* **2018**, *77*, 413. [CrossRef]

61. Cocean, G.; Cocean, P. An Assessment of Gorges for Purposes of Identifying Geomorphosites of Geotourism Value in the Apuseni Mountains (Romania). *Geoheritage* **2016**, *9*, 71–81. [CrossRef]

Article

Quantitative and Qualitative Assessment of the "El Sexmo" Tourist Gold Mine (Zaruma, Ecuador) as A Geosite and Mining Site

Paúl Carrión-Mero [1,2,*], Oscar Loor-Oporto [3], Héctor Andrade-Ríos [3], Gricelda Herrera-Franco [4], Fernando Morante-Carballo [5], María Jaya-Montalvo [3], Maribel Aguilar-Aguilar [3], Karen Torres-Peña [3] and Edgar Berrezueta [6,*]

[1] Centro de Investigaciones y Proyectos Aplicados a las Ciencias de la Tierra (CIPAT), ESPOL Polytechnic University, Escuela Superior Politécnica del Litoral, ESPOL, Campus Gustavo Galindo Km 30.5 Vía Perimetral, P.O. Box 09-01-5863 Guayaquil, Ecuador

[2] Facultad de Ingeniería en Ciencias de la Tierra, ESPOL Polytechnic University, Escuela Superior Politécnica del Litoral, ESPOL, Campus Gustavo Galindo Km 30.5 Vía Perimetral, P.O. Box 09-01-5863 Guayaquil, Ecuador

[3] Bira Bienes Raíces S.A. (BIRA S.A.), Av. Alfonso de Mercadillo, P.O. Box 071350 Zaruma, Ecuador; oslooro@gmail.com (O.L.-O.); hecandra@gmail.com (H.A.-R.); mjaya@espol.edu.ec (M.J.-M.); maesagui@espol.edu.ec (M.A.-A.); kdtorres@espol.edu.ec (K.T.-P.)

[4] Facultad de Ciencias de la Ingeniería, Universidad Estatal Península de Santa Elena (UPSE), Avda. principal La Libertad—Santa Elena, La Libertad 240204, Ecuador; grisherrera@upse.edu.ec

[5] Facultad de Ciencias Naturales y Matemáticas, Geo-recursos y Aplicaciones GIGA, ESPOL Polytechnic University, Escuela Superior Politécnica del Litoral, ESPOL, Campus Gustavo Galindo Km 30.5 Vía Perimetral, P.O. Box 09-01-5863 Guayaquil, Ecuador; fmorante@espol.edu.ec

[6] Instituto Geológico y Minero de España (IGME), C/Matemático Pedrayes 25, 33005 Oviedo, Spain

*   Correspondence: pcarrion@espol.edu.ec (P.C.-M.); e.berrezueta@igme.es (E.B.); Tel.: +593-99-826-5290 (P.C.-M.)

Received: 6 December 2019; Accepted: 26 February 2020; Published: 10 March 2020

Abstract: Zaruma is host to the "El Sexmo" tourist mine, the galleries of which extend below the city, and its exploitation dates back to precolonial times. The mining boom created important development in the area, but informal mining also emerged causing environmental issues and safety problems. This study presents a qualitative and quantitative assessment of the "El Sexmo" Tourist Mine in the context of its potential as a tourism geosite and mining site. The methodological stages included: (i) The process and systematization of the general mine information and its surroundings; (ii) the assessment of the geological and mining interest of the mine, through GAM and Brilha method; and (iii) description and proposal of action strategies through Delphi analysis and a Strengths, Weaknesses, Opportunities, and Threats (SWOT) matrix. Based on the results of the quantitative evaluation, the high values in the educational, scientific, and tourist aspects of the two applied methodologies, show the mine as a potential geosite and mining site with added cultural value. In addition, the quantitative assessment in correspondence with the qualitative analysis, allowed to propose improvement strategies to take advantage of the geological resources and mining identity of the area, as an alternative that strengthens the infrastructure of the mine and consolidates the geotouristic development of the area.

Keywords: geotourism; geosite; mining site; Zaruma; El Sexmo; tourist mine

---

1. Introduction

Mining activity generates positive impacts on the economy and development of surrounding communities around the influence area of mineral deposits, it also produces negative effects on the economic, ecological, and aesthetic aspects [1–4]. Once mineral exploitation ceases, these productive zones become abandoned areas, and different post-mining reclamation methods are proposed to rescue these areas [5]. Outstanding initiatives include solar power plants in inactive open-pit mines (sustainable energy) [1], theme parks (creation of green areas) [6], underground museums [7], and tourist mines (geological, mining, and cultural heritage) [8].

Some tourist mines represent an example of mining geo-heritage due to the peculiar characteristics that can be used for education, research, and geotourism, and in particular, these places retain historical issues that are closely related to such as cultural aspects, thus constituting a means to approach mining activity in a sustainable way.

Around the world, there are mines that are part of UNESCO (United Nations Educational, Scientific and Cultural Organization) World Cultural Heritage and UNESCO Global Geoparks declared regions. An example of World Cultural Heritage mines are the salt mines of Wieliczka and Bochnia (Poland), whose galleries have hundreds of kilometers with works of art, underground chapels, and statues sculpted in salt that reflect the mining activity developed in Europe from the 13th to the 20th century [9]. In 1978, the Wieliczka salt mine was declared a UNESCO cultural heritage site and later, in 2013, the Bochnia mine was added to the UNESCO World Heritage List as an extension of the Wieliczka salt mine inscription [9,10]. On the other hand, an example of a mine belonging to the UNESCO Global Geoparks is the diamond mine of the Yimengshan Geopark (China) where the largest kimberlite-type diamonds in Asia were extracted during the active phase of the mine. In addition, the mine was the source of the first primary diamond discovered in China [11]. Minas Gerais, Mariana, Tiradentes, Sabara, and Diamantina in Ouro Preto (Brazil), as well as the El Teniente mine in Sewell town (Chile) are some of the gold mining references [12].

Geo-heritage or geological heritage has been defined by Brilha [13] as those elements of in situ and ex situ geodiversity (abiotic elements of nature's diversity) that have exceptional scientific value. Brilha [14] made a list of the main types of geo-heritage in 'UNESCO Global Geoparks' (UGGs) based on the Information Available at the UNESCO Website (127 geoparks' websites), some types of geo-heritage in UNESCO Global Geoparks (UGG) are geomorphology, glacial and ice age, karst, landforms, mining, palaeontological, stratigraphical, tectonics, and volcanological types. Examples of geoparks linked to mining geo-heritage are: Parco Geominerario della Sardegna (Italy), the Tuscan Mining Park (Italy), Comarca Minera (Mexico), Idrija (Slovenia), and central Catalonia (Spain). Mining geo-heritage is a concept directly related to geo-heritage [15] and is part of one of the specific anthropogenic geo-heritage categories [12].

In Ecuador, the term 'geological heritage' or 'geo-heritage' was recently unknown. However, in the wake of the creation of the Ecuadorian Geopark Committee (CEG) in February 2019 and the official declaration of the Imbabura Geopark as a UNESCO Global Geopark in April 2019, which is the first geopark in Ecuador [16] and the seventh geopark to be recognized in South America, the dissemination of the subject has gained increased attention in Ecuador in recent years. Several geopark project initiatives have emerged, such as the Tungurahua Volcano Geopark project, Napo Sumaco Geopark project, Santa Elena Geopark project [17], Galapagos, Puyango Petrified Forest [18], Jama Pedernales, and the "Ruta del Oro" (Gold Route) project [15].

Diverse authors have assessed sites with geological interest using methods based on parameters or criteria (scientific value, educational value, cultural value, scenic/aesthetic values, ecological value, potential use value, recreational value, degradation risk, economic value, tourism value, protection, and functional value, etc.) [19–26].

From the known evaluation methods, two quantitative methods have been selected in this study, considering the different evaluation approaches: (i) The geosite assessment model (GAM), based on a preliminary physical evaluation of geosites and (ii) the Brilha method, which evaluates elements of geodiversity. For instance, the Vrdnik mine, an abandoned coal mine consisting of 26 underground mine shafts up to a depth of 280 m, was a geosite evaluated by the GAM (geosite assessment method) and was part of the first Serbian national park, Fruška Gora Mountain [27]. Another example is the Sierra Mágina Natural Park, located in the south-center of Jaén, Spain, where five sites were evaluated with the Brilha method. The evaluation included Sierra Mpage Karst, which obtained moderate ratings in terms of scientific (300 points), didactic (220 points), and tourism interests (205 points), with a low degradation risk (75 points), which qualifies it as a site of geological interest [28].

In a previous phase of the Academic Research Project, the Spanish Inventory of Geological Places of Interest (IELIG, acronym in Spanish) methodology was used for the consideration of a list of potential geosites and mining sites identified in the Zaruma-Portovelo area (27 sites were previously assessed). The "El Sexmo" Tourist Mine obtained the highest score for its scientific, academic, and tourist interests [15]. With a desire to strengthen the assessment of this site of geological and mining interest, the GAM and Brilha method are used.

Is it possible that the approach and promotion of a tourist mine through enhancement and development strategies of technical and integral mining with consideration of Sustainable Development Goals (SDGs) can become a reference methodology or process for geotourism in the context of a geopark project?

Considering the problem currently (subsidence due to illegal underground mining activity) facing Zaruma and looking for an alternative that enhances the geomining values and its relationship with culture, the aim of this study is to evaluate the "El Sexmo" Tourist Mine as an agent of geotourism and development linked to Sustainable Development Goals (SDGs) through the detailed assessment of its geological and mining interest based on internationally recognized methodologies to proposal of strategies that optimize their use in the context of geological and mining heritage.

1.1. Study Zone

The mining city of Zaruma is located in the southwest of Ecuador, in the upper part of El Oro Province (Figure 1a,b). Zaruma has a strong mining tradition, and it was declared a cultural heritage site of the Ecuadorian State in 1999 [29]. Since 1998, Zaruma has been on UNESCO's tentative list of candidates as a World Heritage Site [30]. Zaruma has a characteristic architectural style (19th century wooden buildings) and lies in a geomorphological landscape featured by steep slopes (Figure 1c). The foundation of the city was on underground galleries due to the existence of an important epithermal-type gold deposit [31]. The "El Sexmo" Tourist Mine is linked to the underground gallery system (Figure 1b,d), whose exploitation ceased because the ore extraction was no longer secure when it transformed into an urban area. Since 2005, the mine has become a tourist complex [32] and it is the oldest mine in Ecuador, whose exploitation dates back to precolonial times. Zaruma city distinguishes the mine for its cultural and geological wealth, and it is an icon of the beginnings of the mining activity in Ecuador. Currently, the area is at high geological risk as it has suffered from subsidence since 2016 due to illegal underground mining activity [33].

1.2. "El Sexmo" Tourist Mine History

The "El Sexmo" Tourist Mine is a mining monument (Figure 1d), a synonym of history, culture, and development. The remarkable structural aspect of the galleries provides evidence of the working conditions of the miners of the 15th century [34]. During the 1880–1896 period, the company Great Zaruma Gold Mining had estimated reserves of 8.40 g/t Au and 5.90 g/t Ag in the Sexmo–Miranda system [35]. Currently, a segment of 500 m of the mine is used for tourism development.

The origin of the "El Sexmo" Tourist Mine dates back five centuries in the mining history of Ecuador, when the golden bonanza season in Zaruma took place in the second half of the 16th century and the first of the 17th century [36]. The gold boom in the region marked this epoch to such an extent that it made the city of Zaruma one of the main economic sources of Spanish America and the royal audience of Quito [30]. At that time, the Inca Empire established the Zaruma–Portovelo district,

and history mentioned that the gold extracted from Zaruma's mines was the reward to rescue the
Atahualpa Emperor [37].

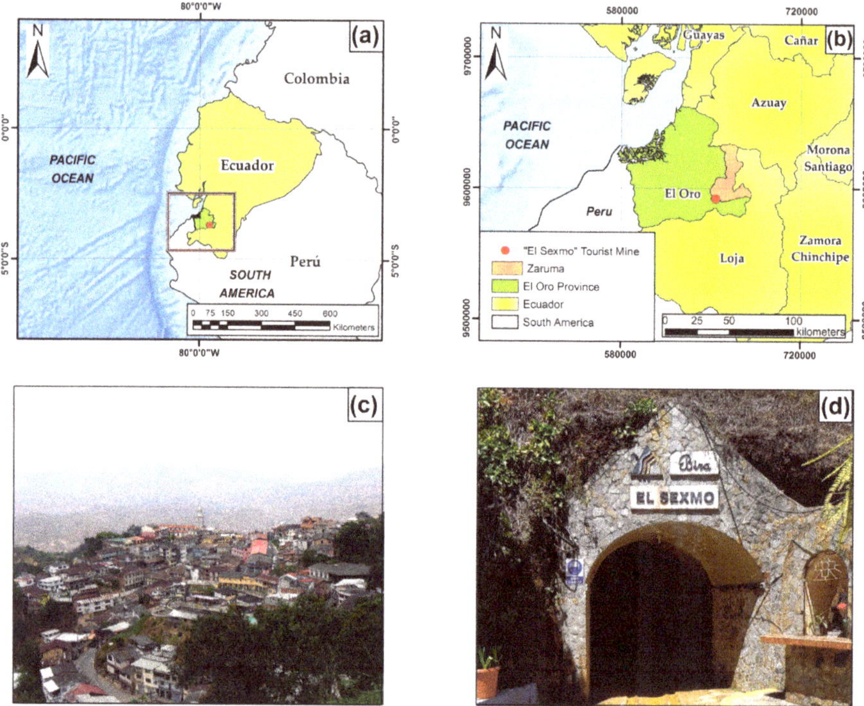

Figure 1. (**a**,**b**) Location of the study area. (**c**) Panoramic view of the Zaruma city. (**d**) "El Sexmo"
Tourist Mine.

During the first colonial period, the mining activity operated under a schedule of rotating labor
days imposed on the indigenous community; this was known as the system of "mitas" and was
proposed by Toledo in 1573 [38–40]. As a result of this cycle, the increase in mineral production gave
rise to a historic event in one of Zaruma's mines, the discovery of "pepa de oro", a gold nugget, whose
weight was more than three pounds. The gold nugget was given as a gift to Philip II King of Spain,
and in gratitude, the King reduced the one-fifth real tax, a percentage that was paid from the extraction
of precious metals, to one-sixth (sexmo, in Spanish). The tax was one-sixth of the profits obtained
from the deposits and the origin of the mine´s name, i.e., "El Sexmo" [36]. However, extreme labor
conditions and the proliferation of epidemics became obstacles to the progress of mining and led to the
first crisis in the district [36].

In the late 19th century and during the first half of the 20th century, with the exploration of the
German geologist Teodoro Wolf in 1876 [41], investment capital from foreign companies began in
Zaruma. In 1880, the first company, the English Zaruma Gold Mining Co. Limited, was in charge of
the exploitation of the "El Sexmo" Tourist Mine [36]. Then, the South American Development Co.
(SADCO) remained in the country from 1896 to 1950 [42–44]. Later, "Compañía Industrial Minera
Asociada" (CIMA, acronym in Spanish), whose shareholders were the Zaruma municipality (52%) and
corporate workers (48%), oversaw the mining operations. However, the latter declared bankruptcy in
1978 [42–45]. Currently, due to the depletion of minerals available for safe extraction, the "El Sexmo"
Tourist Mine is inactive, and the company "Bira Bienes Raíces S.A." (BIRA S.A., acronym in Spanish)
turned it into a tourist mine in 2005 [32].

1.3. "El Sexmo" Tourist Mine Services Offered to Visitors

The services offered by the mine to its visitors are the following: (i) Video of the "El Sexmo" Tourist Mine history, (ii) Underground Tour, (iii) Gift Shop, (iv) Special Exhibitions, (v) Coffee Shop, (vi) Hanging Bridge, and (vii) Viewpoint. It is an amazing tourist experience that helps tourists learn about the mining history of the area, to see objects related to former mining activities, and to learn about its geological context of mineralization, with the main attraction consisting in the underground tour (Figure 2).

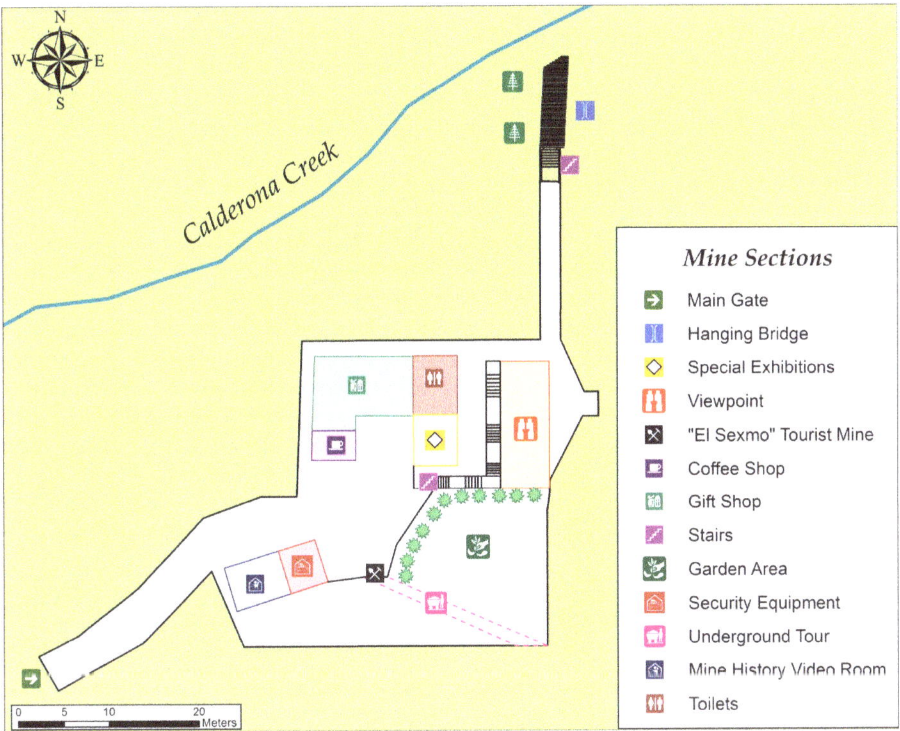

Figure 2. Map of the "El Sexmo" Tourist Mine.

The tour begins in a room where it is possible to see a 10 min long video of the history of the mine and where it is possible to admire photos from 1890 to 1900. After the mine history video, the staff provides personal protective equipment for underground tours and gives a talk about mine safety measures. Visitors then navigate inside a 500 m long underground tunnel. This is on slippery ground due to a constant trickle of water, and there are numerous dummies simulating the work of the miners (Figure S1a,b in the Supplementary Material) and other artisanal and small-scale gold mining elements.

In the underground tour of the mine, it is possible to observe different geological elements such as (i) supergenic alteration on tabular and centimetric quartz veins (Figure S2a in the Supplementary Material) hosting possible disseminated gold and other base metal minerals, (ii) sections of the weathered orebodies characterized by green and brown patinas of Cu and Fe oxide minerals (Figure S2b in the Supplementary Material), (iii) highly fractured andesite host rock (Figure S2c in the Supplementary Material), and finally (iv) formation of stalactites of approximately 30 cm in length generated by leaching of pyrites in contact with water and oxygen (Figure S2d in the Supplementary Material).

Some minerals such as quartz, pyrite, tiny fragments of gold extracted from the mine, and other minerals of the area are exhibited on the superficial tourist complex (Figure S3a–c in the Supplementary

Material), where tourists can buy souvenirs made by local people (Figure S3d in the Supplementary Material). Finally, visitors can enjoy a coffee shop (Figure S4a in the Supplementary Material), experience the view of the valley below and of the surrounding mountain features (Figure S4b in the Supplementary Material), and see flora species of the area through a small hanging bridge (Figure S4c in the Supplementary Material).

2. Geological Setting

The "El Sexmo" Tourist Mine is part of an intermediate sulphidation vein system of Au ± Ag ± Cu [31], have been related to propylitic, argillic, silica, and sericitic alterations type [42,46–48], and that is regionally located in the extension of the southwestern segment of the Miocene metallogenic belt (mineral deposits from the Azuay–El Oro district) in Ecuador [49] (Figure 3a). This district contains tertiary granitoids [50], such as the Cangrejos–Zaruma intrusive complex [49]. The complex has an ESE tendency near the Piñas–Portovelo regional fault, and it is hosted in volcanic rocks of the Oligocene–Early Miocene Saraguro Group.

Within the Azuay–El Oro district, there are several epithermal mineral deposits of the Miocene, such as Zaruma–Portovelo [51,52] (Figure 3b), where there is important structural control of the Palestina and Piñas dextral faults. The faults have a NW–SE trend, where the Palestina fault marks the northern limit of the Zaruma–Portovelo mineral field, and the Piñas fault comprises the southern limit and separates the Saraguro group from the El Oro metamorphic complex [53,54]. Consequently, the structural dynamics have favored the formation of dilating spaces for the location of vein assemblages (Figure 3b), where gold mineralization occurs [46,55] and associated minerals (pyrite, chalcopyrite, sphalerite, galena, bornite, hematite, tetrahedrite, molybdenite, quartz, and calcite) identified in different studies [55–57]. These vein systems have been established in three mineralized domains: N–S, NE–SW, and NW–SE, and the Sexmo fault represents one of the structures of the NE–SW domain [42].

The geology within the study area is formed of metamorphic, igneous, and sedimentary rocks ranging from the Precambrian-Paleozoic to the Quaternary age [55]. The oldest rocks are made up of Tahuín group metamorphic rocks [46], separated from the Andesites and basaltic andesites of the Celica formation (Cretaceous) by the Piñas fault [58]. The rhyolites and rhyolitic tuffs of the Tarqui formation (Mio-Pliocene) unconformably overlain all these units, followed by numerous subvolcanic rhyolite dikes and stocks (Riolitas de Zaruma Urcu, Quaternary) which crosscut the volcanic units. Finally, along the Amarillo and Calera rivers are the Quaternary alluvial and colluvial deposits [59] (Figure 3b).

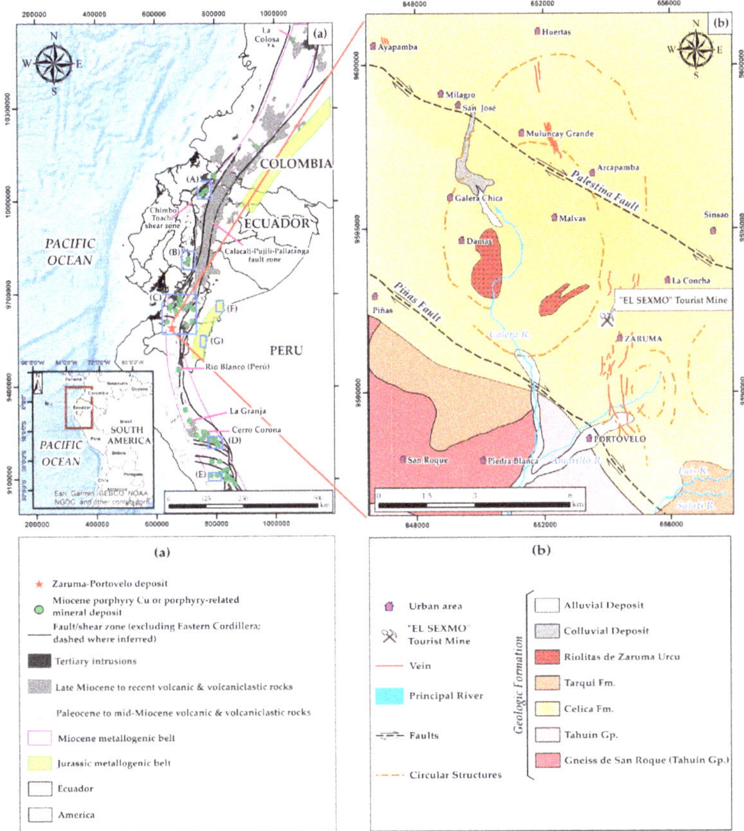

Figure 3. (**a**) A simplified metallogenic map of the NW of the South American margin and location of the Zaruma–Portovelo deposit. Abbreviations: (A) Imbaoeste district (Cuellaje, Junín (Llurimagua)); (B) Bolívar district (Telimbela, Balzapamba); (C) Azuay-El Oro districts (Chaucha, Gaby-Papa Grande, Quimsacocha, Tres Chorreras, El Mozo, Cangrejos, Portovelo-Zaruma); (D) Cajamarca mega-dristric (incl. Yanacocha, Michinquillay); (E) Quiruvilca district (incl. Lagunas Norte); (F) Pangui distric (Mirador, Fruta del Norte); (G) Nambija district, (**b**) The geological setting of the Zaruma–Portovelo mining field and its main structures, veins, geological formations, and the location of the "El Sexmo" Tourist Mine. Adapted from [42,49].

3. Methodology

The methodology used in this study comprised three stages (Figure 4): (i) A detailed inventory of the historical, cultural, tourist, and geological information of the study area; (ii) assessment of the interest and importance of the "El Sexmo" Tourist Mine with two different evaluation methods; and (iii) definition of strategies that optimize the use of the mine using the Delphi and strengths, weaknesses, opportunities, and threats (SWOT) analysis.

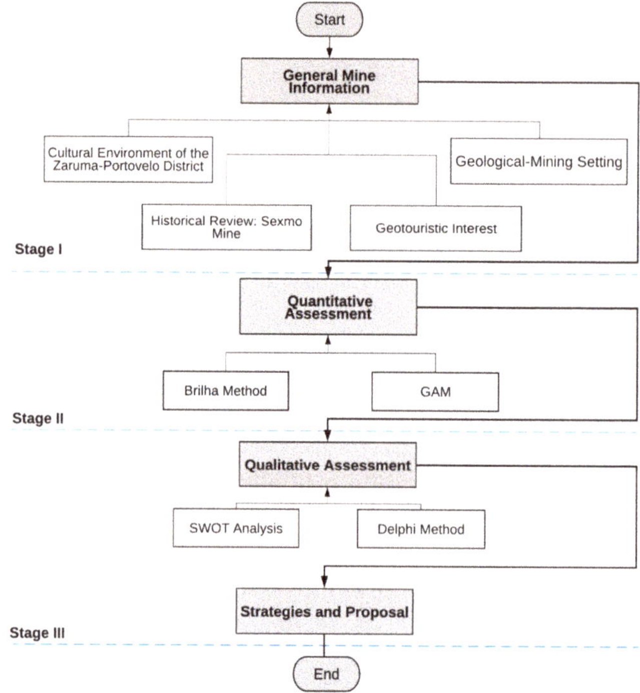

Figure 4. Flowchart of the methodology used in this study.

3.1. First Stage: General Mine Information

In the first stage, the researchers carried out an integral analysis of the characteristics of the mine studied. The process consisted of the collection of historical and cultural information (e.g., [15,30,32,35–37,39,41,43,45]). In addition, the mining geological context of the mine and its setting was defined based on previous studies [31,42,46,49–55] as a basis for subsequent assessment. The potential geotouristic interest presented by the study area was described. Specifically, this phase comprised the use of databases and synthesis of the information in scientific publications, divulgation work, project reports, data collection through interviews, and surveys administered to native people in the industry, former miners, etc. Finally, the study included a review of the regional and local geology of the district of the "El Sexmo" Tourist Mine, mineral deposits, mineralization, and the geological context of the area.

3.2. Second Stage: Quantitative Assessment

Within this phase, assessments were made of the study site using two methodologies, the GAM [26] and Brilha method [13], carried out by three anonymous geoscientific experts with knowledge in geology and mining and professional experience in the area. The information obtained was stored digitally to facilitate its data processing. Next, the parameters that consider each one of the methods used are detailed by means of a schematic and summary presentation of these.

3.2.1. The Geosite Assessment Model (GAM)

The GAM assesses the potential geosite based on two groups of indicators: Main and additional. The first group comprises three indicators: Scientific/educational (VSE), scenic/aesthetic (VSA), and protection (VPr) values. The second group includes functional (VFn) and tourist (VTr) values. In total, there are 12 subindicators of main values and 15 subindicators of additional values that are rated from

0 to 1, according to the parameters in Table 1, with the GAM assessment resulting from the application of Equation (1):

$$\text{GAM} = \text{Main values (VSE + VSA + VPr)} + \text{Additional values (VFn + VTr)}. \quad (1)$$

The results of the GAM assessment are illustrated in a matrix, where the X axis corresponds to the main values and the Y axis to the additional values. This matrix is divided into nine fields (zones) and, according to the rating, the geosite is included in the GAM matrix for interpretation [26].

Table 1. The geosite assessment model (GAM) showing the indicators of the main values and the additional values [26].

Geosite Assessment Model (GAM)			
	Indicators/Subindicators		**Indicators/Subindicators**
Main Values	**Scientific/Educational values (VSE)**	**Additional Values**	**Functional values (VFn)**
	1. Rarity 2. Representativeness 3. Knowledge on geo-scientific issues 4. Level of interpretation		1. Accessibility 2. Additional natural values 3. Additional anthropogenic values 4. Vicinity of emissive centres 5. Vicinity of important road networks 6. Additional functional values
	Scenic/Aesthetic values (VSA)		**Touristic values (VTr)**
	1. Viewpoints 2. Surface 3. Surrounding landscape and nature 4. Environmental fitting of sites		1. Promotion 2. Organized visits 3. Vicinity of the visitors' center 4. Interpretative panels 5. Number of visitors 6. Tourism infrastructure 7. Tour guide service 8. Hostel service 9. Restaurant service
	Protection values (VPr)		
	1. Current condition 2. Protection level 3. Vulnerability 4. Suitable number of visitors		

3.2.2. The Brilha Method

This method allows valuation by establishing: (i) Scientific Value (SV), (ii) Use Educational Potential (UEP), (iii) Potential Tourism Use (PTU), and (iv) Degradation Risk (DR). Each criterion is represented by different indicators that are rated from 1 to 4. The final evaluation of each criterion is the weighted sum of the indicators based on their respective scores and a predefined variable weight, as shown in Table 2 [13].

Table 2. Criteria and indicators used for the quantitative assessment of geosites [13].

Brilha Method			
Indicators/Subindicators		**Values**	**Weight**
Scientific Value (SV)			
A. Representativeness			30
B. Key locality			20
C. Scientific knowledge			5
D. Integrity		1–4	15
E. Geological diversity			5
F. Rarity			15
G. Use limitations			10
Total			**100**
Use Educational Potential (UEP) and Potential Tourism Use (PTU)		**Values**	**Weight**
UEP	**PTU**		**UEP** / **PTU**
A. Vulnerability			10 / 10
B. Accessibility			10 / 10
C. Use limitations			5 / 5
D. Safety			10 / 10
E. Logistics			5 / 5
F. Density of population		1–4	5 / 5
G. Association with other values			5 / 5
H. Scenery			5 / 15
I. Uniqueness			5 / 10
J. Observation conditions			10 / 5
K. Didactic potential	K. Interpretative potential		20 / 10
L. Geological diversity	L. Economic level		10 / 5
	M. Proximity of recreational areas		5
Total			**100** / **100**
Degradation Risk (DR)		**Values**	**Weight**
A. Deterioration of geological elements			35
B. Proximity to areas/activities with potential to cause degradation			20
C. Legal protection		1–4	100
D. Accessibility			15
E. Density of population			10
Total			**100**

The total assessment of the risk of degradation, as shown in Table 3, aims to establish priorities in the action plan of any site and to propose strategies for necessary improvement [13].

Table 3. Considering the final value, the degradation risk (DR) can de classified into three classes: Low, moderate, and high [13].

Total Weight	Degradation Risk
<200	Low
201–300	Moderate
301–400	High

3.3. Third Stage: Qualitative Assessment

In this stage, the researchers applied a Delphi method [60,61] to collect, analyze, and understand information on the current potential of the mine and to design improvement plans. The process included interviews with four anonymous experts in the area. Moreover, the experts completed a questionnaire based on five aspects: (i) Quality of services offered by the geological site, (ii) the aesthetic appearance, (iii) prominent qualities, (iv) proposed improvement, and (v) geosite deterioration.

The ideas collected were used as a reference for a SWOT analysis. Next, a SWOT analysis [62] of the "El Sexmo" Tourist Mine and its surroundings was applied. The SWOT analysis was developed to determine the potential of the area in a more ambitious future project and to propose initiatives for the efficient and effective use of the mine and its environment. This stage was carried out with the participation of members of the public and private sectors of the area of interest. Finally, as a product of this third phase, specific alternatives for optimizing the "El Sexmo" Tourist Mine were defined. This was performed as an interpretation and analysis of the processes described above and, in some cases, as an estimate of improvements that could occur.

4. Results

4.1. General Mine Information

Data of the annual visits to the mine (2014–2017) showed a clear upward trend in the number of visits, and the peak number of visits was reached in 2016 (Figure 5). Registration of local and international visitors showed that "El Sexmo" Tourist Mine had more than 35,000 visits (on average about 9000 visits per year) during the last four years. The estimation indicated the average visit duration in the mine per tourist was an hour and a half (25 min of introductory talk, 45 min of visiting inside the mine, and 20 min of sightseeing).

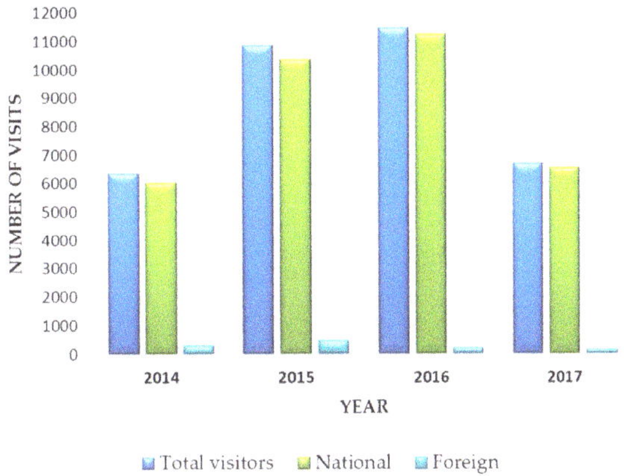

Figure 5. Annual visit data of the "El Sexmo" Tourist Mine.

Although the number of current visits to the mine and their duration indicated that the mine is a relevant tourist destination in the area, it is possible to improve these figures through an increase in the supply of possible activities to extend the duration of the visits. However, the undermining and sinking of a school in Zaruma [33] in December 2016 provided evidence of the damage caused by illegal mining in the area. This incident led to the decree of the State of Exception of the site, and the tourist mine had to close from September to December 2017. Hence, the monthly trend of visits changed negatively that year (Figure 6).

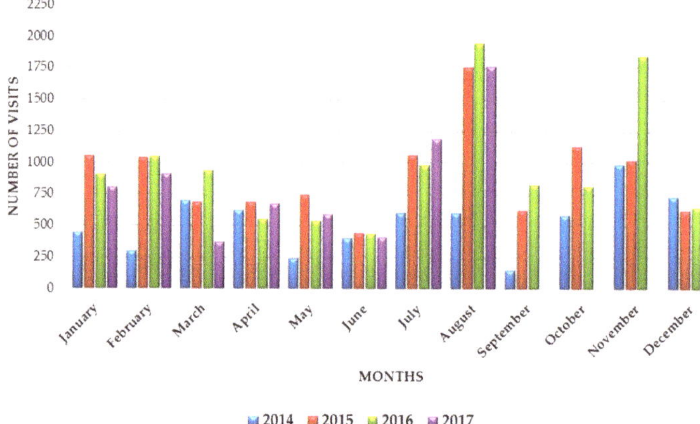

Figure 6. Monthly visit data of the "El Sexmo" Tourist Mine.

The information processing of the mine and its setting, besides serving as a basis for applying the geological-mining interest assessment methods, focused on dissemination materials. Specifically, schemes/posters/panels were made with summarized information on the historical evolution of the mine. Moreover, the information about the geography and geology of the mining area (Figure 7) was incorporated.

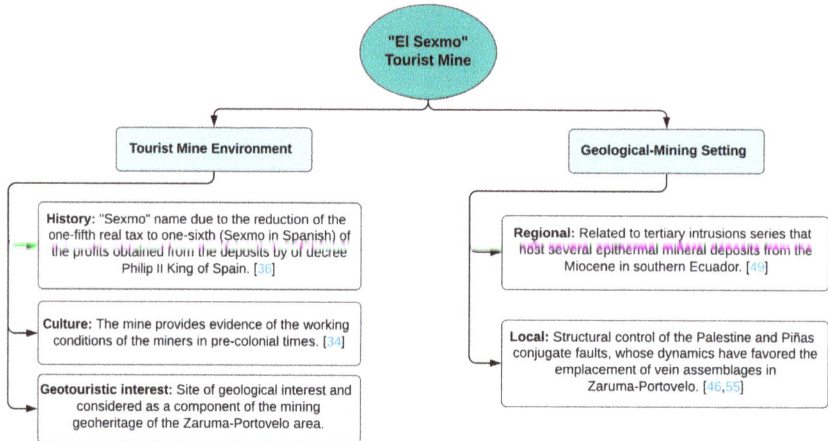

Figure 7. Summary scheme of the characteristics of the "El Sexmo" Tourist Mine and its setting.

4.2. Quantitative Assessment: GAM and Brilha

This section shows the main characteristics of the geological and mining interest site, considering the scientific, educational, and tourist value, degradation risk, and the scenic/aesthetic value, according to parameters assessed in the GAM [26] and Brilha method [13] (Table 4).

Table 4. Description of the main indicators assessed by the GAM and Brilha method.

Values	Indicators/Subindicators	Description
Scientific Value	Representativeness	The mine is a symbol of artisanal mining identity of Zaruma city, recognized as the first mine of the area. It has unique geological characteristics, highlighting the rosary-style veins belonging to part of a regional vein system: Nicole, Gaby, and Octubrina, of hydrothermal origin with a predominance of sulphides and the presence of gold, belonging to the Zaruma–Portovelo intermediate sulphuration deposit. Its average orientation is N-S, associated with the trend of regional fracture of the NW–SE direction (Palestina and Piñas Fault). It is also possible to visualize the city and its unique geomorphological landscape with very steep slopes.
	Geological diversity	It has a high scientific interest; the mineralogical aspect stands out as the primary geological interest, whereas the intrusive, geomorphological, hydrological, petrological, tectonic, landscape, and artisanal exploitation techniques are secondary interests.
	Key locality	The site of geological interest is a good example of mineralization in the south of the country, near the mine is the trend of mineralization formed by the structural dynamics of the Piñas and Palestine faults, which have favored the formation of dilating spaces for the location of vein assemblages with gold content.
	Scientific knowledge	There are studies by universities that have been published in national and international impact journals under different themes (geological, economic, tourism, and cultural).
	Integrity	The good state of conservation of the site allows us to visualize the veins present with mineralization; at the beginning of the tour ("bocamina"), the mine presents stabilization with concrete and drains, which prevent detachment by water and fracture action in the massif.
	Rarity	The regional mineralogical environment is identified as epi-mesothermal; however, in the study area, the minerals present indicate a uniquely epithermal environment. The site of geological interest is the only one recognized as a founder of artisanal mining in Ecuador. This mine has characteristics of the development of artisanal mining in the precolonial times, as well as stories of Spanish mining.
	Use limitations	The mine allows studies of scientific interest after requesting the necessary permits. It is a site of geotouristic interest that collaborates and maintains a relationship with the scientific community.
Potential Educational and Touristic Uses	Vulnerability	The site is in proximity to areas with illegal mining, an anthropic activity that has caused subsidence events in Zaruma, such as the sinking of the "La Inmaculada Fe y Alegría" school and surrounding houses
	Accessibility	The site is located less than 900 m from a Zaruma urban center by paved road (Estimated duration: 3 min per car and 15 min on foot).
	Use limitations	The mine tour is free (no need to pay an entrance fee); operating hours: Monday–Sunday (8:00 to 16:00), and the site has no limitations for use by students and tourists.
	Safety	The site provides personal protective equipment such as a helmet and rubber boots for underground tours; the mine entrance is reinforced concrete and is drained to ensure its stability. Additionally, there is an account mobile phone coverage, and the mine is located less than 1.5 km from emergency services.
	Logistics	Lodging and restaurants for groups of 50 persons are less than 15 km away from the site.
	Density of population	The site is located in a municipality with 37.14 inhabitants/km^2 [63].
	Association with other values	An illustrative example of the cultural heritage of Zaruma town is the "El Sexmo" Tourist Mine; the site reflects mining activity carried out in precolonial times, and it is the oldest mine in Ecuador. The mine is named after the king, due to the historic event of a reduction of the fifth real tax by one-sixth by decree of Philip II King of Spain in gratitude for the discovery of "pepa de oro" (gold nugget), whose weight was more than three pounds. In addition, the site shows the good side of mining as one of the initiatives of reclamation in a post-mining environmentally friendly and sustainable way.

Table 4. *Cont.*

Values	Indicators/Subindicators	Description
	Scenery	The site is currently used as a tourism destination in national campaigns such as field visits by universities (e.g., ESPOL Polytechnic University) by students of mining and geology engineering, in addition to visits organized by groups of students from local secondary and elementary schools.
	Uniqueness	The site shows unique and uncommon features in Ecuador due to the mine being an artisanal and small-scale gold mining example that keeps aspects of the cultural heritage of the mining area.
	Observation conditions	All geological elements are observed to be in good condition.
	Didactic potential	After the mine history video, visitors navigate inside a 500 m long underground tunnel on slippery ground along the railway that hauled tons of ore out of the mine, wearing equipment used by geologists and miners, attempting to understand the difficult work of miners (artisanal and small-scale gold mining in SW Ecuador) as they learn about the mining history of the area, objects related to former mining activities, and the geological context of hydrothermal alterations in one type of rock (igneous rock). The information can be easily understood by students with different educational levels.
	Geological diversity	The site has more than three types of geodiverse elements that occur in the site (mining, intrusive, geomorphological, hydrological, petrological, tectonic, landscape, and artisanal exploitation techniques).
	Interpretative potential	The site presents artisanal and small-scale gold mining elements in a very clear and expressive way to all members of the public.
	Economic level	The average of the Zaruma Canton of unsatisfied basic needs (NBI) referring to the poor population is 75.9% and that of the nonpoor population 24.1%. In other words, nine out of 10 parishes have a poor population in excess of 60%, and one parish in a nonpoor population exceeds 60% [63].
	Proximity of recreational areas	The site is located less than 5 km from recreational areas and tourist attractions (Zaruma Center Historic, Zaruma Municipal Museum, Shrine of the Virgen del Carmen, Zaruma Urcu hill, Calvario hill, a coffee shop, and Zaruma's traditional meals).
Degradation Risk (DR)	Deterioration of geological elements	There is a possibility of deterioration of secondary geological elements.
	Proximity to areas/activities with potential to cause degradation	The site is located less than 1 km from a potential degrading area/activity.
	Legal protection	The mine lacks legal support, which ensures sustainability and validity, but it is located within a new area of exclusion of mining activity delimited under the urban area of the city of Zaruma according to the Ministerial Agreement 2019–0050-AM.
Scenic/ Aesthetic values	Viewpoints	The mine has a main gate with a sign displaying its name that can be seen from at least two vantage points.
	Surface	The underground tunnel is 500 m long, and the surface zone of the mine (mine history video room, gift shop, hanging bridge, viewpoint, and coffee shop) has an approximate area of 1300 m^2.
	Surrounding landscape and nature	Tourists can enjoy a spectacular view of the valley below and of the surrounding mountains.
	Environmental fitting of sites	The site fits perfectly with the natural environment of the study area.

4.2.1. GAM Results

This section shows the evaluation of the "El Sexmo" Tourist Mine with the GAM [26]. The main values (scientific–natural) show a high degree in its VSE (3/4) and VPr indicators (3.5/4) and moderate in VSA (2.5/4), with a low value in the surface subindicator due to the limited size of the geosite (Table 5). An additional (anthropogenic) value has high scores in its VFn indicators (5.75/6) and VTr (6.25/8). Considering the total of the main and additional values, this methodology places the mine in the Z_{33} field, as shown in the matrix of Figure 8.

Table 5. Evaluation scores for each indicator of the main and additional values of the GAM [26] applied to the "El Sexmo" Tourist Mine.

Geosite Assessment Model (GAM)					
Values	Indicators/Subindicators	Score	Values	Indicators/Subindicators	Score
	Scientific/Educational values (VSE)			**Functional values (VFn)**	
	1. Rarity	0.5		1. Accessibility	1
	2. Representativeness	0.75		2. Additional natural values	1
	3. Knowledge on geoscientific issues	1		3. Additional anthropogenic values	1
	4. Level of interpretation	0.75		4. Vicinity of emissive centres	1
	(VSE) Total	**3.00**		5. Vicinity of important road network	0.75
				6. Additional functional values	1
MAIN VALUES	**Scenic/Aesthetic values (VSA)**		**ADDITIONAL VALUES**	**(VFn) Total**	**5.75**
	1. Viewpoints	0.5		**Touristic values (VTr)**	
	2. Surface	0		1. Promotion	1
	3. Surrounding landscape and nature	1		2. Organized visits	0.75
	4. Environmental fitting of sites	1		3. Vicinity of the visitors' center	0.25
	(VSA) Total	**2.5**		4. Interpretative panels	0.5
	Protection (VPr)			5. Number of visitors	0.5
	1. Current condition	1		6. Tourism infrastructure	0.75
	2. Protection level	1		7. Tour guide serve	0.75
	3. Vulnerability	0.5		8. Hostel service	1
	4. Suitable number of visitors	1		9. Restaurant service	0.75
	(VPr) Total	**3.5**		**(VTr) Total**	**6.25**

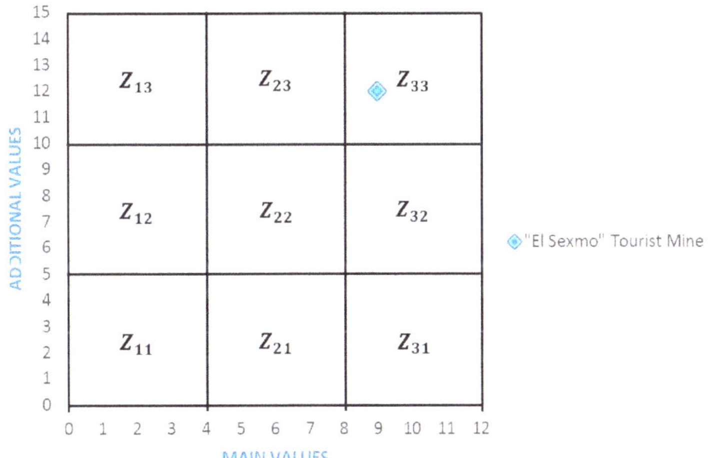

Figure 8. Position of the "El Sexmo" Tourist Mine as a result of the assessment with the GAM [26].

4.2.2. The Brilha Method Results

The parameters evaluated by the Brilha method [13] indicated high values for the scientific (340/400 points), tourism (340/400 points), and educational (330/400 points) indicators. However, the value obtained for the parameters of the degradation risk indicator (250/400 points) indicated a moderate degradation risk, as detailed in Table 6.

Table 6. Criteria indicators used for the quantitative assessment of the "El Sexmo" Tourist Mine by [13].

Brilha Method			
Indicators/Subindicators		**Values**	**Weight**
Scientific Value (SV)			
A. Representativeness		4	120
B. Key locality		2	40
C. Scientific knowledge		4	20
D. Integrity		4	60
E. Geological diversity		4	20
F. Rarity		4	60
G. Use limitations		2	10
Total			**340**

Potential Educational and Touristic Uses		**Values**		**Weight**	
UEP	**PTU**			**UEP**	**PTU**
A. Vulnerability		3		30	30
B. Accessibility		3		30	30
C. Use limitations		4		20	20
D. Safety		4		40	40
E. Logistics		4		20	20
F. Density of population		1		5	5
G. Association with other values		4		20	20
H. Scenery		4		20	60
I. Uniqueness		3		15	30
J. Observation conditions		4		40	20
K. Didactic potential	K. Interpretative potential	4	4	80	40
L. Geological diversity	L. Economic level	4	1	40	5
	M. Proximity of recreational areas	4			20
Total				**330**	**340**

Degradation Risk (DR)	**Values**	**Weight**
A. Deterioration of geological elements	2	70
B. Proximity to areas/activities with potential to cause degradation	1	20
C. Legal protection	3	60
D. Accessibility	4	60
E. Density of population	4	40
Total		**250**

4.3. The Delphi Method Results

Based on the questionnaires, the "El Sexmo" Tourist Mine and its surroundings have relevance as a place of visit due to the location, access, available services, and, mainly, the great historical importance. Moreover, the results highlighted the geological and mining interest. However, the anthropic activities and natural phenomena have increased the risk of degradation. As the mine is an example of recovery of mining spaces for geotourism, it is necessary to consider actions for preserving this heritage. Both cultural and historical aspects provided an opportunity for sustainable development of the sector (geotourism). The proposed ideas were the basis for the development of the SWOT analysis.

4.4. The SWOT Matrix

The SWOT matrix was developed from the data obtained from the Delphi method and the focus group of community members and municipal authorities during the workshop "Geotourism Perspectives in the upper part of El Oro province, Zaruma–Portovelo". The workshop allowed the use of participatory methodologies for recognizing "El Sexmo" as a site of great historical and cultural relevance with great potential if promotional strategies are carried out. Furthermore, a recommendation for the mine protection plan arose, as its conservation is vital to disseminate and maintain the site representativeness of the mining identity. The SWOT summary of the study area is presented in the Table 7.

Table 7. Strengths, weaknesses, opportunities, and threats (SWOT) matrix of the "El Sexmo" Tourist Mine.

	Strengths	Weaknesses
Internal Environment / **External Environment**	S_1. "El Sexmo" Tourist Mine, a symbol of mining identity. S_2. Strategic location. S_3. It has a variety of services of a tourist mine such as free underground tours, free surface tours, gift shop, special exhibitions, and a mine history video. S_4. Area of geological-mining interest (geosite potential).	W_1. The parking service and roads of the geosite and the place are limited. W_2. The mine does not present a virtual tour, and there is no official website of the site. W_3. There is a lack of a recreational park for children as a learning tool. W_4. Limited educational and scientific publications.
Opportunities	**Strategies: Strengths + Opportunities**	**Strategies: Weaknesses + Opportunities**
O_1. Support the creation of a geopark. O_2. Use the mine as an example of environmental awareness projection for cultural and natural conservation through tourism. O_3. Invest in basic services and road infrastructure. O_4. Exploit the mine as an alternative for the scientific/educational interest, tourism, and economic development of the area.	$S_1.O_1$. Bolster the recognition of several geosites in the study area and the creation of a geopark proposal. $S_1.O_2$. Boost an innovation project for the mine to be considered an official geosite mining site. $S_4.O_1.O_3$. Adapt the surrounding conditions for the repowering of the geosite. $S_3.O_1.O_4$. Promote the "El Sexmo" Tourist Mine as a technical, social, environmental, and cultural example.	$W_1.O_3$. Redesign the road infrastructure taking advantage of underground spaces as parking spaces for visitors. $W_3.O_4$. Complement the services offered by the mine, implementing spaces that encourage child participation and scientific/educational development. $W_4.O_4$. Promote the site (geosite) through informative and scientific publications that involve the scientific–social–business nexus.
Threats	**Strategies: Strengths + Threats**	**Strategies: Weaknesses + Threats**
T_1. High degradation because it is located near illegal mining activity, which has caused damage to the urban infrastructure. T_2. Deterioration due to natural events. T_3. The high density of abandoned underground spaces with the possibility of becoming tourist attractions in the study area would cause the mine to lose its uniqueness. T_4. It lacks full legal support, which ensures its sustainability and validity.	$S_1.T_1.T_4$. Protect the tourist mine as an identity of Zaruma in the inventory of the Tourism and Heritage Ministry. $S_2.T_2$. Generate studies/stabilization works for the mine due to its natural conditions. $S_1.T_3$. Promote a program of innovation strategies that add value to the site.	$W_1.T_3$. Take advantage of abandoned underground spaces for parking. $W_2.T_3$. Create a web page with virtual visitors to strengthen tourism activity. $W_4.T_1.T_2$. Develop scientific publications to identify the possible threats to anthropic and natural events.

4.5. Initiatives and Proposals

As a global result of the assessments and analysis of the "El Sexmo" Tourist Mine, the researchers proposed some specific and general initiatives that could be carried out in the study area. These include the following:

- The official presentation of the evaluation work of the "El Sexmo" Tourist Mine as a site with potential for local authorities (Zaruma City Council) for the achievement of official recognition as a UNESCO Global Geopark. In addition, the strategies derived from the SWOT analysis will be delivered in detail as a reference for developing initiatives to take advantage of this important resource.
- The definition of complementary activities to those currently offered at the "El Sexmo" Tourist Mine. Specifically, the creation of a gold-panning zone in the facilities, creation of informative panels of the geological and mining interest of the mine and the locality, and the organization of periodical visits of students from schools and colleges. A general estimate of the increase in the average tourist visit duration time, with the materialization of the proposals described above, would reach 100% (from one and a half to three hours).
- The development of a detailed inventory and evaluation proposal of other locations with potential to be considered as geosites (e.g., Cerro de Arcos) and mining sites (e.g., Magner Turner Mineralogical Museum) in the Zaruma and Portovelo municipalities [15]. A general estimate of the increase in the average tourist view visit duration in the area, with the inclusion of two additional points of interest would mean having a visit offered for one full day.
- Scientific support for the initiative to create a geopark that would be called "Ruta del Oro" (Gold Route) that has its starting point at the "El Sexmo" Tourist Mine, which also includes 27 sites of

geological and mining interest in its proposal. This initiative would be based on previous studies that already addressed this possibility (e.g., [15,64]). This initiative would mean being able to offer a broader visit to achieve at least a couple of days' worth of visitation possibilities, with the corresponding economic benefit for the involved stakeholders.

5. Interpretation of Results and Discussion

Based on the methodology for performing the qualitative and quantitative assessment to define "El Sexmo" Tourist Mine as a geosite or mining site, the study allows exposing the geological, historical, and cultural environment that surrounds the mine. Moreover, it addresses the need for alternative routes of economic development. Carrión Mero et al. [15] carried out a previous assessment of a group of geological places of interest (27 sites), where the "El Sexmo" Tourist Mine obtained the highest score for its scientific, academic, and tourist interest and highlighting the mineralogical and historical attraction of the gold mining operation. Thus, the assessment work presented in this study with the GAM [26] and Brilha method [13] (Table 5) presents similar results to the estimates obtained by [15] using the García–Cortés assessment methodology [19].

Despite having high additional values (functional and tourist), it is essential to monitor these values to analyze the consequences of tourism development by increasing the number of visits and assessing the possible threats to the mine, such as the deterioration of the mine and the natural environment in which it is located. Thus, the data would provide information for planning according to the load capacity and reality of the site.

The high scores obtained for the SV, PEU, and PTU with the Brilha method [13] show that the study area represents an important geological feature with aesthetic relevance, which makes it a destination for research and scientific knowledge at an international level (Table 6). Regarding the total assessment of the DR, it presents a moderate risk of degradation due to natural or anthropic events, with little legal protection and proximity to areas with illegal mining, where the secondary galleries of the mine could be used as an alternate access, contributing even more to the problems of illegal exploitation that the study area is currently experiencing. As there is a high probability of man-induced degradation, strategies that reduce the risk and enhance the mine as an alternative for geotouristic development are essential. In addition, through the assessment methodology, one of the lowest values obtained was that of the economic level of the sector. This parameter influences the quality of tourism services provided by Zaruma city and its surrounding areas, which could be reflected in the number of tourists.

The results of the assessments with the GAM and Brilha method provided data for the Delphi qualitative analysis where geoscientific experts highlighted the importance of the mine as an example of a mine closure plan and the recovery of abandoned underground spaces. Moreover, the analysis suggested that the geotourism alternative would boost the initiative of the "Ruta del Oro" (Gold Route) Geopark project.

Considering the point of view of the community and the previous analysis of experts, the SWOT matrix provided information to propose strategies that complement the services offered by the mine (Table 7). Among those services, the study highlights the implementation of spaces aimed at child participation as a didactic tool and the creation of a gold-panning zone (activity of the mining culture of the area), as well as the installation of information panels of geological and mining interest. Moreover, another idea is to take advantage of one of the main weaknesses of the place (limited parking spaces) by using inactive tunnels as parking lots.

Finally, it is essential to incorporate a virtual visit that strengthens and increases the influx of the tourism sector to disseminate information of the site nationally and internationally and to consider the registration of annual visits (Figures 5 and 6).

6. Conclusions

The methodologies for the assessment of geological and mining points of interest allowed for a correct characterization and evaluation of the study area under different parameters that include high values for the educational, scientific, and tourist aspects. Therefore, the "El Sexmo" Tourist Mine has high value as a geological and mining heritage, which makes it an icon of the geotourism of the sector. These characteristics add a cultural and positive value to the mining activity as a solution to the existing problem. The researchers analyzed the types of anthropic threats (illegal mining in urban areas) and natural threats (landslides and deterioration of the massif by water leaks) that the geosite faces to reinforce aspects of protection and conservation.

As the study area was classified as a geosite and mining site of considerable relevance, the SWOT and Delphi analyses were fundamental techniques. The "El Sexmo" Tourist Mine was highlighted as a symbol of the mining identity of Ecuador and a point of geological and mining interest. However, the analysis presented the illegal mining activity, road infrastructure, and limited scientific and tourist dissemination as weaknesses. Under these conditions, the solution strategies proposed are as follows: (i) Promote the "Ruta del Oro" (Gold Route) Geopark project, (ii) include the mine as a recognized geosite, (iii) to redesign the road infrastructure of the sector, and (iv) implement a disclosure system through scientific manuscripts and virtual advertising.

Based on the general report of annual visits of approximately 9000 visitors per year, this study and its proposals for protection and improvement could increase the number of visitors to approximately 12,000 people per year. Consequently, the results of this study will boost tourism, knowledge, and economic activity, which will promote the sustainable development of geotourism in the sector. Finally, given the current limited strategies of the mine, a modification in the communication routes could give excellent results in the educational and geotouristic sectors.

Supplementary Materials: The following are available online at http://www.mdpi.com/2079-9276/9/3/28/s1, Figure S1: Services in "El Sexmo" Tourist Mine. (a,b) Underground tour, Figure S2: Underground tour. (a) Quartz vein with copper sulphides lixiviated to copper sulphates (green hue). (b) Rock massif with presence lixiviation of oxides and hydroxides of iron and copper sulphides. (c) Wall rock (andesite) with fractures. (d) Mine tunnel with stalactites (iron oxides and hydroxides stalactites), Figure S3: Superficial tourist complex. (a–c) Special exhibitions and (d) Gift Shop, Figure S4: Superficial tourist complex. (a) Coffee Shop, (b) Viewpoint, and (c) Hanging bridge.

Author Contributions: All authors (P.C.-M., O.L.-O., H.A.-R., G.H.-F., F.M.-C., M.J.-M., M.A.-A., K.T.-P., and E.B.) wrote the paper and participated in the design and performed the methods, evaluation, discussion, and conclusion of this research article. All authors have read and agreed to the published version of the manuscript.

Funding: This research was funded by the "Geoparque Ruta del Oro" academic research project by ESPOL University and BIRA S.A. Company.

Acknowledgments: This work has been possible thanks to support from BIRA Bienes Raices S.A., Zaruma Rotary Club, and Zaruma Municipality, we thank Gabriela Zambrano and María del Cisne Chuico for their help. We also would like to thank four anonymous reviewers for their constructive comments and the editorial office for the editorial handling.

Conflicts of Interest: The authors declare no conflict of interest.

References

1. Mert, Y. Contribution to sustainable development: Re-development of post-mining brownfields. *J. Clean. Prod.* **2019**, *240*, 118212. [CrossRef]

2. Cao, X. Regulating mine land reclamation in developing countries: The case of China. *Land Use Policy* **2007**, *24*, 472–483. [CrossRef]

3. Kuter, N. Ulusal Mermer ve Taş, Onarım Teknikleri Sempozyumu, Isparta. Terk Edilmiş Açık Ocak Maden Sahalarında Doğaya Yeniden Kazandırma Çalışmaları. 2014, pp. 354–367. Available online: http://ormanweb.isparta.edu.tr/mermerteknik/belgeler/BildirilerKitabi_v2.pdf (accessed on 3 March 2020).

4. Mancini, L.; Sala, S. Social impact assessment in the mining sector: Review and comparison of indicators frameworks. *Resour. Policy* **2018**, *57*, 98–111. [CrossRef]

5. Orman ve Su işleri Bakanlığı. *Maden SAhaları Rehabilitasyon Eylem Planı 2014–2018*; OSİB: Ankara, Turkey, 2014.

6. Belediyesi, Selçuklu. Selçuklu Kanyon Park 2007. Available online: http://www.selcuklu.bel.tr/ilcemiz/detay/305/selcuklu-kanyon-park.html (accessed on 23 March 2019).

7. Garofano, M.; Govoni, D. Underground geotourism: A historic and economic overview of show caves and show mines in Italy. *Geoheritage* **2012**, *4*, 79–92. [CrossRef]

8.   Williams, P. World heritage caves and karst. *IUCN Gland* **2008**, *57*. Available online: https://portals.iucn.org/library/sites/library/files/documents/2008-037.pdf (accessed on 23 March 2019).

9.   Zięba, K. Reales minas de sal de Wieliczka y Bochnia: Museo subterráneo, pero aún una mina. *Rev. Del Patrim. Mund.* **2017**, *84*, 44–47.

10.   Saltmine, Wieliczka. "Wieliczka" Salt Mine 2017. Available online: http://www.wieliczka-saltmine.com/ (accessed on 4 July 2019).

11.   Cai, Y.; Wu, F.; Han, J.; Chu, H. Geoheritage and Sustainable Development in Yimengshan Geopark. *Geoheritage* **2019**, *11*, 991–1003. [CrossRef]

12.   Mata-Perelló, J.; Carrión, P.; Molina, J.; Villas-Boas, R. Geomining Heritage as a Tool to Promote the Social Development of Rural Communities. In *Geoheritage: Assessment, Protection and Management*; Reynard, E., Brilha, J., Eds.; Elsevier: Amsterdam, The Netherlands, 2017; pp. 167–177. [CrossRef]

13.   Brilha, J. Inventory and quantitative assessment of geosites and geodiversity sites: A review. *Geoheritage* **2016**, *8*, 119–134. [CrossRef]

14.   Brilha, J. Geoheritage and geoparks. In *Geoheritage. Assessment, Protection, and Management*; Brilha, J., Ed.; Elsevier: Amsterdam, The Netherlands, 2018; pp. 323–335. [CrossRef]

15.   Carrión Mero, P.; Herrera Franco, G.; Briones, J.; Caldevilla, P.; Domínguez-Cuesta, M.; Berrezueta, E. Geotourism and local development based on geological and mining sites utilization, Zaruma-Portovelo, Ecuador. *Geosciences* **2018**, *8*, 205. [CrossRef]

16.   Sánchez-Cortez, J.L. Conservation of geoheritage in Ecuador: Situation and perspectives. *Int. J. Geoherit. Parks* **2019**, *7*, 91–101. [CrossRef]

17.   Herrera, G.; Carrión, P.; Briones, J. Geotourism potencial in the context of the geopark project for the development of Santa Elena province, Ecuador. *Wit Trans. Ecol. Environ.* **2018**, *217*, 557–568.

18.   Jaramillo, J.P.; García, T.; Bolaños, M. Bosque Petrificado de Puyango y sus alrededores: Inventario de lugares de interés geológico. *Rev. Científica Geolatitud* **2017**, *1*, 18.

19.   García-Cortés, A.; Carcavilla, L.; Díaz-Martínez, E.; Vegas, J. Inventario de Lugares de Interés Geológico de la Cordillera Ibérica. 2012. Available online: http://www.igme.es/patrimonio/Informe%20Ib%C3%A9rica%20Final.pdf (accessed on 5 July 2019).

20.   Pereira, P.; Pereira, D.I.; Alves, M.I. Geomorphosite assessment in Montesinho natural park (Portugal). *Geogr. Helv.* **2007**, *62*, 159–168. [CrossRef]

21.   Reynard, E.; Fontana, G.; Kozlik, L.; Scapozza, C. A method for assessing the scientific and additional values of geomorphosites. *Geogr. Helv.* **2007**, *62*, 148–158. [CrossRef]

22.   Erhartič, B. Geomorphosite assessment. *Acta Geogr. Slov.* **2010**, *50*, 295–319. [CrossRef]

23.   Feuillet, T.; Sourp, E. Geomorphological heritage of the Pyrenees National Park (France): Assessment, clustering, and promotion of geomorphosites. *Geoheritage* **2011**, *3*, 151–162. [CrossRef]

24.   Kubalíková, L. Geomorphosite assessment for geotourism purposes. *Czech J. Tour.* **2013**, *2*, 80–104. [CrossRef]

25.   Dmytrowski, P.; Kicinska, A. Waloryzacja geoturystyczna obiektów przyrody nieożywionej i jej znaczenie w perspektywie rozwoju geoparków. *Probl. Ekol. Kraj.* **2011**, *29*, 11–20.

26.   Vujičić, M.D.; Vasiljević, D.A.; Marković, S.B.; Hose, T.A.; Lukić, T.; Hadžić, O.; Janićević, S. Preliminary geosite assessment model (GAM) and its application on Fruška Gora Mountain, potential geotourism destination of Serbia. *Acta Geogr. Slov.* **2011**, *51*, 361–376. [CrossRef]

27.   Petrović, M.; Vasiljević, D.; Vujičić, M.; Hose, T.; Marković, S.; Lukić, T. Geoparque global y análisis candidato-comparativo del geoparque de la montaña Papuk (Croacia) y la montaña Fruška Gora (Serbia) utilizando el modelo GAM. *Carpathian J. Earth Environ. Sci.* **2013**, *8*, 105–116.

28.   Álvarez-Jurado, Á. El papel del patrimonio geológico en la gestión de espacios naturales protegidos. *Apl. Sierra Mágina* **2018**, 35–43. Available online: http://tauja.ujaen.es/jspui/handle/10953.1/6431 (accessed on 20 March 2019).

29.   Culturaypatrimonio. Available online: http://picultural.culturaypatrimonio.gob.ec/ (accessed on 20 March 2019).

30.   UNESCO World Heritage Centre. Available online: http://whc.unesco.org/en/tentativelists/6089 (accessed on 18 March 2019).

31.   Spencer, R.M.; Montenegro, J.L.; Gaibor, A.; Perez, E.P.; Mantilla, G.; Viera, F.; Spencer, C.E. The Portovelo-Zaruma mining camp, SW Ecuador: Porphyry and epithermal environments. *Seg Newsl.* **2002**, *49*, 8–14.

32. Herrera, G.; Carrión, P.; Jimenez, S.; Mina, A. Oportunidades para el turismo de patrimonio geológico y minero a partir de modificaciones recientes al marco legal en el Ecuador, El caso de la mina El Sexmo. In *Patrimonio Minero y Sustentabilidad: Propuestas y Experiencias de Reutilización*; López, M., Pérez, L., Eds.; Ediciones Universidad del Bío-Bío-CYTED: Concepción, Chile, 2014; pp. 167–177.

33. Vangsnes, G.F. The meanings of mining: A perspective on the regulation of artisanal andsmall-scale gold mining in southern Ecuador. *Extr. Ind. Soc.* **2018**, *5*, 317–326.

34. Carrión, P.; Ladines, L.; Morante, F.; Blanco, R.A. Diagnóstico de la situación geomecánica y de contaminación de Zaruma y Portovelo (Ecuador). In *Actas del IV Congreso Internacional Sobre Patrimonio Geológico y Minero*; Sedpgym, V.S., Ed.; Aragón, España, 2003; pp. 283–295. Available online: http://www.sedpgym.es/18-publicaciones/actas-congresos/78-actas-del-iv-congreso-internacional-sobre-patrimonio-geologico-y-minero-teruel-2004 (accessed on 20 March 2019).

35. Barrantes, C. *Reservas Probadas de Mineral y su Valor in Situ*; CIMA: Portovelo, Ecuador, 1977.

36. Murillo, R. *ZARUMA, Historia Minera. Identidad en Portovelo*; ABYA-YALA: Quito, Ecuador, 2000; Volume 1.

37. Poma, V. *Historia Orense: Documentos de Zaruma*; P&C DOS MIL 3: Machala, Ecuador, 1992. Available online: https://issuu.com/historiografiaorense/docs/documentos_de_zaruma (accessed on 20 March 2019).

38. Levillier, R. *Gobernantes del Perú: Cartas y Papeles: Siglo XVI*; Sucesores de Rivadeneira (S.A.): Madrid, España, 1921.

39. Ramón, G. Loja y Zaruma: Entre las minas y las mulas, 1557–1700. *Rev. Ecuat. Hist. Económica* **1990**, *7*, 111–143.

40. Lane, K. Unlucky strike: Gold and labor in Zaruma, Ecuador, 1699–1820. *Colonial Lat. Am. Rev.* **2004**, *13*, 65–84. [CrossRef]

41. Rengel, J. *Desarrollo Nacional de la Minería Ecuatoriana*; Escuela Superior Politécnica del Litoral: Guayaquil, Ecuador, 1985.

42. Bonilla, W. Metalogenia del Distrito minero Zaruma-Portovelo, República del Ecuador. Ph.D. Thesis, Universidad de Buenos Aires, Buenos Aires, Argentina, 2009.

43. Astudillo, C. *El Sudor del Sol: Historia de la Minería Orense*; LaTierra, Ed.; Corporación de Ediciones la Tierra: Quito, Ecuador, 2007.

44. Velásquez-López, P.C.; Veiga, M.M.; Hall, K. Mercury balance in amalgamation in artisanal and small-scale gold mining: Identifying strategies for reducing environmental pollution in Portovelo-Zaruma, Ecuador. *J. Clean. Prod.* **2010**, *18*, 226–232. [CrossRef]

45. Sandoval, F. Small-scale mining in Ecuador. *Environ. Soc. Found.* **2001**, *75*. Available online: https://pubs.iied.org/pdfs/G00720.pdf (accessed on 20 March 2019).

46. Van Thournout, F.; Salemink, J.; Valenzuela, G.; Merlyn, M.; Boven, A.; Muchez, P. Portovelo: A volcanic-hosted epithermal vein-system in Ecuador, South America. *Miner. Depos.* **1996**, *31*, 269–276. [CrossRef]

47. Meyer, C.; Hemley, J.J. Wall rock alterations, 166–235. In *Geochemistry of Hydrothermal ore Deposits (HG Barnes, ed.)*; Winston Inc.: New York, NY, USA, 1967; p. 670.

48. Heald, P.; Foley, N.K.; Hayba, D.O. Comparative anatomy of volcanic-hosted epithermal deposits; acid-sulfate and adularia-sericite types. *Econ. Geol.* **1987**, *82*, 1–26. [CrossRef]

49. Schütte, P.; Chiaradia, M.; Barra, F.; Villagómez, D.; Beate, B. Metallogenic features of Miocene porphyry Cu and porphyry-related mineral deposits in Ecuador revealed by Re-Os, 40 Ar/39 Ar, and U-Pb geochronology. *Miner. Depos.* **2012**, *47*, 383–410. [CrossRef]

50. Schütte, P.; Chiaradia, M.; Beate, B. Petrogenetic evolution of arc magmatism associated with Late Oligocene to Late Miocene porphyry-related ore deposits in Ecuador. *Econ. Geol.* **2010**, *105*, 1243–1270. [CrossRef]

51. PRODEMINCA. Depósitos epitermales en la Cordillera Andina. In *Evaluación de Distritos Mineros del Ecuador*; UCP PRODEMINCA Proyecto MEM BIRF: Quito, Ecuador, 2000; Volume 2, pp. 36–55.

52. PRODEMINCA. Depósitos porfídicos y epi-mesotermales relacionados con intrusiones de las Cordilleras Occidental y Real. In *Evaluación de Distritos Mineros del Ecuador*; UCP PRODEMINCA Proyecto MEM BIRF: Quito, Ecuador, 2000; Volume 4, pp. 36–55.

53. Aspden, J.A.; Bonilla, W.; Duque, P. *The El Oro Metamorphic Complex, Ecuador: Geology and Economic Mineral Deposits*; British Geological Survey: Nottingham, UK, 1995. [CrossRef]

54. Riel, N.; Martelat, J.E.; Guillot, S.; Jaillard, E.; Monié, P.; Yuquilema, J.; Mercier, J. Fore arc tectonothermal evolution of the El Oro metamorphic province (Ecuador) during the Mesozoic. *Tectonics* **2014**, *33*, 1989–2012.

55. Berrezueta, E.; Ordóñez-Casado, B.; Bonilla, W.; Banda, R.; Castroviejo, R.; Carrión, P.; Puglla, S. Ore petrography using optical image analysis: Application to Zaruma-Portovelo deposit (Ecuador). *Geosciences* **2016**, *6*, 30. [CrossRef]
56. Paladines, A.P.; Rosero, G. *Zonificación Mineralogénica del Ecuador*; Laser Editores S.A.: Quito, Ecuador, 1996.
57. Vikentyev, I.; Banda, R.; Tsepin, A.; Prokofiev, V.; Vikentyeva, O. Mineralogy and formation conditions of Portovelo-Zaruma gold-sulphide vein deposit, Ecuador. *Geochem. Mineral. Petrol.* **2005**, *43*, 148–154.
58. Chiaradia, M.; Fontboté, L.; Beate, B. Cenozoic continental arc magmatism and associated mineralization in Ecuador. *Miner. Depos.* **2004**, *39*, 204–222. [CrossRef]
59. La Misión Británica y la Dirección General de Geología y Minas. *Hoja Geológica Zaruma 38 CT-NVI-E escala 1:100,000*; Instituto de Investigación Geológico y Energético, Ministerio de Recursos Naturales y Energéticos, Dirección General de Geología y Minas: Quito, Ecuador, 1980.
60. McKenna, H.P. he Delphi technique: A worthwhile research approach for nursing? *J. Adv. Nurs.* **1994**, *19*, 1221–1225. [CrossRef] [PubMed]
61. Hasson, F.; Keeney, S. Enhancing rigour in the Delphi technique research. *Technol. Forecast. Soc. Chang.* **2011**, *78*, 1695–1704. [CrossRef]
62. Dyson, R.G. Strategic development and SWOT analysis at the University of Warwick. *Eur. J. Oper. Res.* **2004**, *152*, 631–640. [CrossRef]
63. INEC. Instituto Nacional de Estadística y Censos 2010. Available online: https://www.ecuadorencifras.gob.ec/censo-de-poblacion-y-vivienda/ (accessed on 1 January 2020).
64. Berrezueta, E.; Domínguez-Cuesta, M.J.; Carrión, P.; Berrezueta, T.; Herrero, G. Propuesta metodológica para el aprovechamiento del patrimonio geológico minero de la zona Zaruma-Portovelo (Ecuador). *Trab. Geol.* **2006**, *26*, 103–109.

Article

"Perugia Upside-Down": A Multimedia Exhibition in Umbria (Central Italy) for Improving Geoheritage and Geotourism in Urban Areas

Laura Melelli

Department of Physics and Geology, University of Perugia, 06123 Perugia, Italy; laura.melelli@unipg.it; Tel.: +39-07-5584-9579

Received: 17 July 2019; Accepted: 13 August 2019; Published: 17 August 2019

Abstract: Multimedia materials represent a promising approach to the promotion of geoheritage. Despite geology being normally associated with natural environments, new tendencies are noted towards better knowledge of the "geological reason" for the selection of a location and the development of urban settlements. The urban environment is, in fact, a perfect laboratory for opening the scientific topics to a broad audience. In this paper, the experience of a geological exhibition organized in the city of Perugia (Umbria, central Italy) is discussed, highlighting the SECRET (SEe and CREaTe) for creating an effective dissemination activity. Panels, interactive tools, laboratories, and trekking tours outside the museum are the main activities, which hosted more than eight thousand visitors in a few months. Moreover, the exhibition was the starting point for ongoing projects on geotourism in the city, with important consequences in terms of visibility and financial return.

Keywords: geotourism; geoheritage; urban geology

---

1. Introduction

The common idea of geology as a scientific discipline restricted to the natural environment is quite widespread and consolidated. However, increasing attention to the geological investigation of urban areas is growing in the scientific community [1–3]. The establishment of a city always has a geological reason. The situation and the site are the initial starting points. The situation or position is the geographical location related to the surrounding areas, being fundamental for communications, economic relations, and cultural exchanges with other communities. In other words, the position refers to how a place is related to other cities or productive places [4]. The site conditions set the direct relations within the environmental context [4]. The topographic conditions (slope angle values in relation to the possibility of defending against external attacks) as well as the proximity of rivers or the sea and the availability of underground water are the most important criteria for site selection. Moreover, the bedrock composition should support the building material and the possibility to create hypogean cavities for a large number of uses (drainage or water supply, food storage, underground passages, shelters in case of war). The geomorphological conditions, in particular, the evolution of a site in relation to landslides or flooding events, establishes the possibility for the urban fabric to extend in the surrounding areas. A large part of scientific literature is focused on natural hazards in cities [5]. Floods or droughts [6] and their increasing effects due to climate change [7] are one of the topics in this area. Other specific and more local natural hazards, such as volcanic or seismic events, also affect urban areas [8–10].

Presently, an opposite trend is growing in the scientific and administrative environments: The geotouristic approach, where the geological context is a new and promising resource for the touristic and didactic issues in urban areas. Geotourism is the branch of tourism focused on activities, products, and services related to Earth sciences [11,12] where the subject is the geological component of the natural environment and social context with a high scientific, educational and cultural value. The prefix "geo-" includes "geology, geomorphology", and the natural resources of the landscape, landforms, fossil beds, rocks and minerals, with an emphasis on appreciating the processes that are creating and created such features [11]. Geotourism links the geology as a scientific discipline using objective criteria and scientific methods to tourism, which needs subjective criteria and aesthetic components [13]. Geotourism is the most efficient approach for exporting the scientific contents of the Earth sciences to a wider audience, characterized by a wide spectrum of ages and cultural backgrounds. Geoheritage is the cornerstone of geotourism, that is a category of heritage where the geological component is relevant. Where some areas show characteristics of uniqueness in both the scientific and cultural aspects, they are selected and classified as geosites [14] and geomorphosites [15–17]. A geosite is the best expression of geoheritage but geosites are not always present in some areas and, moreover, their definition is not simply objective. Therefore, in order to export the knowledge derived from Earth sciences to a wider public, it is essential to find the geological component of a landscape also in common features and daily experiences.

Aside from the definitions surrounding the subjects of geotourism, the real challenge is how to communicate this heritage and most of all, how to make it a recreational activity. A huge amount of scientific papers and research activities are devoted to these methods and represent the main vehicle of dissemination for cultural geoheritage [18,19]. This tendency has had exponential growth since 2001 [20]. However, several problems arise for dissemination including the technical scientific language, the geological time scale (millions of years) and the spatial scale varying from extensions of thousands of kilometers to the microscopic scale [13]. Another problem is the high heterogeneity in the tourists involved in geotourism. Differences in age, cultural level, and physical capability may be a serious obstacle for successful dissemination [13]. Finally, the area of interest of geotourism does not equally cover all the branches of knowledge related to Earth sciences. Geomorphology, volcanology, and paleontology are the most exploited in dissemination activities [13] since these subjects investigate more than others the macroscopic effects of geodynamics and are linked to the most fascinating aspects of the geology, recalling spectacular and impressive natural events.

Introducing the idea of the geological component in a city as a strong point of tourist activity is not easy. The traditional approach in visiting and getting to know a city, both for tourists and educational purposes, is generally starting from a historical framework. The geographical introduction, if it is present, it is reduced to a brief paragraph. Moreover, the link between the geographical setting and the human presence is absent in most cases. Improving the geological heritage should be the basis for introducing people to a city. The morphological and hydrographic arrangement is a direct consequence of the geological evolution of the area. The time span considered is much broader, but it is essential information for understanding where, why, and how the local populations made their choices in order to exploit resources and oppose the limits of the territory. To date, the geological parameter in cities is perceived as a risk. Where the geological heritage in situ is not present or well evident, as in some urban areas, a good compromise is represented by the ex situ items, such as museum collections. In dissemination activities, the museum with permanent and temporary exhibitions are one of the most successful possibilities [21]. Nevertheless, except for dinosaurs, volcanoes, and earthquakes, geological matter is not very interesting for non-specialists [21].

In order to stress the idea of the exhibition as a good tool for dissemination activities related to Earth sciences, a geological exhibition was organized in Perugia in 2017 (Umbria, central Italy, Figure 1A).

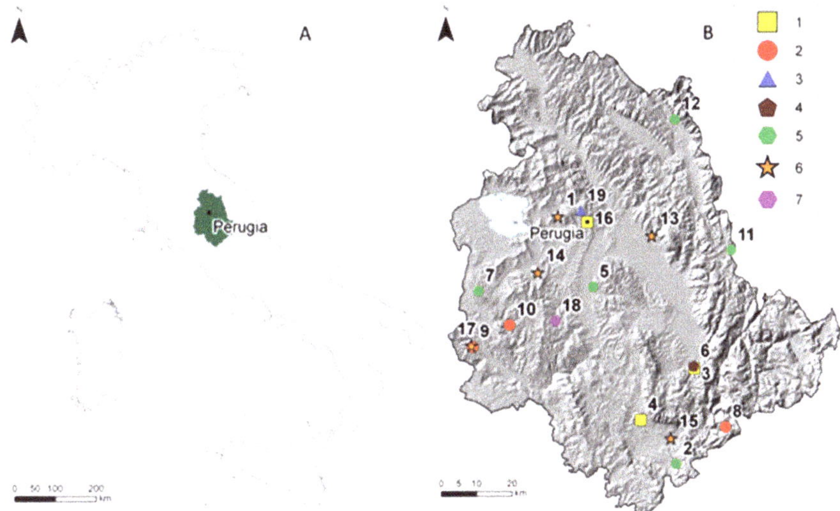

Figure 1. (**A**) Location map: the Umbria region in Italy with the Perugia city. (**B**) The Umbria region with the scientific museum already present on the regional territory and dedicated to natural sciences and Earth sciences. The different symbols represent the specialization: (1) Didactics, (2) Geology, (3) Hydrogeology, (4) Mine, (5) Natural Science, (6) Paleontology, (7) Volcanology. The numbers inside the figure refer to Table 1: (1) Antiquarium Museum (Corciano, PG), (2) Civic Museum of Natural History (Stroncone, Terni), (3) Earth Science Laboratory of Spoleto (Spoleto, PG), (4) GeoLab (San Gemini, TR), (5) GSN Gallery of Natural History (Casalina, PG), (6) Morgnano Mines Museum (Spoleto, PG), (7) Museum of Natural History and of Territory (Città della Pieve, PG), (8) Museum of the Apennines (Polino, TR), (9) Museum of the Geological Cicles (Allerona, TR), (10) Museum of the Territory (Parrano, TR), (11) Naturalistic Museum of Colfiorito Park (Colfiorito, PG), (12) Naturalistic Museum of Cucco Mt. and Earth Science (Costacciaro, PG), (13) Paleontological Museum (Assisi, PG), (14) Paleontological Museum (Pietrafitta, PG), (15) Paleontological Museum (Terni, TR), (16) TerraLab (Perugia, PG), (17) The Botanic Palaeontology Centre of the Fossil Forest in Dunarobba, (Allerona, TR), (18) Volcanological Park of San Venanzo (San Venanzo, TR), Water Museum (Perugia, PG). The abbreviation PG is for Perugia, the abbreviation TR is for Terni, Perugia and Terni are the two provinces of the Umbria region.

Umbria is a region with strong evidence of a connection between topography, morphology, and geology, and so is an excellent test area for such dissemination activities. Nineteen museums, with permanent exhibitions focused on some aspects of Earth sciences are already present in the regional territory (Figure 1B). Five of them are focused on paleontological heritage and as many on a wider naturalistic aspect where the geological component is only a part of the exhibition. Three museums are devoted to general aspects of local geology, while one is dedicated to mining activity, one to volcanology and another one to hydrogeology. All these museums offer occasional didactic laboratories but only the remaining three museums have permanent laboratories and exhibitions for didactic purposes. In Table 1, the museums are listed with their specific vocations.

Although Umbria is a small region, it can, therefore, count on a good number of initiatives aimed at divulging geological data. However urban geology has never been the subject of dissemination activities in the region. This paper illustrates the first attempt to do that in Perugia, one of the most important hotspots for cultural initiatives in central Italy and offering a large number of aspects useful for research on urban geology. This paper illustrates in detail the scientific background, the dissemination techniques and the results of this experience.

Table 1. Museums present in Umbria. The numbers refer to Figure 1B. Type: D) Didactics, G) Geology, H) Hydrogeology, M) Mine, NS) Natural Science, P) Paleontology, V) Volcanology. The abbreviation PG is for Perugia, the abbreviation TR is for Terni, Perugia and Terni are the two provinces of the Umbria region.

N.	Name	Location	Type
1	Antiquarium Museum	Corciano (PG)	P
2	Civic Museum of Natural History	Stroncone (TR)	NS
3	Earth Science Laboratory of Spoleto	Spoleto (PG)	D
4	GeoLab	San Gemini (TR)	D
5	GSN Gallery of Natural History	Casalina (PG)	NS
6	Morgnano Mines Museum	Spoleto (PG)	M
7	Museum of Natural History and of Territory	Città della Pieve (PG)	NS
8	Museum of the Apennines	Polino (TR)	G
9	Museum of the Geological Cicles	Allerona (TR)	G
10	Museum of the Territory	Parrano (TR)	G
11	Naturalistic Museum of Colfiorito Park	Colfiorito (PG)	NS
12	Naturalistic Museum of Cucco Mt. and Earth Science Laboratory	Costacciaro (PG)	NS
13	Paleontological Museum	Assisi (PG)	P
14	Paleontological Museum	Pietrafitta (PG)	P
15	Paleontological Museum	Terni	P
16	TerraLab	Perugia	D
17	The Botanic Palaeontology Centre of the Fossil Forest in Dunarobba	Allerona (TR)	P
18	Volcanological Park of San Venanzo	San Venanzo (TR)	V
19	Water Museum	Perugia	H

2. The "Perugia Upside-Down" Exhibition: An Example of Best Practice

Perugia is the capital city of the Umbria region, (central Italy) and is located on a triangular-shaped hill with an areal extent of about 27 km². The maximum altitude value is about 493 m a.s.l. with a minimum of ca. 200 m along the Tiber River valley, at the bottom of the hill (Figure 2). The hill is distributed along five main ridges spreading from the highest altitude toward NE, E, SSE, SW, and W, separated by several small rivers. The hill of Perugia is made of sediments derived from fluvial and/or lacustrine environments, widespread in the area during the Pliocene and Pleistocene.

In these periods, an extensional tectonic phase, still acting, affected the area and the morphological result of this phase are several intermountain basins bordered by normal faults [22,23]. Perugia is located along the western edge of the Tiberino Basin, the largest basin in Umbria (about 1800 km²) and one of the largest in Central Italy (Figure 2B). The bedrock of the hill of Perugia is made of clastic sediments of different sizes, from blocks and gravels to sands and clays, transported by the rivers flowing from the surrounding mountains and then deposited on the bottom of the intermountain basins. In addition, a drainage network of rivers, swamps, and lakes was widespread along the plain areas inside the basins, covering with new sediments and reshaping the previous deposits. The sedimentary sequence, dated in Perugia from Early to Middle Pleistocene, has variable thicknesses from few to hundreds of meters and is defined as "Perugia Unit" (Figure 2B). The unit is divided into some litofacies according to sedimentary and paleoenvironmental principles. In each of these litofacies some deposits prevail. In the Volumni Litofacies, present in the downtown of the city (lower Pleistocene), conglomerates and sand are prevalent. The extensional tectonic stress is still acting with the result that the morphological evolution is very dynamic [24,25]. Along the borders of the intermountain basins, the sedimentary sequences are faulted and eroded, resulting in gentle hilly areas. The topographic arrangement, with a higher altitude, compared to the lowlands of the alluvial plain, often covered by stagnant water, guaranteed a healthier environment. In the same time, the gentle slope values along the flanks of the hills allowed an easier connection with roads and cities in comparison to the steep mountain areas [26]. The most important historical cities in Umbria are located on the top of these sedimentary hills and their position and sites are a clear consequence of the geological history of the area. Perugia is a perfect example of this condition and a good test site for urban geology and for the scientific communication of this topic.

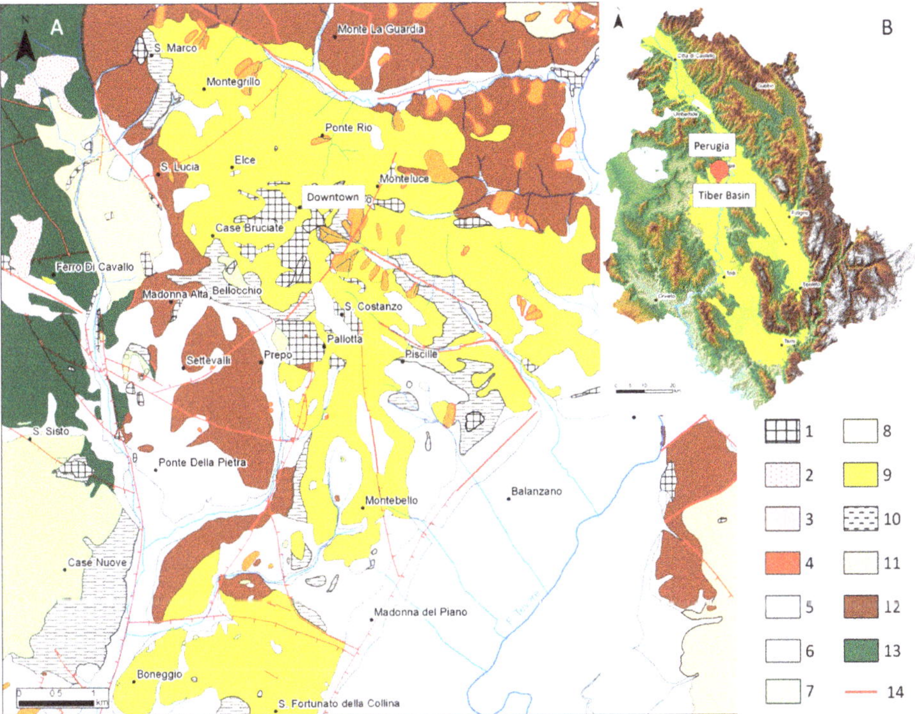

Figure 2. (**A**) The geological map of the Perugia city area. (1) Anthropic deposits, (2) Debris (Holocene), (3) Colluvial deposits (Holocene), (4) Landslides, (5) Alluvial deposits (Holocene), (6) Alluvial terrace (Holocene), (7) Perugia Unit, Ellera Litofacies (upper–medium Pleistocene), (8) Perugia Unit, Pian di Massiano Litofacies (medium Pleistocene), (9) Perugia Unit, Volumni Litofacies (lower Pleistocene), (10) Perugia Unit, Ferrini Litofacies (lower Pleistocene), (11) Solfagnano Unit (lower Pleistocene), (12) Terrigenous Complex (Burdigalian–Tortonian), (13) Limestone Complex (upper Trias–lower Miocene). (**B**) The Perugia city in Umbria region and inside the limits of the Tiberino Basin (in yellow).

"Perugia Upside-Down: When the Geology Describes the City" is the title of an exhibition developed by the Department of Physics and Geology of the University of Perugia, inaugurated on 10 November 2017. The exhibit location is the POST Museum (Perugia Officina della Scienza e della Tecnica—Perugia Science and Technology Laboratory, http://www.perugiapost.it), the most important and visited a scientific museum in the city. The exhibition lasted until the spring of 2018 (Figure 3).

Figure 3. The poster used to promote the exhibition. The title was "Perugia Upside-Down: When the Geology Tells the City" (the reproduction of the image is allowed by POST).

2.1. Methodology

Exhibition is a common and useful practice in order to attract people [27,28], but the results in terms of the number of visitors or the level of satisfaction are not always encouraging [13]. Problems are related to the traditional method of exposition (samples and description). Taking inspiration from previous experiences, to which others have been added over time as highlighted in the references list [29–34] some already tested items were proposed in the exhibition. One of the most successful approaches is to highlight and illustrate some of the stones used for historical buildings [35,36]. The identification of petrographic characteristics is the starting point to expand the information linked to the paleoenvironmental conditions. Thus, the building stones are snapshots of the geological history of the surrounding areas and, because of their position, visible on the most important buildings of our cities, are instruments always openly available. In addition, some samples show palaeontological features, thus not only the sedimentological or mineralogical data are present but also other added values, introducing a wide range of geological aspects [37,38]. In order to understand the geological composition of lithotypes outcropping under the cities or in the surrounding areas, the building stones are used as a point of interest for urban trekking and they are one of the best expressions of the geoheritage present in urban areas [12]. Historical buildings, bridges or industrial constructions are viewpoint geosites. The considerable height of towers, belfries, industrial sheds (especially if associated with a surrounding topographic arrangement with lower altitude values) offers a unique opportunity to admire the landscape and understand the morphology and spatial location of a city [39,40]. In addition, other aspects are properly used for touristic and didactic purposes. The geomorphological evolution perhaps represents one of the most intriguing cases because, traveling through time, rewinds the morphological evolution and reveals the past, the present and the future landscape [41,42]. The paleontological heritage, if it was found in the urban area, could be an excellent topic for an exhibition [43,44]. Moreover, when characteristics concerning geology are combined with other fields, such as archaeology, the sites where these characteristics are present at the same time can be excellent targets for geotourism and thematic exhibitions [26,45].

Once the geological aspects are known, the next step is to identify the best solution to disseminate the content. To translate the urban geology from a scientific perspective to well-understood information, some criteria must be satisfied [13]. First of all, is the time interval. Geology is a science that takes into account timescales of up to hundreds of millions of years (the Earth system has been evolving since approximately from 4.5 billion years ago) while the human experience covers at most a few millennia. For common people, thinking in terms of ancient times generally means to enlarge the time perspective up to a few hundred years. The morphological evolution is perceived as something related to an unchangeable system where only the catastrophic events (earthquakes, volcanic eruptions, tsunami, landslides) suggest that the Earth is a dynamic planet. The dynamic equilibrium controlling the surface processes modeling the Earth surface is invisible to the human eyes. This is one of the most relevant difficulties when, for example, the perceived risk is lower than the real risk during natural hazard events. Therefore, in order to communicate the geological evolution of an area, it is fundamental to underline the time spans in relation to human life. The second problem is the four-dimensional perspectives, necessary to a geologist to understand features and events. A geologist often needs to consider a landscape in 3D. In addition, a fourth dimension is needed, considering the structure under the topographic surface too. This means having the skills to consider the landscape from a geographical perspective on the surface of the Earth and keeping the visualization vertical, imagining the removal of the topographic surface as if it was only a thin layer. This skill is not common for people with different knowledge, and therefore one of the greatest efforts that must be made to make scientific communication effective is to introduce a tourist or a student into a "bird's eye" view and then take them below the Earth's surface in the fourth dimension. The third problem is the scale. Geology includes patterns and processes that range from the infinitely large to the infinitely small. To understand Earth dynamics, the observations embrace a spatial framework going from the solar system and beyond until the microscopic observation of the structure of minerals. The challenge is to make clear that these scales are the opposite sides of the same coin and to join the information deriving from different approaches in a unique way. The fourth drawback is the language. Every experience related to scientific communication should translate the scientific language in a common way, using few but unavoidable rules: Concise and without technical terms but exhaustive, in other words, simple but not simplistic. To find the best compromise between complete information avoidance to being incomprehensible and boring is not so obvious. Many experiences attempt to avoid the problem using a glossary, but this is a false solution. It is quite rare that in dissemination activity people are so intensely involved as to seek out clarification each time it is necessary, consulting a glossary. The first reaction is to read a text without fully understanding it. The fifth point, which is specific for urban geology, is to never forget that in the cities the geological evolution is strictly related to human settlement, so never separate the naturalistic aspect from the anthropic one. People may be interested in the natural environment, but they become even more interested if this environment is something that has an impact on their everyday life.

To try to get the best result, the exhibition was structured with a basis of panels in the museum but with several parallel activities with the aim of encouraging visitors to become active subjects, both in the museum and outside: tools, laboratories, trekking tours. To overcome the problems related to the disclosure of a scientific subject listed above, this exhibition was prepared in a synergy between researchers, museum workers, and designers. The researchers have devised the thread of the information flow and prepared the text, the figures, and the theoretical basis for the tools. Moreover, they prepared the laboratories and led the trekking around the cities. The staff museum built the infrastructures to house the material. Most of all, they provided an irreplaceable contribution in simplifying the scientific character of the texts and figures. The designers created the graphics and organized texts and figures on the panels.

Although panels may be the most boring aspect of an exhibition, the cooperation with designers and staff museum guaranteed an amazing and effective final product. The panels were designed following some criteria (Figure 4). The upper section was devoted to the title and to a progressive number showing the path to be followed. At the bottom of the panels, only graphics were present to not force the visitor to bend down. In the middle part, the text was separated in columns with a logical idea, which imposed the public to read the contents going from the left to the right. On the left, only the fundamental concepts were summarized, then moving toward the right side of the panel, other peculiar information was added. The aim was to introduce the visitor to the topic described on the panel, presenting information step by step and giving them the possibility to decide when to finish reading, without losing important information. Figures and photos were always present. Some supplementary boxes were included for explaining technical words or particular geological concepts.

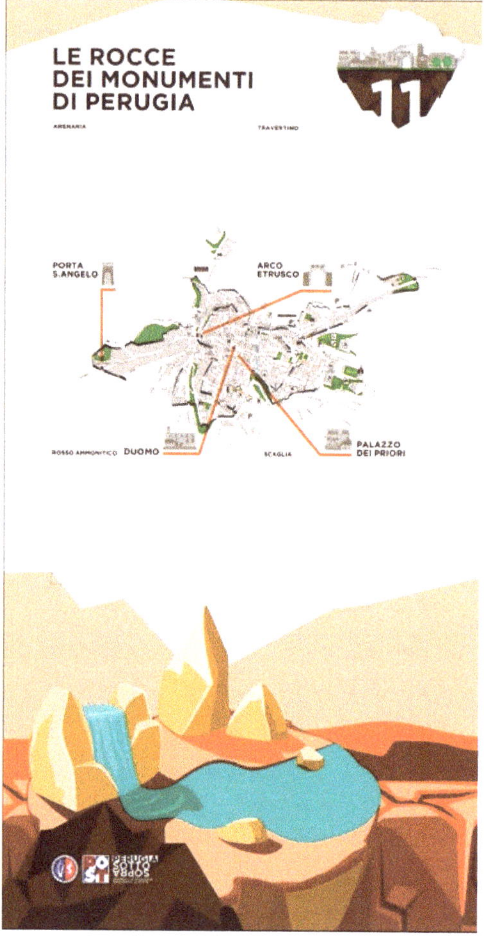

Figure 4. An example of a panel with the division of the space. In the upper part and at the bottom the graphics are present. In the middle part, the text and figures. In this example, the white squared contain the showcases with the rock samples referred to some historical buildings in Perugia. From the left corner on the top and proceeding clockwise: sandstone, travertine, limestone, Rosso Ammonitico Formation. The title of this panel is "Rocks and Monuments in Perugia" (the reproduction of the image is allowed by POST).

Multimedia tools interrupted the path of the exhibition, guided by the numbered panels. Transparent and illuminated showcases contained samples of rocks and terrain. Videos with real images and paleo-environments reconstructed with digital techniques were broadcast continuously. Moreover, some interactive tools invited visitors to create their own experience with the different geological components. In the opening period of the exhibition, some laboratories in the museum and outside were organized devoted to scholarships. Urban trekking completed the offer with the possibility for people of all ages and cultural levels to observe the places that they were introduced to in the exhibition within the city. The results and methods of this approach are described below.

2.2. Results

2.2.1. The Panels

The exhibition was structured in five sections all included in the main hall of the museum (Figure 5), each one devoted to a particular aspect of urban geology present in Perugia with a theoretical scheme following an initial introduction to the geological history and then moving toward some more particular aspects.

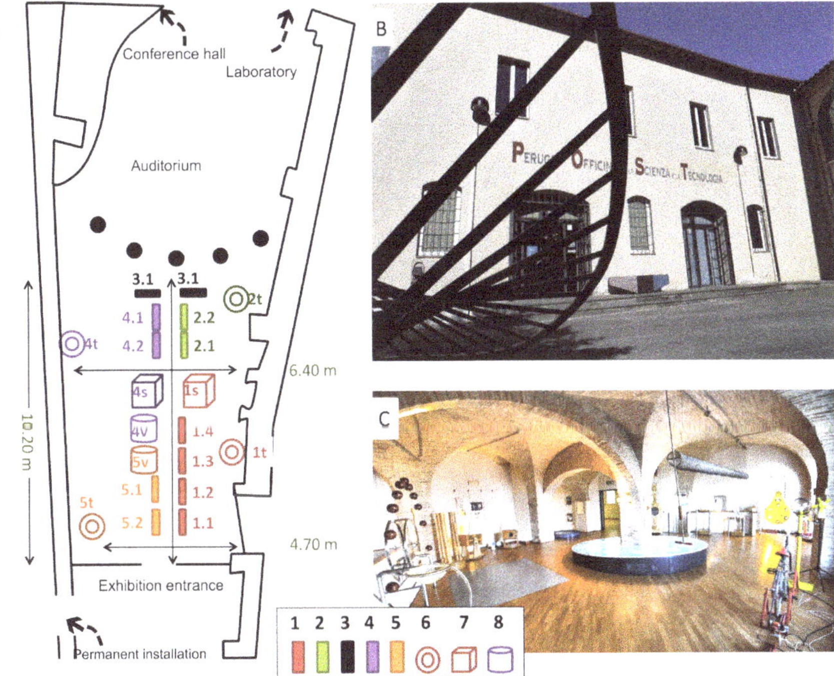

Figure 5. (**A**) Map of the POST museum. In addition to the room where the exhibition was installed, the museum has a room with a permanent installation, an auditorium, a conference hall and a room for the laboratories. (1) Panels of geological section, (2) panels of geomorphological section, (3) panels of the human presence, (4) panels of building stones section, (5) panels of paleontological section, (6) tools: (1t) 3D puzzle, (2t) ARSandbox, (4t) optical microscope, (5t) rhino skull model. (7) Showcases: (1s) boxes with conglomerates, sand, clay, (4s) boxes with samples of travertine, limestone, and sandstone. (8) Videos: (4v) video of building stones, (5v) video of Pleistocene paleoenvironments. (**B**) the entrance of POST museum, (**C**) The room with the permanent installation (photo by POST, use allowed by POST).

The first section was assigned to the geology, followed by the second one, where the geomorphology was the topic. The third section illustrated the relationship between human presence and geological context, while the fourth section was dedicated to building stones. The fifth section was used to house the paleontological heritage found in the city and to explain the fauna present in the Pleistocene (Figure 6).

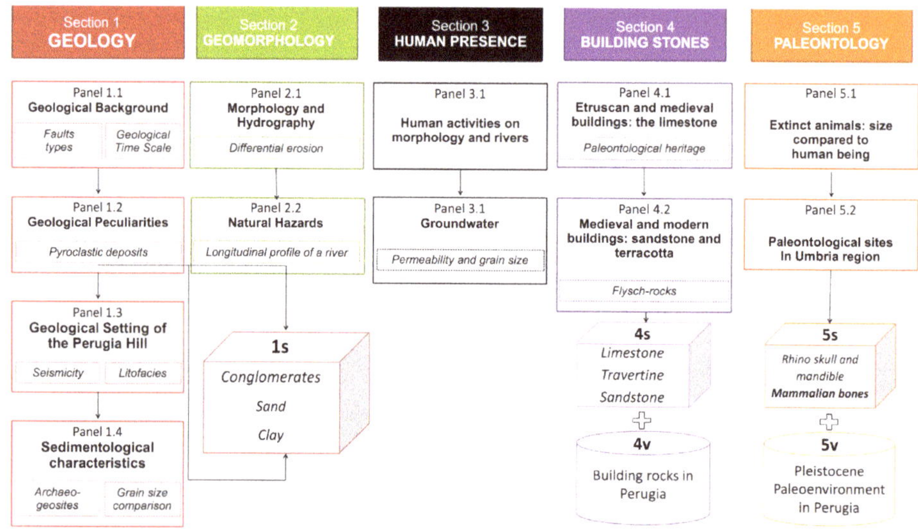

Figure 6. The list of the panels divided according to the different sections: geology, geomorphology, human presence, building stones, paleontology. The topic of each panel is under the number. The subjects of the in-depth boxes are specified in the frames with the dashed lines. The cubes represent the showcases (s) while the cylinders the videos (v).

The geological section (Section 1) proposed four panels where the bedrock composition and the geological evolution of the area were summarized (Figure 7).

In the panel 1.1 the geological background summarized, with a geological time scale, the entire geological history of Umbria region from the oldest rocks dated about 250 My up to now. Two in-depth boxes better explained what a fault is and the principles of stratigraphy. The panel 1.2 had the aim to dispel some "false myths" still deep-rooted in the popular culture of the place. In particular local traditions identified some mountains close to Perugia as ancient volcanoes. This information is still present in some websites, pointing out the poor communication between academic institutions and local people. The in-depth box tried to explain what is a true pyroclastic deposit. The panel 1.3 was focused on the geological setting of the Perugia hill with two in-depth boxes. The first one explained the concept of litofacies due to the fact that the sedimentary sequence outcropping in the city is organized in several litofacies. The second box illustrated the relationship of the area with the seismicity of the central Apennines. Although Perugia is located in an area with low seismic risk, moving eastward, the Apennines record events with high magnitude and thus the effects of seismic shocks are evident in the city too and affect, mostly from a psychological point of view, a large part of the citizenry. The first section ended with the panel 1.4 where the sedimentological characteristics of the deposits are detailed. Three showcases contained conglomerates, sand clay with a reference scale beside each box. Visitors were able to observe the difference in size between the various deposits. On the panel, one in-depth box suggested some archaeological sites in Perugia were these different deposits might be observed and introduce the concept of archaeo-geosite.

The second geomorphological section was split in only two panels. The first one (2.1) explains the relationship between morphology, hydrography, and the geological arrangement. The typical landscape of Perugia, divided into ridges and rivers, has been interpreted with a geological approach. Due to the fact that some landscape particularities in the Perugia slopes are due to the different grain size of the deposits, the concept of differential erosion is detailed in a box. In the second panel (2.2) the attention was focused on the mass wasting and fluvial processes acting on the area with an analysis of related natural hazards. River erosion is the main cause of landslides, mostly along the headwater drainage divide close to the downtown. Therefore, the in-depth box explains the concept of the longitudinal profile of a river and the tendency to an equilibrium state, gained through erosion and sedimentation activities.

Figure 7. The exhibition along Section 1 geology and Section 2 geomorphology (on the left) and Section 3 human presence (on the right, photo by POST, use allowed by POST).

The third section is on the human presence and the two panels reveal the topographic surface changes made by humans over the centuries to prevent landslides or for the construction of important historic buildings. The definition of a "morphological false" is present in panel 3.1, to explain some characteristic areas in the downtown, and was very appreciated by visitors. The panel 3.2 highlights the ancient water supply methods. In the downtown a large number of historical wells and tanks, from the Etruscan (from V to I century B.C.) and medieval periods are present. Due to the sedimentary grain size sequence, the oldest part of the city has a huge amount of underground water reserve even today. In the panel the concept of porosity and permeability is detailed.

The mineralogical section is the fourth one and it was dedicated to the building stones. In fact, in Perugia, there is a very close relationship between some historical periods (Etruscan and Roman, medieval and the passage between the XIX and XX centuries) and the use of specific lithotypes for the construction of the main religious and civil buildings. The panel 4.1 shows the use of travertine in the Etruscan period (Etruscan walls) and of limestone in the medieval one, while the panel 4.2 highlights the use of sandstone in the medieval walls and of terracotta, derived from the clay present at the bottom of the hill, on the most recent historical buildings (beginning of XX century). The in-depth box reveals the paleontological heritage hidden on the façade of some important buildings in the downtown and that several tiles are made of Rosso Ammonitico Formation (Toarciano). The name of this formation, well widespread on the regional territory, derives from the high content of ammonite fossils. Four showcases contained many samples of travertine, limestone, and sandstone. A video with subtitles, close to the showcases, evidenced the use of these lithotypes on the most famous religious and civil buildings in the downtown of Perugia and the natural environments where these sedimentary rocks originate.

Finally, the last palaeontological section illustrates the mammal fauna of central Italy in the Pleistocene (Figure 8). One of the most important results of the exhibition was to show for the first time the mammal fossils (Pliocene and Pleistocene) discovered in the past century on the Perugia hill, with a well-preserved rhino skull usually not visible to the public. In this section, a video was present too (Figure 9). With surface mesh digital techniques some contemporary places in the city were overlaid with the moving images of mammal fossils in order to show the palaeoenvironmental conditions in the Pleistocene.

Figure 8. One of the panels in the paleontological section. On the desktop the several parts of the model of the rhino skull are visible (photo by L. Melelli, use allowed by POST).

Figure 9. The palaeontological section: the real rhino skull is in the showcase on the right, on the left the video 5v with a frame representing the merge between a present landscape of Perugia and a digital reconstruction of the lake present in the area in the Pleistocene with some mammals moving along the shore (photo by L. Melelli, use allowed by POST).

2.2.2. The Tools

In each section, a tool invited the visitors to be an active subject of the exhibition (Figures 5 and 10).

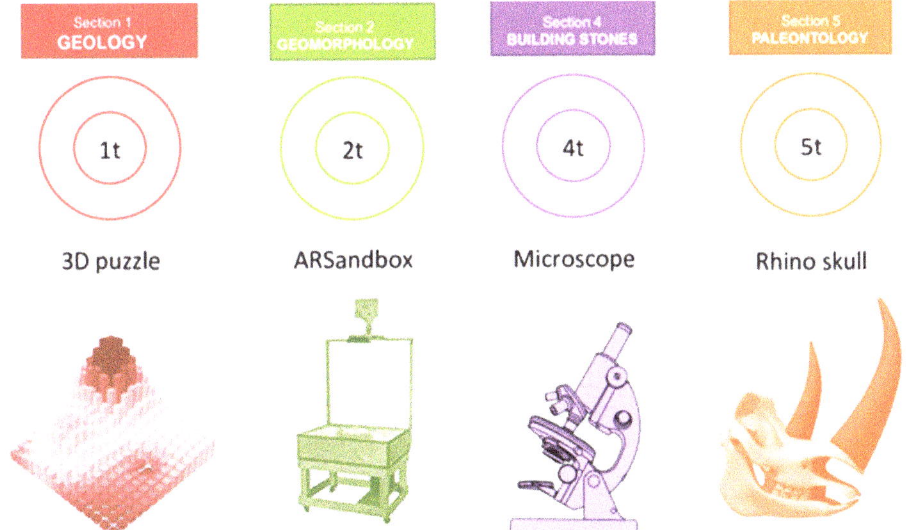

Figure 10. The tools in each section. The symbols with the numbers refer to Figure 5. 1t is the 3D puzzle in the geological section, 2t is the ARSandbox in the geomorphological section, 4t is the optical microscope in the section dedicated to building stones, 5t is the model of rhino skull in the paleontological section.

In the geological section, to help the visitor understand the spatial distribution of the lithotypes a 3D puzzle of the area was created (Figure 11). The first step was to extract some contour lines from a digital elevation model of the hill of Perugia (cell size 5 × 5 m). Then the polygons of the geological complexes were overlaid. Finally, only for the downtown area, the polygons of the watersheds are added where the drainage divide of the main rivers flowing on the city center converges. A 3D printer, analyzing the vector data, created the plastic model of the Perugia hill and surrounding area. Different colors were associated with the geological complexes while the plastic was cut along some boundaries corresponding to the limits between different lithological complexes or along drainage divides. Then some labels were available to be added to the puzzle and to identify the symbolic places of the city. In order to help the visitors, a poster in front of the plastic model was present with the names of the places printed on the labels.

In the geomorphological section, an augmented reality (AR) sandbox was installed (https://arsandbox.ucdavis.edu) allowing the 3D visualization of virtual topographic surfaces (Figure 12).

In particular, topographic contour lines and an elevation color map were visualized, and the water flow was simulated. The visitor, by hand-shaping the sand in the box, could modify the topographic surface and try to reproduce the morphology of the area.

In the mineralogical section, an optical microscope and thin sections of the main rocks present in the exhibition were made available to visitors (Figure 13). Each thin section was illustrated by a card where the petrographic and paleontological characteristics present in the thin section were detailed and highlighted. Beside the microscope, a hand lens was available for observing the macroscopic petrographic characteristics.

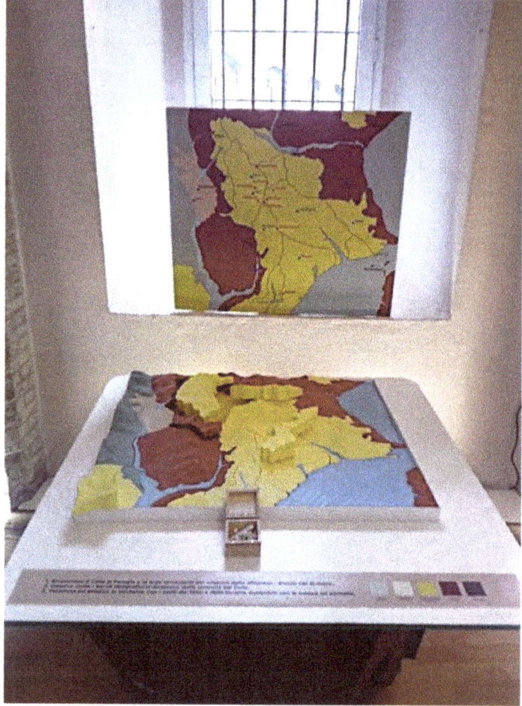

Figure 11. The 3D puzzle in the geological section. On the desktop, the model created with the 3D printer is available to visitors. The little box on the desktop contains the labels with the place names to be arranged on the model while the legend details the meaning of the different colors corresponding to the lithotypes. The poster hung in front of the window has the aim to help the visitors in doing this activity and represents the model in plain view with the watershed boundaries and the place names already put in order (photo by L. Melelli, use allowed by POST).

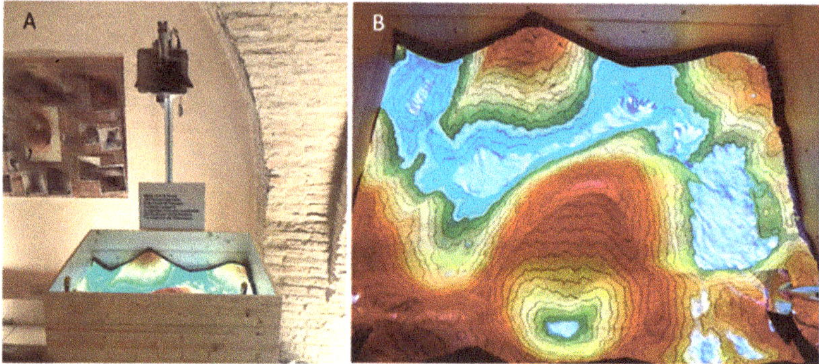

Figure 12. The ARSandbox in the geomorphological section. (**A**) The sandbox with the full equipment. (**B**) The surface of the model with the color ramp projected on the sand. The cold colors (blue one) refer to the lowest altitude, the heights increase going from green to yellow and brown for the highest altitude values. The contour lines are projected too. It is possible to observe in the hollowed areas the water effect (photo by L. Melelli, use allowed by POST).

Figure 13. The optical microscope in the mineralogical section. On the left, the thin sections are available together with the instruction manual (photo by L. Melelli, use allowed by POST).

Finally, in the palaeontological section, a rhino skull was reproduced with a 3D printer and divided into some pieces along the morphological limits. Visitors were invited to put together the pieces to reconstruct the entire skull and better understand the shape and the function of the different pieces.

2.2.3. Laboratories and Trekking

During the regular time schedule for museum visits, some laboratories were organized. The laboratories were mainly dedicated to schools (Figure 14).

Figure 14. One of the laboratories prepared for the exhibition. In particular, this laboratory was dedicated to the paleontological section. (**A**) A school group working on the field to observe the rocks on an outcrop of limestone. (**B**) The laboratory's activity for creating the shape of some ammonites with modeling paste and for observing the morphological characteristics (photo by G. Margaritelli, use allowed by the author and by POST).

According to normal school planning, Earth sciences are focused on natural locations. In these laboratories, the aim was to introduce the cities as geological environments. Children, teenagers, and young people live daily in their cities, and most of their educational and recreational experiences are connected to urban infrastructures and places. For this reason, it is fundamental to exploit what each city can offer to bring young people closer to Earth sciences. Among the activities offered, the AR Sandbox appeared to be the most attractive tool. The key to understanding the scientific content is the augmented reality component. Contour lines and a terrain color ramp were projected on the virtual topography and movement was tracked using a Microsoft Kinect 3D camera. Placing an object at a particular height above the sand surface, a virtual rain is simulated, and water flowed over the landscape. Some fundamental topographic attributes, such as slope angle, could be visualized and easily modeled and modified. By connecting the slopes to the flow direction and accumulation may facilitate the understanding of drainage network modeling. Moreover, the AR Sandbox allows the capturing of photographs of the surface morphology at different times during use, rebuilding the sequence of events that modify the virtual landscape and offering the opportunity to follow its evolution over time. The strong point of this tool is that visitors can interact with the virtual topography by providing the SECRET "SEe and CREaTe" [46] for effective scientific communication. During the exhibition, weekly workshops were organized for schools of all levels and adult people (Figure 15A). Moreover, the material presented in the exhibition represented an important resource to be used in the activities of dissemination and information about the degrees in geology offered by the University of Perugia to different schools in the city.

Figure 15. Some images of the activities organized during the exhibition. (**A**) A conference in the auditorium (see Figure 5), (**B**) trekking in the downtown to show the paleontological heritage on the building stones (photo by M. Coli and http://www.circolosanmartino.unipg.it, use allowed by the author and by POST).

Moving outside of the exhibition and remembering the information acquired inside the museum allowed visitors to complete their experience and to consolidate their cultural experience. The idea was to propose trekking tours in four dimensions (Figure 15B).

Two dimensions were presented walking along a path and referring to a map for improving the sense of direction and spatial arrangement of places. The third dimension was the perspective observable from scenic viewpoints. Being a hilly city, Perugia offers several scenographic standpoints. Moreover, Perugia has two opposite landscapes, the steep and uninhabited scenery along the northern area and the gentle and urban one on the opposite side. This contrast is a good starting point for recalling geological and geomorphological aspects, such as tectonics, differential erosion, and fluvial and gravitational processes.

One of the most successful trekking routes was from the POST Museum up to the top of the downtown. There were six stops in total: one focused on the fluvial processes and natural phenomena, two on the anthropic modifications of natural morphology, two on building stones, and the last was run underground and exploited one of the most important Etruscan wells, the most important archaeological evidence of the ancient human presence on the hill related to water resources research. Trekking experiences represent the key to effective scientific communication. People could see, touch, look for, and most of all, connect an abstract idea to something tangible. Moreover, they could apply a scientific subject to daily life and acquire the capability to observe the urban environment from a different perspective. During the trekking tours, visitors were entertained above all by "fossil hunting". None of them, despite having lived in Perugia for decades, had ever noticed that on the facade of the city's main church, fossils of ammonites were present (Figure 16). This hints that the idea of the geologist obliged to search for scarce and rare fossils in natural environments is outdated, suggesting it is sufficient simply to observe our surroundings, especially those of historical buildings.

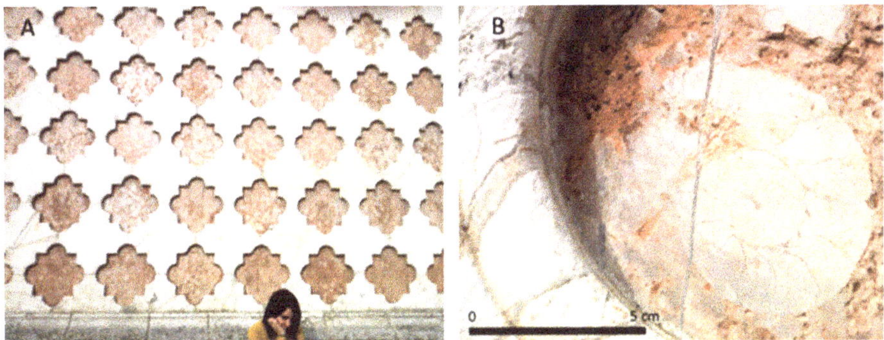

Figure 16. The palaeontological heritage on the building stones: (**A**) The façade of the San Lorenzo Cathedral, the most important church in the Perugia city, also in Figure 15. The limestone shows two colors, pink and white, for a better aesthetic result. (**B**) One ammonite inside the tiles.

2.3. Discussion

Urban geotourism is a promising approach to disseminating Earth sciences to a wide audience. Urban areas guarantee several advantages compared to natural environments. Cities with relevant historical and artistic contexts are generally already structured for needs related to tourism. The connection between human activities and the original natural environment, both in past centuries and in the present day, is well evident. Cities are places where digital tools (Wi-Fi and electronic devices, such as smartphones and tablets) are, in most cases, already structured and available for free [47] so that in urban areas the dissemination activities are facilitated and encouraged in order to increase the tourist flow. Several approaches are already tested in several cities in the world [48]. São Paulo in Brasil [49], Mexico City [30], London (http://londonpavementgeology.co.uk), Lisbona in Portugal [31], Brno city in Czech Republic [50], Belgrad in Serbia [34], Shiraz city in Iran [32] are only some examples. In Italy the geotouristic approach in urban areas has been already tested in some important cities. Rome [41,42,51,52], Milan [29], Genoa [53], Naples [54], Turin [55].

The "Perugia Upside-Down" exhibition was the first experience of geotourism dedicated to urban geology in the city of Perugia. 8046 people, 3915 of whom were students, visited the exhibition. This number is a good result for the city and an excellent outcome for the POST Museum, which is dedicated exclusively to scientific topics. Panels, real samples in showcases, videos, and multimedia tools are the avenues chosen to involve the public in themes present in the exhibition. Didactic laboratories and urban trekking are an incisive answer to "force" the visitors in moving out from the museum and discovering the contents of the exhibition in the real world. Moreover, Perugia, if compared with other cities like Roma or Milan, has the great opportunity to be in a hilly environment. Trekking activities may exploit several scenic views and the geomorphological experience could be much more interesting and richer.

The exhibition, despite good results, made clear some critical issues. The structure and content of the panels fully satisfied visitors. However, the number of panels and the large amount of information within them has made it difficult for younger visitors to understand. The texts should be written with non-technical terms, but in particular, they should be extremely brief. Although the tools obtained the best results in terms of involvement, two of them have raised some problems. In particular, the optical microscope showed significant limits. The managing of the several mechanical and optical components of the microscope requires a specialist beside the visitors. Although an explanatory sheet was next to the microscope, the comprehension of the thin section was not always clear. The location of the microscope was a mistake too, being along the path and without a dedicated corner where the visitors could observe the thin section comfortably and without feeling rushed. An alternative method, like a screen connected to the microscope with predefined focus, guided views and only some controlled rotation of the objects could be an alternative and better solution. The 3D model of the rhino skull was not always easy to manage for the visitors. The model was divided into some parts, according to a morphological principle. When the visitors found the sections already divided on the desk, it was very difficult to put the model together again. A detailed guide with the instructions listed step by step and figures of each component could facilitate a better understanding of the procedure. The most successful tools were the 3D puzzle and the ARSandbox. In both cases, no difficulty was identified. The visitors presented themselves as both amused and interested. These results confirm that when the dissemination activity satisfies the SECRET (SEe and CREaTe) for good communication, it goes beyond the limits imposed by the scientific nature of the content. The 3D puzzle is particularly worthwhile for obtaining awareness of geographical space and acquiring the ability to orientate places and put them in topological relation. The third dimension of the model facilitates the understanding of the distribution of altitude values. Observing and touching the distribution of slope values makes it possible to link some theoretical concepts, such as river erosion and the connection with slope evolution. In addition, the lithotypes being highlighted with different colors, it is possible to explain the influence of structural factors on superficial morphology. The ARSandbox is efficient in communicating the concepts of geomorphological processes, in particular, where the runoff is the main focus. The contour lines being visualized together with a color ramp make the sandbox a perfect visualization tool in the modeling of the real world with topographic maps. Moving the sand, the visitors modify the topographic surface and control the topographic attributes like slope, aspect, and curvature. The superimposition of the water flow effect shows the interaction between river drainage network and topography. The advantage of ARSandbox is the strong interaction opportunity presented to the visitors with the tool, mostly effective with young people and children.

Finally, for urban geoheritage promotion, the trekking experience turned out to be extremely positive. Visitors were invited to express their opinion and the results were extremely positive. Once again, to combine the daily experiences in the real world with theoretical concepts seems to be the key for effective dissemination of urban geological phenomena. Despite this, if compared with other similar experiences, the trekking activity could be improved. If urban areas offer some advantages in using digital technologies, this possibility should be strongly exploited. Where digital technologies empower the tools for geotourism, new approaches and potentialities are growing. This is the case of the mobile application technology developed for Lausanne [55], Turin [54], and Rome [55]. In the "Perugia Upside-Down" trekking activities the structure of the trekking was the traditional one with a guide speaking in front of the point of interest. This simple solution is not the most charming and the introduction of a mobile application is strongly recommended.

3. Conclusions

In 2017, looking for the best practice to transfer knowledge from a scientific or technical community to a broader audience in an urban environment, the Department of Physics and Geology in the University of Perugia organized an exhibition. The idea was to open decades of data collected by geologists, archaeologists, historians, and architects to citizens and tourists. The exhibition was structured in panels, interactive tools, laboratories, and trekking within the city. In this video: https://www.youtube.com/watch?v=oDng-kPKvpw, it is possible to take a virtual tour of the exhibition. The experience, despite good results, highlighted some critical issues. In the panels, the text could be further shortened and simplified. Some tools turned out not to be suitable for an exhibition for educational purposes or, more precisely, not without some precautions that simplify their use. Trekking in urban areas could be more effective if supported by digital devices that expand the information.

Starting from the "Perugia Upside-Down" experience, new projects started in order to improve geotourism. SILENE (a LIDAR system for exploring the Palazzone necropolis remote sensing and geology for enhancing archaeological sites) is a project with the aim of promoting the Etruscan necropolis of Palazzone in Perugia, that is undoubtedly one of the most valuable Etruscan burial sites in Central Italy [26,45]. More than two hundred tombs are present in the necropolis, all dug at different levels within the deposits of the Perugia hill. The perimeter walls are real "three-dimensional geological sections", allowing the observation of the sediments from various orientations. The project revealed to the visitors the paleogeographic environment of the Perugia hill through the sedimentary structures present in the deposits suggesting the importance of the Necropolis as an archeo-geosite where historical-artistic value and geological importance are combined. GPS and digital surveying (LIDAR—laser imaging detection and ranging) together with a drone appropriately equipped for carrying out aerial surveys, allowed topographic maps, orthophotos and a detailed digital model assisting in the production of virtual images and tours. The results are visible on http://www.silenepg.it.

The experience of urban trekking during the exhibition suggests us to exploit digital techniques to better involve people in consuming information and obtaining a completely satisfying experience. For this reason, a second project is being developed, named HUSH (hiking in urban scientific heritage). Mixing science, technology, and augmented reality, HUSH will show the naturalistic and geological heritage hidden in the city along several urban trekking routes. The recent advancements in augmented reality technologies create the basis for the development of immersive and customized touristic experiences (abstract HUSH). The last ongoing project is HUSH Underground that is a section of HUSH dedicated to the underground cavities present in the downtown area of Perugia. The common starting point of all these projects is the geological heritage hidden in the city of Perugia. To this day, geology in Perugia has been linked to the hydrogeological instability affecting large areas close to the downtown. With this new approach, geotourism could be a precious resource and a unique opportunity not only for future research but for didactic and cultural purposes with significant commercial and administrative impacts.

Funding: This research was funded by the Dipartimento di Fisica e Geologia, Università degli Studi di Perugia, project title "PERUSIAE (PERUgia StratIgraphy, geoArcheology and landscapE): a multidisciplinary reappraisal of the geological assessment of Perugia Hill)", RicBAs2014, awarded to Laura Melelli.

Acknowledgments: The author would like to thank POST Museum (http://www.perugiapost.it) for the contribution in the creation of the structure and organization of the exhibition. The staff of the communication agency "Le Fucine Art & Media" (https://www.lefucine.it) are authors for the design of the panels and for the poster used for the advertising. Marco Cherin (https://www.unipg.it/personale/marco.cherin) is the author of the paleontological section. The Sabrina Nazzareni (https://www.unipg.it/personale/sabrina.nazzareni) is the author of the description of the thin sections used with the optical microscope.

Conflicts of Interest: The author declares no conflict of interest.

References

1. Legget, R.F. *Urban Geology, Canadian Building Digest, 113*; National Research Council: Ottawa, ON, Canada, 1969; p. 4.
2. Legget, R.F. *Cities and Geology*; McGraw-Hill: New York, NY, USA, 1973; p. 624.
3. Karrow, P.F.; White, O.L. (Eds.) *Urban Geology of Canadian Cities*; Special Paper 42; Geological Association of Canada: St. John's, NL, Canada, 1998; p. 500. ISBN 0-919216-62-5.
4. Gisotti, G. *La fondazione delle città. Le scelte insediative da Uruk a New York*; Carocci Ed.: Roma, Italy, 2016; p. 559.
5. Bathrellos, G.D.; Skilodimou, H.D.; Chousianitis, K.; Youssef, A.M.; Pradhan, B. Suitability estimation for urban development using multi-hazard assessment map. *Sci. Total Environ.* **2017**, *575*, 119–134. [CrossRef] [PubMed]
6. Güneralp, B.; Güneralp, I.; Liu, Y. Changing global patterns of urban exposure to flood and drought hazards. *Glob. Environ. Chang.* **2015**, *31*, 217–225. [CrossRef]
7. Christenson, E.; Elliott, M.; Banerjee, O.; Hamrick, L.; Bartram, J. Climate-related hazards: A method for global assessment of urban and rural population exposure to cyclones, droughts, and floods. *Int. J. Environ. Res. Public Health* **2014**, *11*, 2169–2192.
8. Carreño, M.L.; Cardona, O.D.; Barbat, A.H. Urban seismic risk evaluation: A holistic approach. *Nat. Hazards* **2007**, *40*, 137–172. [CrossRef]
9. Chester, D.K.; Degg, M.; Duncan, A.M.; Guest, J.E. The increasing exposure of cities to the effects of volcanic eruptions: A global survey. *Glob. Environ. Chang. Part B Environ. Hazards* **2000**, *2*, 89–103. [CrossRef]
10. Orsi, G.; Di Vito, M.A.; Isaia, R. Volcanic hazard assessment at the restless Campi Flegrei caldera. *Bull. Volcanol.* **2004**, *66*, 514–530. [CrossRef]
11. Dowling, R.K.; Newsome, D. *Geotourism*; Routledge: London, UK, 2006.
12. Reynard, E.; Fontana, G.; Kozlik, L.; Scapozza, C. *Geoheritage: Assessment, Protection, and Management*; Elsevier: Amsterdam, The Netherlands, 2017.
13. Garofano, M. Challenges in the popularization of the earth sciences: Geotourism as a new medium for the geology dissemination. *Anuário Do Inst. De Geociências* **2012**, *35*, 34–41. [CrossRef]
14. Reynard, E.; Kaiser, C.; Martin, S.; Regolini, G. A method for assessing 'scientific' and 'additional values' of geomorphosites. *Geogr. Helv.* **2007**, *62*, 148–158. [CrossRef]
15. Pelfini, M.; Bollati, I. Landforms and geomorphosites ongoing changes: Concepts and implications for geoheritage promotion. *Quaest. Geogr.* **2014**, *33*, 131–143. [CrossRef]
16. Reynard, E.; Coratza, P.; Regolini-Bissig, G. *Geomorphosites*; Verlag Dr. Friedrich Pfeil: Munich, Germany, 2009.
17. Reynard, E.; Panizza, M. Geomorphosites: Definition assessment and mapping. An introduction. *Géomorphologie Relief Process. Environ.* **2005**, *11*, 177–180.
18. Cayla, N. An overview of new technologies applied to the management of geoheritage. *Geoheritage* **2014**, *6*, 91–102. [CrossRef]
19. Cayla, N.; Martin, S. Digital geovisualisation technologies applied to geoheritage management. *Geoheritage* **2018**, 289–303. [CrossRef]
20. Ruban, D.A. Geotourism—A geographical review of the literature. *Tour. Manag. Perspect.* **2015**, *15*, 1–15. [CrossRef]
21. Hose, T.A. European geotourism–geological interpretation and geoconservation promotion for tourists. In *Geological Heritage: Its Conservation and Management*. Barettino, D., Wimbledon, W.A.P., Gallego, E., Eds.; Instituto Tecnologico Geominero de Espana: Madrid, Spain, 2000; pp. 127–146.
22. D'Agostino, N.; Giuliani, R.; Mattone, M.; Bonci, L. Active crustal extension in the central Apennines (Italy) inferred from GPS measurements in the interval 1994–1999. *Geophys. Res. Lett.* **2001**, *28*, 2121–2124. [CrossRef]
23. Della Seta, M.; Melelli, L.; Pambianchi, G. Relief, intermontane basins and civilization in the umbria-marche apennines: Origin and life by geological consent. In *Landscapes and Landforms of Italy*, 1st ed.; Soldati, M., Marchetti, M., Eds.; Springer: Cham, Switzerland, 2017; pp. 317–326.
24. Melelli, L.; Pucci, S.; Saccucci, L.; Mirabella, F.; Pazzaglia, F.; Barchi, M. Morphotectonics of the Upper Tiber Valley (Northern Apennines, Italy) through quantitative analysis of drainage and landforms. *Rendiconti Lincei* **2014**, *25*, 129–138. [CrossRef]

25. Pucci, S.; Mirabella, F.; Pazzaglia, F.; Barchi, M.R.; Melelli, L.; Tuccimei, P.; Saccucci, L. Interaction between regional and local tectonic forcing along a complex Quaternary extensional basin: Upper Tiber Valley, Northern Apennines, Italy. *Quat. Sci. Rev.* **2014**, *102*, 111–132. [CrossRef]

26. Bizzarri, R.; Melelli, L.; Cencetti, C. Archaeo-geosites in urban areas: A case study of the etruscan Palazzone Necropolis (Perugia, central Italy). *Alp. Mediterr. Quat.* **2018**, *31*, 1–12.

27. De Wever, P.; Guiraud, M. Geoheritage and museums. *Geoheritage* **2018**, 129–145. [CrossRef]

28. Zwoliński, Z.; Hildebrandt-Radke, I.; Mazurek, M.; Makohonienko, M. Existing and proposed urban geosites values resulting from geodiversity of Poznań City. *Quaest. Geogr.* **2017**, *36*, 125–149. [CrossRef]

29. Pelfini, M.; Bollati, I.; Giudici, M.; Pedrazzini, T.; Sturani, M. Urban geoheritage as a resource for Earth science education: Examples from Milan metropolitan area, the teaching approach to geosciences in early Italian High school: Recent trends and new perspectives. *Rend. Online Soc. Geol. It.* **2018**, *45*, 71.

30. Palacio-Prieto, J.L. Geoheritage within cities: Urban geosites in Mexico City. *Geoheritage* **2015**, *7*, 365–373. [CrossRef]

31. Rodrigues, M.L.; Machado, C.R.; Freire, E. Geotourism routes in urban areas: A preliminary approach to the Lisbon geoheritage survey. *Geoj. Tour. Geosites* **2011**, *8*, 281–294.

32. Habibi, T.; Ponedelnik, A.A.; Yashalova, N.N.; Ruban, D.A. Urban geoheritage complexity: Evidence of a unique natural resource from Shiraz city in Iran. *Resour. Policy* **2018**, *59*, 85–94. [CrossRef]

33. Portal, C.; Kerguillec, R. The shape of a city: Geomorphological landscapes, abiotic urban environment, and geoheritage in the western world: The example of parks and gardens. *Geoheritage* **2018**, *10*, 67–78. [CrossRef]

34. Petrović, M.D.; Lukić, D.M.; Radovanović, M.; Vujko, A.; Gajić, T.; Vuković, D. "Urban geosites" as an alternative geotourism destination-evidence from Belgrade. *Open Geosci.* **2017**, *9*, 442–456. [CrossRef]

35. Brocx, M.; Semeniuk, V. Building stones can be of geoheritage significance. *Geoheritage* **2019**, *11*, 133–149. [CrossRef]

36. Da Silva, C.M. Urban geodiversity and decorative arts: The curious case of the "Rudist Tiles" of Lisbon (Portugal). *Geoheritage* **2019**, *11*, 151–163. [CrossRef]

37. Borghi, A.; D'atri, A.; Martire, L.; Castelli, D.; Costa, E.; Dino, G.; Groppo, C. Fragments of the Western Alpine chain as historic ornamental stones in Turin (Italy): Enhancement of urban geological heritage through geotourism. *Geoheritage* **2014**, *6*, 41–55. [CrossRef]

38. De Wever, P.; Baudin, F.; Pereira, D.; Cornée, A.; Egoroff, G.; Page, K. The Importance of geosites and heritage stones in cities—A review. *Geoheritage* **2017**, *9*, 561–575. [CrossRef]

39. Migoń, P.; Pijet-Migoń, E. Viewpoint geosites—Values, conservation and management issues. *Proc. Geol. Assoc.* **2017**, *128*, 511–522. [CrossRef]

40. Mikhailenko, A.V.; Ruban, D.A. Geo-heritage specific visibility as an important parameter in geo-tourism resource evaluation. *Geosciences* **2019**, *9*, 146. [CrossRef]

41. Del Monte, M. Aeternae urbis geomorphologia—Geomorphology of Rome, aeterna urbs. In *B Landscapes and Landforms of Italy*, 1st ed.; Soldati, M., Marchetti, M., Eds.; Springer: Cham, Switzerland, 2017; pp. 339–350.

42. Pica, A.; Vergari, F.; Fredi, P.; Del Monte, M. The aeterna urbs geomorphological heritage (Rome, Italy). *Geoheritage* **2016**, *8*, 31–42. [CrossRef]

43. Dominici, S.; Cioppi, E. Evolutionary theory and the Florence paleontological collections. *Evol. Educ. Outreach* **2012**, *5*, 9. [CrossRef]

44. Murriello, S. The palaeontological exhibition: A venue for dialogue. *Public Underst. Sci.* **2015**, *24*, 86–95. [CrossRef]

45. Melelli, L.; Bizzarri, R.; Baldanza, A.; Gregori, L. The Etruscan "Volumni Hypogeum" Archeo-Geosite: New sedimentological and geomorphological insights on the Tombal complex. *Geoheritage* **2016**, *8*, 301–314. [CrossRef]

46. Pelfini, M.; Fredi, P.; Bollati, I.; Coratza, P.; Fubelli, G.; Giardino, M.; Liucci, L.; Magagna, A.; Melelli, L.; Padovani, V.; et al. Developing new approaches and strategies for teaching physical geography and geomorphology: The role of the Italian association of physical geography and geomorphology (AIGeo). *Rend. Online Della Soc. Geogr. Ital.* **2018**, *45*, 119–127. [CrossRef]

47. Reynard, E.; Pica, A.; Coratza, P. Urban geomorphological heritage. An overview. *Quaest. Geogr.* **2017**, *36*, 7–20. [CrossRef]

48. Del Lama, E.A.; Bacci, D.D.L.C.; Martins, L.; Garcia, M.D.G.M.; Dehira, L.K. Urban geotourism and the old centre of São Paulo city, Brazil. *Geoheritage* **2015**, *7*, 147–164. [CrossRef]

49. Kubalíková, L.; Kirchner, K.; Bajer, A. Secondary geodiversity and its potential for urban geotourism: A case study from Brno city, Czech Republic. *Quaest. Geogr.* **2017**, *36*, 63–73. [CrossRef]
50. Pica, A.; Luberti, G.M.; Vergari, F.; Fredi, P.; Del Monte, M. Contribution for an urban geomorphoheritage assessment method: Proposal from three geomorphosites in Rome (Italy). *Quaest. Geogr.* **2017**, *36*, 21–36. [CrossRef]
51. Del Monte, M.; Fredi, P.; Pica, A.; Vergari, F. Geosites within Rome city center (Italy): A mixture of cultural and geomorphological heritage. *Geogr. Fis. E Din. Quat.* **2013**, *36*, 241–257.
52. Sacchini, A.; Imbrogio Ponaro, M.; Paliaga, G.; Piana, P.; Faccini, F.; Coratza, P. Geological landscape and stone heritage of the Genoa Walls Urban Park and surrounding area (Italy). *J. Maps* **2018**, *14*, 528–541. [CrossRef]
53. Garofano, M.; Govoni, D. Underground geotourism: A historic and economic overview of show caves and show mines in Italy. *Geoheritage* **2012**, *4*, 79–92. [CrossRef]
54. Gambino, F.; Borghi, A.; d'Atri, A.; Gallo, L.M.; Ghiraldi, L.; Giardino, M.; Valentino, D. TourinStone: A free mobile application for promoting geological heritage of Torino (NW Italy). In Proceedings of the Congresso SIMP-SGI-SOGEI-AIV, Pisa, Italy, 4–6 September 2017; p. 171.
55. Pica, A.; Reynard, E.; Grangier, L.; Kaiser, C.; Ghiraldi, L.; Perotti, L.; Del Monte, M. GeoGuides, urban geotourism offer powered by mobile application technology. *Geoheritage* **2018**, *10*, 311–326. [CrossRef]

Article

Iceland, an Open-Air Museum for Geoheritage and Earth Science Communication Purposes

Federico Pasquaré Mariotto [1,*], Fabio Luca Bonali [2,3] and Corrado Venturini [4]

[1] Department of Human and Innovation Sciences, Insubria University, Via S. Abbondio 12, 22100 Como, Italy
[2] Department of Earth and Environmental Sciences, University of Milan Bicocca, Piazza della Scienza 1, 20126 Milan, Italy; corrado.venturini@unibo.it
[3] CRUST-Interuniversity Center for 3D Seismotectonics with Territorial Applications, 66100 Chieti Scalo, Italy
[4] Department of Biological, Geological and Environmental Sciences, Bologna University, Via Zamboni 67, 40126 Bologna, Italy; fabio.bonali@unimib.it
* Correspondence: pas.mariotto@uninsubria.it; Tel.: +39-0332-218931

Received: 7 December 2019; Accepted: 28 January 2020; Published: 2 February 2020

Abstract: Iceland is one of the most recognizable and iconic places on Earth, offering an unparalleled chance to admire the most powerful natural phenomena related to the combination of geodynamic, tectonic and magmatic forces, such as active rifting, volcanic eruptions and subvolcanic intrusions. We have identified and selected 25 geosites from the Snæfellsnes Peninsula and the Northern Volcanic Zone, areas where most of the above phenomena can be admired as they unfold before the viewers' eyes. We have qualitatively assessed the selected volcano–tectonic geosites by applying a set of criteria derived from previous studies and illustrated them through field photographs, unmanned aerial vehicle (UAV)-captured images and 3-D models. Finally, we have discussed and compared the different options and advantages provided by such visualization techniques and proposed a novel, cutting-edge approach to geoheritage promotion and popularization, based on interactive, navigable Virtual Outcrops made available online.

Keywords: Iceland; geosite; faults; fractures; dykes; geoheritage; Earth Science communication

---

1. Introduction

Geological heritage, better known as geoheritage, has been discussed and reviewed in a number of papers published over the last two decades [1–7]. Another key concept is geodiversity, which can be defined as "the variety of rocks, minerals, fossils, landforms, sediments and soils, together with the natural processes which form and alter them" [8]. Geoheritage comprises elements of geodiversity, which have scientific, cultural, educational value and can be promoted, popularized and protected through geoscience museums [9–11], geoparks [12–14] and geotourism [15–19].

The conservation of geoheritage, aimed at preserving the natural diversity of significant geological and geomorphological features, is a key activity documented and reviewed in a number of works [20–24].

Geoheritage is tightly associated with geological heritage sites, or geosites. These may preliminarily be defined as those parts of the geosphere that are important in terms of the understanding of Earth's history; as far as their social relevance is concerned, they can be regarded as geological or geomorphological objects that may have a scientific, cultural, historical, aesthetic, social and economic value [25]. According to other works [26,27] on this topic, geosites are "geological objects or fragments of the geological environment exposed on the land surface, thus, accessible for visits and studies". As both geology and geomorphology are included within the concepts of geodiversity, geoheritage and geoconservation, geosites encompass both geological and geomorphological features [28].

The most thorough classification of geosites was proposed in a milestone paper [29], in which the biological natural heritage was subdivided into the following categories: geological, geomorphological, structural, tectonic, mineralogical, paleontological, petrographic, sedimentological, stratigraphical and others. Complex geosites are a combination of two or more of the above categories. Geosites may be single outcrops, caves, quarries, mines, individual volcanic landforms and tectonic structures [27]. Based on their importance, they can be of global, national, regional or local relevance [30]. Geosites can be further classified [26] in terms of their spatial appearance (circumscribed sites like outcrops, linear features and aerially extended features such as peaks) and their dynamic state (inactive features/processes vs. active ones).

Over the last two decades, several authors have attempted to qualitatively and quantitatively assess the quality of geosites using a range of criteria. Most assessment efforts make reference to the scientific value [31]. This is made up of four subcriteria: rarity, representativeness, integrity [4,29,32] and also the level of current scientific knowledge about the geosite, reflected by the existence of published scientific studies [7]. Representativeness pertains to the exemplarity of a geosite in terms of the geological processes (active or inactive) represented there. Rarity is related to the uncommonness of the geosite at the regional or global level [32].

In addition to the scientific value, other values, defined as "additional" [33,34], can be individuated and assessed: ecological, cultural, aesthetic, economic and educational. The cultural value is composed of four subcriteria: religious, historical, artistic or literary and geohistorical importance [32]. Among the above additional values, particularly worthy of mention is the educational one, which may be defined [7] as the combination of the following elements: didactic relevance (how easily a geosite's features might be understood by nonexperts), accessibility, safety and current exploitation of the geosite for education-related activities (excursions and guided tours). Moving on to the description of the area we selected for our work, Iceland (Figure 1a) is widely regarded as a natural laboratory, offering a seemingly endless variety of geosites. They include places where geothermal-related phenomena can be admired (geyser eruptions at Geysir, Figure 1b) and major landforms, such as presently active rift zones (the Thingvellir, majestic rift valley in SW Iceland, Figure 1c) and table mountains (e.g., the impressive Herdubreid, Figure 1d) and fluvial features like waterfalls (Gulfoss in SW Iceland, Figure 1e) and gigantic glaciers (such as Vatnajökull, the largest glacier in Europe, Figure 1f). Most of the above geological objects have a specific geoheritage-related importance and, at the same time, can be used for geoscience education and communication. Thanks to its stunning variety of volcanic, tectonic, fluvial and glacial features, of which the ones above are only a tiny, although meaningful, portion, Iceland is one of the top locations for geotourism at the worldwide level [35]. As geotourism may have profound impacts on the geodiversity of the country, awareness is being raised among the local scientific community about the need to strengthen geoconservation and foster sustainable geotourism. This may be accomplished through the establishment of new geoparks [36], following the example of the two UNESCO Global Geoparks (UGGp), Katla and Reykjanes, already present in the country. In this regard, among the recent events aimed at promoting geoconservation policies and values in Iceland, particularly worthy of mention is the VIII International ProGEO Symposium held in Reykjavík in September 2015 [37].

This paper is aimed at illustrating two major Icelandic geoheritage areas and showing their relevance for geosite popularization and promotion, as well as for educational and geoscience communication goals. We have selected these particular locations because, having carried out scientific research there over the last decade, we have had a chance, during our numerous field surveys, to work at hundreds of sites that may be considered relevant in terms of geoheritage. The two areas are (Figure 1a): A) the Snaefellsnes Peninsula (Figures 2 and 3), comprising the world-famous Snæfellsjökull volcano, well-exposed subvolcanic bodies and related sheet swarms, as well as tectonically guided alignments of Holocene volcanic cones and B) the Northern Volcanic Zone (NVZ, Figure 4), where the following volcano–tectonic elements can be observed: (i) a textbook example of a triple–junction interaction between an onshore transform fault and an active rift system; (ii) the Theystareykir Fissure Swarm

(ThFS), an active rift zone characterized by a central volcano, several major faults and a great number of eruptive fissures and (iii) the Krafla Fissure Swarm (KFS), another major rift zone also marked by the diffuse presence of eruptive fissures and dominated by the active Krafla Volcano.

Figure 1. (**a**) Volcanic systems of Iceland, modified after [38]. RPR: Reykjanes Peninsula Rift, WVZ: Western Volcanic Zone, EVZ: Eastern Volcanic Zone, CIVZ: Central Iceland Volcanic Zone and NVZ: Northern Volcanic Zone. The circles indicate central volcanoes belonging to active volcanic zones. GOR: Grímsey Oblique Rift, HFZ: Húsavík–Flatey Transform Zone, DAL. Dalvík Zone and KOR. Kolbeinsey Ridge. (**b**) Geyser eruption in Geysir. (**c**) The Thingvellir National Park. (**d**) The Herðubreið table mountain. (**e**) The Gullfoss Waterfall. (**f**) The southeasternmost section of the Vatnajökull Glacier, viewed from Höfn (SE Iceland).

With the purpose of showcasing a total of 25 geosites selected in the two areas, we made use of the following tools: (i) field photographs, some of which have been reworked by highlighting the most relevant features [6]; (ii) highly detailed images, captured by unmanned aerial vehicles (UAVs) and (iii) 3-D models of field outcrops, based on individual, UAV-captured pictures, further elaborated by using Structure-from-Motion photogrammetry techniques (SfM). The importance of UAV-based techniques as innovative and helpful tools for Earth Science research has been widely demonstrated in the last decade, as documented by Bonali et al. [39] and Fallati et al. [40].

Moreover, we have qualitatively evaluated the identified geosites by using three of the above cited criteria (scientific, cultural and educational) wherever they could be applied; such criteria were considered as most suitable for enabling us to carry out a preliminary, and by no means exhaustive, assessment. In two instances, we attempted to judge the aesthetic value as well. We need to point out that no attempt has been made to make a numerical assessment, as opposed to what has been done in other works [4,18,19,32,33], which were exclusively focused on the evaluation of geosites and geomorphosites.

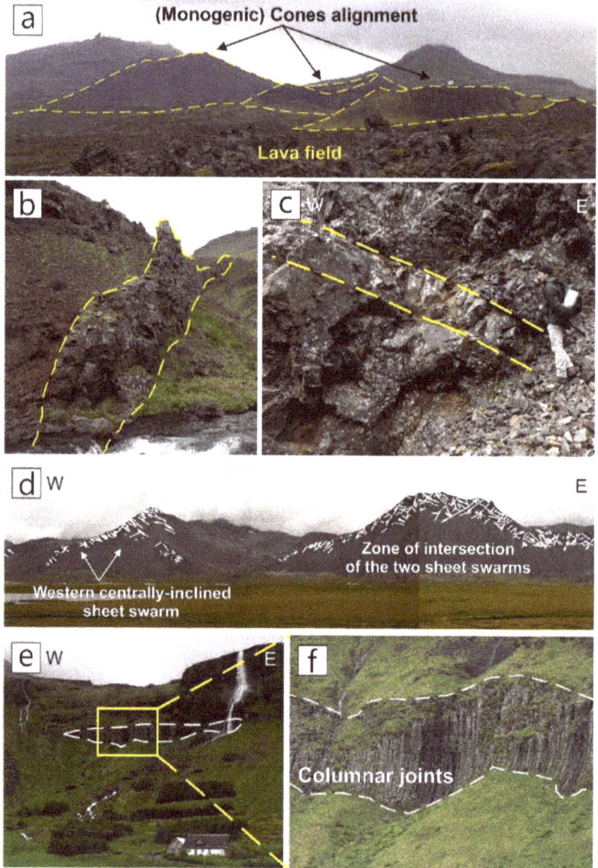

Figure 2. Field photographs of geosites in the Snæfellsnes Peninsula. (**a**) The alignment of monogenetic cones is a clear indicator of the direction of the underlying dyke feeding the cone growth. (**b**) A 1.2-m-thick, steeply-dipping dyke. (**c**) A shallow-dipping sheet (person for scale). (**d**) Panoramic view of the sheet swarms, cross-cutting each other in the right-hand side of the picture. (**e**) The typical appearance of a sill, concordant with the bedding of the overlying and underlying lava units (house for scale). (**f**) Detail of the columnar joints formed during cooling of the magma.

2. Geological Setting of the Two Selected Areas

Iceland is the result of the combination of hot spot and mid-ocean ridge magmatism [41–44]. Its geological setting is characterized by the widespread presence of Neogene and Pleistocene basalts, bordering the active rift system that cuts through the island from SW to NNE (Figure 1a). All eruptions that occur on the island are associated with volcanic systems [45], which are made of swarms of faults, extension fractures and basaltic volcanoes; all have been formed due to plate–pull-induced mid-oceanic ridge activity and also as a result of magma upwelling from the Icelandic mantle plume. There are 30 presently-active volcanic systems, 40–150 km long and 5–20 km wide; all host central volcanic edifices. Analogous systems, nowadays extinct, are found in Iceland's eroded Neogene and Pleistocene lava successions; 15 of these have been mapped and about 40 have been identified and located [45]. Extinct systems provide crucial information that, in turn, enables gaining insight into the possible 3-D structure and evolution of presently active ones and interpreting their dynamics. The following sections provide a brief geological and tectonic background on the two areas selected for their geoheritage-related importance, as well as for their suitability for geoscience communication purposes.

2.1. The Snaefellsnes Peninsula

The 80-km-long and 10–30-km-wide Snaefellsnes Peninsula (Figure 3a) is home to three volcanic systems: (1) Snæfellsjökull Volcano (Figure 3b) on the peninsula's westernmost tip, whose last eruptions are dated to 1750 BP; (2) the Lýsuhóll volcanic system, active in Holocene times and (3) the Ljósufjöll volcanic center, the easternmost of the three volcanic systems, with the latest eruptions dating back to 960 AD [46].

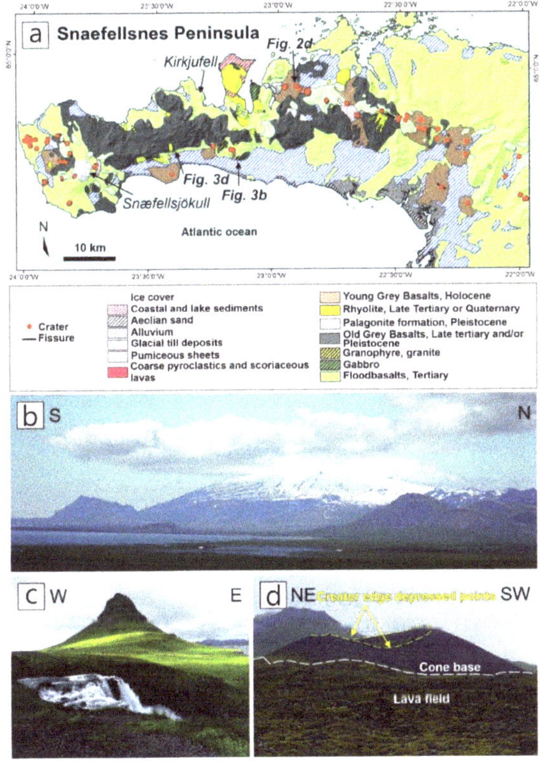

Figure 3. (a) Geological map of the Snæfellsnes Peninsula, modified after [47]. (b) Panoramic view of glacier-capped, 1446-m-high Snæfellsjökull stratovolcano. (c) Kirkjufell, or "Church Mountain". (d) Textbook example of a monogenetic volcano (scoria cone), with indication (yellow arrows) of the crater rim's depressed points, useful for assessing the trend of the underlying magma-feeding fracture.

The rock successions cropping out across the peninsula are mainly basaltic lavas and palagonites, with ages ranging from the late Neogene to Holocene times (Figure 3a).

The effects of glacial erosion have unveiled three extinct magmatic complexes: the Midhyrna and Lysuskard intrusions near the peninsula's southern coastline and the Kolgrafarmùli intrusion along its northern side; all intrusions were injected into Neogene basalts (Figure 3a). A fairly recent paper [47] focused on the central and eastern sector of Snaefellsnes, providing a detailed structural description of vertical dykes, inclined sheets and two shallow magma bodies; respectively, the Midhyrna gabbroic intrusion and the Lysuskard granophyric one (Figure 2).

2.2. The Northern Volcanic Zone

The Northern Volcanic Zone (NVZ) has been active since 8–9 Ma [48,49] and is composed of the following five N–S-striking rift zones (also called "fissure swarms"): Theystareykir, Krafla, Fremrinámar, Askja, and Kverkfjoll (Figure 4a). Each of these active rifts is made of 5–20-km-wide and 60–100-km-long swarms of extension fractures, dip–slip faults and a main volcano.

The Theystareykir Fissure Swarm (ThFS, Figure 4) is 10-km-wide and is cut by several N–S-striking normal faults (some with decametric dip–slip offsets) and by a huge number of extension fractures; it is also marked by the presence of the Theistareykjaburnga shield volcano [50,51]. The most prominent tectonic structure in the ThFS is the Gudfinnugja Fault (GF), a Holocene dip–slip fault that represents the westernmost edge of the rift system. This active rift zone is extensively covered by lava flows from Theistareykjaburnga, dated to about 14.5 ka BP [52,53]; the last eruption (so far) at this edifice led to the emplacement of the 'Theystareykjahraun' lava flows (younger than 2.4 ka BP).

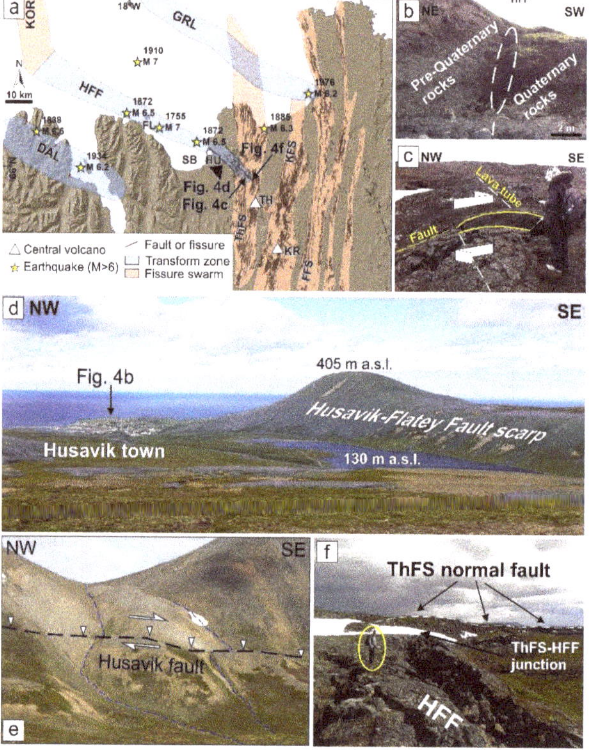

Figure 4. Tectonic setting of northeastern Iceland and main features of the onshore segment of the Husavik–Flatey Fault (HFF), modified after [54,55]. (**a**) The mid-Atlantic Ridge is offset by the Húsavík–Flatey Fault (HFF), as well as by the Grímsey (GRL) and Dalvík (DAL) Lineaments. Orange stripes locate volcano–tectonic rift zones, also called "fissure swarms"; the three westernmost ones are referred to as ThFS = Theystareykir Fissure Swarm, KFS = Krafla Fissure Swarm and FFS = Fremrinámar Fissure Swarm. Recent volcanoes are highlighted with white triangles. KR = Krafla Volcano and TH = Theistareykjaburnga Volcano. (**b**) Outcrop of the HFF along the coast, just north of Husavik (HU). Photo A. Tibaldi. (**c**) Western onshore termination of the HFF, across Husavik. Notice the huge, 270m-high fault scarp. (**d**) Right-lateral offset of a lava tube along the easternmost segment of the HFF. Photo A. Tibaldi. (**e**) Right-lateral offset of gullies along the HFF. (**f**) The triple junction between the HFF and the westernmost fault of the ThFS (person for scale).

Another major tectonic feature in the NVZ is the Husavik–Flatey Fault (HFF), a major seismically active transform fault with a total length (comprising its offshore and onshore sections) of about 100 km [56]. The length of its emerged (onshore) section is 25 km [38,57]. This fault has been active at least during the last 7 Ma, producing a total estimated right-lateral displacement of 60 km [57]. Along with this impressive strike–slip component, the HFF has produced also a major dip–slip one; its total vertical displacement may amount to as much as 1.4 km/m [58].

What is more striking about the emerged section of the HFF is that it displays a unique tectonic structure similar to the ones that are documented along mid-oceanic ridges; what is normally hidden beneath sea level is fully visible here: the NW–SE-trending HFF connects with the N–S-trending Gudfinnugja normal fault, forming a textbook example of an emerged triple junction, which has been the subject of extensive field studies in the recent past [54,55,59–64].

The third area within the NVZ, suitable for illustrating several examples of volcanic and volcano–tectonic processes, is the Krafla Fissure Swarm (KFS), extending about 50 km towards the north and about 40 km towards the south from the Krafla Volcano (location in Figure 4a). This world-famous edifice is characterized by an 8-km-wide caldera that was formed following a paroxysmal eruption around 100 ka BP. Since the time of its formation, the caldera has expanded about 2 km in an E–W direction, owing to active processes of plate spreading and rifting. Within the KFS, it is possible to observe a great number of extension fractures and eruptive fissures, formed during repeated rifting events over the last 10 ka [39].

3. Methodologies for Image Collection and Processing

3.1. General Overview

Field observations and data collection are essential for geoscience research, teaching, outreach and popularization; however, very often, the study areas are characterized by difficult logistic conditions, due to the outcrops and landforms not being easily reachable [65]. In order to overcome these limitations, remotely captured images and derived models (3-D Digital Outcrop Models—DOMs, orthomosaics and Digital Surface Models—DSMs) have been increasingly used by the scientific community over the last decade, allowing for extensive collection of geological data in previously inaccessible areas [66]. In particular, the past few years have seen the steady development of unmanned aerial vehicles (UAV). These have become popular for scientific purposes, owing to the following reasons: (i) The most convenient data acquisition procedures and schedules can be chosen. (ii) The flight height can be adjusted so as to obtain the required (usually very high) spatial resolution. (iii) Flights can be repeated on a daily basis or also several times per day. (iv) Vertical rock cliffs, normally inaccessible to direct examination, can be examined in great detail. (v) Data acquisition is substantially less expensive than in the case of high-resolution satellite imagery. (vi) UAVs can be equipped with several types of sensors, designed for specific purposes. Thanks to the above, UAVs have been employed, over the past decade, to enhance knowledge of different types of geohazards, ranging from seismic [67,68] to landslide [69–73] to volcanic [74–77] and flood hazards [78,79]. With regard to the study of active tectonics and volcano–tectonics, focused on geological objects like those documented in the present work, after the first attempts with balloons [80], UAV-captured images have become a major option in the study of active faults [81–88]. In most of these works, the use of UAVs has been integrated by Structure-from-Motion (SfM) photogrammetry [89–91], a powerful technique to augment traditional methods used to gather outcrop data. Therefore, the combination of UAV-digital image collection and SfM photogrammetry has been increasingly applied to geological and environmental research [39,40]. In order to accomplish the goals of the present work, we have used the above techniques both for collecting images of the selected geosites, as well as constructing 3-D DOMs, which may be considered "Virtual Outcrops (VOs)" [92,93]. VOs can be defined as follows: "a 3-D representation of surface geology, but it does not contain subsurface information" [94]. This cutting-edge technique can be used for achieving the following overall purposes, all relevant to the present paper: (i) characterize, illustrate and assess geosites; (ii) communicate Earth Science by explaining ongoing, active geological processes and (iii) engage the younger generation, usually very attracted to innovative and interactive communication tools.

3.2. UAV Selection and Use

For the sake of the present work, we chose multirotor vehicles (Figure 5a,b), as they can be remotely controlled (e.g., Figure 5c), are characterized by very stable hovering, can be easily transported in the field and are less expensive than hybrid and fixed-wing models (e.g., Figure 5d,e). In addition to the above, multirotor models can fly at very low heights, thus, obtaining greater field resolution; most importantly, take-off and landing operations are smoother than for fixed-wing models, and this can prove to be crucial, especially when operating in difficult logistic terrains, such as lava flow outcrops or remote beach areas [39,40].

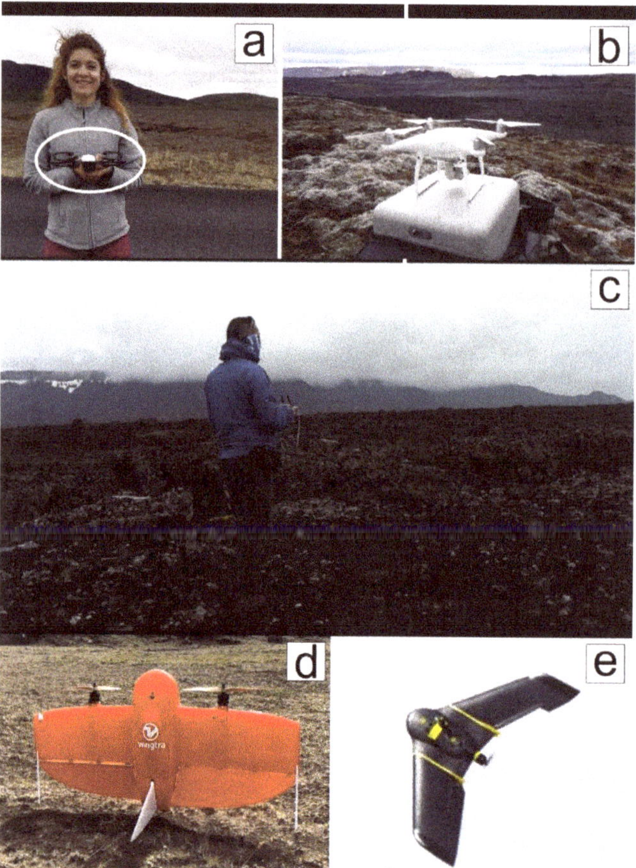

Figure 5. (**a**) The DJI Spark; person for scale. (**b**) The DJI Phantom 4 PRO. (**c**) Unmanned aerial vehicle (UAV) pilot remotely controlling the DJI Spark. (**d**) Hybrid UAV type; picture courtesy of Joël Ruch. (**e**) Fixed-wing UAV model.

Considering the above, we selected two different types of multirotor UAVs: the DJI Phantom 4 PRO quadcopter and the smaller DJI Spark. The DJI Phantom 4 PRO (Figure 5b) is a 1.375-kg vehicle equipped with a 20 megapixel camera, EXIF information (Exchangeable Image file Format) and GPS coordinates, provided by the integrated satellite positioning systems GPS/GLONASS (refer to the WGS84 datum); it records 4K videos at 100 fps and supports a micro SD card with a maximum capacity of 128 GB. In this case, flight time is approximately 30 min. This model provides constantly stable flights thanks to the integrated GPS system, including position holding, altitude locking and stable hovering. Its transmission distance is up to 7 km. The DJI Spark (Figure 5a) is a 0.300-kg vehicle, also known as the "selfie drone", equipped with a 12 megapixel camera, EXIF information (Exchangeable Image file Format) and GPS coordinates. In this case, flight time is approximately 16 min. Owing to its small size and low weight, this model is pretty useful for field research and 3-D DOM reconstruction focused on outcrops located in remote areas. The high resolution guaranteed by both instruments is a key asset for image collection, SfM processing, 3-D DOM and VO reconstruction for teaching, outreach and research purposes.

3.3. UAV-Based Structure-from-Motion: Data Collection (Part I)

Our workflow aimed at 3-D DOM construction can be broken down in two parts (Figure 6a): (i) Part I pertains to data collection (digital image gathering and setup of ground control points—GCPs). (ii) Part II has been dedicated to data processing and model reconstruction. The first step has been devoted to defining the area to be surveyed and planning the details of the flight missions, such as path orientation. In doing this, care must be taken in considering wind direction, which may affect UAV flight performance. As the surveyed geological objects are situated in very remote areas, we made use of the smaller DJI Spark, managed through the DJI GO App. The UAV was manually controlled by the pilot for the entire duration of the mission; as shown in Figure 6b, a number of paths, parallel to each other, were planned. As suggested in recent works [39,95–97], UAV-captured photos should have an overlap of 90% along single paths and 80% in a lateral direction (e.g., Figure 6b), so as to obtain a better alignment of the images and reduce distortions on the resulting orthomosaics. During image collection for the goals of the present work, field photographs were taken from a height of 30 m by flying the drone at a speed of 2 m/s and obtaining an overlap consistently in a range of 90%–85% along paths and 80%–75% in a lateral direction; images were captured every 2 s (equal time interval mode) and in optimal light conditions suitable for the camera ISO range (100–1600). This was done to minimize the motion blur, avoid the rolling shutter effect and achieve well-balanced camera settings (exposure time, ISO and aperture); in this way, ensuring sharp and correctly exposed images (e.g., [98]).

Moreover, to reduce shadows around elevated features, we operated the drones when the sun was straight overhead (at zenith). Luckily, during the summer period in Iceland (July), the sun angle variation is minimal, and the exposure to sunlight is almost uniform during the central part of the day. In order to allow for the co-registration of datasets and the calibration of models resulting from SfM processing (e.g., [89,91,99,100]), World Geodetic System (WGS84) coordinates of, at least, four artificial ground control points (GCPs) were fixed near every corner of each surveyed area (an additional one was selected in the central part); this procedure helped to reduce the "doming" effect resulting from SfM processing. The GCPs were chosen as well-visible natural targets: stones, lava flow edges and piercing points along fractures; they were surveyed by using single frequency receivers in RTK configuration (with centimetric accuracy).

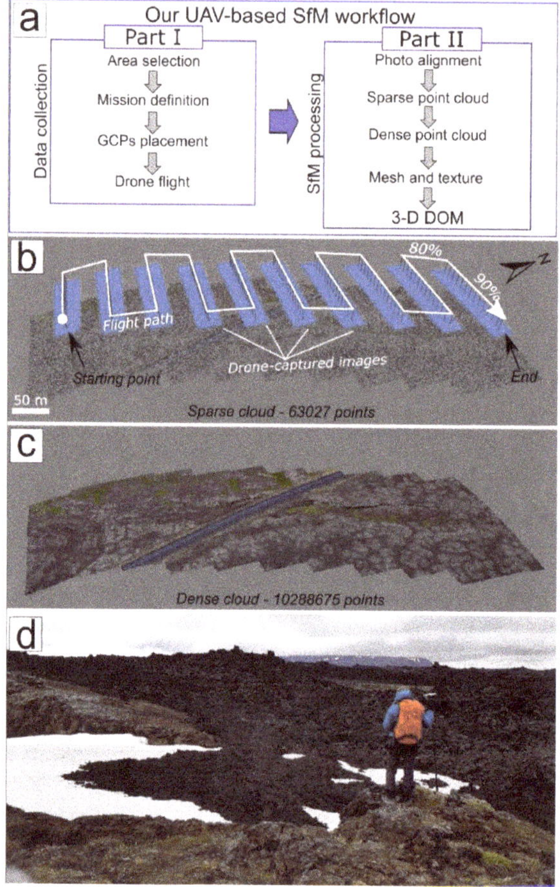

Figure 6. (**a**) Flow chart illustrating the different steps that led to the production of 3-D digital outcrop models (DOMs) using UAVs and Structure-from-Motion (SfM)-dedicated software. (**b**) Example of typical UAV flight paths, with indication that the suggested overlap among pictures is sparse. (**c**) Dense cloud generated by the SfM software. Computed camera positions are shown as blue rectangles and black lines indicate pitch angle and camera orientation. (**d**) Operator collecting ground control points (GCPs).

3.4. UAV-Based Structure-from-Motion: Data Processing and Model Reconstruction (Part II)

Part II has been dedicated to data processing aimed at 3-D DOM and Virtual Outcrop (VO) construction; the collected images were processed through Agisoft Metashape (Agisoft LLC, 11 Degtyarniy per., St. Petersburg, Russia, 191144), a commercial Structure-from-Motion software (SfM). This application has been increasingly used for both UAV and field-based SfM reconstructions, owing to its user-friendly interface, intuitive workflow and high quality of point clouds [101]. The SfM technique allowed us to identify matching features in different photos and combine them to create a sparse and a dense cloud (Figure 6b,c), an orthomosaic, a DSM and, eventually, 3-D DOMs as final products [89,102]. The steps leading to model construction are shown in Figure 6a; further details are provided hereunder. The first step was to obtain an initial low-quality photo alignment, only considering measured camera locations. Thereafter, we excluded the photos with quality value <0.5 (or out of focus) from further processing, as suggested by the software's user manual. Following this initial quality check, ground control points (GCPs) were added to all photos, where available, so as to: (i) scale and georeference the point cloud (and thus, the resulting model); (ii) optimize extrinsic parameters, such as estimated camera locations and orientations and (iii) improve the accuracy of the final model. Images were then realigned, camera locations and orientations were better established and the sparse point cloud was computed by the software. The next phase consisted in reconstructing the dense point cloud (e.g., Figure 6d) from the sparse point one, using a mild-depth filtering and medium quality settings. The 3-D DOMs were finally created by generating the Tiled model (mesh and texture) through the Agisoft Metashape software. The resulting 3-D DOMs are characterized by a high-resolution texture always in the range of 1.0–1.5 cm/pixel.

4. Geosites in the Two Selected Icelandic Areas: Description and Assessment

4.1. Geosites in the Snæfellsnes Peninsula

The two most iconic geosites in the peninsula, which are worthy of mention also for their geotourism-related significance, are Snæfellsjökull Volcano and Kirkjufell. The former (Figure 3a,b), located at the westernmost edge of Snæfellsnes, is a majestic, snow and ice-capped composite edifice that rises to almost 1500 m a.s.l. This geosite can be evaluated by applying most of the above outlined criteria. In fact, it can be assessed in terms of its aesthetic value, which can be defined as the combination of two factors: visibility and structure [32]. A geosite's visibility is greater when it can be clearly spotted from multiple viewpoints and also from a considerable distance. Structure, on the other hand, refers to the fact that features with a vertical development, such as isolated peaks, are generally perceived as the nicest [103]. This volcano fits both criteria, as it is an isolated peak that, being located at the tip of a peninsula, is perfectly visible from tens of kilometers away. As regards the scientific value, two subcriteria can be applied: firstly, the volcano was the subject of extensive scientific research (e.g., [104]) and, secondly, it is highly representative of the combination of volcanic and glacial processes. As far as the criterion "artistic or literary importance" is concerned, the volcano has also a cultural value; in fact, Snæfellsjökull was made famous by Jules Verne, who included it in his best-known science-fiction novel, *Journey to the Center of the Earth* (1864). Finally, this geosite has a major educational value, being easily accessible and located within Snæfellsjökull National Park (established in 2001), where guided tours to its glacier, as well as recreational/educational activities, are organized on a regular basis. Kirkjufell ("Church Mountain" in Icelandic) is a stunning, 468-m-high peak found on the northern coast of the peninsula, which, along with the beautiful waterfalls, is one of the most photographed natural features in Iceland (Figure 3a,c). This geosite has a considerable aesthetic value: firstly, it has a peculiar morphologic structure, clearly visible from a considerable distance (although not comparable to the much higher Snaefellsjökull Volcano). It is also very representative of glacial processes that, all over Iceland, carved thick piles of basaltic lava flows, to the point of creating impressive erosion-related landforms, such as Kirkjufell and many others. However, as opposed to its neighboring volcano, this geosite has not been the subject of scientific research in the past. In regard to its educational value, all the subcriteria can be applied: it is perfectly accessible, the processes that led to its formation can be easily understood and there are plenty of possibilities to take part in guided tours.

Along the northern coast of the peninsula, around 20 km east of Kirkjufell, another geosite that is worth highlighting is a scoria cone with a basal diameter of 550 m (Figure 3d), which is a perfect example of a continuous crater rim with two depressed points. The line ideally connecting the two depressed points is considered [105–107] to be parallel to the fracture in the substrate along which magma was rising, leading to the growth of the cone. Another way to define the most probable orientation of near-surface magma paths is to analyze the alignment of pyroclastic cones. In this respect, Tibaldi et al. [47] documented that 51 pyroclastic cones, clustered in groups of three or more (such as in Figure 2a), and mostly elliptical in plain view and are aligned in an approximately E–W direction, consistent with the trend of the whole Snaefellsnes Peninsula. As regards the assessment of this geosite, the scientific criterion that can be applied here is representativeness; this is a very clear example of the appearance of such scoria cones all over the country, and in many other volcanic regions of the world. With regard to the educational value, this site is easily accessible, but its didactic relevance is rather low, as it would be difficult to explain to nonexperts the topic of magma paths and their relation to the geometry of scoria cones, as well as to the alignment of multiple monogenetic edifices.

Along the central-southern coast of the peninsula, a very complex system of regional dykes crops out; dykes are major subvolcanic features [108] that are responsible for feeding fissure eruptions [74] and flank eruptions at volcanoes [109]. In some cases, they can also induce the destabilization of volcanic edifices, potentially leading to lateral collapse [110]. The outcrop in Figure 2b represents a geosite that is suitable for showing the geometry of dykes in the field.

On the southern side of the Senaefellsnes Peninsula are located the Midhyrna and Lysuskard intrusions, very clear examples of extinct magma chambers that once stored magma feeding eruptions at the surface. Of particular interest are the swarms of inclined sheets surrounding the two intrusions. Sheets are subvolcanic bodies that channel magma from a deep reservoir to the surface. In the field, they are clearly distinguishable from dykes, as the latter are almost always subvertical or vertical, whereas sheets are always dipping at a low angle. Iceland is one of the places on Earth where sheet swarms can be better observed and studied [108]. At Snaefells, as is the case in other locations in Iceland and at the Isle of Skye [111–115], the sheets are inclined towards the subvolcanic bodies from which they were injected. The outcrop in Figure 2c represents a geosite that may be functional for explaining the geometrical appearance (dipping at a shallow angle) of sheets in the field that are different from those of dykes (which are vertical or subvertical). In the central-southern portion of the peninsula, the 384 sheets that were mapped are arranged in a particular fashion around the Midhyrna and Lysuskard subvolcanic intrusions [47]. The two aligned ridges in Figure 2d can be considered a unique geosite (laterally extended for 3.8 km), which is key to pointing out a rather complicated process hardly ever observed in subvolcanic geology: the inclined sheets in the western part of the photograph preferentially dip to the E at a shallow angle towards the location of the Lysuskard composite doleritic–granopyric intrusion, which represents the magma chamber off which they were injected. In the central sector of Figure 2d, there is the intersection of two different swarms of sheets (highlighted as thin white lines), which dip towards either the W or the E. In the eastern zone (not included in Figure 2d), the sheets dip preferentially towards the Mydhirna doleritic intrusion. Another feature that is frequently possible to view in the Snaefellsnes Peninsula is represented by sills, subvolcanic intrusions with a subhorizontal attitude [108], concordant with the underlying and overlying host rocks (geosite in Figure 2e), especially worthy of attention is the presence of well-developed columnar joints that are formed during the cooling of magma (geosite in Figure 2f). At the scientific level, all the above geosites are highly representative of the subvolcanic processes that are common in volcanic regions, such as Iceland. All have been the subject of scientific research, leading to publications [47,116]. Moreover, they are easily accessible and, with the sole exception of the volcano–tectonic process leading to a sheet swarm intersection (Figure 2d), all could be suitable for teaching the basics of subvolcanic geology to a public of nonexperts.

4.2. Geosites in the Northern Volcanic Zone

Moving from W to E, the first geological feature in the NVZ is the aforementioned Husavik–Flatey Fault (HFF), an amazing example of an oceanic transform fault that can be observed along its emerged prolongation (Figure 4a). The HFF, together with the Grimsey and Dalvik lineaments, compose the so-called Tjornes Fracture Zone, which connects the NVZ to the Kolbeinsey Ridge (Figure 1a). The HFF has an impressive appearance in the field, separating pre-Quaternary from Quaternary volcanites (Figure 4b) and offsetting structures, such as lava tubes (Figure 4c), in a dextral sense. The clearest exposure of the fault is the one in Figure 4d, where the sheer size of the fault plane is visible in its completeness; here, the location of Husavik town a short distance away from this gigantic tectonic element provides an eerie reminder of the seismic hazard the town is prone to. Another view of the dextral displacement along the fault is given in Figure 4e, where the fault clearly offsets gullies and water divides. Figure 4f depicts the above-explained triple junction: the field photograph captures one of the few tectonic interactions, visible on Earth's surface, between a transform fault (the HFF) and a rift system (the ThFS), whose northwesternmost fault (the Gudfinnudgja Fault) can be seen in the distance, with its about 30-m vertical offset [61].

The five tectonic geosites that have been introduced above are all of the linear type, as they are related to a strike–slip fault. They can be considered active geosites, as the HFF has produced four historical earthquakes in the last 200 years [61], and displacements along any sectors of this major fault may take place in the future. Their scientific value is considerable due to the following reasons: Firstly, they are highly representative of the appearance of a major strike–slip structure in the field. Secondly, they belong to a fault that has been documented in a great deal of high-profile publications, as mentioned above. They are also extremely rare at the scale of Iceland, because they belong to an oceanic transform fault that extends onshore, a process that takes place only in this portion of the island. It is worth noting that one of the five geosites in Figure 4 is rare also at the worldwide level—the textbook-example of an emerged triple junction shown in Figure 4f.

With regard to the educational value of the geosites, all are suitable for explaining the activity of a strike–slip fault. Particularly worthy of notice in this respect are the geosites in Figure 4c,d. The former is a very good example of the movements along a strike–slip fault, which are easy to visualize and understand thanks to the presence of a displaced lava tube. The gigantic fault plane in Figure 4d, on the other hand, is key to explaining the existence of a dip–slip component of movement superimposed on the strike–slip one. The geosites in Figure 4b,d are easily accessible, as opposed to the other three, which would require potential visitors to walk long distances across a harsh volcanic landscape. To our knowledge, none of the above geosites have been the focus of educational activities, probably also on account of the difficulty to access them. The second selected area in the NVZ is the Theystareykir Fissure Swarm (ThFS), marked by the presence of geosites that are representative of a number of active volcano–tectonic processes, such as faulting and fissuring, as well as the development of central volcanoes and associated geothermal areas.

Figure 7a is the geological map of a portion of the ThFS about 15 km south of the triple junction; here, the older volcanic and volcanoclastic units pre-date the Last Glacial Maximum (LGM), whereas the younger ones were emplaced after the last deglaciation. In this area, we selected a few geosites, among which a geothermal area (Figure 7b), home to several pools of hot mud. Another geosite is a 30-m-wide and 300-m-long volcano–tectonic graben, described in detail in a recent paper [39]; this extensional structure (Figure 8a) is bounded by two main normal faults, striking NNE–SSW, affecting 2.4-ka-old lavas. Another image (Figure 8b) documents the offset (12 m) produced along one of the many dip–slip faults that compose this active rift system; a third UAV-captured image, depicting a fracture field (Figure 8c), enables observing a set of extension fractures, roughly parallel to each other, affecting older, pre-LGM lavas and marked by dilation amounts > 40 cm (in the range of 40 to 120 cm). A close-up field photograph (Figure 8d) visualizes one of the thousands of extension fractures affecting post-LGM lavas (with dilation from a few centimeters to about 40 cm); here, clear "piercing points" can be spotted, suitable for assessing the vector of fracture opening and the amount of dilation. Finally, Figure 8e shows the Stórihver recent volcanic cone, with a crater of 60 m in diameter.

Regarding the assessment of the six geosites (Figures 7b and 8a–e), all have a major scientific relevance. In fact, they belong to an active rift that has been intensively investigated in the last three decades, as documented by several high-profile scientific publications, some of which are mentioned above. Moreover, all the selected geosites are highly representative of active processes, both geothermal and tectonic ones, the latter leading to the development of dip–slip faults and extension fractures. As opposed to the geothermal and tectonic geosites, which are all representative of presently active processes, the volcanic geosite displayed in Figure 8e is the result of a localized eruption that resulted in a monogenetic volcanic edifice that is nowadays extinct. With regard to rarity, none of the geosites can be regarded as uncommon, because most of Iceland is pervasively cut by faults and fractures and dotted with monogenetic cones in response to ongoing crustal extension and hot spot-related volcanism.

The above geosites could be undoubtedly used for didactic purposes, as they enable us to explain active extensional processes that can be easily understood thanks to the favorable exposure of the outcrops. However, the educational value of four out of five of the considered geosites is hampered by their limited accessibility; apart from the graben in Figure 8a, whose location is reachable by car, all the others are found in remote areas; moreover, the floor of the ThFS is riddled with gaping fractures and holes, which make the area relatively unsafe for nongeologists.

In order to overcome these limitations, we produce a series of 3-D models, as illustrated hereunder. The first example (Figure 9a) is a 3-D view of the 30-m-wide graben (cut by the road), previously shown in Figure 8a, which represents a typical effect of extensional tectonics across the ThFS.

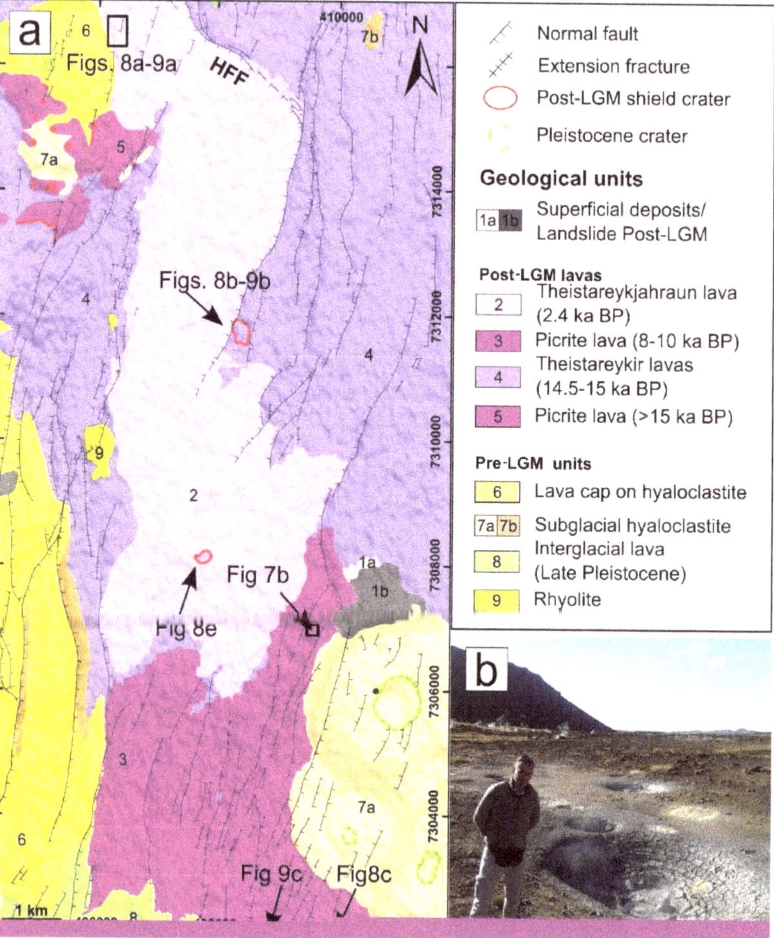

Figure 7. (**a**) Geological map of a portion of the Theystareykir Fissure Swarm (ThFS) about 15 km south of the triple junction, modified after [39]. (**b**) Pools of hot mud in a geothermal area within the ThFS. LGM = Last Glacial Maximum.

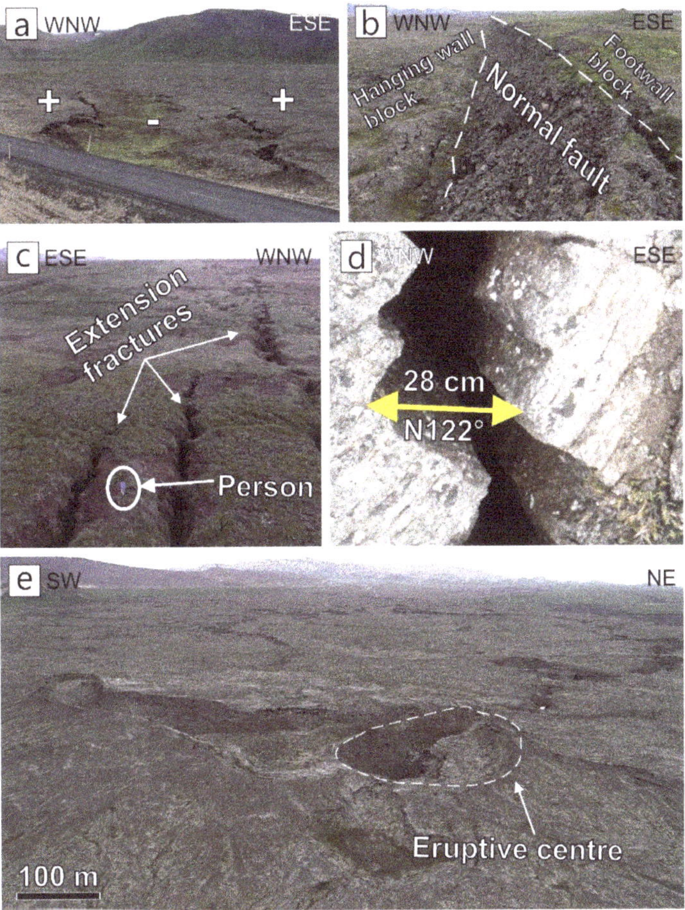

Figure 8. Field and UAV-taken photographs in the ThFS. (**a**) A graben, with indication of the low-lying floor of the extensional structure. (**b**) Field picture documenting the magnitude of the offset (12 m) at a dip–slip fault. (**c**) The result of extensional fracturing (image courtesy of Fabio Marchese). (**d**) Close-up photographs of an extension fracture. Piercing points and amount of dilation are indicated by the yellow arrow. (**e**) Aerial view of the flat-lying floor of the rift zone. The crater of a recent eruptive center in the foreground (named Stórihver) is highlighted with a white dashed line (image courtesy of Fabio Marchese).

The 3-D model visualizes, with exceptional detail, the two sets of opposite-dipping normal faults that border the graben, as well as the low-lying floor of the volcano–tectonic structure. The model is also instrumental in highlighting that tectonic subsidence across the graben floor has developed in a differential fashion, as attested by the fault system to the WNW (upper part of the figure), marked by a greater offset than its counterpart to the ESE. The second 3-D model (Figure 9b) portrays the above-illustrated dip–slip fault (Figure 8b); here, the geometry of a recent, active dip–slip fault with a steeply-dipping fault plane separating two horizontal surfaces (the top of the footwall block and the hanging wall block, respectively) can be viewed in great detail. The third model (Figure 9c), aimed at providing a clearer picture of the effects of extensional tectonics, is focused on a segment of the previously shown fracture field (maximum opening 3 m) affecting pre-LGM lava units (Figure 8c).

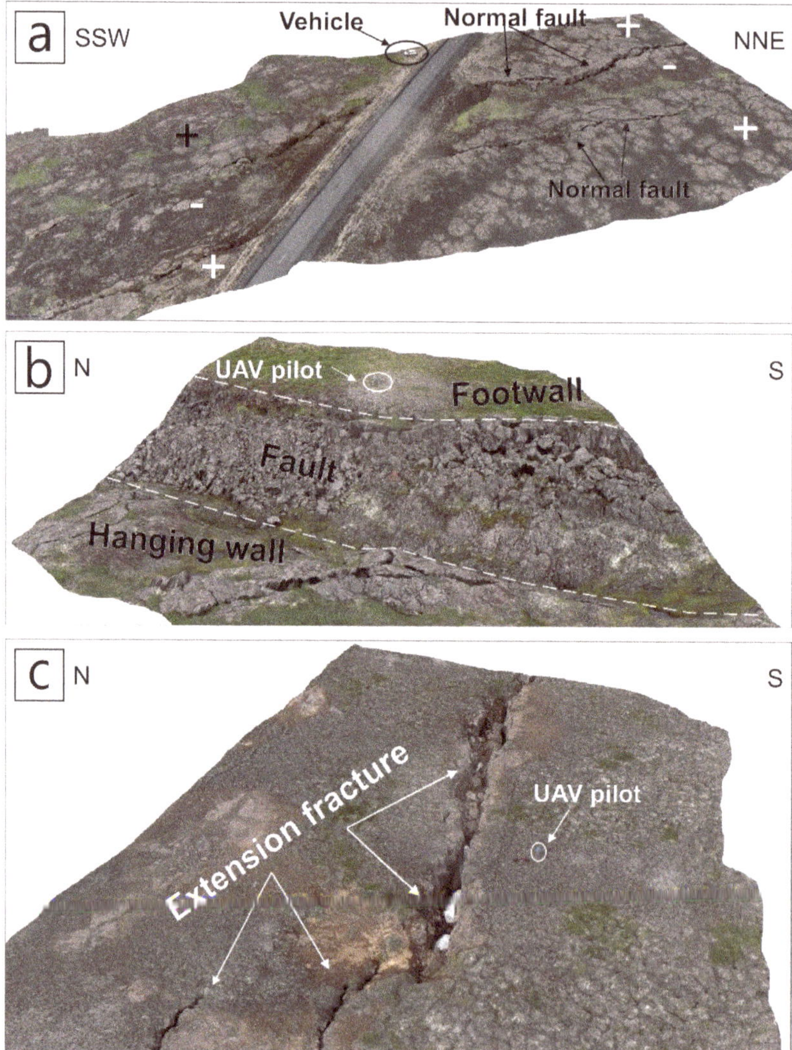

Figure 9. 3-D models of three geosites in the ThFS. (**a**) The graben in Figure 8a is represented here in a much more detailed fashion. Note the elongated area crossed by the road and lying at a lower altitude than the surrounding topographic surface. The vehicle for scale aids in understanding the size of this volcano–tectonic structure. (**b**) This is an improvement of the previous Figure 8b in terms of enabling the viewer to understand the process of dip–slip faulting. (**c**) The effects of extensional fracturing are clearly visible, as well as the amount of dilation across the fractures (UAV pilot for scale).

The third area in the NVZ is a segment of the northern portion of the Krafla Fissure Swarm (KFS) at a location that is north of the Krafla central volcano. The geological map in Figure 10a enables us to observe that, as opposed to the ThFS, the KFS is marked by the presence of historical lava fields, as well as both historical and pre-LGM volcanic centers.

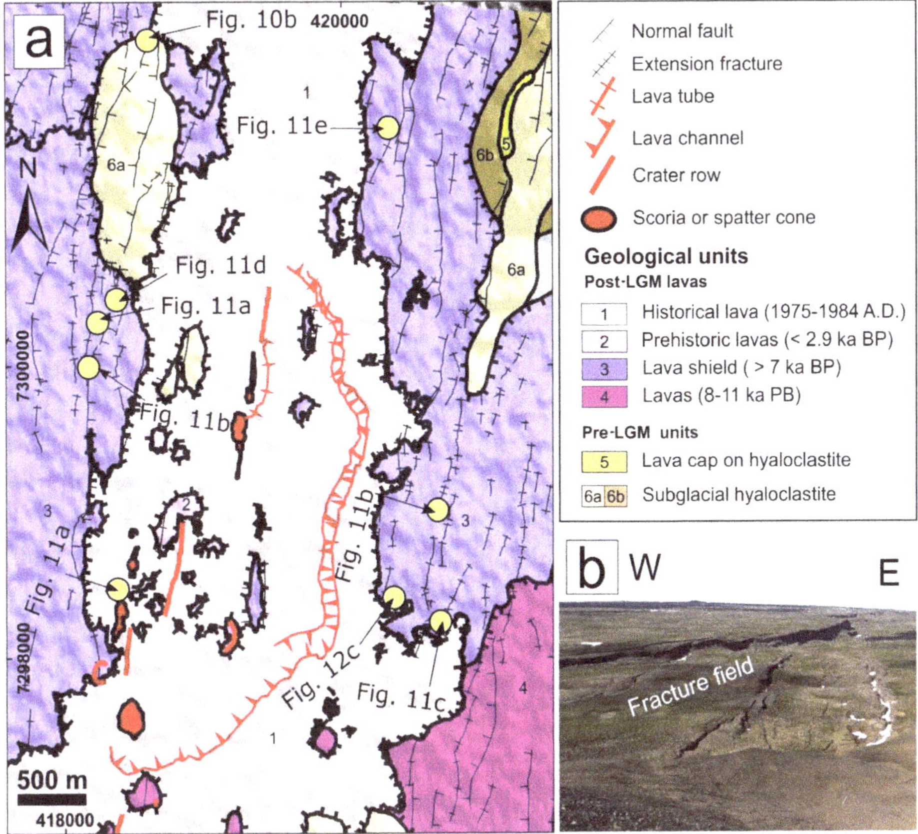

Figure 10. (**a**) Geological map of a portion of the Krafla Fissure Swarm (KFS). (**b**) A field picture visualizing a graben, different from the one portrayed in Figures 8a and 9a, as the faults bordering it have a curvilinear arrangement (the maximum width of the graben is 90 m).

In particular, as reported by Hjartardóttir et al. [117], two major rifting episodes took place within the KFS in the last 1140 years: the 1724–1729 "Mývatn rifting episode" and the instrumentally recorded "Krafla rifting episode" (better known as "the Krafla Fires"), which occurred from 1975 to 1984. During both episodes, there were periods of strong earthquake activity and motions along a number of faults (accommodating the widening and subsidence along tectonic graben). During the "Krafla Fires", the continuous emplacement of dykes resulted in the opening of eruptive fissures, which, in turn, led to lava fountaining and the outpouring of lava flows.

We individuated six geosites in this area: The first one, displayed in Figure 10b, is a textbook-example of a recent volcano–tectonic graben affecting post-LGM lava units bordered by two dip–slip faults that diverge from a common point (highlighted by a yellow circle in Figure 10a). The graben floor, with a maximum width of 90 m, is affected by active stretching, as testified to by the development of an extension fracture field. The other volcanic and tectonic geosites in the KFS are the following: a cluster of recent monogenetic volcanoes (scoria cones), two of which are visible in the background and one at the center of the image (Figure 11a). The larger cone in the foreground (350 m × 150 m) was formed in 1984, at the end of the "Krafla Fires" eruptive cycle [118].

Especially notable is a very recent pahoehoe lava flow, which was outpoured by the crater in the foreground. Another geosite is one of the typical extension fractures (with dilation between 1 and

1.5 m) formed within the lavas older than 7 ka (Figure 11b). As is the case in the ThFS, in the KFS, the wider extension fractures are found within the older lava units. In Figure 11c, a field photograph documents a historical lava flow (emitted during the "Krafla Fires") coming from the left-hand side of the image, which partially infilled a 2-m-wide extension fracture. Figure 11d is a UAV-captured image, offering a chance to take a look at a geosite composed of the combination of a volcano and an extension fracture field. In the background of the image is the Hituholar volcanic edifice (500 m of basal diameter in E–W direction), made of hyaloclastites at the base and scattered pillow lavas in its upper portion. From its southern base, and extending southward, a fracture field cuts both through 12-ka-old lava units and the volcanic edifice. Finally, in Figure 11e, another UAV-captured image enables us to observe a geosite made of the two main extension-related structures that characterize active rift zones such as the KFS: These are N–S-trending dip–slip faults (marked by important offsets) and N–S to NNE-trending extension fractures, whose main characteristics are the absence of vertical displacement and the presence of a major dilation compatible to the regional extensional regime. The individual fault and the fractures composing the geosite in Figure 11e are approximately parallel and spaced about 130 m from each other.

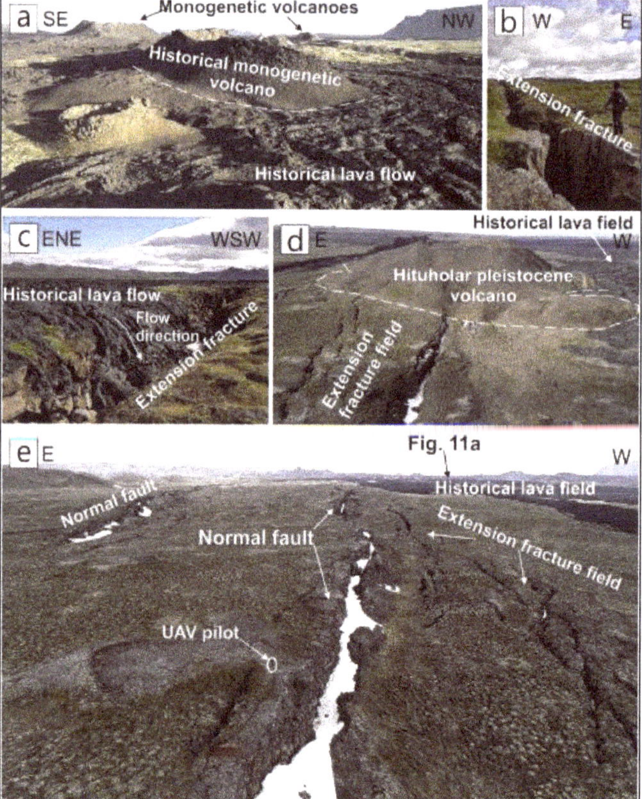

Figure 11. (**a**) UAV-captured image of a number of recent monogenetic volcanoes: two in the background and one at the center of the image (image courtesy of Fabio Marchese). (**b**) One of the extension fractures in the area. (**c**) A beautiful example of a recent lava flow. (**d**) UAV-taken photograph with the Hituholar volcanic edifice and a fracture field (image courtesy of Luca Fallati). (**e**) Aerial image with dip–slip faults and extension fractures (image courtesy of Luca Fallati).

As far as the assessment of the KFS geosites is concerned, all have a considerable scientific value, attested by the several research efforts and publications dedicated to this active rift zone and the recent volcanic episodes that took place there. All are representative of volcanic and tectonic processes within an active rift system. However, none can be considered rare, for the same reasons that were cited above in regard to the ThFS geosites.

Their educational relevance is generally high, though special mention has to be made to the two geosites in Figure 11a,c. The former is suitable for explaining a number of volcanic features (the geometry of monogenetic cones and the morphology of a recent lava flow) that may be easily understood also by nonexperts. The geosite in Figure 11c enables documenting of the interaction between volcanic (a lava flow) and tectonic processes (an extension fracture). However, as these geosites are located in remote areas (relatively unsafe as well, due to the presence of a great number of fractures and holes), their accessibility is limited, and their educational value is, therefore, greatly diminished.

As was the case for some of the geosites in the ThFS, in this case, we created some 3-D models: The first one (Figure 12a) depicts a 100 m × 200 m area marked by the presence of a set of very long and wide (as much as 4 m) extension fractures cutting through 12-ka-old lavas. The 3-D model in Figure 12b depicts a major N–S-trending dip–slip fault (with a 10-m vertical offset) in a similar way as in Figure 9b. Finally, the model in Figure 12c is aimed at capturing a rather common, yet spectacular, occurrence in active rifts: a historical, basaltic, pahoehoe lava flow cascading into a 2.3-m-wide extension fracture. The viewer has a chance to take a look at the so-called ogives, particular structures (common in basaltic, pahoehoe lavas) produced by the bending of the crust during the movement of the underlying, still-molten lava; the convexity of the ogives points in the direction of the lava flow.

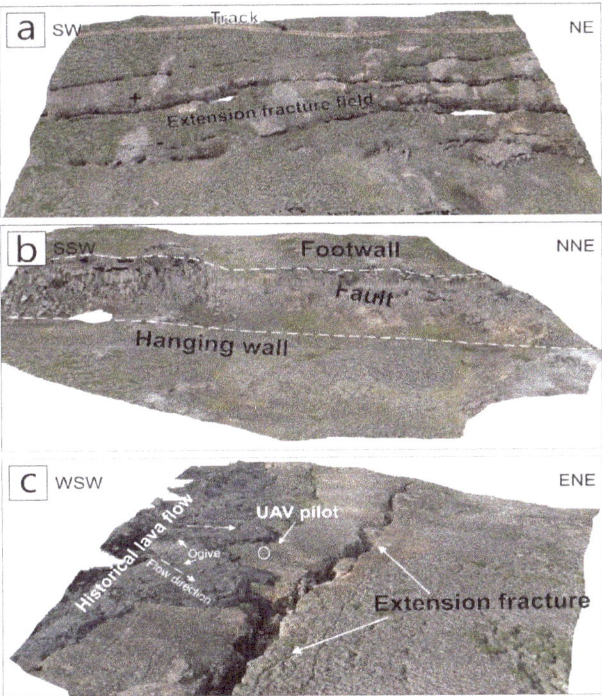

Figure 12. (**a**) 3-D models of volcano–tectonic structures in the KFS. (**a**) A set of very long and wide extension fractures. (**b**) The geometry of a N–S-trending dip–slip fault. (**c**) A historical pahoehoe lava flow cascading into an extension fracture. See text for a detailed explanation.

5. Discussion

5.1. Geosite Assessment

Our qualitative assessment of the geosites has been performed by following the guidelines contained in several key papers devoted to geoheritage and geotourism [4,6,7,15,19,25–27,29–31,33,119]. As a result of our work, it is possible to highlight that only one geosite, Snæfellsjökull Volcano, is characterized by all the values that we selected as most appropriate for the purpose of our work. Most of the criteria that have been used to assess the scientific value, and the additional values of Snæfellsjökull, have been applied to Kirkjufell, which, however, lacks a cultural value and has not been the subject of relevant scientific publications. These two geosites have been assessed also in terms of their aesthetic value, based on their visibility and prominent structures. In regard to the other sites in the Snæfellsnes Peninsula, all have a considerable scientific value and a potentially high educational value, enhanced by their accessibility and safety.

All the geosites in the NVZ, located in the three selected areas, have major scientific value, in that they are representative of tectonic and volcanic processes (most of which are active) documented in several scientific publications. In addition to that, the geosites along the HFF are very rare at the scale of Iceland, and the triple junction (Figure 4f) is extremely uncommon at the worldwide level.

With reference to the educational value, all of the considered geosites in the NVZ share the advantage of being suitable for teaching and communication purposes. In fact, they are outstanding outcrops or flat-lying surfaces, where volcanic and tectonic processes are visible in such a way as to be understood by the lay public as well. However, with the exception of two geosites along the HFF (Figure 4b,d) and one in the ThFS (Figure 8a), most are located in areas that are difficult to reach. Therefore, their educational value is diminished and it has been necessary to come up with strategies to communicate them, making them accessible in a "virtual" fashion, as illustrated in the following paragraph.

5.2. Geosite Visualization, Fruition and Communication

The selected geosites are situated in two Icelandic "milestone" areas: the Snæfellsnes Peninsula and the Northern Volcanic Zone. In describing the selected geosites, which are outcrops, geological objects and features at scales from a few centimeters to kilometers, we have made use of: (i) field photographs, (ii) highly detailed images captured by unmanned aerial vehicles (UAVs) and (iii) 3-D models elaborated by way of Structure-from-Motion techniques (SfM).

In the following section, we will highlight the advantages offered by each of the above types of visualization techniques. Starting with the first geoheritage area, the Snæfellsnes Peninsula, it is worth pointing out that, when studying the area back in 2012, our research team hadn't experimented with UAV-based techniques yet. However, we believe that the field photographs provided above are effective in capturing some key examples of the Icelandic geoheritage. In particular, while Figure 3b,c display milestone geosites famous all over the world, Figures 3d and 2a are suitable for introducing the concepts of monogenetic cone morphology and alignment as indicators of the direction of a feeding fracture below the surface; this, in turn, is the expression of the tectonic regime at the time of the development of the cones. In this particular case, it cannot be denied that a UAV-captured image would have made the observation of these geosites much clearer for the viewer. Regarding the other four field photographs in Figure 2 aimed at showing the geometry of subvolcanic intrusions in the field, in this case, a UAV-captured image wouldn't be effective. In fact, a high-resolution field photograph taken from a distance of about two kilometers (such as the one in Figure 2d) is capable of communicating the concept of a sheet swarm intersection in a much more effective fashion.

With regard to the Northern Volcanic Zone, our research team has studied this area since 2015, a time period over which we have progressively used more and more UAV-captured images for visually describing outcrops and interpreting volcano–tectonic processes. The first figure, aimed at showcasing the importance of the Husavik–Flatey Fault (HFF), is composed exclusively of field photographs, two of which (Figure 4b,c) are the perfect choice in terms of showing details of the fault at the outcrop scale, including an example (Figure 4c) of right-lateral offset across a lava tube. With regard to Figure 4d, the selected picture, taken several kilometers away from the HFF fault plane, represents the best choice for communicating two geological concepts at the same time: the considerable size of the vertical offset along a giant transform fault like the HFF (the result of transtensional tectonics) and the level of seismic risk affecting the nearby town of Husavik, indicated by an arrow in the background. Figure 4e is another field-taken picture, useful to explaining the effects of strike–slip faulting on landforms. Here, we highlighted the fault trace by way of a black dashed line, as well as the offset of river gullies (with light-blue dashed lines).

In order to show the intersection between the HFF and the westernmost fault of the ThFS, the field picture in Figure 4f was taken from a perspective that allows us to "intuitively guess" the different orientations of the two faults and their connection in the location indicated by the black arrow. Unfortunately, we do not have any suitable UAV-captured image, nor do we have a 3-D model that would have made the contents of this geosite much more compelling and easy to communicate.

As we have surveyed other locations in the ThFS by using UAVs, Figure 8 provides a combination of field-based and drone-taken photographs. Figure 8a is an oblique, UAV-captured perspective that enables the viewer to take a look at the effects of extensional tectonics in the form of the development of a graben. Figure 8b is a field picture taken from the top of the footwall block of a dip–slip fault. The obliquity of this point of view is helpful in illustrating the geometry of this type of rifting-produced normal fault. Figure 8c is another UAV-captured image, which has the advantage of showing the scale of the process of extensional fracturing. As already underscored, in this case, extensional forces across these fractures have been at work for a very long time, as expressed by the extreme degree of dilation; on the contrary, only a close-up photograph like the one in Figure 8d can help visualize the process of extensional fracturing across very recent extensional fractures with centimetric dilation. A UAV-captured photograph like the one in Figure 8e, on the contrary, is the best option to offer a view across the flat-lying floor of the rift zone, punctuated by recent lava flows and scattered eruptive centers such as the one in the foreground.

As underscored in the previous paragraph, the educational value of the geosites in the ThFS is hampered by their limited accessibility. To make the most of the didactic relevance of the geosites illustrated in Figure 8a–c, we produced 3-D models by following the workflow outlined in Chapter 3. Figure 9 shows that 3-D models are even more effective than individual UAV-taken pictures in explaining volcano–tectonic concepts. Figure 9a represents a step forward in showing the structure of a graben. Here, the presence of the road that crosses this extensional structure aids in communicating the topic of geological hazards affecting man-made structures. Additionally, Figure 9b can be regarded as an improvement of the previous Figure 8b in terms of visualizing the process of dip–slip faulting. The same goes for Figure 9c, which allows us to view the effects of extensional fracturing and the extent to which the rifting processes have been opening the fractures over time (this is made possible by the scale, represented by the UAV pilot). With the purpose of enhancing the popularization and fruition of these three geosites in the ThFS, we uploaded three "Virtual Outcrops" [92–94] on the web. 3-D DOMs can also be navigated through Virtual Reality techniques, as described in recent contributions [95,97,120,121]. The geosite in Figure 9a is accessible in Virtual Outcrop format at [122]. The 3-D model in Figure 9b is accessible in Virtual Outcrop format at [123]. The geosite in Figure 9c is available as Virtual Outcrop at [124].

The online fruition of these Virtual Outcrops is greatly enhanced by the possibility to rotate the images and zoom in to examine the outcrops' tiniest details. This way, it is possible to accomplish two key goals: firstly, to popularize the selected geosites, making them "accessible" and, thus, enhancing their educational value and, secondly, to communicate Earth Science topics to a potentially wider audience, including younger people usually very keen on technological and interactive applications.

With regard to the geosites in the Krafla Fissure Swarm (KFS), we also used a combination of three types of visualization techniques: Figure 10b is a field photograph that, taken from a volcanic ridge, provides a clear view of a graben, which is different from the one in the ThFS, as the faults bordering it have a curvilinear arrangement. All the geosites in Figure 11 have been illustrated through the combination of field pictures and UAV-captured photographs. Figure 11a, taken from the top of a slope, aids in showing two of the main features of active rift zones: monogenetic volcanoes and their effusive eruptive products, or lava flows. Figure 11b displays the amount of dilation produced by extensional fracturing, whereas Figure 11c is aimed at visualizing, from a short distance, the "encounter" between a lava flow and an extension fracture, a rather common occurrence in active rift zones where fracturing, faulting and lava outpouring are all guided by the combination of the regional, extensional regime and the presence of magma upwelling from underneath. A UAV-captured image (Figure 11d) is more helpful than a field picture in showing the coexistence of two different products of extensional tectonics: the growth of volcanic edifices (here represented by the Pleistocene-age Hituholar Volcano) and the development of extension fractures. Figure 11e is another example of how a UAV-taken photograph can make the difference in showing volcano–tectonic processes in a rift zone. The scale, represented by the UAV pilot standing on top of the fault's footwall block, helps to point out the much more pervasive effects produced by faulting, in comparison to extensional fracturing.

Finally, Figure 12 is dedicated to 3-D models that, as already discussed, can play a paramount role in Earth Science communication. For example, Figure 12a is aimed at highlighting the presence of a man-made structure (the track) parallel to the trend of this fracture field. Figure 12b, just like the previously commented Figure 9b, works well to illustrate the geometry of a dip–slip fault and the displacement along it. Figure 12c provides a different perspective of the interaction between magmatic and tectonic processes. Here, it is possible to vividly illustrate (an added value with respect to Figure 11c) the movement of the lava flow, and also the structures produced by the lava, slowly creeping towards the gaping fracture.

As underscored for the ThFS, in order to enhance the popularization and fruition of three geosites in the KFS, we published three "Virtual Outcrops". The geosite in Figure 12a is available in Virtual Outcrop format at [125]. The 3-D model in Figure 12b is available in Virtual Outcrop format at [126]. The geosite in Figure 12c is accessible in Virtual Outcrop at [127].

6. Conclusions

Among all the areas that compose the Icelandic "open-air geological museum", we selected the Snaefellsnes Peninsula and the Northern Volcanic Zone (NVZ), which are home to a wide gamut of subvolcanic and volcano–tectonic outcrops and landforms, many of which can be considered potential geosites, as they reflect the multifaceted variety of the Icelandic geoheritage. Our purpose has been to document, describe and assess a number of geosites based on a set of criteria chosen from previous research efforts focused on geoheritage, geoconservation and geotourism. We have selected a total of 25 geosites, 8 in the Snæfellsnes Peninsula and 17 in the NVZ (five along the HFF, six in the ThFS and six in the KFS).

The qualitative assessment we performed can be summarized by pointing out that the majority of the geosites in both areas have a high scientific value. However, only the geosites in the Snæfellsnes Peninsula may be regarded as having an overall educational value, thanks to their accessibility; in the case of Snæfellsjökull Volcano and Kirkjufell, there is also the possibility, offered to visitors, to take part in guided tours and other educational activities. Of the 17 geosites in the NVZ, only three are easily accessible, and this hampers their educational value.

The promotion of geosites, especially those which are not easily accessible, can be achieved through an accurate work of illustration, description and popularization. To accomplish these goals, we have made use of a range of visualization techniques, from field photographs to highly detailed images captured by unmanned aerial vehicles (UAVs) and 3-D models of field outcrops produced by means of Structure From-Motion (SfM) photogrammetry. At the same time as showcasing the selected geosites, we discussed and compared the advantages provided by the different types of image-taking techniques, from traditional ones such as field photographs to more advanced ones such as UAV-based photographs and 3-D models. Finally, with the aim of making it possible, for potential users, to interact with the geosites through Virtual Reality techniques, we have uploaded six Virtual Outcrops online.

This may represent a novel, cutting-edge approach to improve geoheritage popularization and geoscience communication, allowing for the engagement of a wider audience, with special reference to potential end-users from the younger generation.

Author Contributions: Original manuscript preparation, F.P.M., F.L.B. and C.V.; figure preparation, F.L.B.; supervision, F.P.M. All authors have read and agreed to the published version of the manuscript.

Funding: The present work was carried out in the framework of ILP Task Force II, "Structural and rheological constraints on magma migration, accumulation and eruption through the lithosphere" (2015–2019). It has also been carried out in the framework of ESA Project Nr. 38829 and MIUR project ACPR15T4_00098–Argo3D (http://argo3d.unimib.it/).

Acknowledgments: We are deeply grateful to three anonymous referees for their insightful comments and suggestions. We also acknowledge Alessandro Tibaldi for his feedback during the writing of this manuscript. This article is also an outcome of Project MIUR–Dipartimenti di Eccellenza 2018–2022. Agisoft Metashape is acknowledged for photogrammetric data processing. Fabio Marchese and Luca Fallati are acknowledged for ongoing scientific collaboration focused on UAV-based SfM techniques. Finally, this paper is an outcome of GeoVires, the Virtual Reality Lab for Earth Sciences, hosted at the Department of Earth and Environmental Sciences, University of Milan Bicocca, U4, Piazza della Scienza 4, 20126 Milan, Italy.

Conflicts of Interest: The authors declare no conflict of interest.

References

1. Eberhard, R. *Pattern and Process: Towards a Regional Approach to National Estate Assessment of Geodiversity*; Technical Series No. 2; Eberhard, R., Ed.; Australian Heritage Commission and Environment Forest Taskforce: Canberra, Australian; Environment Australia: Canberra, Australian, 1997.

2. Brocx, M.; Semeniuk, V. Geoheritage and geoconservation history, definition, scope and scale. *J. R. Soc. West. Aust.* **2007**, *90*, 53–87.

3. Asrat, A.; Demissie, M.; Mogessie, A. Geoheritage conservation in Ethiopia: The case of the Simien mountains. *Quaest. Geogr.* **2012**, *31*, 7–23. [CrossRef]

4. Fassoulas, C.; Mouriki, D.; Dimitriou-Nikolakis, P.; Iliopoulos, G. Quantitative assessment of geotopes as an effective tool for geoheritage management. *Geoheritage* **2012**, *4*, 177–193. [CrossRef]

5. Wimbledon, W.A.P.; Smith-Meyer, S. *Geoheritage in Europe and Its Conservation*; Wimbledon, W.A.P., Smith-Meyer, S., Eds.; ProGEO: Oslo, Noruega, 2012; p. 405.

6. Bruno, D.E.; Crowley, B.E.; Gutak, J.M.; Moroni, A.; Nazarenko, O.V.; Oheim, K.B.; Ruban, D.A.; Tiess, G.; Zorina, S.O. Paleogeography as geological heritage: Developing geosite classification. *Earth Sci. Rev.* **2014**, *138*, 300–312. [CrossRef]

7. Brilha, J. Inventory and quantitative assessment of geosites and geodiversity sites: A review. *Geoheritage* **2016**, *8*, 119–134. [CrossRef]

8. Dudley, N. *Guidelines for Applying Protected Area Management Categories*; Dudley, N., Ed.; IUCN: Gland, Switzerland, 2008.

9. Reis, J.; Póvoas, L.; Barriga, F.J.A.S.; Lopes, C. Science education in a museum: Enhancing Earth Sciences literacy as a way to enhance public awareness of geological heritage. *Geoheritage* **2014**, *6*, 217–223. [CrossRef]

10. Pasquaré Mariotto, F.; Venturini, C. Strategies and tools for improving Earth Science education and popularization in museums. *Geoheritage* **2017**, *9*, 187–194. [CrossRef]

11. Venturini, C.; Pasquaré Mariotto, F. Geoheritage promotion through an interactive exhibition: A case study from the Carnic Alps, NE Italy. *Geoheritage* **2019**, *11*, 459–469. [CrossRef]

12. De Grosbois, A.M.; Eder, W. Geoparks—A tool for education, conservation and recreation. *Environ. Geol.* **2008**, *55*, 465–466. [CrossRef]

13. Mckeever, P.; Zouros, N.; Patzak, M. The UNESCO global network of national geoparks. In *Geotourism. The Tourism of Geology and Landscape*; Newsome, D., Dowling, R.K., Eds.; Goodfellow Publishers Ltd.: Oxford, UK, 2010; pp. 221–230.

14. Pásková, M.; Zelenka, J. Sustainability management of unesco global geoparks. *Sustain. Geosci. Geotourism* **2018**, *2*, 44–64. [CrossRef]

15. Panizza, M.; Piacente, S. Geomorphosites and geotourism. *Revista Geográfica Acadêmica* **2008**, *2*, 5–9.

16. Newsome, D.; Dowling, R.K. *Geotourism: The tourism of Geology and Landscape*; Goodfellow Publishers Ltd.: Oxford, UK, 2010.

17. Burek, C.V. The role of LGAPs (Local Geodiversity Action Plans) and Welsh RIGS as local drivers for geoconservation within geotourism in Wales. *Geoheritage* **2012**, *4*, 45–63. [CrossRef]

18. Kubalíková, L. Geomorphosite assessment for geotourism purposes. *Czech J. Tour.* **2013**, *2*, 80–104. [CrossRef]

19. Suzuki, A.; Takagi, H. Evaluation of geosite for sustainable planning and management in Geotourism. *Geoheritage* **2018**, *10*, 123–135. [CrossRef]

20. Cleal, C.J. Geoconservation—What on Earth are we doing? In *Regional Conference on Geoconservation: Geological Heritage in the South-European Europe Book of Abstracts*; Hlad, B., Herlec, U., Eds.; Environmnetal Agency of the Republic of Slovenia: Ljubljana, Slovenia, 2007; p. 25. Available online: http://arsis.net/circular/ProGEO-Abstract.pdf (accessed on 22 February 2010).

21. Burek, C.V.; Prosser, C.D. The History of Geoconservation: An introduction. *Geol. Soc. London Spec. Pub.* **2008**, *300*, 1–5. [CrossRef]

22. Prosser, C.; Murphy, M.; Larwood, J. *Geological Conservation: A Guide to Good Practice*; English Nature: Peterborough, UK, 2011.

23. Prosser, C.D. Our rich and varied geoconservation portfolio: The foundation for the future. *Proc. Geol. Assoc.* **2013**, *124*, 568–580. [CrossRef]

24. Gordon, J.E. Geoheritage, Geotourism and the cultural landscape: Enhancing the visitor experience and promoting geoconservation. *Geosciences* **2018**, *8*, 136. [CrossRef]

25. Reynard, E. Geosite. In *Encyclopedia of Geomorphology*; Goudie, A.S., Ed.; Routledge: London, UK, 2004.

26. Ruban, D.A. Quantification of geodiversity and its loss. *Proc. Geol. Assoc.* **2010**, *121*, 326–333. [CrossRef]

27. Ruban, D.A.; Kuo, I. Essentials of geological heritage site (geosite) management: A conceptual assessment of interests and conflicts. *Nat. Nascosta* **2010**, *41*, 16–31.

28. Crofts, R.; Gordon, J.E. Geoconservation in protected areas. In *Protected Area Governance and Management*; Worboys, G.L., Lockwood, M., Kothari, A., Feary, S., Pulsford, I., Eds.; ANU Press: Canberra, Australia, 2015; pp. 531–568.

29. Grandgirard, V. L'évaluation des géotopes. *Geol. Insubr.* **1999**, *4*, 59–66.

30. Zorina, S.O.; Silantiev, V.V. Geosites, classification of. In *Encyclopedia of Mineral and Energy Policy*; Springer: Berlin/Heidelberg, Germany, 2014. [CrossRef]

31. Lima, F.; Brilha, J.; Salamuni, E. Inventorying geological heritage in large territories: A methodological proposal applied to Brazil. *Geoheritage* **2010**, *2*, 91–99. [CrossRef]

32. Reynard, E.; Fontana, G.; Kozlik, L.; Scapozza, C. A method for assessing "scientific" and "additional values" of geomorphosites. *Geogr. Helv.* **2007**, *62*, 148–159. [CrossRef]

33. Coratza, P.; Giusti, C. Methodological proposal for the assessment of the scientific quality of of geomorphosites. *Il Quat.* **2005**, *18*, 307–313.

34. Coratza, P.; Panizza, M. (Eds.) *Geomorphology and Cultural Heritage*; Memorie Descrittive Della Carta Geologica d'Italia; ISPRA: Rome, Italy, 2009; p. 87.

35. Ólafsdóttir, R.; Tverijonaite, E. Geotourism: A systematic literature review. *Geosciences* **2018**, *8*, 234. [CrossRef]

36. Ólafsdóttir, R.; Dowling, R. Geotourism and Geoparks. A tool for geoconservation and rural development in vulnerable environments: A case study from Iceland. *Geoheritage* **2014**, *6*, 71–87. [CrossRef]

37. Ásbjörnsdóttir, L. The VIII international ProGEO symposium in reykjavík, 2015. *Geoheritage* **2018**, *10*, 139–142. [CrossRef]

38. Einarsson, P.; Sæmundsson, K. Earthquake epicenters 1982–1985 and volcanic systems in Iceland: A map. In *Í Hlutarins Eðli, Festschriftfor Þorbjörn Sigurgeirsson*; Menningarsjóður, R., Sigfússon, Þ., Eds.; Menningarsjodur: Reykjavik, Iceland, 1987.

39. Bonali, F.L.; Tibaldi, A.; Marchese, F.; Fallati, L.; Russo, E.; Corselli, C.; Savini, A. UAV-based surveying in volcano-tectonics: An example from the Iceland rift. *J. Struct. Geol.* **2019**, *121*, 46–64. [CrossRef]

40. Fallati, L.; Polidori, A.; Salvatore, C.; Saponari, L.; Savini, A.; Galli, P. Anthropogenic Marine Debris assessment with Unmanned Aerial Vehicle imagery and deep learning: A case study along the beaches of the Republic of Maldives. *Sci. Total Environ.* **2019**, *693*, 133581. [CrossRef]

41. Jacoby, W.; Gudmundsson, A. Hotspot iceland: An introduction. *J. Geodyn.* **2007**, *43*, 1–5. [CrossRef]

42. Einarsson, P. Plate boundaries, rifts and transforms in Iceland. *Jokull* **2008**, *58*, 35–58.

43. Jakobsdóttir, S.S.; Roberts, M.J.; Guðmundsson, G.B.; Geirsson, H.; Slunga, R. Earthquake swarms at Upptyppingar, North-east Iceland: A sign of magma intrusion? *Stud. Geophys. Geod.* **2008**, *52*, 513–528.

44. Mjelde, R.; Digranes, P.; van Schaack, M.; Shimamura, H.; Shiobara, H.; Kodaira, S.; Næss, O. Crustal structure of the outer Vøring Plateau, offshore Norway, from ocean bottom seismic and gravity data. *J. Geophys. Res.* **2001**, *106*, 6769–6791. [CrossRef]

45. Gudmundsson, A. Dynamics of volcanic systems in Iceland: Example of tectonism and volcanism at juxtaposed hot spot and mid-ocean ridge systems. *Ann. Rev. Earth Planet. Sci.* **2000**, *28*, 107–140. [CrossRef]

46. Flude, S.; Burgess, R.; McGarvie, D.W. Silicic volcanism at Ljósufjöll, Iceland: Insights into evolution and eruptive history from Ar–Ar dating. *J. Volcanol. Geotherm.* **2008**, *169*, 154–175. [CrossRef]

47. Tibaldi, A.; Bonali, F.L.; Pasquaré Mariotto, F.; Rust, D.; Cavallo, A.; D'Urso, A. Structure of regional dykes and local cone sheets in the Midhyrna-Lysuskard area, Snaefellsnes Peninsula (NW Iceland). *Bull. Volcanol.* **2013**, *75*, 764–780. [CrossRef]

48. Young, K.D.; Jancin, B.; Orkan, N.I. Transform deformation of tertiary rocks along the Tjornes Fracture Zone, North Central Iceland. *J. Geophys. Res.* **1985**, *90*, 9986–10010. [CrossRef]

49. Bergerat, F.; Angelier, J. Immature and mature transform zones near a hot spot: The South Iceland Seismic Zone and the Tjornes Fracture Zone (Iceland). *Tectonophysics* **2008**, *447*, 142–154. [CrossRef]

50. Opheim, J.O.; Gudmundsson, A. Formation and geometry of fractures, and related volcanism, of the Krafla fissure swarm, northeast Iceland. *Geol. Soc. Am. Bull.* **1989**, *101*, 1608–1622. [CrossRef]

51. Garcia, S.; Dhont, D. Structural analysis of the Húsavík-Flatey Transform Fault and its relationships with the rift system in Northern Iceland. *Geodin. Acta* **2005**, *18*, 31–41. [CrossRef]

52. Slater, L.; McKenzie, D.; Gronvold, K.; Shimizu, N. Melt generation and movement under Theistareykir, NE Iceland. *J. Petrol.* **2001**, *42*, 321–354. [CrossRef]

53. Stracke, A.; Zindler, A.; Salters, V.J.M.; McKenzie, D.; Blichert-Toft, J.; Albarède, F.; Gronvold, K. Theistareykir revisited. *Geochem. Geophys. Geosyst.* **2003**, *4*, 8507. [CrossRef]

54. Tibaldi, A.; Bonali, F.L.; Einarsson, P.; Hjartardóttir, Á.R.; Pasquarè Mariotto, F.A. Partitioning of Holocene kinematics and interaction between the Theistareykir Fissure Swarm and the Husavik-Flatey Fault, North Iceland. *J. Struct. Geol.* **2016**, *83*, 134–155. [CrossRef]

55. Bonali, F.L.; Tibaldi, A.; Pasquaré Mariotto, F.; Saviano, D.; Meloni, A.; Sajovitz, P. Geometry, oblique kinematics and extensional strain variation along a diverging plate boundary: The example of the northern Theistareykir Fissure Swarm, NE Iceland. *Tectonophysics* **2019**, *756*, 57–72. [CrossRef]

56. Metzger, S.; Jonsson, S.; Danielsen, G.; Hreinsdottir, S.; Jouanne, F.; Giardini, D.; Villemin, T. Present kinematics of the Tjornes Fracture Zone, North Iceland, from campaign and continuous GPS measurements. *Geophys. J. Int.* **2013**, *192*, 441–455. [CrossRef]

57. Gudmundsson, A. Infrastructure and evolution of ocean-ridge discontinuities in Iceland. *J. Geodyn.* **2007**, *43*, 6–29. [CrossRef]

58. Gudmundsson, A.; Brynjolfsson, S.; Jonsson, M.T. Structural analysis of a transform fault-rift zone junction in North Iceland. *Tectonophysics* **1993**, *220*, 205–221. [CrossRef]

59. Fjäder, K.; Gudmundsson, A.; Forslund, T. Dikes, minor faults and mineral veins associated with a transform fault in North Iceland. *J. Struct. Geol.* **1994**, *16*, 109–119. [CrossRef]

60. Magnúsdòttir, S.; Brandsdòttir, B. 2011. Tectonics of the þeistareykir fissure swarm. *Jokull* **2011**, *61*, 65–79.

61. Pasquaré Mariotto, F.; Bonali, F.L.; Tibaldi, A.; Rust, D.; Oppizzi, P.; Cavallo, A. Holocene displacement field at an emerged oceanic transform-ridge junction: The Husavik-Flatey Fault—Gudfinnugja Fault system, North Iceland. *J. Struct. Geol.* **2015**, *75*, 118–134. [CrossRef]

62. Tibaldi, A.; Bonali, F.L.; Pasquaré Mariotto, F.A. Interaction between transform faults and rift systems: A combined field and experimental approach. *Front. Earth Sci.* **2016**, *4*, 33. [CrossRef]

63. Tibaldi, A.; Bonali, F.L.; Pasquaré Mariotto, F.A.; Russo, E. Interplay between inherited rift faults and transcurrent structures: Insights from analogue models and field data from Iceland. *Glob. Planet. Chang.* **2018**, *171*, 88–109. [CrossRef]

64. Tibaldi, A.; Bonali, F.L.; Pasquaré Mariotto, F.A.; Ranieri Tenti, L.M. The development of divergent margins: Insights from the North Volcanic Zone, Iceland. *Earth Planet. Sci. Lett.* **2019**, *509*, 1–8. [CrossRef]

65. Jordan, B.R. A bird's-eye view of geology: The use of micro drones/UAVs in geologic fieldwork and education. *Geol. Soc. Am. Today* **2015**, *25*, 50–52. [CrossRef]

66. Pavlis, T.L.; Mason, K.A. The new world of 3D geologic mapping. *Geol. Soc. Am. Today* **2017**, *27*, 4–10. [CrossRef]

67. Zekkos, D.; Manousakis, J.; Athanasopoulos-Zekkos, A.; Clark, M.; Knoper, L.; Massey, C.; Archibald, G.; Greenwood, W.; Hemphill-Haley, M.; Rathje, E.; et al. Structure-from-Motion based 3D mapping of landslides & fault rupture sites during 2016 Kaikoura earthquake reconnaissance. In Proceedings of the 11th U.S. National Conference on Earthquake Engineering, Integrating Science, Engineering & Policy, Los Angeles, CA, USA, 25–29 June 2018.

68. Yao, Y.; Chen, J.; Li, T.; Fu, B.; Wang, H.; Li, Y.; Jia, H. Soil liquefaction in seasonally frozen ground during the 2016 Mw6. 6 Akto earthquake. *Soil Dyn. Earthq. Eng.* **2019**, *117*, 138–148. [CrossRef]

69. Gong, J.H.; Wang, D.C.; Li, Y.; Zhang, L.H.; Yue, Y.J.; Zhou, J.P.; Song, Y.Q. Earthquake induced geological hazard detection under hierarchical stripping classification framework in the Beichuan area. *Landslides* **2010**, *7*, 181–189. [CrossRef]

70. Rathje, E.M.; Franke, K. Remote sensing for geotechnical earthquake reconnaissance. *Soil Dyn. Earthq. Eng.* **2016**, *91*, 304–316. [CrossRef]

71. Brook, M.S.; Merkle, J. Monitoring active landslides in the Auckland region utilising UAV/structure-from-motion photogrammetry. *Jpn. Geotech. Soc. Spec. Publ.* **2019**, *6*, 1–6. [CrossRef]

72. Cignetti, M.; Godone, D.; Wrzesniak, A.; Giordan, D. Structure from motion multisource application for landslide characterization and monitoring: The champlas du col case study, sestriere, North-Western Italy. *Sensors* **2019**, *19*, 2364. [CrossRef]

73. Warrick, J.A.; Ritchie, A.C.; Schmidt, K.M.; Reid, M.E.; Logan, J. Characterizing the catastrophic 2017 Mud Creek landslide, California, using repeat structure-from-motion (SfM) photogrammetry. *Landslides* **2019**, *16*, 1201–1219. [CrossRef]

74. Müller, D.; Walter, T.R.; Schöpa, A.; Witt, T.; Steinke, B.; Gudmundsson, M.T.; Dürig, T. High-resolution digital elevation modeling from TLS and UAV campaign reveals structural complexity at the 2014/2015 holuhraun eruption site, Iceland. *Front. Earth Sci.* **2017**, *5*, 59. [CrossRef]

75. Darmawan, H.; Walter, T.R.; Brotopuspito, K.S.; Nandaka, I.G.M.A. Morphological and structural changes at the Merapi lava dome monitored in 2012–2015 using unmanned aerial vehicles (UAVs). *J. Volcanol. Geotherm. Res.* **2018**, *349*, 256–267. [CrossRef]

76. Favalli, M.; Fornaciai, A.; Nannipieri, L.; Harris, A.; Calvari, S.; Lormand, C. UAV-based remote sensing surveys of lava flow fields: A case study from Etna's 1974 channel-fed lava flows. *Bull. Volcanol.* **2018**, *80*, 29. [CrossRef]

77. De Beni, E.; Cantarero, M.; Messina, A. UAVs for volcano monitoring: A new approach applied on an active lava flow on Mt. Etna (Italy), during the 27 February–02 March 2017 eruption. *J. Volcanol. Geotherm. Res.* **2019**, *369*, 250–262. [CrossRef]

78. Hashemi-Beni, L.; Jones, J.; Thompson, G.; Johnson, C.; Gebrehiwot, A. Challenges and Opportunities for UAV-based digital elevation model generation for flood-risk management: A case of Princeville, North Carolina. *Sensors* **2018**, *18*, 3843. [CrossRef]

79. Langhammer, J.; Vackova, T. Detection and mapping of the geomorphic effects of flooding using UAV photogrammetry. *Pure Appl. Geophys.* **2018**, *175*, 3223–3245. [CrossRef]

80. Johnson, K.; Nissen, E.; Saripalli, S.; Arrowsmith, J.R.; McGarey, P.; Scharer, K.; Williams, P.; Blisniuk, K. Rapid mapping of ultrafine fault zone topography with structure from motion. *Geosphere* **2014**, *10*, 969–986. [CrossRef]

81. Angster, S.; Wesnousky, S.; Huang, W.L.; Kent, G.; Nakata, T.; Goto, H. Application of UAV photography to refining the slip rate on the Pyramid Lake fault zone, Nevada. *Bull. Seismol. Soc. Am.* **2016**, *106*, 2. [CrossRef]

82. Deffontaines, B.; Chang, K.J.; Champenois, J.; Fruneau, B.; Pathier, E.; Hu, J.-C.; Lu, S.-T.; Liu, Y.C. Active interseismic shallow deformation of the Pingting terraces (Longitudinal Valley–Eastern Taiwan) from UAV high-resolution topographic data combined with InSAR time series. *Geomat. Nat. Hazards Risk* **2016**, *8*, 120–136. [CrossRef]

83. Jiao, Q.; Jiang, W.; Zhang, J.; Jiang, H.; Luo, Y.; Wang, X. Identification of paleoearthquakes based on geomorphological evidence and their tectonic implications for the southern part of the active Anqiu–Juxian fault, eastern China. *J. Asian Earth Sci.* **2016**, *132*, 1–8. [CrossRef]

84. Bi, H.; Zheng, W.; Ren, Z.; Zeng, J.; Yu, J. Using an unmanned aerial vehicle for topography mapping of the fault zone based on structure from motion photogrammetry. *Int. J. Remote Sens.* **2017**, *38*, 2495–2510. [CrossRef]

85. Gao, M.; Xu, X.; Klinger, Y.; van der Woerd, J.; Tapponnier, P. High-resolution mapping based on an Unmanned Aerial Vehicle (UAV) to capture paleoseismic offsets along the Altyn-Tagh fault, China. *Sci. Rep.* **2017**, *7*, 8281. [CrossRef] [PubMed]

86. Wu, F.; Ran, Y.; Xu, L.; Cao, J.; Li, A. Paleoseismological study of the Late Quaternary slip-rate along the South Barkol Basin Fault and its tectonic implications, Eastern Tian Shan, Xinjiang. *Acta Geol. Sin. Engl. Ed.* **2017**, *91*, 429–442. [CrossRef]

87. Rao, G.; He, C.; Cheng, Y.; Yu, Y.; Hu, J.; Chen, P.; Yao, Q. Active normal faulting along the Langshan Piedmont Fault, North China: Implications for slip partitioning in the western Hetao Graben. *J. Geol.* **2018**, *126*, 99–118. [CrossRef]

88. Trippanera, D.; Ruch, J.; Passone, L.; Jónsson, S. Structural mapping of dike-induced faulting in Harrat Lunayyir (Saudi Arabia) by using high resolution drone imagery. *Front. Earth Sci.* **2019**, *7*, 168. [CrossRef]

89. Westoby, M.J.; Brasington, J.; Glasser, N.F.; Hambrey, M.J.; Reynolds, J.M. "Structure-from-Motion" photogrammetry: A low-cost, effective tool for geoscience applications. *Geomorphology* **2012**, *179*, 300–314. [CrossRef]

90. Chesley, J.T.; Leier, A.L.; White, S.; Torres, R. Using unmanned aerial vehicles and structure-from-motion photogrammetry to characterize sedimentary outcrops: An example from the Morrison Formation, Utah, USA. *Sediment. Geol.* **2017**, *354*, 1–8. [CrossRef]

91. James, M.R.; Robson, S.; d'Oleire-Oltmanns, S.; Niethammer, U. Optimising UAV topographic surveys processed with structure-from-motion: Ground control quality, quantity and bundle adjustment. *Geomorphology* **2017**, *280*, 51–66. [CrossRef]

92. Xu, X.; Aiken, C.L.; Nielsen, K.C. Real time and the virtual outcrop improve geological field mapping. *Eos Trans. Am. Geophys. Union* **1999**, *80*, 317–324. [CrossRef]

93. Tavani, S.; Granado, P.; Corradetti, A.; Girundo, M.; Iannace, A.; Arbués, P.; Muñozb, J.A.; Mazzoli, S. Building a virtual outcrop, extracting geological information from it, and sharing the results in Google Earth via OpenPlot and Photoscan: An example from the Khaviz Anticline (Iran). *Comput. Geosci.* **2014**, *63*, 44–53. [CrossRef]

94. Trinks, I.; Clegg, P.; McCaffrey, K.; Jones, R.; Hobbs, R.; Holdsworth, B.; Holliman, N.; Imber, J.; Waggott, S.; Wilson, R. Mapping and analysing virtual outcrops. *Vis. Geosci.* **2005**, *10*, 13–19. [CrossRef]

95. Gerloni, I.G.; Carchiolo, V.; Vitello, F.R.; Sciacca, E.; Becciani, U.; Costa, A.; Riggi, S.; Bonali, F.L.; Russo, E.; Fallati, L.; et al. Immersive virtual reality for earth sciences. In Proceedings of the 2018 Federated Conference on Computer Science and Information Systems (FedCSIS), Poznan, Poland, 9–12 September 2018; IEEE: Piscataway, NJ, USA, 2018; pp. 527–534.

96. Varvara, A.; Nomikou, P.; Pavlina, B.; Pantelia, S.; Bonali, F.L.; Lemonia, R.; Andreas, M. The Story Map for Metaxa Mine (Santorini, Greece): A unique site where history and volcanology meet each other. In Proceedings of the 5th International Conference on Geographical Information Systems Theory, Applications and Management (GISTAM 2019), Heraklion, Crete, Greece, 3–5 May 2019; pp. 212–219.

97. Krokos, M.; Bonali, F.L.; Vitello, F.; Antoniou, V.; Becciani, U.; Russo, E.; Marchese, F.; Fallati, L.; Nomikou, P.; Kearl, M.; et al. Workflows for virtual reality visualisation and navigation scenarios in Earth Sciences. In Proceedings of the 5th International Conference on Geographical Information Systems Theory, Applications and Management, GISTAM 2019, Heraklion, Crete, Greece, 3–5 May 2019; pp. 297–304.

98. Vollgger, S.A.; Cruden, A.R. Mapping folds and fractures in basement and cover rocks using UAV photogrammetry, Cape Liptrap and Cape Paterson, Victoria, Australia. *J. Struct. Geol.* **2016**, *85*, 168–187. [CrossRef]

99. James, M.R.; Robson, S. Straightforward reconstruction of 3D surfaces and topography with a camera: Accuracy and geoscience application. *J. Geophys. Res.* **2012**, *117*, F03017. [CrossRef]

100. Turner, D.; Lucieer, A.; Watson, C. An automated technique for generating georectified mosaics from ultra-high resolution unmanned aerial vehicle (UAV) imagery, based on structure from motion (SfM) point clouds. *Remote Sens.* **2012**, *4*, 1392–1410. [CrossRef]

101. Burns, J.H.R.; Delparte, D. Comparison of commercial structure-from-motion photogrammetry software used for underwater three-dimensional modeling of coral reef environments. *Int. Arch. Photogramm. Remote Sens. Spat. Inf. Sci. ISPRS Arch.* **2017**, *42*, 127–131. [CrossRef]

102. Stal, C.; Bourgeois, J.; De Maeyer, P.; De Mulder, G.; De Wulf, A.; Goossens, R.; Hendrickx, M.; Nuttens, T.; Stichelbaut, B. Test case on the quality analysis of structure from motion in airborne applications. In *32nd EARSeL Symposium: Advances in Geosciences*; European Association of Remote Sensing Laboratories (EARSeL): Mykonos, Greece, 2012.

103. Grandgirard, V. Géomorphologie, protection de la nature et gestion du paysage. Thèse de Doctorat en Géographie, Université de Fribourg, Fribourg, Switzerland, 1997; p. 210. Available online: https://www.persee. fr/doc/karst_0751-7688_2000_num_35_1_2461_t1_0064_0000_2 (accessed on 31 January 2020).

104. Kokfelt, T.F.; Hoernle, K.; Lundstrom, C.; Hauff, F.; van den Bogaard, C. Time-scale for magmatic differentiation at the Snaefellsjokull central volcano, western Iceland: Constraints from U-Th-Pa-Ra-disequilibria in post-glacial lavas. *Geochem. Cosmochem. Acta* **2009**, *73*, 1120–1144. [CrossRef]

105. Corazzato, C.; Tibaldi, A. Fracture control on type, morphology and distribution of parasitic volcanic cones: An example from Mt. Etna, Italy. *J. Volcanol. Geotherm. Res.* **2006**, *158*, 177–194. [CrossRef]

106. Bonali, F.L.; Corazzato, C.; Tibaldi, A. Identifying rift zones on volcanoes: An example from La Réunion Island. *Indian Ocean. Bull. Volcanol.* **2011**, *73*, 347–366. [CrossRef]

107. Tibaldi, A.; Bonali, F.L. Intra-arc and back-arc volcano-tectonics: Magma pathways at Holocene Alaska-Aleutian volcanoes. *Earth Sci. Rev.* **2017**, *167*, 1–26. [CrossRef]

108. Gudmundsson, A.; Pasquarè Mariotto, F.A.; Tibaldi, A. Dykes, Sills, Laccoliths, and Inclined Sheets in Iceland. In *Physical Geology of Shallow Magmatic Systems*; Breitkreutz, C., Rocchi, S., Eds.; Springer: Cham, Germany, 2014.

109. Scudero, S.; De Guidi, G.; Gudmundsson, A. Size distributions of fractures, dykes, and eruptions on Etna, Italy: Implications for magma-chamber volume and eruption potential. *Sci. Rep.* **2019**, *9*, 4139. [CrossRef]

110. Tibaldi, A.; Pasquaré Mariotto, F.; Papanikolaou, D.; Nomikou, P. Discovery of a huge sector collapse at the Nisyros volcano, Greece, by on-land and offshore geological-structural data. *J. Volcanol. Geotherm. Res.* **2008**, *177*, 485–499. [CrossRef]

111. Pasquaré Mariotto, F.; Tibaldi, A. Structure of a sheet–laccolith system revealing the interplay between tectonic and magma stresses at Stardalur Volcano, Iceland. *J. Volcanol. Geotherm. Res.* **2007**, *161*, 131–150. [CrossRef]

112. Tibaldi, A.; Vezzoli, L.; Pasquaré Mariotto, F.; Rust, D. Strike-slip fault tectonics and the emplacement of sheetelaccolith systems: The Thverfell case study (SW Iceland). *J. Struct. Geol.* **2008**, *30*, 274–290. [CrossRef]

113. Tibaldi, A.; Pasquaré Mariotto, F. A new mode of inner volcano growth: The "flower intrusive structure". *Earth Planet. Sci. Lett.* **2008**, *271*, 202–208. [CrossRef]

114. Tibaldi, A.; Pasquaré Mariotto, F.; Rust, D. New insights into the cone–sheet structure of the Cuillin Complex, Isle of Skye, Scotland. *J. Geol. Soc.* **2011**, *168*, 689–704. [CrossRef]

115. Bistacchi, A.; Tibaldi, A.; Pasquaré Mariotto, F.; Rust, D. The association of cone-sheets and radial dykes: Data from the Isle of Skye (UK), numerical modelling, and implications for shallow magma chambers. *Earth Planet Sci. Lett.* **2012**, *339–340*, 46–56. [CrossRef]

116. Gianelli, G. Cumulus textures of the Midhyrna layered intrusion. *Boll. Soc. Geol. Ital.* **1972**, *91*, 419–438.

117. Hjartardóttir, A.R.; Einarsson, P.; Bramham, E.; Wright, T.J. The Krafla fissure swarm, Iceland, and its formation by rifting events. *Bull. Volcanol.* **2012**, *74*, 2139–2153. [CrossRef]

118. Thordarson, T.; Larsen, G. Volcanism in Iceland in historical time: Volcano types, eruption styles and eruptive history. *J. Geodyn.* **2007**, *43*, 118–152. [CrossRef]

119. Kubalíková, L. Assessing geotourism resources on a local level: A case study from southern Moravia (Czech Republic). *Resources* **2019**, *8*, 150. [CrossRef]

120. Bonali, F.L.; Tibaldi, A.; Russo, E.; Pasquaré Mariotto, F.; Marchese, F.; Fallati, L.; Savini, A.; Vitello, F.; Becciani, U.; Sciacca, E. Immersive virtual reality for studying volcano-tectonic features: A case study from the northern active rift zone of Iceland. In *Geophysical Research Abstracts*; EGU2019-1159. EGU General Assembly 2019; EGU: Vienna, Austria, 2019; Volume 21.

121. Bonali, F.L.; Russo, E.; Savini, A.; Marchese, F.; Fallati, L.; Nomikou, P.; Varvara, A.; Kyriaki, D.; Di Mauro, B.; Colombo, R.; et al. Learning outcomes from the EGU 2018 Public Engagement grant "Shaping geological 3D virtual field-surveys for overcoming motor disabilities". In *Geophysical Research Abstracts*; EGU2019-8118, EGU General Assembly 2019; EGU: Vienna, Austria, 2019; Volume 21.

122. Sketchfab. Available online: https://skfb.ly/6PKqL (accessed on 8 January 2020).
123. Sketchfab. Available online: https://skfb.ly/6PKqR (accessed on 8 January 2020).
124. Sketchfab. Available online: https://skfb.ly/6PK7U (accessed on 9 January 2020).
125. Sketchfab. Available online: https://skfb.ly/6PKqx (accessed on 9 January 2020).
126. Sketchfab. Available online: https://skfb.ly/6PK7G (accessed on 11 January 2020).
127. Sketchfab. Available online: https://skfb.ly/6PJIM (accessed on 11 January 2020).

 *resources*

Article

MoGeo, a Mobile Application to Promote Geotourism in Molise Region (Southern Italy)

Francesca Filocamo, Gianluigi Di Paola *, Lino Mastrobuono and Carmen M. Rosskopf

Department of Biosciences and Territory, University of Molise, Contrada Fonte Lappone, 86090 Pesche (IS), Italy;
francesca.filocamo@gmail.com (F.F.); lino.mastrobuono@ordinegeologimolise.it (L.M.);
rosskopf@unimol.it (C.M.R.)
* Correspondence: gianluigi.dipaola@unimol.it; Tel.: +39-0874-404168

Received: 18 February 2020; Accepted: 9 March 2020; Published: 12 March 2020

Abstract: Geotourism represents a powerful and new form of sustainable tourism that has rapidly expanded worldwide over the last decades. To promote it, the use of digital and geomatic tools is becoming of increasing importance. Especially mobile information represents one of the most efficient and smart ways to bring geotourism closer to a wide audience. This applies in particular to rural and inner areas, where the exploitation of geoheritage can represent a crucial resource for eco-friendly and sustainable tourism development. With the aim to promote geotourism on a regional scale, we have implemented a mobile devise application for Molise region, tested in the Alto Molise area. This application, called MoGeo App, aims at providing diversified geotourism information that combines geologic attractions (geosites and geologic itineraries) with other possible tourist attractions (other sites of natural and cultural interest), to respond to differentiated interests and needs of a wide audience. Besides geotourism purposes, the structure of MoGeo App can be used also for other purposes such as educational targets, by adapting contents and language. It appears to be a flexible, easily updatable digital tool, adaptable to various target groups, as well as other regional contexts, both inside and outside of Italy.

Keywords: geology-based tourism; geosites; geoheritage; cultural heritage; web-GIS; smartphone; Alto Molise

---

1. Introduction

Geotourism, understood as a form of tourism that specifically focuses on geology and landscape [1–6], represents a powerful and relatively new form of sustainable tourism [7,8]. It has rapidly expanded over the last decades [4–9] all around the world [5,7,10] and become a substantial part of the overall tourism offer [4], as well as an important research direction (e.g., [6,10]).

Geotourism focuses on geoheritage and therefore on geosites that are the most essential part of it [5,9,11–13]. It represents an important alternative or integration to more traditional forms of tourism, such as sun and sand tourism, and cultural tourism. Furthermore, it can become an important economic resource for countries and regions that are characterized by a rich natural heritage and great geodiversity (e.g., [8,14–18]). Especially in rural and inner areas, the exploitation of geoheritage can represent a crucial resource for eco-friendly and sustainable tourism development. Here, traditional and mass tourist destinations are generally scarce or lacking, and major tourist attractions are typically related to the geodiversity and naturalness of the landscape [19–23]. It is in such areas that the geo-landscape and the related geodiversity, biodiversity and cultural values [5,24,25] become important drivers for the local and regional economy.

Geotourism, both in a pure sense and characterized by the integrated fruition of geological sites and other places of interest (natural, historical, archaeological, etc.) [20,26,27], can be of high interest for various target groups [23] and especially for families. Families, in fact, must often turn to tourist forms and offers that take into account differentiated interests and needs of family members, including simple enjoyment and relaxation, as well as the desire to receive stimuli to learn something about and better appreciate the natural and cultural landscape [28–30].

The promotion, organization and exploitation of tourist offers now make increasing use of digital information sources (especially tourism websites) [31–33], not only in relation to the choice of tourist offer, travel and accommodation organisation, etc., but also in relation to the description and illustration of tourist attractions. This is particularly true also for geotourism [33]. Especially in the field of geoheritage conservation, management and promotion, the use of internet, digital and geomatic tools is becoming of increasing importance [33–37], even if geotourism is still being promoted little online [33]. Among such digital tools, especially the use of mobile phone devises and related applications to receive tourist information has experienced very recently a wide and rapid diffusion [38–41]. However, despite of the elevated potential and specific strengths of digital mobile tools (easy to transport, multi-sensorial, etc.) [42], there is still little use in the field of geotourism promotion and tour guiding [33,42–45]. Especially the scarce diffusion of app-based mobile tour guides (AMTG) [46] is surely at least partially due to the possible limitations related to the use of mobile phone applications [44], which must be carefully considered and best reduced. While keeping in mind such limitations, it is clear that mobile phones are already or will quickly be the most important interface between visitors/tourists and tourism contexts.

In geotorism field [34], three main types of applications can be distinguished: (1) applications that are based on georeferencing and mapping of geotourism assets, taking advantage in particular of recent developments in web mapping and mobile data access of maps (e.g., [42]); (2) applications that return 3D models based on photogrammetry, laser scanning or real-time observations of natural phenomena through a webcam (e.g., [47]); (3) applications that make interpretations using Augmented Reality (AR), a process that enriches discovery through digital media or virtual reality technologies creating a virtual universe that helps to imagine everything (e.g., [48]). These typologies can also be combined among them and coexist together [34].

Convinced that mobile apps can strongly support the promotion of geotourism, especially in rural and inner, less urbanized areas, we have implemented a mobile phone application that is illustrated in this paper. This application refers to the first type and aims at providing diversified "geotourism" information that includes not only the geologic attractions (geosites and geologic itineraries) but also other possible tourist attractions (other sites of interest) to respond in this way to different interests and needs of users, especially of families.

The application we propose should operate on a regional scale or on smaller areas. To develop it, we have chosen the Molise region. This region offers on a relative small area a representative view of the major geological-geomorphological and landscape features that typically characterize the central-southern Apennines [18]. It is characterized by a high geodiversity, and a regional inventory of geosites is already available [49]. However, the divulgation of geosites in the Molise region is only scarcely developed and restricted to some notions about geosites in leaflets directly distributed by the regional "Service for Tourist Promotion and Relations with Molisans in the World", and the online information provided by some institutional websites [50–52]. Regarding the promotion of geotourism at the regional or sub-regional scale, specific products are totally lacking. Some associations, essentially the ones promoting respectively the rocky spurs, so-called Morge, in Molise region, the Guardiaregia-Campochiaro Oasis and the Collemeluccio-Montedimezzo Alto Molise Biosphere Reserve [53–55], provide only short information about some geosites and only in Italian, and/or promote trekking activities and itineraries that include some of them. Besides this, some scientific and popular publications (e.g., [18,20,21,56,57]) have been produced with the aim of promoting geosites and geotouristic itineraries of specific areas of the Molise region, such as the Matese area, the Mt. Mainarde-Alto Volturno area and the Alto Molise area.

To start our project, we have selected the Alto Molise area (AM, Figure 1a), one of the seven major physiographic units in which the Molise Region has been subdivided mainly based on geological and orographic-hydrographic characteristics [21,49,58]. The main reasons for this choice are: the Alto Molise area has a small surface area, but is characterized by (i) a geological-geomorphological context that is representative for the history of the central-southern Apennines [18], and (ii) a high geodiversity and a significant number of geosites of different scientific interest. The Alto Molise area is also rich in natural protected areas, as well as important architectural, historical and archaeological sites, and retains important traces of ancient agro-pastoral traditions and crafts. In fact, the data collected during the regional geosite inventory activities [49] and several studies carried out to promote its geotouristic exploitation [18,20,57] have allowed to point out also its geographical features and traditions, as well as historical, archaeological and faunistic-floristic aspects, highlighting its rich natural and cultural heritage [57]. Therefore, the knowledge already available makes the Alto Molise area a good starting point to develop the application.

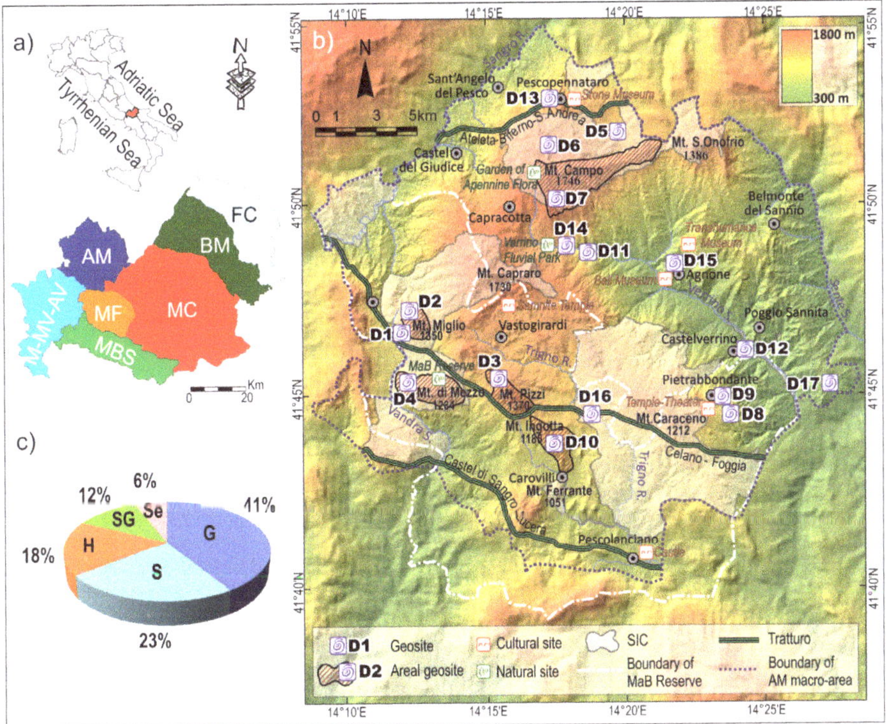

Figure 1. (**a**) The Molise region and its subdivision in seven macro-areas. M-MV-AV = Mainarde-Monti di Venafro-Alto Volturno, MBS = Matese-Conca di Boiano-Sepino, MF = Montagnola di Frosolone, AM = Alto Molise, MC = Molise Centrale, BM = Basso Molise, FC = Fascia Costiera; (**b**) Altitude and drainage network map of the Alto Molise macro-area with the location of the 17 assessed geosites and the other sites of interest; (**c**) Pie chart illustrating the assessed types of geosites and related percentages, based on primary scientific interests (G = Geomorphology, S = Stratigraphy, SG = Structural Geology, Se = Sedimentology, H = Hydrogeology).

2. The Test Area

The Alto Molise area (AM, Figure 1a) has a surface area of 452 km^2 and is located in the northwestern sector of the Molise region [20,21]. It is a macro-area particularly rich in geosites, in total 17 (Figure 1b, Table 1) of the 100 assessed through the regional geosite inventory carried out by the Department of Biosciences and Territory of the University of Molise in partnership and on behalf of the Molise Region [18,49]. During this inventory project, the Molise geosites have been assessed by using a quantitative method that allowed to determine their representativeness, rarity, the scenic-aesthetic, historical-archeological-cultural, and vulnerability values. Based on the assessed values, the used method allowed to calculate for each geosite the so-called "Intrinsic Value of the Site of Geological Interest" (IVSGI) [18], corresponding to the weighted sum of representativeness, rarity, and scenic-aesthetic values.

Table 1. Code, name, main scientific interests, and estimated relevance of the Alto Molise geosites.

ID	Name	Scientific Interests	Relevance
D1	Capo di Vandra spring	Hydrogeology	Regional
D2	Mt. Miglio monocline	Geomorphology	Regional
D3	Mt. Pizzi deep-seated gravitational slope deformation	Geomorphology	Regional
D4	Mt. di Mezzo morphostructure	Geomorphology	Regional
D5	Rio Verde springs	Hydrogeology	Regional
D6	Rocky ledge of the S. Luca Hermitage	Geomorphology	Regional
D7	Mt. Campo –Mt. S. Nicola monocline	Geomorphology Structural geology	National
D8	I Colli Agnone Flysch outcrop	Stratigraphy Sedimentology	Regional
D9	Morge of Pietrabbondante	Stratigraphy Geomorphology	Regional
D10	Mt. Ingotta anticline	Structural geology Geomorphology Paleontology	Regional
D11	Verrino Stream spring	Hydrogeology	Local
D12	Agnone Flysch outcrop of Castelverrino village	Stratigraphy Sedimentology	Regional
D13	Pescopennataro fault planes	Structural geology	National
D14	Verrino Stream waterfalls	Geomorphology	Regional
D15	Cogoli walls of Agnone village	Stratigraphy Sedimentology	Regional
D16	Vomero Resurgence	Geomorphology	Regional
D17	Limestones with pyrite nodules of Reggia Guardia	Sedimentology	Regional

The Alto Molise geosites, based on their primary scientific interest, refer mainly to the geomorphology or stratigraphy type (Figure 1c), although many of them have multiple scientific interests (for example Geomorphology-Structural geology, Table 1). The estimated relevance of the Alto Molise geosites [57] is mostly regional, national or local, instead, for three of them (Table 1). Most of the Alto Molise geosites (14) are already included in the Italian Geosites Inventory managed by ISPRA [52].

The Alto Molise area is characterized by a mainly mountainous and hilly landscape, whose major peaks are Mt. Campo (1746 m) and Mt. Capraro (1730 m) (Figure 1b), and by a few low-lying flat areas, such as the valley floors of the Trigno and Sangro rivers, which are the major watercourses in this area (Figure 2). Most of the outcropping bedrock is part of a sedimentary succession referred to the Molise pelagic basin domain, Oligocene to Late Miocene in age, which is interposed between the Apennine [59] and the Apulia carbonate platforms [60,61]. This succession is mainly composed of four geolithological units (see Figure 2): Varicolored clays and marls (10), Limestones and marls (9), Clayey marls and limestones (7) and Siliciclastic deposits (6). Units 7 and 9 form the main mountain ridges, and Unit 10 the hilly areas surrounding the mountain ridges. Finally, Unit 6 crops out widespread in the valley incisions of the Sangro River and Verrino Stream (Figure 2).

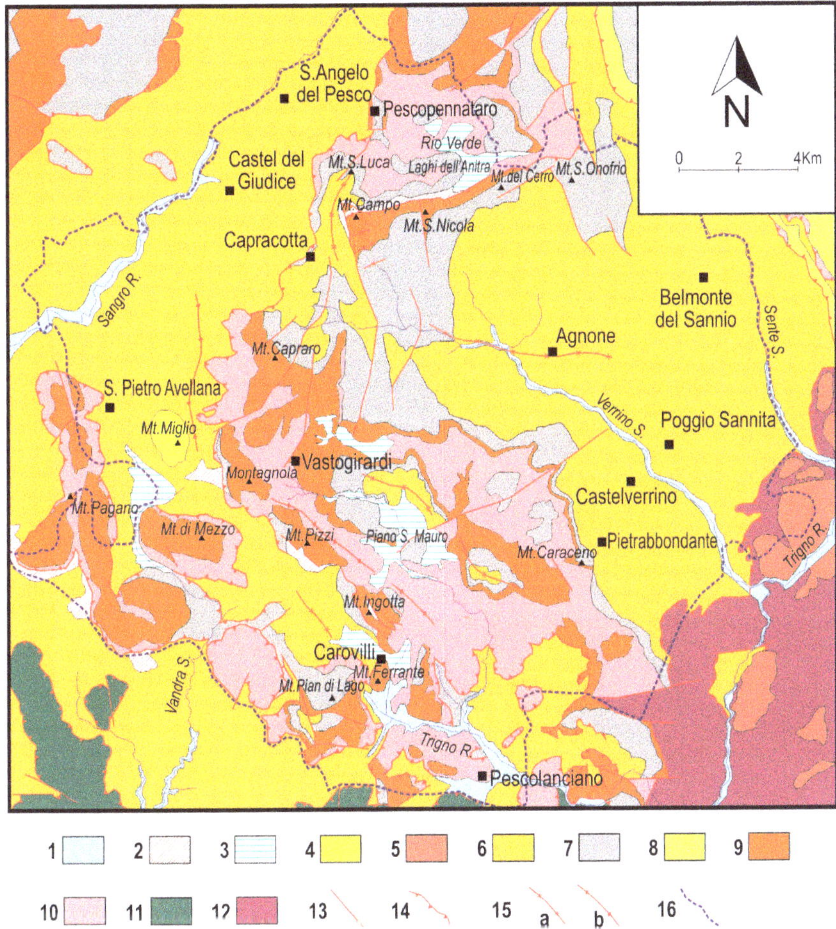

Figure 2. Geological sketch map of the Alto Molise area (modified from Vezzani et al. [62]). (1) Fluvial deposits (Holocene), (2) Slope debris (Late Pleistocene-Holocene), (3) Fluvial-palustrine deposits (Late Pleistocene-Holocene), (4) Limestones and polygenic conglomerates (Early-Middle Pliocene), (5) Siliciclastic deposits (S. Bartolomeo Flysch, Middle-Late Miocene), (6) Siliciclastic deposits (Agnone Flysch, Late Miocene); (7) Clayey marls and limestones (Marne ad Orbulina Formation, Late Miocene), (8) Marly limestones, marls and limestones (Tufillo Formation, Middle–Late Miocene), (9) Limestones and marls (Gamberale-Pizzoferrato Formation, Middle Miocene), (10) Varicolored clays and marls (Oligocene-Early Miocene), (11) Limestones and marly limestones (Frosolone Units, Late Cretaceous–Late Miocene), (12) Clays, marly clays and limestones (Sannio Units, Late Cretaceous–Early Miocene), (13) main faults, (14) main thrusts, (15) folds: a. syncline, b. anticline, 16) boundary of the Alto Molise macro-area.

The actual geological-structural setting of the Alto Molise area is the result of tectonics that acted from Late Miocene onwards. From the Messinian to the Middle Pliocene [63,64], the area was involved in thrusting that led to the tectonic juxtaposition of the Oligocene-Miocene stratigraphic units of the Molise basin on the Late Miocene Agnone Flysch. Then, from Late Pliocene to Early Pleistocene [61,63,64], the compressive structures were cut by strike slip and normal faults that acted from the Middle Pleistocene onwards according to a NE-SW direction of maximum extension.

As a result of this complex geological and tectonic history, the Alto Molise landscape is strongly dominated by structural landforms [20,57], especially monocline and anticline reliefs, often markedly asymmetrical, such as those forming Mt. Miglio, La Montagnola, Mt. Pizzi and Mt. Ingotta (Figure 2), typically aligned in the NW-SE direction. Slope processes and related landforms are widespread. Major phenomena are large rock falls with related talus slopes affecting the steep structural carbonate slopes (such as the one present along the western slope of Mt. Campo), together with complex landslides and phenomena of accelerated water erosion in the surrounding hilly areas. Where the tectonic juxtaposition of rigid carbonate rocks on plastic Miocene siliciclastic deposits has occurred, deep-seated gravitational deformations, as those affecting Mt. Pizzi, are also documented [65]. Furthermore, karst landforms are widespread where carbonate rocks crop out and are mainly represented by exokarst forms, such as karren and dolines, but also by some endokarst landforms, such as the Vomero Resurgence (Table 1).

From the bioclimatic point of view, the Alto Molise area is part of the temperate region [66], characterized by marked differences in winter and summer temperatures, precipitations concentrated in winter months, and summer aridity. Because of its climate conditions and geological-geomorphological features, the Alto Molise area is characterized by a high richness in fauna and flora, and related biodiversity. Its predominant forest vegetation is characterized by a high degree of naturalness [67], indicating that the evolution of the forest ecosystems is controlled especially by natural processes and only marginally influenced by human activities.

The high naturalistic value of the Alto Molise area is strengthened by the presence of numerous protected areas (Table 2) that occupy approximately 299.5 km^2, equal to 66% of its total surface area. Among these, a special mention deserves the Collemeluccio-Montedimezzo Alto Molise Unesco Man and Biosphere Reserve (Figure 1b), a large part of which falls in the Alto Molise territory.

Table 2. The Alto Molise natural protected areas.

Site Type	Code and/or Name of Protected Natural Area	Surface (km^2)
EUAP	Riserva Naturale Orientata Collemeluccio	4.22
	Riserva Naturale Orientata Montedimezzo	3.08
non EUAP	Foresta Demaniale Regionale Bosco Pennataro	3.45
	Foresta Demaniale Regionale Monte Capraro	1.95
	Foresta Demaniale Regionale Bosco S. Martino e Cantalupo	2.15
	Oasi Legambiente Selva Castiglione	3.00
ZPS	IT7221132 Monte di Mezzo	3.13
	IT7221131 Bosco di Collemeluccio	5.00
SIC	IT7218213 Isola della Fonte della Luna	8.67
	IT7218217 Bosco Vallazzuna	2.92
	IT7211120 Torrente Verrino	0.93
	IT7212133 Torrente Tirino (Forra)–M. Ferrante	1.45
	IT7218215 Abeti Soprani–M. Campo–M. Castelbarone–Sorgenti del Verde	30.33
	IT7212134 Bosco di Collemeluccio–Selvapiana–Castiglione–La Cocuzza	62.39
	IT7212124 Bosco di M. di Mezzo–M. Miglio–Pennataro–M. Capraro–M. Cavallerizzo	39.54
MaB	Collemeluccio-Montedimezzo Alto Molise Biosphere Reserve	252.68 [1]

[1] The total surface of the MaB Reserve is indicated, including the portion that falls outside the Alto Molise macro-area.

From the cultural point of view, the Alto Molise area offers various tourist attractions, first some archaeological sites, such as the Temple-theater complex of Pietrabbondante and the Sanctuary of Vastogirardi (Figure 1b). There are also several villages and small towns with nice historical centres, like Agnone, famous for its craftmanship of bell casting, and Capracotta, well known for its cross-country skiing area. Furthermore, this area hosts rich evidence of agro-pastoral traditions that have contributed to the shaping of its cultural landscape, represented per excellence by the *thòlos*, characteristic stone shelters used by shepherds, and the *tratturi* (Figure 1b), i.e., ancient pastoral transhumance paths, also called drove roads [68]. Since the last decades, the tratturi have become increasingly the subject of projects and studies [68,69] aimed at their recovery and fruition. Recently, they have been included in the national catalogue of historical rural landscapes [70,71], while the transhumance, the agropastoral practice of seasonal droving of livestock along migratory routes (tratturi) in the Mediterranean and in the Alps (Austria, Greece and Italy), was entered in 2019 in the representative list of the intangible cultural heritage of Humanity [72].

3. Materials and Methods

3.1. The Contents of the Application

The central contents of the proposed mobile application, hereinafter also called MoGeo App or simply App, are obviously the geosites and related geological itineraries.

The selection of geosites to be included in the App was based on the criteria safety, accessibility, scenic-aesthetic qualities, and interpretative potential that are in essence the selection criteria proposed by Brilha [73] for the qualitative assessment of sites suitable for geotourism use. Based on the data acquired during the Molise geosites inventory and other studies, the suitability of the 17 geosites was assessed by attributing a value for each criterion using scores (1—low; 3—medium; 5—high).

Safety and accessibility are considered indispensable criteria for geosite selection, to ensure their safe and unconditioned access and tourist use. For areal geosites, the accessibility and safety of both panoramic and on-site viewpoints were assessed.

The scenic-aesthetic qualities of geosites, i.e., the visual appeal and the natural beauty of a site, as well as the aesthetic qualities of the surrounding natural landscape, are considered of great importance to attract people to geosites [23,74–76]. Moreover, they can facilitate the interpretation of geosites, stimulate the curiosity of visitors and their desire to understand the geological-geomorphological features of geosites as well as the processes that underlie their genesis and evolution.

Despite the importance of aesthetics in tourism, according to Kirillova et al. [77], basic questions of tourist aesthetic judgment are still under-explored. These authors provide an important contribution to the understanding of "what makes a destination beautiful" by identifying and investigating nine themes and related dimensions of tourist aesthetic judgment in the context of both nature-based and urban tourist destinations. Another important aspect implicated in the aesthetic judgment is highlighted by Mikhailenko et al. [78] who propose a simple aesthetic-based classification of geological structures in outcrops based on their pattern, so as to take into account visions and attitudes of visitors and help evaluating the attractiveness of geosites.

To assess the scenic-aesthetic qualities of geosites, we have considered the following indicators: shape, vegetation, naturalness/anthropic modifications, chromatic variety and contrasts, and uniqueness. These indicators comprise some of the themes identified by Kirillova et al. [77], especially the shape. The latter, in particular, is closely related to the landscape setting and geological structure and includes the pattern sensu Mikhailenko et al. [78].

The interpretative potential of geosites is considered essential for disseminating geological information to non-geologists. It is closely related to the capacity of a geosite to be easily understood by law people and, therefore, to its representativeness, i.e., to its capacity to illustrate geological elements and processes.

The choice of other contents (Figure 3) to be included in MoGeo App was guided by the idea of enhancing and promoting not only geosites and related geo-itineraries, but also the overall naturalistic and cultural contexts of the selected areas. This is also in the awareness that the connection between natural and cultural heritage can represent a strength and a push factor for geotourism promotion, by offering richer and more varied experiences to visitors, who are perhaps not experts in geology or geology lovers, but simply families or lovers of landscape, nature and culture.

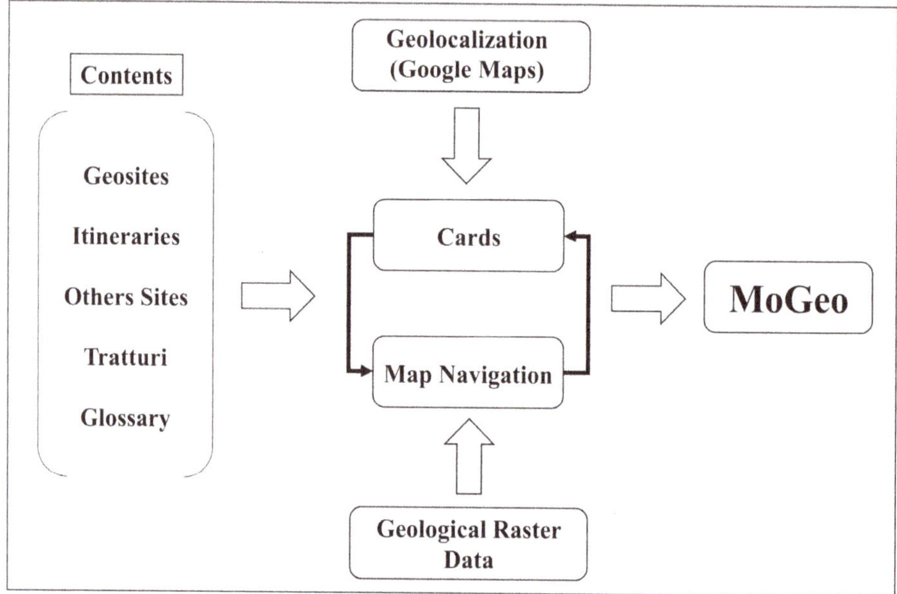

Figure 3. Flow Chart of the MoGeo App.

Therefore, the contents included in MoGeo App consist of:

- descriptive cards of selected geosites
- a glossary of scientific terms
- geological itineraries
- a simplified geo-lithological sketch map
- Other sites of cultural and/or naturalistic interest
- the tratturi

To allow the use of the App to both Italian and foreign tourists, all inserted titles and texts have been written in two languages, Italian and English.

For each selected geosite, a descriptive card was prepared that contains all essential information for understanding and for autonomously visiting the site. The inserted information was largely extracted from the Molise geosites inventory data archive [49,58] and other studies conducted about geosites/geotourism in the Alto Molise area [18,20,57].

Each card contains, in addition to the description of the geosite, information on the main geoscience interests. To select the main geoscience interests to insert in the cards, we did not simply consider the major scientific interests of geosites (Figure 1c and Table 1), but geological themes that could be of wider and perhaps environmental interest. For example, we considered the geoscience interest "Landscape instability" (see section results and Table 3), which is closely linked to the issues of natural hazard and risk and, in turn, to global issues such as the climate change. To ease the disclosure and interpretation of geosites, the descriptive cards were enriched with specific illustrative material (photos, sections, geological sketches, 3D schemes, etc.). Furthermore, to allow an optimal appreciation of each geosite, the best on-site and panoramic viewpoints are indicated.

Table 3. List of selected geosites. ID, scores obtained for each criterion, type of proposed observation points, main geoscience interests, and other natural and cultural interests. H = Hydrogeology, Se = Selective erosion, L = Landscape instability, Pc = Paleoclimate; St = Stratigraphy, Sed = Sedimentology, T = Tectonics, Pe = Paleoenvironment, Pa = Paleontology.

ID	S [1]	A [2]	SE [3]	IP [4]	O [5]	P [6]	MGI [7]	Other Interests
D1	5	3	3	5	X		H	Flora, vegetation and fauna
D2	5	5	5	5		X	Se–H	Flora, vegetation and fauna—Human history
D3	5	5	5	3	X	X	Se–L	Flora, vegetation and fauna
D4	5	5	5	5		X	Se–Pc	Flora, vegetation and fauna
D5	5	5	5	5	X		H	Flora, vegetation and fauna—Human history
D6	5	5	3	5	X		Se–St	Flora, vegetation and fauna—Human history
D7	5	5	5	5	X	X	Se–L–St–Sed	Flora, vegetation and fauna—Agro-pastoral tradition
D8	5	5	3	5	X	X	St–Sed	
D9	5	5	5	3	X	X	Se–St–Sed–T	
D10	5	3	5	5	X	X	Se–Pe–Pa	Flora, vegetation and fauna
D12	5	5	5	5	X	X	Se–St–Sed	
D13	5	5	5	5	X	X	Se–T–Pa	
D14	5	3	5	5	X	X	H	Flora, vegetation and fauna
D15	5	5	3	5	X	X	Se–St–Sed	Human history, Architecture and Handicraft

[1] Safety; [2] Accessibility; [3] Scenic-aesthetic qualities; [4] Interpretative potential; [5] On-site view; [6] Panoramic view; [7] Main geoscience interests.

We tried to use the simplest language possible to be clear even for non-geologists. However, to safeguard the scientific rigor, it was not possible to exclude certain scientific terms, so we included among the contents a glossary of scientific terms.

Considering that the use of geological itineraries and viewpoints is an important tool in geotourism activities [23,79], we have created also some itineraries (four at the moment). These itineraries allow the joint visit of different geosites and involve all selected geosites. Furthermore, to attract a large audience, they follow the main road system. For each itinerary, we have detailed the path and the sequence and location of stops. Based on logistic conditions and facilities (presence of parking areas, rest areas, etc.), the best on-site and panoramic viewpoints were selected. For all stops, descriptive cards were drawn up.

Panoramic viewpoints become particularly important as they allow the observation of sites represented by large landforms, which can be best appreciated from a distance. The importance of panoramic viewpoints, especially where on-site views of geosites are not useful or available has led to conceptualize a specific category of geosites: the viewpoint geosites [80,81]. According to these authors, the location of panoramic viewpoints has to take into account not only the quality of the site view they allow (clarity of features, good light, good visibility, etc.), but also the environmental context surrounding the target, as well as the conditions and pattern of the standpoint.

In the choice of our panoramic viewpoints, also some of the aforementioned criteria for viewpoint geosites have been considered. So we selected standpoints that satisfy the following conditions: easily accessible (normally located along the main road), not placed in private properties, with good safety conditions, with null to minimum anthropogenic degradation, preferentially inserted in a context characterized by few human interventions and a medium to high degree of naturalness, not covered by dense vegetation, and allowing the view of geological/geomorphological features that in many cases stand out (for color contrasts, vertical elevation, etc.) from the surrounding landscape. Furthermore, where possible, we selected several panoramic points for the same geosite to allow the specific observation of different/separate elements of the geosite. The preferential location of panoramic viewpoints along main roads was also guided by the need to guarantee the mobile phone coverage.

To best explain in the application the geosites from these panoramic views, we prepared panoramic photos clearly visible at the screen size, partially "retouched" to better put in evidence the features to be observed, and simplified sketches to illustrate the geological setting and the relief features.

To sustain geological information given in the cards, we realized a geolithological sketch map in the GIS environment, mainly based on the geological data extracted from the Geological Map of

Molise in scale 1:100,000 [62]. Data on tectonics features, which may be excessively complex for an audience of non-geologists, were not included in the map.

The other sites included in the App were selected by considering not only the best known cultural and natural sites/areas of Alto Molise, but also lesser known sites, to give a comprehensive overview of landscape, flora, vegetation and fauna, traditions, history, and archaeology of this territory. A simple and concise information card was produced for each of these sites containing, where available, the link to the official webpage of the site for further information.

Finally, the traces of the three tratturi that cross the Alto Molise territory were inserted in the App, as they represent important and distinctive landscape elements. These traces were extracted from the GIS project implemented during the Molise regional geosites inventory [49]. A general presentation of this theme and a descriptive card for each tratturo were prepared.

3.2. The Implementation of the Application

To develop our App, we considered the following expected characteristics and performance requirements:

- The application should be of easy and linear use. Contents should be accessible both by following a drop-down menu and by starting from an interactive map;
- All sites and itinerary stops, as well as the position of the user, should be geolocalized on an interactive map;
- The application should be usable on any kind of Android device, such as smartphones with varying screen sizes and tablet computers;
- The application should be fast and should not require a high storage capacity of mobile devises;
- The application should be easily updatable and expandable, by adding features such as tools or other data.

MoGeo App has been designed as a hybrid mobile app [45], a combination of web and native mobile applications, in which the cartographic part interfaces directly with the host operating system (in our case Android), while the information cards are inserted on a remote responsive website, optimized for mobile. The web pages are loaded via WebView, a native component, and displayed on the device as if they were themselves native. The advantage of this approach is to view the contents both from MoGeo App and directly from the Web Browser by connecting to the specific website. Through the geolocalization, it is possible to know the position both of the user and the single site, the distance between them, and the shortest way to reach the site.

In detail, MoGeo App resides in two different virtual spaces:

- The cartographic part is developed through the Android Studio Software. This is a fully integrated development system, created and made freely available by Google [82] for the development of new applications. In particular, three different map types are used (Google street and satellite, and geological maps) to localize the geo-touristic data related to different information cards (Figure 3).
- The other part includes the cards containing the information collected for all contents (Figure 3). All information has been transcribed in HTML format by using a Content Management System (CMS), called WordPress, and has been stored on an Italian Web Platform that AlterVista [83] makes freely available for all users registered on this platform.

To avoid speed and performance reductions of the application, information cannot be stored on the mobile devise, but is simply uploaded by using an Internet connection.

4. Results

4.1. Characteristics of Selected Geosites

Fourteen of the 17 Alto Molise geosites have been included in the App (Table 3), based on scores obtained for selection criteria safety, accessibility, scenic-aesthetic qualities and interpretative potential.

Geosites D11, D16 and D17 (Table 1) were excluded due to their medium to low scores in safety and accessibility: geosite D11 due to medium scores both in safety and accessibility, and geosites D16 and D17 (respectively a karst cave whose access is restricted to experts with caving equipment and a landslide area, see Table 1) due to their low and medium scores respectively in accessibility and safety.

The 14 selected geosites have achieved a high score in safety as they can be observed in complete security conditions, with nil to minimum risk for visitors. Nine geosites can be appreciated by both on-site and panoramic views (Table 3), three and two, instead, only by on-site and panoramic views, respectively.

None of the selected geosites present any use limitation due to access permissions. Furthermore, most of the selected geosites have obtained a high score in accessibility because they do not have access difficulties and are easily reached by paved roads. The geosites Capo di Vandra spring and Verrino Stream waterfalls (D1 and D14 in Table 3) are characterized only by medium scores in accessibility, because the first site can only be reached by completing the last part of the route by foot or with an off-road vehicle, while the second site, which is located within the Verrino Fluvial Park, can only be reached on foot. In addition, a medium store in accessibility marks the geosite Mt. Ingotta anticline (D10 in Table 3) as it is necessary to walk a path on foot to reach the on-site view for observing in detail the fossiliferous strata.

All selected geosites, except two, got a high score in interpretative potential, as they can be easily understood by a large audience, in particular also by lay people without a geological background. The two geosites with medium scores in interpretative potential (D3 and D9, Table 3) were anyway selected because they are characterized by high scenic-aesthetic qualities.

Finally, most of the selected geosites have achieved a high score in scenic-aesthetic qualities, while four of them (D1, D6, D8 and D15) reached a medium score, but were selected because of their high interpretative potential.

The geoscience interests of selected geosites range between Hydrogeology and Landscape instability (Table 3), with the latter being the most represented. Other interests of geosites are Flora, vegetation and fauna, Agro-pastoral tradition, Human history, Architecture, and Handicraft (Table 3).

4.2. How to Access Information Using MoGeo App

MoGeo App, which is downloadable by using the link https://geositi.altervista.org/download or the QR code in Figure 4, offers a simple and rapid way to reach information about contents included. Once you open the Homepage (Step 1, Figure 4), information can be accessed through two separate ways: the drop-down menu and the interactive map.

By staying on the Homepage, a drop-down menu can be opened that allows consulting the list of contents (Step 2, Figure 4). By making a selection on one of these contents, for example geosites, the user is redirected to the relative list of geosites (Step 3, Figure 4). By making a further selection on the latter, it is possible to reach the page of the specific site. In addition, at the top of the map view there are three buttons that allow respectively to return to the Homepage, to open the legend and to switch to satellite mode.

Starting from the interactive map, information can be accessed through the sites/places of interest that are localized on it by means of classic Google Maps textures and indicated with different colors according to the type of content (Step 2, Figure 4). By clicking on one of them, a popup opens that contains first information about the site, such as a photo and the name (Step 3, Figure 4). By clicking on the latter, the user is redirected to the information card that contains the description and illustration (text, photos, schemes, etc.) prepared for the specific site (Step 4, Figure 4).

4.3. Contents of MoGeo App

The content Geosites includes a general presentation and the information cards prepared for the 14 geosites. The presentation card (Figure 5) provides basic notions about the meaning of the term geosite and a short information on the Alto Molise geosites. The information cards provide information about the origin, geological-geomorphological features, and main geoscience interests of the geosite, as well as about the age of rock formations involved. All cards are enriched with illustrations, especially photos (Figure 5), geological sections and 3D schemes. The latter have been included mainly to explain some specific geological features such as various types of faults (Figure 5).

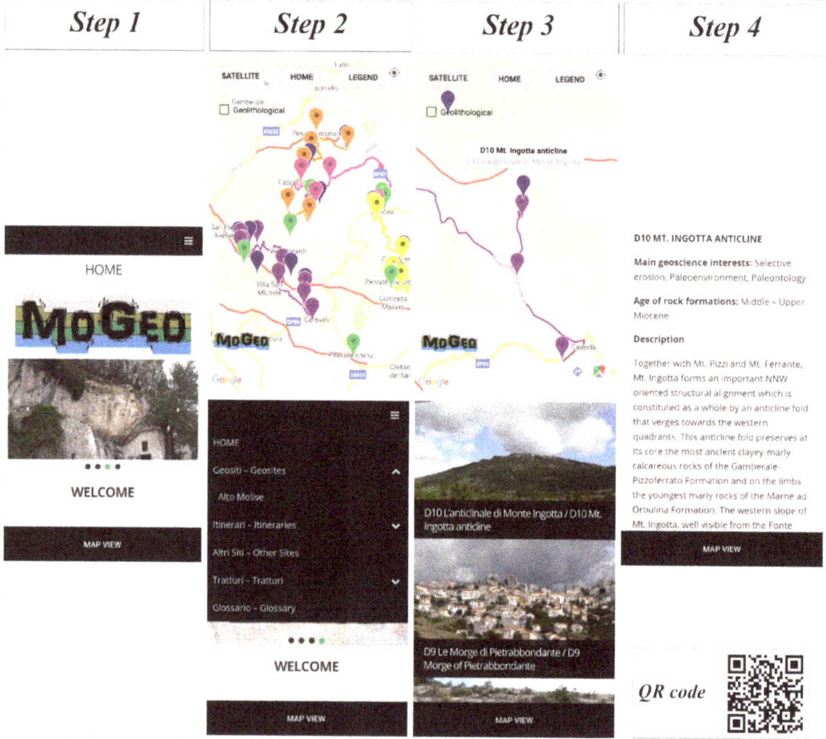

Figure 4. The main structure of the MoGeo App. Step 1: the Homepage that addresses to the two types of data access; Step 2: the interactive map and the drop-down menu access; Step 3: the opening of a pop-up after clicking a site on the interactive map, and the opening of a specific list of sites by selecting a single content from the drop-down menu; Step 4: the information card of a geosite. The figure contains also the QR code to download the MoGeo App.

The content Itineraries give access to the list of created geological itineraries (I1–I4, Table 4 and Figure 6), a card that provides a short information on each itinerary, and the descriptive cards that illustrate each single stop. Two of these itineraries (I1 and I2) mainly develop in the northern sector of the Alto Molise area, the other two (I3 and I4) mainly in the southern sector. They are made of a variable number of stops, from a minimum of 5 (I2, Table 4) to a maximum of 9 stops (I4, Table 4), and embrace several geoscience interests (Table 4) that allow visitors to deepen certain geological topics. Stops are mostly very easy to access, as the itineraries run largely along the main roads. Walking paths are needed only in some cases, precisely to reach the first stop of I2 itinerary, located within the Verrino Fluvial Park, and the third and ninth stops of I4 itinerary (Table 4). Itineraries I1, I2 and I4 partly cross protected natural areas (Table 4).

Figure 5. Sequence of screenshots extracted from MoGeo App that illustrate various aspects of the content Geosites.

Table 4. Main characteristics of the created four itineraries.

Itinerary	Stop	Geosite	O [1]	P [2]	Geoscience Interests	Route	Natural Areas
I1 Capracotta- Rio Verde Springs	1	D7		X	Selective erosion Tectonics Landscape instability Paleoenvironment Hydrogeology	Main roads	IT7218215 SIC
	2	D7		X			
	3	D7	X	X			
	4	D6	X				
	5	D13	X				
	6	D5	X				
I2 Capracotta- Agnone	1	D14	X		Selective erosion Landscape instability Hydrogeology Stratigraphy Sedimentology	Path Main roads	Verrino Fluvial Park
	2	D7	X				
	3	D7		X			
	4	D7		X			
	5	D15	X				
I3 Pietrabbondante- Agnone	1	D9		X	Stratigraphy Sedimentology	Main roads	
	2	D9	X				
	3	D8	X				
	4	D12	X				
	5	D15		X			
	6	D15	X				
I4 Carovilli- Capo di Vandra	1	D10		X	Selective erosion Landscape instability Paleoclimate Paleoenvironment Hydrogeology	Main roads Paths	MAB Reserve Collemeluccio-Montedimezzo Alto Molise
		D3		X			
	2	D10		X			
	3	D10	X				
	4	D3		X			
	5	D3		X			
	6	D2		X			
	7	D2		X			
	8	D4		X			
	9	D1	X				

[1] On-site view; [2] Panoramic view.

Regarding the content Other sites (Figure 7), we have included for now nine sites, three of naturalistic and six of cultural interest (Figure 1), which are well distributed throughout the Alto Molise territory. The sites of naturalistic interest are the Garden of Apennine flora of Capracotta, the Collemeluccio–Montedimezzo Alto Molise MaB Reserve and the Verrino Fluvial Park. The sites of cultural interest are the Samnite Temple of Vastogirardi, the Temple-theater complex of Pietrabbondante, the Bell Museum of Agnone, the Stone Museum of Pescopennataro, the Castel of Pescolanciano, and the Museum of Transhumance of Agnone. Information cards illustrate with a synthetic text and some photos major features of each site (Figure 7).

Regarding the content Tratturi, all three major drove roads that cross the Alto Molise territory, the tratturi Ateleta–Biferno, Castel di Sangro–Lucera and Celano–Foggia (Figure 7), were included in our App. The traces of the first are visible in the northern sector of the Alto Molise area, while the traces of the other two are preserved in the southern sector. Also for the tratturi, in addition to the single information cards (Figure 7), a general presentation card was compiled. This card contains information on the transhumance and related paths, i.e., the tratturi, and allows to have an overview about this ancient agro-pastoral practice and all major tratturi that cross the Molise territory.

Figure 6. Sequence of screenshots extracted from MoGeo App that illustrate various aspects of the content Itineraries.

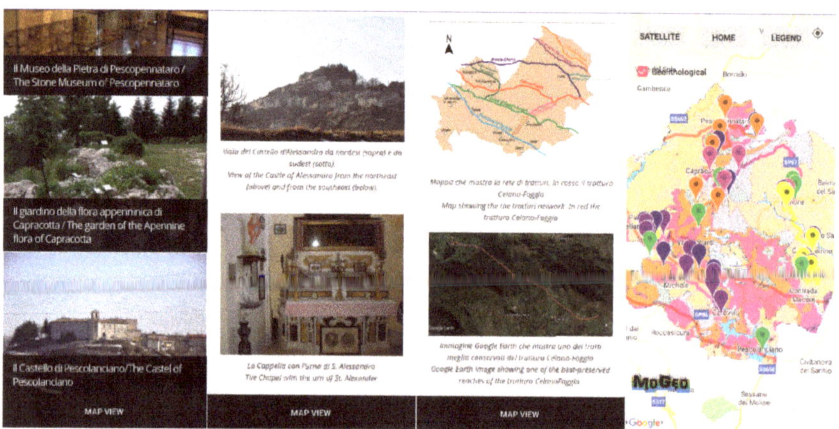

Figure 7. Sequence of screenshots extracted from MoGeo App that illustrate various aspects related to contents Other sites, Tratturi and Geo-lithological map.

Two further products were included in MoGeo App: A geolithological map and a glossary of scientific terms. The geolithological map (Figure 7) allows visualizing each site in its specific geological context, contributing to a better understanding of the geological information provided by the cards. The glossary of scientific terms, instead, can help not only to better understand information provided by cards for geosites, but also represents a useful tool for tourists who wish to deepen their knowledge about geological elements/topics. It allows easy research on specific scientific terms included in the information cards.

5. Discussion and Conclusions

MoGeo App represents a smart tool that allows the access to specific geotouristic information about sites/places of geological, natural and cultural interest geolocalized on Google earth maps and satellite views, by using a small mobile device, such as a smartphone.

It offers a simple and rapid way to reach information and, thanks to its hybrid nature, can be defined as a user-friendly tool, because of the velocity with which it elaborates the commands of the user and the accuracy of its geolocation functionality. In fact, this App has been developed considering that not all users have sophisticated and high-performing mobile devises and/or high knowledge and manual skills in using their smartphone. Obviously, some difficulty or limitation can arise from the diversity of mobile devices and related graphic performance capabilities.

In addition, MoGeo App is based on an archive of information that can be updated, modified and enriched with new information and cartographic tools. In this case, the only limitation of the information that can be accessed is related to the computing power of the device.

The developed application is suitable for guiding individual visits (perhaps even involving only single sites) and tours that are selected among the offered itineraries or developed by users according to their specific geo-naturalistic and cultural interests and time available. Therefore, it is able to meet the needs and preferences of a large audience, including families as well as hikers, amateur geologists and admirers of natural beauty, encouraging them to visit these places often forgotten by the media.

Besides its use during visits/excursions, MoGeo App can also be useful for analyzing preventively the areas of interest, especially for defining the best road connections and acquiring information about the geographical and geological contexts that characterize the area and the individual sites that can be visited. In Italy, there are some examples of applications created for geotourism purposes in other regional areas (e.g., [84,85]). By comparing MoGeo App with other applications, it is possible to observe some differences. One of the main differences is that MoGeo allows to process together different types of information, organized in separate categories, not just geosite information. In fact, great importance assumes the possibility of accessing with a single application different contents (geological, natural, cultural, etc.) that can be geolocated on interactive maps, by using mapping service portals such as Google Maps. Also the drop-down menu appears extremely useful, as it allows to reach the different contents without using the map. In this way, the individual user can choose the way to receive information according to his needs.

Other strengths of this application are its light and smart structure. In particular, the information is acquired directly remotely without downloading data to the smartphone. This allows to avoid speed and performance reductions of the application.

However, MoGeo App also has some limitations. Being developed for smartphones/tablets, due to the reduced size of the screens, the scale of images cannot be of elevated detail and resolution. Another limitation is that, for now, MoGeo does not include the use of Augmented Reality (AR), which could be developed in the future.

To create our application, we chose to use only the Android operating system, because it is the most widespread in Europe and in Italy (in December 2019, 81.4% of Italian smartphone users are Android [86]) and provides for free both the possibility to publish the application on its store and to use the API (Application Programming Interface) to connect to the tools of Google Maps. However, in the future we would like to create a version for IOS, so as to reach all smartphone users.

During the phase of design, MoGeo App has been tested remotely and directly in the field on a group of a dozen people without geological background to get a first quick feedback on the ease of use and efficiency of the application, as well as on the curiosity towards and liking of proposed contents. By questioning the group used as for the test, it was possible to verify a large appreciation of contents and a positive judgment about the lightness of the application (~3 Mb), which allows a quick download on the mobile phone, even on less technologically advanced smartphones. Furthermore, the logic and structure of the application resulted in being intuitive for everyone. In particular, no one had difficulties in finding information both on the geosites and the territory in general, managing to create a personal visit itinerary. There was a widespread appreciation of the cards for the clarity and simplicity of contents. The possibility of a combined consultation of different contents (cards) was also judged positively. In the field, the application works quite well and the function that allows to reach a given site using the Google Map tools incorporated in MoGeo has been particularly appreciated. Only in a few cases, reception problems were found due to the local weakness or lack of the cell phone signal, above all in the high mountain areas. Wind and Iliad network users had more problems than TIM and Vodafone network users.

Further surveys on the reception and use of MoGeo App can certainly help improve it and are scheduled. In the future, we intend to conduct more systematic tests involving a group of families, to better consider the needs and interests of this specific target group and try to shorten the distance between contents for experts and lay people. This will be done by collecting through questionnaires the interest, appreciation and/or criticism expressed by families, as well as the difficulties encountered in using the App and understanding of contents; in short, all the useful tips to improve the App.

The contents of MoGeo App are specifically designed for geotourism purposes, but its structure can be used for promoting and disseminating geoheritage to different target groups, by adapting the cards' content and the language. For example, contents for educational targets or dedicated only to children/teenagers (considering that digital tools help attract and engage this kind of audience) can be devised.

In conclusion, MoGeo App appears to be a flexible, updatable digital tool that can support different contents and can be adapted to various target groups and to other regional contexts, both inside and outside of Italy.

Download MoGeo App: https://geositi.altervista.org/download or QR code in Figure 4.

Author Contributions: Conceptualization, G.D.P., F.F. and C.M.R.; methodology, G.D.P., F.F., L.M. and C.M.R.; software, L.M.; validation, G.D.P., F.F.; investigation, F.F.; data curation, F.F. and C.M.R.; writing—original draft preparation, G.D.P., F.F., L.M. and C.M.R.; writing—review and editing, G.D.P. and C.M.R. All authors have read and agreed to the published version of the manuscript.

Funding: This research received no external funding.

Acknowledgments: The authors wish to thank the anonymous reviewers whose suggestions greatly improved the manuscript.

Conflicts of Interest: The authors declare no conflict of interest.

References

1. Pralong, J. Geotourism: A new form of tourism utilising natural landscapes and based on imagination and emotion. *Tour. Rev.* **2006**, *61*, 20–25. [CrossRef]
2. Dowling, R.K. Geotourism's global growth. *Geoheritage* **2010**, *3*, 1–13. [CrossRef]
3. Hose, T.A. 2012 Editorial: Geotourism and Geoconservation. *Geoheritage* **2012**, *4*, 1–5. [CrossRef]
4. Dowling, R.K. Global Geotourism—An emerging form of sustainable tourism. *Czech J. Tour.* **2013**, *2*, 59–79. [CrossRef]
5. Newsome, D.; Dowling, R. Geoheritage and Geotourism. In *Geoheritage: Assessment, Protection, and Management*; Reynard, E., Brilha, J., Eds.; Elsevier: Amsterdam, The Netherlands, 2018; pp. 305–321.
6. Ruban, D.A. Geotourism—A geographical review of the literature. *Tour. Manag. Perspect.* **2015**, *15*, 1–15. [CrossRef]
7. Newsome, D.; Dowling, R.; Leung, Y. The nature and management of geotourism: A case study of two established iconic geotourism destinations. *Tour. Manag. Perspect.* **2012**, *2*, 19–27. [CrossRef]
8. Dowling, R.; Newsome, D. *Handbook of Geotourism*; Edward Elgar Publishing Ltd.: Cheltenham, UK, 2018; p. 520.
9. Reynard, E. Scientific research and tourist promotion of geomorphological heritage. *Geogr. Fis. Dinam. Quat.* **2008**, *31*, 225–230.
10. Ólafsdóttir, R.; Tverijonaite, E. Geotourism: A Systematic Literature Review. *Geosciences* **2018**, *8*, 234. [CrossRef]

11. Bentivenga, M.; Cavalcante, F.; Mastronuzzi, G.; Palladino, G.; Prosser, G. Geoheritage: The Foundation for Sustainable Geotourism. *Geoheritage* **2019**, *11*, 1367–1369. [CrossRef]

12. Kubalíková, L. Geomorphosite assessment for geotourism purposes. *Czech J. Tour.* **2013**, *2*, 80–104. [CrossRef]

13. Pica, A.; Fredi, P.; Del Monte, M. The Ernici Mountains Geoheritage (Central Apennines, Italy): Assessment of the Geosites for Geotourism Development. *GeoJ. Tour. Geosites* **2014**, *14*, 193–206.

14. Errami, E.; Brocx, M.; Semeniuk, V. *From Geoheritage to Geoparks: Case Studies from Africa and Beyond*; Springer: Cham, Switzerland, 2015.

15. Osipova, E.; Shadie, P.; Zwahlen, C.; Osti, M.; Shi, Y.; Kormos, C.; Bertzky, B.; Murai, M.; Van Merm, R.; Badman, T. *IUCN World Heritage Outlook 2: A Conservation Assessment of all Natural World Heritage Sites*; IUCN: Gland, Switzerland, 2017; p. 92.

16. Marchetti, M.; Soldati, M.; Vandelli, V. The great diversity of Italian landscapes and landforms: Their origin and human imprint. In *Landscapes and Landforms of Italy*; Soldati, M., Marchetti, M., Eds.; Springer International Publishing: Cham, Switzerland, 2017; pp. 7–20.

17. Cappadonia, C.; Coratza, P.; Agnesi, V.; Soldati, M. Malta and Sicily joined by geoheritage enhancement and geotourism within the framework of land management and development. *Geosciences* **2018**, *8*, 253. [CrossRef]

18. Filocamo, F.; Rosskopf, C.M.; Amato, V. A contribution to the understanding of the Apennine landscapes: The potential role of Molise geosites. *Geoheritage* **2019**, *11*, 1667–1688. [CrossRef]

19. Farsani, N.T.; Coelho, C.O.A.; Costa, C.M.M. Rural geotourism: A new tourism product. *Acta Geoturistica* **2013**, *4*, 1–10.

20. Filocamo, F.; Rosskopf, C.M.; Amato, V.; Cesarano, M.; Di Paola, G. The integrated exploitation of the geological heritage: A proposal of geotourist itineraries in the Alto Molise area (Italy). *Rend. Online Soc. Geol. It.* **2015**, *33*, 44–47. [CrossRef]

21. Filocamo, F.; Rosskopf, C.M. The geological heritage for the promotion and enhancement of a territory. A proposal of geological itineraries in the Matese area (Molise, Southern Italy). *Rend. Online Soc. Geol. It.* **2019**, *49*, 142–148.

22. Forleo, M.B.; Giaccio, V.; Giannelli, A.; Mastronardi, L.; Palmieri, N. Socio-economic drivers, land cover changes and the dynamics of rural settlements: Mt. Matese Area (Italy). *Europ. Countrys.* **2017**, *9*, 435–457. [CrossRef]

23. Gordon, J.E. Geoheritage, geotourism and the cultural landscape: Enhancing the visitor experience and promoting geoconservation. *Geosciences* **2018**, *8*, 136. [CrossRef]

24. Reynard, E.; Giusti, C. The landscape and the cultural value of geoheritage. In *Geoheritage: Assessment, Protection, and Management*; Reynard, E., Brilha, J., Eds.; Elsevier: Amsterdam, The Netherlands, 2018; pp. 147–166.

25. Neches, I.M.; Erdeli, G. Geolandscapes and geotourism: Integrating nature and culture in the Bucegi Mountains of Romania. *Landsc. Res.* **2015**, *40*, 486–509. [CrossRef]

26. Cartojan, E.; Di Lisio, A.; Ferretta, C.; Magliulo, P.; Russo, F.; Sisto, M.; Valente, A. Esempi di aree di interesse geoturistico nel territorio Irpino-Sannita (Campania). *Supplemento a Geologia dell'Ambiente* **2011**, *2*, 388–400.

27. Del Monte, M.; Fredi, P.; Pica, A.; Vergari, F. Geosites within Rome city center (Italy): A mixture of cultural and geomorphological heritage. *Geogr. Fis. Din. Quat.* **2013**, *36*, 241–257.

28. Allan, M. Geotourism: Why do children visit geological tourism sites? *Dirasat Hum. Soc. Sci.* **2014**, *41*, 653–661. [CrossRef]

29. Allan, M.; Dowling, R.K.; Sander, D. The motivations for visiting geosites: The case of crystal cave, Western Australia. *GeoJ. Tour. Geosites* **2015**, *16*, 142–153.

30. Dowling, R.; Allan, M. What are geotourists? A case study from Jordan. In *Handbook of Geotourism*; Dowling, R.K., Newsome, D., Eds.; Edward Elgar Publishing: Cheltenham, UK, 2018; pp. 76–86.

31. Noti, E. The Websites of National Tourism Organizations—A challenge of e-marketing. *Angl. J.* **2013**, *2*, 241–249.

32. Bonjisse, B.; Morais, E. Evaluation of Tourism Websites. In *Education Excellence and Innovation Management trough Vision 2020: From Regional Development Sustainability to Global Economic Growth, Proceedings of 29th IBIMA Conference, Vienna, Austria, 3–4 May 2017*; Soliman, K.S., Ed.; IBIMA: King of Prussia, PA, USA, 2017; pp. 3685–3700.

33. Widawski, K.; Rozenkiewicz, A.; Łach, J.; Krzemińska, A.; Oleśniewicz, P. Geotourism starts with accessible information: The Internet as a promotional tool for the georesources of Lower Silesia. *Open Geosci.* **2018**, *10*, 275–288. [CrossRef]

34. Cayla, N. An overview of new technologies applied to the management of geoheritage. *Geoheritage* **2014**, *6*, 91–102. [CrossRef]

35. Ghiraldi, L.; Giordano, E.; Perotti, L.; Giardino, M. Digital tools for collection, promotion and visualisation of geoscientific data: Case study of Seguret Valley (Piemonte, NW Italy). *Geoheritage* **2014**, *6*, 103–112. [CrossRef]

36. Ravanel, L.; Bodin, X.; Deline, P. Using terrestrial laser scanning for the recognition and promotion of high-alpine geomorphosites. *Geoheritage* **2014**, *6*, 129–140. [CrossRef]

37. Cayla, N.; Martin, S. Digital geovisualisation technologies applied to geoheritage management. In *Geoheritage: Assessment, Protection and Management*; Reynard, E., Brilha, J., Eds.; Elsevier: Amsterdam, The Netherlands, 2018; pp. 69–86.

38. Da Silva, A.C.; Da Rocha, S.H.V. M-Traveling: Mobile Applications in Tourism. *Int. J. Infonomics (IJI)* **2012**, *5*, 618–630. [CrossRef]

39. Karanasios, S.; Burgess, S.; Sellitto, C. A classification of mobile tourism applications. In *Global Hospitality and Tourism Management Technologies*; Pablos, P.O., Tennyson, R., Zhao, J., Eds.; Business Science Reference: Hersey, PA, USA, 2011; pp. 165–177.

40. Dickinson, J.E.; Ghali, K.; Cherrett, T.; Speed, C.; Davies, N.; Norgate, S. Tourism and the smartphone app: Capabilities, emerging practice and scope in the travel domain. *Curr. Issues Tour.* **2014**, *17*, 84–101. [CrossRef]

41. Kim, D.; Kim, S. The role of mobile technology in tourism: Patents, Articles, news, and mobile tour app reviews. *Sustainability* **2017**, *9*, 2082. [CrossRef]

42. Pica, A.; Reynard, E.; Grangier, L.; Christian Kaiser, C.; Ghiraldi, L.; Perotti, L.; Del Monte, M. GeoGuides, urban geotourism offer powered by mobile application technology. *Geoheritage* **2018**, *10*, 311–326. [CrossRef]

43. Gambino, F.; Borghi, A.; d'Atri, A.; Gallo, L.M.; Ghiraldi, L.; Giardino, M.; Martire, L.; Palomba, M.; Perotti, L.; Macadam, J. TOURinSTONES: A free Mobile Application for promoting geological heritage in the city of Torino (NW Italy). *Geoheritage* **2019**, *11*, 3–17. [CrossRef]

44. Reynard, E.; Kaiser, C.; Martin, S.; Regolini, G. An application for geosciences communication by smartphones and tablets. In *Engineering Geology for Society and Territory*; Lollino, G., Giordan, D., Marunteanu, C., Christaras, B., Yoshinori, I., Margottini, C., Eds.; Springer: Heidelberg, Germany, 2015; Volume 8, pp. 265–268.

45. Kenteris, M.; Gavalas, D.; Economou, D. Electronic mobile guides: A survey. *Pers. Ubiquit. Comput.* **2011**, *15*, 97–111. [CrossRef]

46. Law, R.; Chan, I.; Wang, L. A comprehensive review of mobile technology use in hospitality and tourism. *J. Hosp. Mark. Manag.* **2018**, *27*, 626–648. [CrossRef]

47. Balzeau, A.; Crevecoeur, I.; Rougier, H.; Froment, A.; Gilissen, E.; Grimaud-Hervé, D.; Mennecier, P.; Semal, P. Applications of imaging methodologies to paleoanthropology: Beneficial results relating to the preservation, management and development of collections. *Comptes Rendus Palevol* **2010**, *9*, 265–275. [CrossRef]

48. Chu, T.H.; Liu, M.L.; Chang, C.H.; Chen, C.W. Using mobile geographic system techniques to develop a location-based tour guiding system based on user evaluation. *Int. J. Phys. Sci.* **2012**, *7*, 121–131.

49. Filocamo, F.; Maglieri, C.; Rosskopf, C.M.; Baranello, S.; Giannantonio, O.; Monaco, R.; Relvini, M.; Iarossi, M. Il censimento e la valorizzazione dei geositi: l'esperienza molisana. *Supplemento a Geologia dell'Ambiente* **2011**, *2*, 135–143.

50. Geositi del Molise. Available online: http://www.regione.molise.it/web/turismo/turismo.nsf/0/2A0005407239EB22C12578180032C505?OpenDocument (accessed on 26 February 2020).

51. I Geositi del Molise. Available online: http://www3.regione.molise.it/flex/cm/pages/ServeBLOB.php/L/IT/IDPagina/382 (accessed on 26 February 2020).

52. Inventario Nazionale dei Geositi. Available online: http://sgi.isprambiente.it/geositiweb/ (accessed on 28 December 2019).

53. Parco Delle Morge Cenozoiche del Molise. Le Morge del Parco. Available online: http://www.parcodellemorge.it/info.php?id=23&tit=Le-Morge-del-Parco (accessed on 26 February 2020).

54. Riserva Regionale Guardiaregia Campochiaro Oasi WWF. I Sentieri. Available online: https://oasiguardiaregia.wordpress.com/i-sentieri/ (accessed on 26 February 2020).

55. Guida Trekking & MTB. Riserva MAB Collemeluccio—Montedimezzo Alto Molise. Available online: http://www.riservamabaltomolise.it (accessed on 29 January 2020).

56. Filocamo, F.; Amato, V.; Rosskopf, C.M. L'itinerario "Le Mainarde—Alto Volturno": Un percorso geoturistico alla scoperta della geologia del settore molisano del Parco Nazionale d'Abruzzo, Lazio e Molise. *Mem. Descr. Carta Geol. d'It.* **2014**, *102*, 131–144.

57. Rosskopf, C.M.; Filocamo, F. *I Geositi Dell'alto Molise*; AGR Editrice—Arti Grafiche La Regione srl: Ripalimosani (CB), Italy, 2014; p. 167.

58. Rosskopf, C.M.; Di Paola, G.; Filocamo, F. Carta di Sintesi dei Geositi Molisani 2014. Available online: http://www3.regione.molise.it/flex/cm/pages/ServeBLOB.php/L/IT/IDPagina/382 (accessed on 28 December 2019).

59. Mostardini, E.; Merlini, S. Appennino centro meridionale. Sezioni geologiche e proposta di modello strutturale. *Mem. Soc. Geol. It.* **1986**, *35*, 177–202.

60. Corrado, S.; Di Bucci, D.; Naso, G.; Damiani, A.V. Rapporti tra le grandi unità stratigrafico-strutturali dell'Alto Molise (Appennino Centrale). *Boll. Soc. Geol. It.* **1988**, *117*, 761–776.

61. Di Bucci, D.; Corrado, S.; Naso, G.; Parotto, M.; Praturlon, A. Evoluzione tettonica neogenico-quaternaria dell'area molisana. *Boll. Soc. Geol. It.* **1999**, *118*, 13–30.

62. Vezzani, L.; Ghisetti, F.; Festa, A. *Carta Geologica del Molise (Scala 1:100.000)*; Selca: Firenze, Italy, 2004.

63. Patacca, E.; Scandone, P.; Bellatalla, M.; Perilli, N.; Santini, U. La zona di giunzione tra l'arco appenninico settentrionale e l'arco appenninico meridionale nell'Abruzzo e nel Molise. *Studi Geol. Carmenti—Volume Speciale* **1991**, *2*, 417–441.

64. Corrado, S.; Di Bucci, D.; Naso, G.; Butler, H.W.R. Thrusting and strike-slip tectonics in the Alto Molise region (Italy): Implications to the Neogene-Quaternary evolution of the central Apennines orogenic system. *J. Geol. Soc. Lond.* **1997**, *154*, 679–688. [CrossRef]

65. Corniello, A.; Santo, A. Geologia e fenomeni gravitativi profondi nell'area dell'alto corso del fiume Trigno. *Geol. Romana* **1994**, *30*, 67–74.

66. Blasi, C.; Michetti, L. Biodiversity and climate. In *Biodiversity in Italy. Contribution to the National Biodiversity Strategy*; Blasi, C., Boitani, L., La Posta, S., Manes, F., Marchetti, M., Eds.; Palombi Editori: Rome, Italy, 2005; pp. 57–66.

67. Marchetti, M.; Marino, D.; Cannata, G. *Relazione Sullo Stato Dell'ambiente Della Regione Molise*; Università degli Studi del Molise: Campobasso, Italy, 2008; p. 536.

68. Minotti, M.; Giancola, C.; Di Marzio, P.; Di Martino, P. Land use dynamics of drove roads: The case of tratturo castel di Sangro-Lucera (Molise, Italy). *Land* **2018**, *7*, 3. [CrossRef]

69. Meini, M.; Di Felice, G.; Petrella, M. Geotourism perspectives for transhumance routes. Analysis, requalification and virtual tools for the geoconservation management of the drove roads in Southern Italy. *Geosciences* **2018**, *8*, 368. [CrossRef]

70. Cardillo, A.; Finodi, A.F. Tratturi dell'Alto Molise/Sheep-tracks in the upper Molise. In *Paesaggi Rurali Storici. Per un Catalogo Nazionale*; Agnoletti, M., Ed.; Editori Laterza: Bari, Italy, 2011; pp. 420–422.

71. Catalogo Nazionale dei Paesaggi Rurali Storici. Available online: https://www.reterurale.it/flex/cm/pages/ServeBLOB.php/L/IT/IDPagina/14372 (accessed on 7 January 2020).

72. Transhumance, the Seasonal Droving of Livestock along Migratory Routes in the Mediterranean and in the Alps. Available online: https://ich.unesco.org/en/RL/transhumance-the-seasonal-droving-of-livestock-along-migratory-routes-in-the-mediterranean-and-in-the-alps-01470 (accessed on 8 February 2020).

73. Brilha, J. Geoheritage: Inventories and evaluation. In *Geoheritage: Assessment, Protection and Management*; Reynard, E., Brilha, J., Eds.; Elsevier: Amsterdam, The Netherlands, 2018; pp. 69–86.

74. Hose, T.A. The significance of aesthetic landscape appreciation to modern geotourism provision. In *Geotourism. The Tourism of Geology and Landscape*; Newsome, D., Dowling, R.K., Eds.; Goodfellow Publishers: Oxford, UK, 2010; pp. 13–25.

75. Migoń, P. Stories behind superlative scenery. *World Herit.* **2012**, *63*, 18–25.

76. Chylińska, D. The role of the picturesque in geotourism and iconic geotourist landscapes. *Geoheritage* **2019**, *11*, 531–543. [CrossRef]

77. Kirillova, K.; Fu, X.; Lehto, X.; Cai, L. What makes a destination beautiful? Dimensions of tourist aesthetic judgment. *Tour. Manag.* **2014**, *42*, 282–293. [CrossRef]

78. Mikhailenko, A.V.; Nazarenko, O.V.; Ruban, D.A.; Zayats, P.P. Aesthetics-based classification of geological structures in outcrops for geotourism purposes: A tentative proposal. *Geologos* **2017**, *23*, 45–52. [CrossRef]

79. Dowling, R.K.; Newsome, D. Setting an agenda for geotourism. In *Geotourism. The Tourism of Geology and Landscape*; Newsome, D., Dowling, R.K., Eds.; Goodfellow Publishers: Oxford, UK, 2010; pp. 1–12.
80. Migoń, P.; Pijet-Migoń, E. Viewpoint geosites-Values, conservation and management issues. *Proc. Geol. Assoc.* **2017**, *128*, 511–522. [CrossRef]
81. Mikhailenko, A.V.; Ruban, D.A. Environment of Viewpoint Geosites: Evidence from the Western Caucasus. *Land* **2019**, *8*, 93. [CrossRef]
82. Goolge. Available online: https://developer.android.com/studio/intro/ (accessed on 8 November 2019).
83. AlterVista. Available online: https://it.altervista.org/ (accessed on 8 November 2019).
84. GEOsiti Friuli Venezia Giulia. Available online: https://play.google.com/store/apps/details?id=mb.divulgando.com.geositifvg&hl=it (accessed on 29 February 2020).
85. Geositi Isola d'Elba. Available online: https://play.google.com/store/apps/details?id=it.netseven.geositielba&hl=it (accessed on 29 February 2020).
86. Kantar. Available online: https://www.kantarworldpanel.com/global/smartphone-os-market-share/ (accessed on 29 February 2020).

resources

Article

Geotourism and the 21st Century–NTOs' Website Information Availability on Geotourism Resources in Selected Central European Countries: International Perspective

Agnieszka Rozenkiewicz [1,*], Krzysztof Widawski [1] and Zdzisław Jary [2]

[1] Department of Regional Geography and Tourism, University of Wrocław, 32 W. Cybulski St.,
    50-205 Wrocław, Poland; krzysztof.widawski@uwr.edu.pl

[2] Department of Physical Geography, University of Wrocław, 34 W. Cybulski St., 50-205 Wrocław, Poland;
    zdzislaw.jary@uwr.edu.pl

* Correspondence: agnieszka.rozenkiewicz@uwr.edu.pl; Tel.: +48-71-375-95-75

Received: 10 November 2019; Accepted: 31 December 2019; Published: 6 January 2020

Abstract: The power of the Internet as a communicative and promotional tool in the contemporary world of tourism is unquestionable. Nevertheless, the context of online information availability referring to geotourism and georesources is very rarely addressed in the academic literature. This article undertakes research into the online information availability on georesources presented on the official websites of the National Tourism Organizations (NTOs) of three selected Central European countries with similar geotourism conditions, namely the Czech Republic, Poland, and Slovakia. Their NTOs underwent a descriptive content analysis in order to highlight the dominating trends in the online presentation of georesources. As concluded in the article, information on geotourism resources available online is rather dispersed, as it is usually presented under divergent umbrella terms. Therefore, measures need to be taken to present a holistic online picture of geoheritage on an international level of availability, where certain pieces of geotourism-related information correspond with each other, accurately applying the system of hyperlinks. The research outcomes and suggestions for the future may find applicable use for various stakeholders of the tourism industry, especially the authorities responsible for different levels of its promotion.

Keywords: geotourism; georesources; Internet

1. Introduction

1.1. The Power of the Internet in the 21st Century World of Tourism

One of the predominant tasks of National Tourism Organizations (NTOs) is the promotion of tourism in certain regions. This promotional activity applies diverse tools and techniques with participation in tourism fairs, media advertising, PR, promotional materials in the form of printed leaflets and catalogues, study tours for journalists, and bloggers' involvement, to name just a few. Their prevailing purpose is evoking the interest of both international and national tourist groups, influencing the tourists' decision-making process regarding the final choice of a particular tourism destination and the undertaken forms of tourism activities, through providing a vast array of choices in accommodation and amenities [1]. Nowadays, printed advertising materials are becoming obsolete [2]. It is the Internet that has gradually become the number one player in the tourism advertising industry since the last decade of the previous century.

Not only has the Internet revolutionized the transfer of tourist information, but it has also triggered a change in tourists' behavior in recent years. A sharp hike in the frequency of using the Internet while planning a journey and making the final decision shows us how much human beings—and thus travelers—have become dependent on modern technology [3,4]. The current situation is the product of many variables, most importantly the Internet's global availability, combined with real-time information updates, as well as the possibility of direct contact with the clients: a reality that has never been known before [5–7].

Nowadays, on the tourism market, as competitive as it is, launching a website cannot be perceived merely as a possibility, but rather as an absolute necessity that needs to be acknowledged by DMOs (Destination Management Organizations) [8,9]. Making use of this distribution channel of information is usually the first step in the modern reality of tourism to promote and commercialize a certain fragment of the tourist space [10]. What is more, due to the Internet, the tourism industry has become a global phenomenon that is available at a reasonable price. The regions promoting their tourist potential online enhance their global market position [11]. On the other hand, it needs to be stated that merely the online presence does not guarantee success [12]. A constant increase in the overall number of tourism-oriented websites causes an increased difficulty in attracting potential tourists, and finally, visiting consumers [13].

1.2. Online Information and Regional Promotion—How to Do It Right?

1.2.1. The Past and the Present—the Spectrum of Change

In the past, a choice of a particular tourism region by a potential tourist was either based on the offer of travel agencies or word of mouth. In the digital era of the 21st century, the Internet is, by far, the most popular source of information. This cyberspace guarantees tourists the ability to make their own individualized choices depending on the tourist information that was made available for them. Information communication technologies have revolutionized the tourism industry since the 1980s [4,14,15]. The websites containing the necessary information for tourists are becoming an indispensable distribution channel of tourist information in general [16,17], and they are the channel that essentially is of the greatest significance these days [18,19]. It has not changed for years now that a destination's website is practically the first information source that tourists consult to find more specific tourist information on the place they wish to visit. In many cases, it is a real game-changer. Based on the exact information found, its availability, and how it is presented online, tourists make their final decisions on the choice of the regions in which they want to spend their vacation time. In the era of such a dynamic development of the information transfer, such as the one enabled by the Internet and social media, communication between a certain region and the institutions responsible for its promotion (NTOs and DMOs) seems to be the key issue. Therefore, it is worth focusing on how the promotion of a region and its resources is carried out online.

The success of a region in terms of its tourism popularity is usually a derivative of an adequately created and user-friendly website. It was as early as at the beginning of the 21st century when Baggio proposed a set of conditions that induce this success. He referred to it as the "Decalogue" [20] (p. 3). The researcher pointed out three basic issues, namely: the necessity of outlining a clear strategy, the aims, and the recipient groups that the offer is aimed at. The elements that enable interaction between the user and the above-mentioned organization should be well-functioning and designed in a user-friendly manner. The website's content should be presented in a way that ensures high levels of acquisition and accuracy, starting from the colors applied, through the font size, to finish off with the correct grammar and text style. The content information provided ought to be credible and essential to encourage repeat online visits. It is best if the website's layout is easy to navigate for all the possible recipient groups. To ensure a high level of content credibility, it should be regularly updated and corrected if needed. It is also necessary to advertise the website widely through both traditional media channels and the Internet.

1.2.2. Advantages and Disadvantages of the Website Information for the Tourism Market

The website's quality is of pivotal importance in brand creation, as indicated by many researchers [21–23]. It is crucial to emphasize that merely the website's existence raises the region's attractiveness and influences its competitiveness on the tourism market [20] (p. 12). Quite often, potential tourists have only a vague idea of the destination they are planning to visit. Hence, online information that is provided easily and interestingly is of unquestionable importance, since it usually has an impact on the final travel destination choice [15,16,24,25].

Undeniably, a website is a key promotional tool [8,26]. In the contemporary world, tourists go online searching for all kinds of information, from practical information, such as the prices, through to the accommodation offer, transport possibilities, entertainment offerings, and regional attractions [27], and the Internet usually is the only source of information acquired [28]. Both DMO-managed and private websites are amongst most frequently visited ones by tourists prior to undertaking their journey [8,29], and their positive impact on the decision-making process is a fact [30–32]. It is also unquestionable how much the Internet has changed the tourism world [33–35]. Tourism belongs to the industry branches that refer to the Internet tools most willingly, with the websites being the most popular information sources, and therefore the most popular promotional tools of tourism regions in general [36,37]. The website's content is as essential as its level of interactivity. Both of these elements increase the attractiveness of the tourism destination for the potential client [38,39]. Thus, a well thought out and adequate website proposal ought to become one of the basic elements of the promotional development strategy in the region [40,41]. From the recipient's perspective, the functionality of the website is not only limited to its content value [42,43], but it is also created by other facilitating tools, such as the navigation system that constitutes the core of its usability [44–46].

The advantages of the online presentation of tourism-related information far outbalance its drawbacks. However, it is noteworthy that there is a risk for the web users that stems from the fact that easily accessible and visually appealing online information more and more frequently replaces the real-life experience. This trend also refers to geotourism resources. On the other hand, regardless of how attractive a certain tourism resource is in reality, if it does not exist in the virtual world, the chance of its occurrence in the tourists' awareness decreases to almost none. This shows how far the dependence on the online information amongst tourists has grown these days.

1.3. Georesources and the Role of Online Presentation

Finally, a question arises: what is the place of geotourism and its resources in the context of online presentation and regional promotion? This form of tourism, even though it is based on the natural values of the landscape that were considered to be important elements of the tourism space for years, received closer attention as a separate form of tourism only in the last decade of the previous century. As highlighted by Hose—one of the pioneer researchers who defined the phenomenon in question—the geological aspect in the concept of geotourism deserves the utmost attention [47,48]. Another issue of great importance is the proper interpretation of the elements of the inanimate nature aided by the tourist facilities, owing to which the tourist gains an insight into the knowledge in geology and geomorphology of the visited place [49,50]. What is more, this knowledge considerably exceeds the basic level of information needed to admire the aesthetics of the observed fragment of the natural space [51]. Newsome and Dowling [52] indicate the importance of taking a sustainable approach to the visited resources. On the one hand, geotourism facilitates need to appreciate their role in the landscape and the evolutional history of the Earth. On the other hand, a sustainable approach protects the geotourism resources by raising the awareness of the tourists, who are interested in getting to know the geology of the place in more detail, when it comes to the necessity of their preservation, to ensure the continuity of the geoheritage for the future generations. The obligation of preserving the heritage of an inanimate nature was also articulated by Hose and Vasiljević [53]. It originated in the idea that is commonly referred to in the academic literature as geoconservation. Its predominant task is the need to protect the geodiversity that, according to Grey [54], can be understood as a set of geological elements amongst which the greatest focus is placed on rocks, minerals, and fossils, as well as on the geomorphological elements such as landforms and physical processes, to finish off with the elements of soil. All of the above is referred to as geoheritage [48,55]. Most of the recent studies in this subject area emphasize the dualistic character of the phenomenon of geotourism [56]. Namely, the in situ geosites are the predominant points of tourist interest; nevertheless, what is growing in popularity are the ex situ elements of geoheritage presented to the wide audience in the form of museum exhibitions. The latter form of presentation of an inanimate nature is characteristic of the urban areas in particular.

So as to avoid any ambiguity in understanding of the basic concepts used in this research paper, its authors have presented them in a visual form (Figure 1) and proposed the following definitions. The notion of *geodiversity* entails all the possible elements related to the inanimate nature, regardless of their scale. Hence, geodiversity comprises both single sites, such as individual rocks and minerals, as well as surface elements, i.e., mountain ranges, uplands, etc. Within the concept of geodiversity, one can differentiate the elements that are valuable witnesses of the natural processes, which are unique in character, or for other reasons that are cognitively important. These elements of an inanimate nature can be defined as *geoheritage*. *Geosites*, on the other hand, are the integral components of geoheritage that due to their smaller scale can be perceived holistically by the public. Most geosites can be classified as *potential resources for geotourism*, whereas only the ones of the greatest scientific value that, at the same time, are adequately prepared to the needs of the tourist traffic (in the form of the existing infrastructure) can be referred to as the *real* or *actual geotourism resources* that constitute the basis for creating the geotourism products.

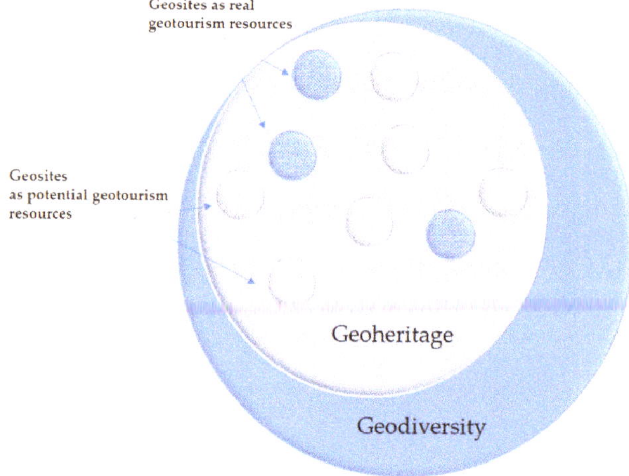

Figure 1. Interrelations between the basic 'geo-concepts'. Source: Authors' own elaboration.

This specific set of tourism resources requires adequate presentation and promotion to ensure the development of geotourism in regions that abound in the resources of an inanimate nature. These processes are happening on different levels—local, regional, national, and international—and require divergent tools. The most popular and the most effective at the same time is the Internet. Therefore, it is worth analyzing how the plethora of the Central European geoheritage resources is perceived by the National Tourism Organizations of chosen countries and how they promote it, so as to ensure becoming an essential factor in generating the geotourism movement in this part of Europe, which only three decades ago came back on the map of the Old Continent's tourism landscape.

2. Materials and Methods

The predominant research aim in this paper was determining the level of accessibility of online information on geoheritage and geotourism on the websites of the NTOs of selected Central European Countries. The choice of the countries applied in the research was made in accordance with the regional division by the International Geographical Union. Additionally, only the countries whose National Tourism Organizations are operating under the auspices of the European Travel Commission, a nonprofit organization responsible for promoting Europe as a tourism destination, were taken into consideration. Altogether, three neighboring countries with similar geoconditions, and hence comparative geotourism resources potential, were chosen to be analyzed: namely Poland, the Czech Republic, and Slovakia (It is worth accentuating the fact that all three countries in question entered the system and market changes at a relatively similar time, and therefore faced comparative challenges. It is the Internet that offers them a chance to fully enter the global market, and to create a truly competitive tourist offer). Their NTOs underwent descriptive content analysis as the qualitative method used to conduct the research. The categories under assessment comprised the following:

- Placement consistency of georesources under the main bookmarks;
- Scope of information available in English, as opposed to the original language of the NTO;
- Embedded search engines' results for the selected search terms, i.e., geotourism, geosite, geopark, geoheritage, geology, landscape, nature.
- Presence and validity of hyperlinks directing the reader to further information on geoheritage;
- Mobile website version and mobile applications' availability;
- Information accuracy and cohesion;
- Language accuracy.

Furthermore, the international perspective included in the article's title was achieved through focusing on the English versions of the NTO's websites, which was based on the assumption that "English is an international language in a global sense" [57]. The online research was carried out between 15 October and 17 November 2019, applying the Google Chrome web search engine.

3. Results

As previously stated, the websites of three NTOs, namely the Polish Tourism Organization, the Czech Tourism Authority, later in the text referred to as Czech Tourism, and the Slovak Tourist Board, underwent the content analysis focused on the geotourism theme. Tables 1–3 show the online research outcomes divided into five main categories under comparison, i.e.,:

- Homepage bookmarks (English version), where the general content is set aside the geotourism-related content;
- Homepage bookmarks and geotourist information in the NTOs' original language version (mutual relations);
- Embedded search engine responses to the selected geotourism keywords;
- Availability of the mobile version of the website and mobile applications in English;
- Language accuracy.

Table 1. Geotourism content analysis of the Polish NTO's website.

Country	Poland	
Official Name of the NTO	Polish Tourism Organization (operating under the supervision of the Ministry of Sport and Tourism of the Republic of Poland)	
NTO's Website Addresses	English version: https://www.polar.l.travel/en Original version: https://www.pols-a.travel/pl	
	General Content	**Geotourism-Related Content**
Homepage Bookmarks (English Version)	NEWS: *Travel inspirations*: 74 website articles containing mostly ideas for theme-inspired journeys from taking part in sport events through discovering the secrets of World War II to visiting famous Hollywood movies' locations, finishing off with culinary tourism experiences and travels suggested by famous bloggers.	*Poland, a place to be in 2020*: The PDF attachment to the article named Poland's Highlights 2019/2010 #Visit Poland #Polandtravel in the initial position highlights that the mining sites of Krzemionki Opatowskie have been recently inscribed on the World Heritage List. A hyperlink to a separate article is provided, where the reader can also find the hyperlink to the official website of the attraction [58]. Unfortunately, the original name "Opatowskie" was misspelled in the promotional materials. *New UNESCO Site*: The article merely mentions that there is a new UNESCO site in Poland providing the most basic information; interestingly, the name of the place is not given, neither are the directions, nor the hyperlink to any further information or article is provided [59]. *#VISITPOLAND Poland by YouTubers*: landscape values of the country are mentioned in chosen videos e.g., hiking adventures in Poland. However, no specific geotourism information is presented.
	CITIES: 18 cities in Poland are described through their characteristic features; hyperlinks to the city's official portals are provided.	Mostly not applicable. One of the exceptions is the description of the vicinity of chosen cities, for instance, the possibility of visiting the "Wieliczka" Salt Mine near Kraków (described in the article named *Kraków—Top attractions*).
	REGIONS: 16 regions of the country according to the administrative division into voivodeships are presented; introductory articles accentuating the predominant tourist attractions are available; hyperlinks directing the reader to the official tourist portals of each region (with the exception of the Greater Poland voivodeship and the Opole voivodeship) are provided; the only invalid hyperlink was the one of the portal of the Lubelskie voivodeship [60]. Two official portals did not have their English versions, i.e., West Pomerania [61] and Świętokrzyskie Voivodeship [62]; nota bene, the hyperlinks directed the reader to the official websites of the voivodeships and not to the official tourism portals as mistakenly suggested. Further information on the Silesian region [63] was available in German only.	Selected georesources are presented in relation to the regional division of the country. Searching for the relevant geotourism information requires initial knowledge on the topic. An example of the region rich in geotourism resources is Lower Silesia. *Lower Silesia (Dolny Śląsk)—the Land of Many Treasures*: includes information on the old mines and underground attractions of the area; the role of the Sudety mountain range is emphasized. Two more examples of regions crucial from the point of view of georesources are *Śląskie* and *Świętokrzyskie*. The first one constitutes an example of the largest industrialized area in the country with its mining facilities of the "Guido" Coal Mine, the Queen Louise Adit, and the Tarnowskie Góry Silver Mine (a UNESCO-listed site) being introduced for the readers in the main article [64]. The majority of the above-mentioned attractions are available for tourists who wish to embark on a journey tracing the Industrial Monuments Route. When it comes to the online presentation of the *Świętokrzyskie* region, geological values of the Świętokrzyskie Mountains and the Pepper Mountains are stressed. The most important tourist geoattractions, namely the flint mines of Krzemionki, the Raj cave, and the Jurassic Park in Bałtów are mentioned in the main article. Overall, the introductory articles describing each region strike a balance between the presentation of the cultural and natural resources at tourists' disposal. Nevertheless, in relation to the natural values of the landscape, the biotic aspects of it seem to overtake the general scope of attention.

Table 1. *Cont.*

Country	Poland
DISCOVER: Tourist information is organized under nine categories: World Heritage Sites, Castles and Palaces, Art and Culture, Holy Places, National Parks and Outdoors, Caves and Mines, Historical Trails, Genealogy Trails, and Shopping in Poland.	*World Heritage Sites*: Altogether as many as 16 places were recognized as UNESCO sites in Poland, six of which are especially important from the point of view of geotourism, namely: Krzemioki, Tarnowskie Góry Silver Mine, Muskauer Park, Wieliczka, and Bochnia Royal Salt Mines. Quite surprisingly, the newest UNESCO site in Poland—Krzemionki, inscribed on the list of the World Heritage Sites only in July 2019—is accompanied by the shortest article of all, and is preceded by other articles, i.e., *Family time travels, Eileen Aldis—A vlogger in Poland*, and pictures of *Travel inspirations* (originally placed under the bookmark called *NEWS*). Hence, in order to reach the main article with 117 words in its word count, the reader has to scroll down to the bottom of the page and then, if interested in further information on the site's exact location etc., has to enter the official website of Krzemionki. *National Parks and Outdoors*: Basic information on the 23 national parks in Poland including a map and introductory articles are available; six national parks do not have the official website versions available in English; three hyperlinks are invalid, i.e., the hyperlinks to the Góry Stołowe National Park, the Roztoczański National Park, and the Gorczański National Park. No information is provided on the fact that the Giant Mountains National Park was established a geopark. *Caves and Mines*: The underground wonders of Poland including the cave systems (9), the military sites (10), mines (17), and other tourist attractions of the type are described shortly. A separate introductory article is devoted to the selected Polish caves, the Queen Louise Adit, the Guido Coal Mine, the "Wieliczka" Salt Mine and its Graduation Tower, the Bochnia Salt Mine, the underground chalk tunnels of Chełm, and the Tarnowskie Góry Silver Mine.
EXPERIENCE: Tourist information is organized under eight categories: Boating and Sailing, Bike Trails, Skiing, Hiking and Climbing, Spa, Polish Cuisine, Polish Traditions, and Spice Up Your Travel.	Four categories, namely *Boating and Sailing, Bike Trails, Skiing, Hiking, and Climbing* focus on the landscape values of Poland, offering numerous possibilities for engaging into active tourism. No information specifically aimed at geotourists is given. Notwithstanding, basic pieces of geotourism-related information appear in the section of *Hiking and Climbing*; for instance, the rock formations of the Table Mountains are mentioned in the article devoted to hiking and climbing in the Sudetes [65]. Another category worth mentioning is *Spa*, especially in the context of geothermal springs. Their geothermal features when perceived as visual attractions can be categorized as geotourism resources [66]. The geothermal springs of the country are briefly described in different articles of the subcategory called *Health Resorts*. The main article distinguishes the geothermal waters of Iwonicz. Nevertheless, without initial knowledge on the topic, the reader has to go through as many as 43 short articles to find more accurate information on the availability of geothermal springs. Unfortunately, there is no English equivalent of the article called *Thermal baths in Poland as a way to deal with the autumn blues*; it is available only in Polish, and the role of the Podhale region in this respect is strongly emphasized [67].

Table 1. *Cont.*

Country	Poland
	Download brochures: 10 documents downloadable as PDF files are available for visitors, i.e., *Poland, Poland: Major tourist attractions, Poland: Cities, Poland: Castles and palaces, Poland: Culture and art, Poland: Health resorts and spas, Tastes of Poland, Poland: religions, rites and traditions, Poland: Wild nature, and Poland: UNESCO World Heritage Sites*. Two brochures offer basic information concentrating mainly on both biotic and abiotic natural resources of Poland. The first one, simply named *Poland*, contains sections titled Sea, Lakes, and Forests. It underlines the uniqueness of the amber coast of the Baltic Sea, the waters of the Masurian lakes, as well as the Białowieża treasures. Moreover, there is information concerning the Polish mountain ranges with the Karkonosze Range, the Tatras, and the Podhale Region, as well as the Bieszczady mountain range. The scenic view opportunities for the photography enthusiasts are described in relation to the Słowiński National Park and its shifting dunes, the Mouth of the Warta River joining the Oder, the Biebrza Marshes, the Bieszczady Mountains, the Białowieża Forest, and the Tatras with their alpine landscape. A separate page is devoted to the Salt Mine in Wieliczka; as the article states, it is a masterpiece created by nature and man. What is more, active tourism opportunities in the picturesque landscapes of the Great Masurian Lakes, the Dunajec Gorges, as well as the underground attractions of the Golden Slope and the Golden Mountains, together with the cave systems of the Kościeliska Valley in the Tatras are described. The second brochure is called *Poland: Major Tourist Attractions* and offers tourist information on 163 places worth visiting, which are grouped according to the administrative division of the country into 16 regions. Chosen examples of geotourist attractions include the shifting dunes in the Słowiński National Park, the seashore cliff of the Wolin Island, the rock labyrinth of the Jagged Rocks (Błędne Skały) in the Table Mountains National Park, the River Prądnik Valley in the Ojców National Park, the Krzemionki flint mines near Bałtów, and many more [68].
PLAN YOUR TRIP: Useful tips regarding the planning process prior to, or during the journey are facilitated through six categories: Traveling, Practical Information, Trip planner, Accommodation, Brochures to Download, General Contact, and Database Search.	The third brochure that is important from the geotourists' perspective is *Poland: Wild Nature*. It mainly focuses on the biosphere; nevertheless, such attractions of the inanimate nature such as the Stołowe Mountains Rock Maze and the Magical Land of Rocks in the Kraków-Częstochowa Upland, the Dunajec River Gorge, the Thermal Baths in Podhale, the Tatra Mountains, and the Bear Cave in the Sudetes are also presented. This brochure is interactive; therefore, it enables the user to visit the linked websites. The brochure entitled *Poland—UNESCO World Heritage Sites* provides a brief, usually a two-page long description of each World Heritage Site (WHS) in the country. However, it needs updating, as no information on the Krzemionki flint mines is provided. Interestingly, the promotional article on the Muskauer Park does include the following information: "this natural phenomenon also covers the area of the Muskauer Arch—the only terminal moraine visible from space" [69]. Nevertheless, no information is given on the fact that the Muskauer Arch is a geopark protected by UNESCO. This brochure is interactive and guarantees easy access to further website information. A brochure providing information on the geothermal springs of the country is named *Poland: Health Resorts and Spas*. A separate short article on the thermal springs of Podhale and Cieplice Śląskie-Zdrój is provided. Another proper name of the resort important in this respect mentioned in the brochure's text is Lądek-Zdrój in the Sudetes. The trip planner enables visitors to customize their trip according to the geographical location, the length of the visit, personal interests, attraction type, number of people, approximate cost, transportation type, etc. No information on the Polish geoparks appears while using the planner with divergent searching options. Geopark, geosite, and geoheritage do not appear as attraction types to choose from. The list of the UNESCO World Heritage Sites does include basic information on the Muskauer Park (Park Mużakowski) in Łęknica, as one of the greatest examples of the 19th-century gardening architecture in Europe. Nevertheless, no information is provided on the fact that the Muskauer Park lies within the landscape park area of the same name, and more importantly from the geotourism perspective, within the Muskauer Arch Geopark area. The latter is the only Polish geopark that is on the list of the Global Geoparks Network under the UNESCO auspices.
WTM: promotional information of the participation of the Polish Tourism Organization in the World Travel Market in London.	Not applicable.

Table 1. *Cont.*

Country	Poland		
Homepage Bookmarks and Geotourist Information (Original Language Version)	Homepage bookmarks are different in Polish than in English; so is the placement of georesources under certain categories. The scope of information on geotourism in Polish is broader. An example of the information missing in an English version is a separate article on the phenomenon of geotourism [70] presenting one geopark of the international rank—the Muskauer Arch Geopark, belonging to the Global Geoparks Network [71], and two geoparks of the national rank, i.e., the St. Anne Mountain Geopark and the Giant Mountains Geopark with the buffer zone. The hyperlinks to more information on the last two geoparks mistakenly direct the reader to the website of the country's Ministry of the Environment, where no further information can be found [72]. Two more geoparks attracting visitors mainly locally are also mentioned (Geopark Kielce and Geopark "Stone Forest in Roztocze"). The hyperlink directing the reader to the information on georesources provided by the Polish Geological Institute is invalid [73]. Even though comprehensive information on geotourism is available there in Polish [74], the English version of the website is unattainable.		

	Keywords	Number of Results	Content
Embedded Search Engine Responses to Selected Search Terms	geotourism	0	none
	geosite	0	none
	geopark	1	*Geopark in Kielce* [75] Centre for Geo-Education, established in 2003 and renamed Geopark Kielce for marketing purposes [76], comprises a network of geotourist and educational facilities presenting the area's georesources under the protection of the geological reserves. Apart from a short article, the hyperlink to the website of the Geopark in Kielce with comprehensive information available in English is provided. No information on the Muskauer Arch Geopark, the Saint Ann's Mountain Geopark, or the Giant Mountains Geopark with the buffer zone is mentioned.
	geoheritage	0	none
	geology	2	The results for the search term 'geology' are as follows: The Bałtowski Jurassic Park (*Experience–Family Fun*; Family Fun is not actually available under the bookmark called Experience on the main website, it only becomes visible after using the embedded search engine. Some places much more clearly related to the topical word in question do not appear as the results at all; for instance, none of the geoparks are linked with the word 'geology', neither are the regions that abound in georesources, such as the Świtokrzyskie voivodeship [77]); The Pieprzowe Mountains and the Queen Jadwiga's Ravine.
	landscape	59	A vast array of articles from which to choose. A potential geotourist needs initial knowledge on the topic to find the relevant information, as some results are only remotely related to the concept in question.
	nature	59	A vast array of articles to choose from. A potential geotourist needs initial knowledge on the topic to find the relevant information, as some results are only remotely related to the concept in question.

Mobile Version of the Website and Mobile Applications	A mobile version of the website is available for both Android and iOS devices. Mobile applications: information on the top 10 mobile applications useful for traveling in Poland is given, i.e., Foodspotting, Accuweather, Google Maps, With XE Currency App, Go2Stop, Voice Translator, Blablacar, PeakFinder, TripAdvisor, and Polska Niezwykła. Especially the last three of them provide selective information on georesources.		
Language Accuracy	Native level of English proficiency in the text, no major mistakes, only a few typos can be found, e.g., Krzemionki "Opiatowskie" instead of Opatowskie [78] or Mużakowski "Psark" instead of Park [78].		

Source: Authors' own elaboration.

Table 2. Geotourism content analysis of the Czech NTO's website.

Country	The Czech Republic	
Official Name of the NTO	The Czech Tourism Authority (operating under the supervision of the Ministry for Regional Development)	
NTO's Website Addresses	English version: https://www.czechtourism.com/home/ Original version: https://www.czechtourism.cz/	
	General Content	**Geotourism-Related Content**
	DESTINATIONS: Comprehensive tourist information together with an interactive map is provided for Prague, Central Bohemia, East Bohemia, West Bohemia, South Bohemia, North Bohemia, Moravia, and Silesia.	A very detailed list of places to visit in the regional division of the country together with the capital city of Prague as a separate unit is given. Finding certain information on geotourism resources requires initial knowledge on their spatial distribution and the recognition of their original names. An example of the region that abounds in geotourism resources is the Bohemian Switzerland in North Bohemia, whose geoheritage is succinctly described [79]. The article that the reader is led to following this access path clearly states that Bohemian Paradise is a UNESCO Geopark. Website navigation is facilitated by the use of hyperlinks, directing the reader to downloadable brochures and guides (available within www.czechtourism.com), as well as to booking.com offering accommodation nearby. Additionally, each article presenting a place to visit is accompanied with the option of showing its location on the map as well as the basic phone and address data. The hyperlinks (usually more than one) for each place under the regional division in most cases enable the reader to reach an external website whose English version is available.
Homepage Bookmarks (English Version)	*CULTURAL HERITAGE:* UNESCO Spiritual Czechia Historical Towns Castles and Chateaux Intangible Heritage Museums and Galleries Architecture Culture and Entertainment Christmas *ACTIVE HOLIDAY:* Tourist information is organized under the umbrella terms of Cultural Heritage, Active Holiday, and Health and Spa. *HEALTH AND SPA:* Medical Spas Medical Tourism Wellness in and out of spas	Geotourism-related content is most easily accessible under the *Active Holiday* bookmark. There are three main categories of active holiday in the Czech Republic proposed, i.e., *Summer Activities*, *Winter Activities*, and the category of *Natural Heritage*. Among other examples of activities for the summer period, visitors are encouraged to enjoy the Bohemian Paradise through hiking [80], biking experiences [81], or having a walk along the gorges of the Bohemian Switzerland [82]. The most important from the point of view of geotourism is the information provided under the bookmark named *Natural Heritage* [83], which is further divided into six subcategories, namely *Mountains, Rivers, and Lakes, Protected Areas, Caves, Observation Towers and Lookout Points*, and finally, *Rock Towns* [84]. The latter is of pivotal importance for geotourism with the Adršpach rock formations, the Bohemian Paradise UNESCO Geopark, Prachov Rocks, Broumovské stěny, Hřensko and its gorges, and Svatoš Rocks being singled out and thoroughly described. There is an interesting system of further categorization of the places into the summer sport locations or nature-related spots that can also make the decision-making process of the tourist easier. Another key bookmark is *Protected Areas*, where practically all the places described have at least one feature to them that can easily qualify them as geoheritage-related, starting from the Pravčická brána (Pravčice Sandstone Gate), which is the biggest natural sandstone arch in Europe, to three geoparks, namely the Bohemian Paradise–UNESCO Geopark, the GeoLoci GeoPark, the Železné hory Geopark. *Health and Spa* with its subcategories is a place to look for further information on geothermal waters. Introductory articles provide very succinct information; therefore, without initial topical knowledge, the reader needs to click on most of the suggestions to find out more on the West Bohemian spa triangle with the Vřídlo hot spring of the Karlovy Vary or the mineral-rich springs of the Teplice and Bečvou Spa, and many more.

Table 2. *Cont.*

Country	The Czech Republic
EVENTS: All sorts of events are presented applying a very useful information filtering system by region, date range and category, i.e., Festivals, Musical and Theatrical events, Exhibitions, Social Events, Events for Children, Sports Events, Folk Crafts and Markets, Gastronomic Events, Historical and Military Events.	Not applicable.
TRAVEL INFO: Tourist information is categorized under Practical Information, Brochures, Useful Mobile Apps, Restaurants, and Accommodation.	*Brochures*: Altogether as many as 50 downloadable PDF documents are at visitors' disposal. The information of a universal character that geotourists can find useful is predominantly placed under the following categories: *Trips from Prague* provides basic information on georesources of the Bohemian Paradise, Karlovy Vary, and the Bohemian Switzerland; *Czech UNESCO Treasures* is a 10-page long section of the document that is devoted to the UNESCO Natural Wonders of the Czech Republic; a separate article focuses on the Bohemian Paradise [85], which was established a UNESCO Global Geopark; a UNESCO Treasures Map is also provided. Single pieces of information on geoheritage also appear in the *Best of the Czech Republic, Active Holiday, Family Holidays, Czech Mountains: Bike and Hike, Regions of the Czech Republic: Top Sights and Attractions, Spas and Health Resorts.*
STORIES: As the main marketing slogan suggests (*Czech Republic Land of Stories*), tourist information is told through several stories. Altogether, there are as many as 20 stories to choose from.	The Bohemian Paradise [86] and the Bohemian Switzerland are yet again brought to the visitors' attention together with the Beskid Mountains as regions of the most outstanding beauty of the natural landscape. Each traveling story is told through a short video encouraging the visit.
TRIPS: The trips offered are divided by region, by the trip type (cycling and hiking trips), and by category (Towns and their Stories, Jewish Bohemia). Under the trip details, the visitor is provided with the story behind the trip, its itinerary, the list of attractions with a short description, as well as with the suggested restaurant and accommodation details.	There are no trips dedicated to geotourists specifically.
TRAVEL PROFESSIONALS: The bookmark provides professionals with the necessary details on undertaking cooperation with Czech Tourism for marketing reasons.	Information provided is not applicable for regular visitors.
Homepage Bookmarks and Geotourist Information (Original Language Version)	Not only are the bookmarks different in the Czech version of the website, but the website itself in the original language of the NTO is not available from the position of the language change. It is a completely different online portal serving a different purpose. Geotourism information is limited in comparison to the English version. There is a hyperlink to a bachelors thesis devoted to the design of the tourist product of the Vysočina Geopark [87].

Table 2. *Cont.*

Country	The Czech Republic		
	Keywords	Number of Results	Content
Embedded Search Engine Responses to Selected Search Terms	geotourism	0	none
	geosite	0	none
	geopark	11	The list below shows the chosen, most accurate results for the search term "geopark", together with their original access path [88]. *GeoLoci GeoPark* (Activities–Active Holiday–Natural Heritage–Protected Areas); *Železné hory Geopark* (Activities–Active Holiday–Natural Heritage–Protected Areas); Bohemian Paradise–UNESCO Geopark (Activities–Active Holiday–Natural Heritage–Protected Areas); *Egeria Geopark* (Activities–Active Holiday–Natural Heritage–Mountains). Apart from the short description of each geopark, the reader is provided with the option of showing the exact location of the place on the map, as well as with the basic contact information, a hyperlink to the place's official website (if in operation), and a hyperlink to download guides and brochures. A hyperlink to search for accommodation in the surrounding area is also given. Some of the search results, for instance "Prague—a magical atmosphere you won't find anywhere else", seem not to have a clear relation to the search term under discussion.
	geoheritage	0	none
	geology	5	The list below presents the exact titles of the articles as the results for the search term "geology": *Železné Hory Geopark;* *Central Bohemian Mountains and Lusatian Mountains;* *Winter in the Czech Republic: Where to Head out for in Search of Warmth and a Good Time;* *Lower Morava Biosphere Reserve;* *Keeping the Children Entertained—Zoos, Science Parks, and Sporting Activities in the Czech Republic* [89]. Not all of the articles mentioned above are linked with the search term directly, and others, such as the previously described geoparks, are ignored.
	landscape	229	A vast array of articles to choose from. A potential geotourist needs initial knowledge on the topic to find the relevant information, as some results are only remotely related to the concept in question.
	nature	311	A vast array of articles to choose from. A potential geotourist needs initial knowledge on the topic to find the relevant information, as some results are only remotely related to the concept in question.
Mobile Version of the Website and Mobile Applications	Mobile version of the website is available for both Android and iOS devices. Mobile Applications: Czech Republic Land of Stories, Jewish Bohemia and Moravia, Czech Republic Wine Trails, Camping Sites in the Czech Republic, Pocket Fennec—Your City Guide, Omio—Travel around the Czech Republic, GoOut: Cultural Events in the Czech Republic, Man About World—Gay Travel Magazine + Guides, Vocabulary Miner, TOP 100 Czech Sights mobile app, PhoneMaps, Prague Trips by Public Transport, Czech Money mobile app, Czech Spa Resorts, Czech Days—The Official Portal, AppsMapper Czech—The Best Android Apps for Czech. Most of the 16 mobile applications available on the website are of universal character; no specific app is oriented on the theme of nature, landscape, or geoheritage.		
Language Accuracy	Native level of English proficiency in the text, minor mistakes that do not influence its understanding. There are several typos to be found, e.g., the Brdy Mountains (double "s" at the end of the word) [90], or "tthe" instead of "the" in the article concerned with the Pravčice Sandstone Gate [91].		

Source: Authors' own elaboration.

Table 3. Geotourism content analysis of the Slovak NTO's website.

Country	Slovakia	
Official Name of the NTO	Slovak Tourist Board (operating under the supervision of the Ministry of Transport and Construction of the Slovak Republic)	
NTO's Website Addresses	English version: https://slovakia.travel/en Original version: https://slovakia.travel/sk	
	General Content	**Geotourism-Related Content**
Homepage Bookmarks (English version)	PLACES TO GO: The best of Slovakia UNESCO Unique Slovakia Tourist regions	The bookmark entitled *The best of Slovakia* emphasizes the natural resources of such areas as the Tatras and Northern Spiš [42], the National Park of Slovenský Raj, the Caves of the Slovak Karst, as well as the Diery (Malá Fatra Mountains). Each short article is accompanied with an option to send it to a friend, print its content, and—elements that are especially useful in the trip planning process—display it on a map and show the street view. Under *the UNESCO Natural Heritage* one can find basic geotourism information in particular on the caves and abysses of the Slovenský karst and the Dobšinská ľadová jaskyňa cave. Using the hyperlinks, the reader is lead to the article called the Caves of the Slovak Karst (originally placed under *The Best of Slovakia*). *Unique Slovakia* lists the following geoattractions that either cannot be found elsewhere or constitute extremely rare examples of geosites typical of Slovakia [43]: The Ochtinská Aragonite Cave; The biggest cave dripstone in Krásnohorská Cave; The rock houses of Brhlovce (the complex of rock dwellings in tuff rocks); The stone balls of Megoňky; The Slovak opal mines. Geotourism resources can be found in practically all of the 21 tourist regions presented on the website; nevertheless, the ones of the greatest importance include the following: the Region of Horehronie, the Tatras and Northern Spiš, the Danubeland (Podunajsko), the Region of Central Považie, the Region of Gemer and Malohont, the Region of Liptov, the Region of Lower Zemplín, the Region of Orava, the Region of Pohronie, the Region of Poiplie, the Region of Šariš, the Region of Southern Spiš, and the Region of Žilina and its environs. Finding more specific information on georesources requires either initial topical knowledge, or the reader has to read through all the introductory articles.
	THINGS TO SEE AND DO: Tourist information is organized under seven categories, i.e., Nature and the Countryside Culture and Sights Spa and Relax Sports and Activities Traditions and Gastronomy Autumn Inspirations Autumn in Slovakia—Good Idea	Amongst the things to see and do in Slovakia in terms of geotourism, the most important is the category named *Nature and the Countryside*, which is further divided into *National Parks, Geoparks, Caves, Water Reservoirs, Botanic Gardens and Zoos*, and *Natural Curiosities*. The categories that clearly stand out here are *Geoparks, Caves, Natural Curiosities*, and *National Parks*. The first one, with its online slogan (*Attractive Geoparks Luring for Hiking*), encourages tourists to undertake active rest in three geoparks of Slovakia, i.e., in Banská Bystrica Geopark, Banská Štiavnica Geopark, and Novohrad-Nógrád Global UNESCO Geopark [44]. The first two are of national importance; the last one was inscribed on the list of Global Geoparks, and is under the patronage of UNESCO [45]. Apart from naming the geosites within their borders and underlying their overall significance in geoheritage protection, the articles devoted to each geopark provide their exact location, including the GPS geographical coordinates. The articles were also equipped with four active icons, namely 'send to a friend', 'print', 'display on map', and 'street view'. There is also a hyperlink directing the reader to the official website of each geopark. It is available in English in all three cases, although sometimes further information is very limited in its range, as in the case of the Banská Bystrica Geopark [46]. The second category that is significant for geotourists is *Caves*. According to the introductory article [47], only eight from the overall number of 6200 caves in Slovakia are adjusted to the visitors' needs, whereas the list of accessible caves in Slovakia, which are available for the visitors as the PDF file as a part of the same article, states that there are as many as 19 show caves in Slovakia [48]. The aforementioned list provides all the necessary tourist information, including the exact GPS location, the spatial distribution on the map, contact details, website address (for 18 caves), and most importantly, availability according to the season. Under *Natural Curiosities*, one can find chosen examples of the Slovakian geosites, e.g., the Bešeňová Travertine Formation, the Craters of Ružbachy in the Spišská Magura Mountains, the Devínska Kobyla Mountain in the Malé Karpaty Mountains, the Diery in the Malá Fatra Mountains, the Súľovské rocks in the Súľovské Vrchy Mountains, or the geyser of Herľany, and many more. A standard description is provided; no external links to the websites with more detailed information are given. Practically, each of the nine national parks in Slovakia, mainly due to the mountainous character of the country, has a wide range of attractions to offer for geotourists. Their website introductory articles are informative (also including practical information). No links to the external official websites of the national parks are provided. A video gallery for two national parks is available; nevertheless, it focuses mainly on the animated aspect of the landscape. What is unprecedented so far is the way of presenting the nearby attractions divided into categories, e.g., *Nature and the Countryside, Culture and Sights, Spa and Relax, Sports and Activities, Traditions and Gastronomy, Autumn Inspirations*, and many more, each marked with an icon [49].

Table 3. *Cont.*

Country	Slovakia
ABOUT SLOVAKIA: The tourist information is further divided into eight categories: Facts, History, Who We Are, Regional Division, Towns, Interesting Information, Practical Information, and Videos.	Mostly not applicable. Information on georesources is of fragmentary character and can be found under the category of *Regional Division*, but it requires initial knowledge on the spatial distribution of the resources. Videos, on the other hand, focus mainly on the visual appeal of the Slovakian natural and cultural landscapes divided into 14 themes.
EVENTS: The Slovakian events are presented as weekend events, nearby events, and regular events. There is also an event calendar with the date, event type, and country as the search filters, as well as the Autumn Palette with events planned for the autumn time.	Not applicable.
TRAVEL ESSENTIALS are further divided into Transport, Accommodation, Map, Brochures, Online Travel Guide, Mobile Applications, Regional Discount Cards, and the category named Focused On, which is further categorized as follows: Conference and Incentives (MICE), Disabled Access Travel in Slovakia, and Gay and Lesbian Travel.	The interactive *Map* tool is extremely useful, as it corresponds to the website's bookmarks and its overall organization of attractions. Using a system of ticks limits the number of choices displayed on the screen. For instance, only the location of geoparks or other geotourism-related sites can be shown [100]. To make use of the *Online Travel Guide*, a potential geotourist needs to possess basic knowledge on the spatial distribution of the Slovakian georesources. Essential travel information on as many as five regions of the country is given, i.e., the Bratislava region, Central Slovakia, Eastern Slovakia, North Slovakia, and Western Slovakia. *Brochures*: Four documents downloadable as PDF files are available for visitors, i.e., *The Most Beautiful Sites in Slovakia, UNESCO World Heritage Treasures, Slovak Spas and Wellness*, and *Slovakia—A Land of Wine*. Geotourism information is mostly the same as described above under the bookmark named *Places to Go* and the subcategories of *The Best of Slovakia* and *UNESCO*, respectively. The brochure presenting the *Slovak Spas and Wellness* tourism options might also be used by geotourists interested in the thermal springs of the Trenčianske Teplice Spa, Bojnice Spa, Rajecké Teplice Spa, Turčianske Teplice Spa, Sklené Teplice Spa, and many more.
JOURNALS ARTICLES: The 10 most beautiful castle ruins in Slovakia; 10 romantic destinations in Slovakia; The Blue eyes of the Slovak mountains; Country holiday experiences in the air and on the water; How our ancestors used to live; How to see a lot and not get too tired; Sheepfolds, farms, and ranches of the Slovak countryside; Specialties of the Slovak meadows and forests; Symbol of the Tatras and other mobile treasures; The beauty preserved in wood; Unique sacral treasures from wood; Wandering through sheepfolds in Liptov; What can Smolenice offer; and Wooden churches of UNESCO World Heritage	Especially the article entitled 'The Blue eyes of the Slovak mountains' can be interesting for geotourists, as it presents the country's most spectacular and picturesque sites of the Tatra Mountains, including the Kmeť ov waterfall in the Nefcerka valley, the Waterfalls of Cold Creek rank, and other sites such as the Herliansky geyser.
PLANNER: an online tool that enables visitors to customize the trip to Slovakia. There is an option to filter the tourist offer by category, region, and season. There are also separate bookmarks giving information on tourist trails, accommodation, and upcoming events. Not all of the information is available in English [101].	The filtering option under *Attractions* offers geotourists a chance to find information further categorized as *Caves* and *Nature Interests*. What is really useful is an accompanying interactive map. What makes it hard to use for the English-speaking user, though, is an inconsistency in translation, as most of the articles are available in Slovak and only some are available in English [102].
Homepage Bookmarks and Geotourist Information (Original Language Version)	The English version of the website is almost exactly the same as the original. Both the main bookmarks and their internal content are repeated. In addition, the introductory articles are neither shortened nor prolonged in English, offering international visitors the same information access as for the domestic ones. The only exception to the rule of the consistent appearance of information in all languages is the travel planner, where one can find information available mainly in the Slovak language [101].

Table 3. *Cont.*

Country	Slovakia		
	Keywords	Number of Results	Content
Embedded Search Engine Responses to Selected Search Terms	geotourism	2	The first hyperlink directs the reader to the general article on geoparks, as it was previously described under *Things to See and Do* [94]. From the three geoparks in Slovakia, the second response to the search term 'geotourism' is the Banská Bystrica Geopark [103]. Information on the existence of the other two geoparks as separate results is not available.
	geosite	2	The first hyperlink directs the reader to the website's homepage in the Slovak language [104], whereas the second one directs readers to the general article on geoparks, as in the case of 'geotourism'.
	geopark	?	Not available for proper comparison, as the embedded search engine of the website is connected with the general Google search engine. Therefore, in most cases, even though relevant information can be found this way, it repeats itself depending on the language of translation and the exact number of results cannot be given, or results unavailable on the website under analysis are provided.
	geoheritage	0	none
	geology	?	As described before in the case of the 'geopark'.
	landscape	?	As described before in the case of the 'geopark'.
	nature	?	As described before in the case of the 'geopark'.
Mobile Version of the Website and Mobile Applications			The mobile version of the website is available for both Android and iOS devices. Altogether, as many as 13 mobile applications are advertised for tourists traveling in Slovakia. Geotourists might be especially interested in the *Košice Region Tourism* app presenting the geosites of the Slovak Paradise and the Slovak Karst, as well as the article titled 'Exciting Underground Adventures in UNESCO-Listed Caves' [105]. Another one that geotourists might want to use is the *High Tatras Region* app, where the tourist routes are accompanied with the necessary description, visuals, and other pieces of practical information. The remaining apps are universal in character and, among others, offer a vast range of functionalities from the *Tourist Map of Slovakia* to an *App for the Road*, helping drivers in case of emergency [106].
Language Accuracy			Native level of English proficiency in the text; no major mistakes to be found. Minor mistakes in translation appear; for instance, in an article concerning geoparks, one comes across the following sentence: "In this reagrd, geoparks have been created to focus on the protection of abiotic nature, scientific research as well as to education to arouse interest of the public" (a typo in the word 'regard', 'education'–a noun that should be replaced with a verb 'to educate', and followed by a linking word 'and'). Another example: "Growing around the world through the growth of geoparks, geotourism focuses not only to flora and fauna, but also to abiotic environment" (the English expression "to focus on" was changed into "to focus to") [94]. At times, examples of translation inconsistency can be found, especially in relation to proper names. It can be confusing for an English-speaking reader. For instance, the original name Dobšinská ľadová jaskyňa is either translated into "Dobšinská ľadová jaskyňa cave", or into "Dobšina Ice Cave" [97]. In addition, a misspelled and pluralized category named "Technical Herritages" appears in the Travel Planner [107].

Source: Authors' own elaboration.

4. Discussion

All three NTOs analyzed provide the readers with information under seven main bookmarks maximum. What is fully understandable is that the organization of content on each website is a part of its overall design and an idea behind it. It is common knowledge that regardless of personal interests and the exact information that visitors want to obtain, the websites should be as easy to navigate as possible. It is also crucial for them to be intuitive and enable the readers to find whatever they are looking for as quickly as possible. Hence, the smaller the number of clicks and the clearer the names of the categories used on the website, the more universally understandable they are in the tourism industry, which is preferable.

Only two steps need to be taken in order to find the geotourist information in the case of the Polish Tourism Organization, where one of the main bookmarks—*Discover*—leads the reader to the article named *Caves and Mines: Underground Attractions* (Figure 2). However, it has to be stated that the above-mentioned are only two examples of attractions within the interest of geotourists.

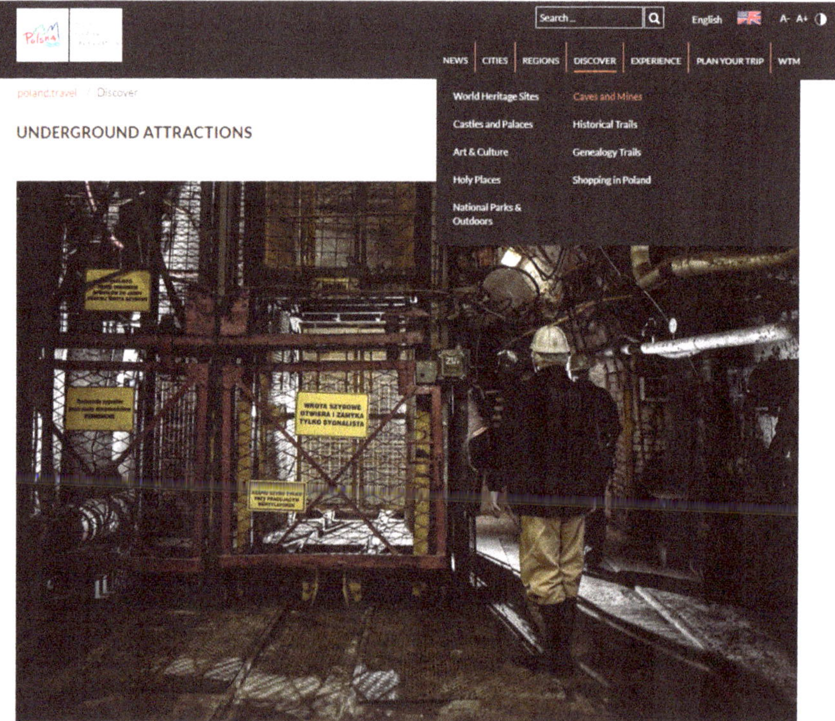

Figure 2. Underground attractions of Poland (a two-step access path: *Discover–Caves and Mines*). Source: [108].

The access path of three steps was required by the Slovak Tourist Board to find a separate article on *Geoparks*, which was most appropriate from the perspective of geoheritage protection and the geotourist offer (Figure 3). At the same time, it was not singled out as an individual subcategory on the websites of the Polish and Czech NTOs, which is not understandable, as all three countries can pride themselves on having geoparks belonging to the Global Geoparks Network, not to mention the ones of the national and local rank.

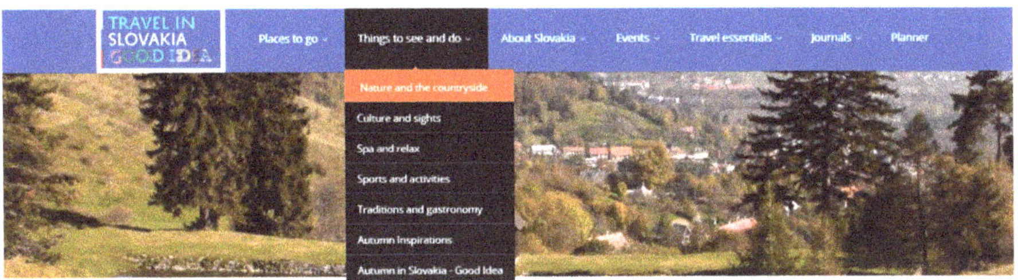

Figure 3. Geoparks (a three-step access path: *Things to See and Do–Nature and the Countryside–Geoparks*). Source: [94].

Most frequently, as many as three to four steps and more were to be taken in order to find more specific geotourist information, as indicated in Tables 1–3. For instance, in all three cases, information on geoheritage is provided in relation to the regional division of the country. Regions and their tourist offer are either distinguished as the main bookmarks, as on the website of the Polish Tourism Organization, or Czech Tourism, where they are referred to as *Regions* and *Destinations* respectively, or as on the website of the Slovak Tourist Board, which they are found within the subcategory of the bookmark titled *Places to Go*. Geotourist information available via this access path requires initial knowledge on the topic. Therefore, considering the overall number of clicks and introductory articles to go through, it is much more likely to be used by a person who is already familiar with the areas in question (Figure 4), making it much less likely for potential inanimate nature-interested travelers to do so. Moreover, taking into consideration the natural values of the landscape, the biotic aspects of it seem to overtake the general scope of attention.

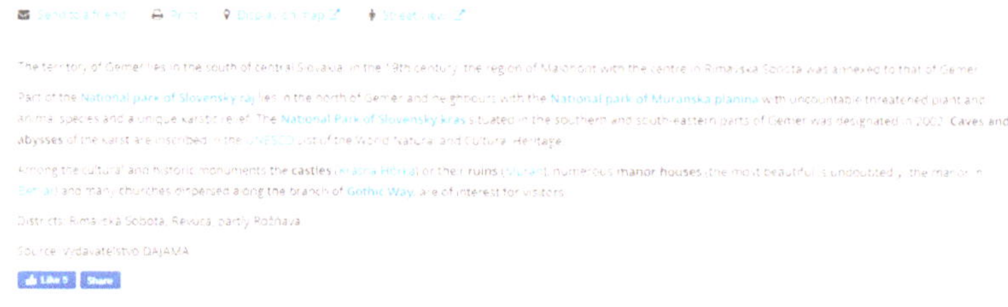

Region of Gemer and Malohont

Location: Banskobystrický kraj, okres Rimavská Sobota, Rimavská Sobota
GPS: N48°22'60" E20°1'6"

Figure 4. National Park of Slovenský raj (a four-step access path: *Places to Go–Tourist Regions–Region of Gemer and Malohont–National Park of Slovenský Raj*). Source: [109].

Another example of similar categories under which more information in question can be found for all three countries are the national parks. Similarly to the regional division of the countries, it requires the reader either to possess initial knowledge on the topic to find the necessary information relatively quickly or to spend some quality time reading through the introductory articles to learn about the georesources that they hold (Figure 5).

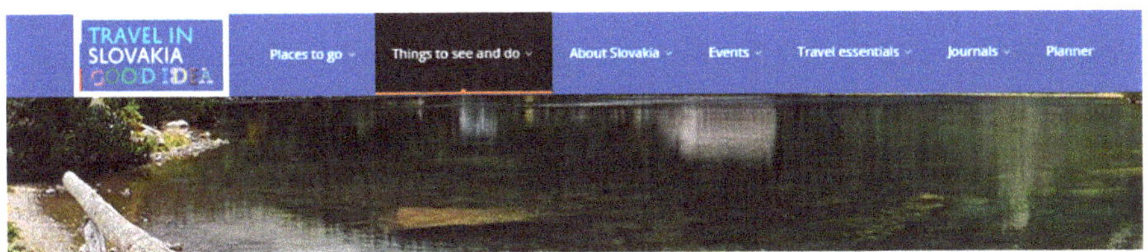

Small area but with nine national parks

Figure 5. A four-step access path: Things to See and Do–Nature and the Countryside–National Parks–National Park of High Tatras). Source: [110].

Additionally, and quite surprisingly, even though all three countries advertise their cultural and natural resources under the auspices of UNESCO, none of them even mentions in the main article that certain geoparks within their area belong to the Global Geoparks Network, and hence can be referred to as UNESCO Global Geoparks. According to the Global Geoparks Network [111], "UNESCO Global Geoparks are single, unified geographical areas where sites and landscapes of international geological significance are managed with a holistic concept of protection, education and sustainable development". As it is a relatively new initiative that originated in 2015, the website's content and the way in which information appears on the page should be updated and follow the newest trends and guidelines for the tourism industry and nature protection. Czech Tourism uses the name "Bohemian Paradise–UNESCO Geopark", but as stated before, not in the main article devoted to the UNESCO sites of the country, and this piece of information is available after knowing the following access path *Activities–Active Holiday–Natural Heritage–Protected Areas–Bohemian Paradise UNESCO Geopark* (Figure 6). In addition, in the case of Poland, even though the Muskauer Park (Park Mużakowski) in Łęknica, as one of the greatest examples of the 19th century gardening architecture in Europe, is advertised under the access path of *Discover–World Heritage Sites–Muskauer Park*, no information whatsoever, even in the form of a hyperlink, is given on the fact that it lies within the

Muskauer Arch Geopark area, which is the only Polish geopark that is on the list of the Global Geoparks Network under the UNESCO auspices. The Slovak Tourist Board seems to have made the information on the Novograd-Nógrád UNESCO Global Geopark most easily accessible via *Things to See and Do–Nature and Countryside–Geoparks—Novohrad Geopark*. It is understandable that the main article on the UNESCO-listed heritage of each country predominantly focuses on the most recognizable lists. Notwithstanding, since selected geoparks function under the same patronage, it is worth at least mentioning this fact or providing a proper hyperlink.

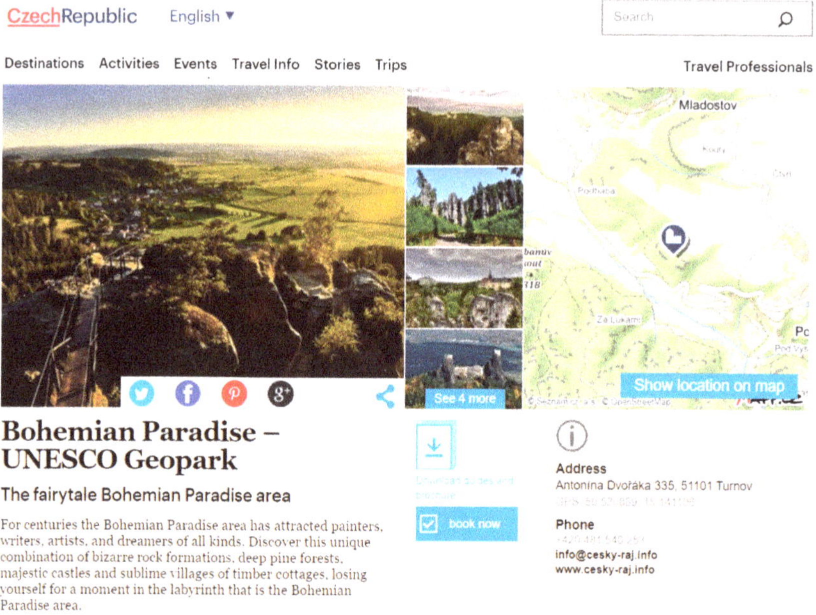

Figure 8. Bohemian Paradise–UNESCO Geopark (a five-step access path: *Activities–Active Holiday–Natural Heritage–Protected Areas–Bohemian Paradise*). Source: [112].

The hyperlinks system of each NTO leading the reader to more detailed geotourism-related information also needs improvement. For instance, in the case of the Polish Tourism Organization's website, when the reader is willing to enter the tourism portal of the West Pomeranian Voivodeship, s/he is directed to the website of the West Pomeranian Marshall's Office with no English language version. Additionally, even though the hyperlinks to the official websites of the Polish national parks are provided, generally speaking, the information that can be found there is usually far from ideal, mainly due to the English language version's unavailability and/or issues in translation. A similar problem appears when reaching the official national parks' websites of the Czech Republic. In contrast, the Slovak Tourist Board does not provide the reader with the hyperlinks to the country's national parks at all. However, it does so for its geoparks. Nevertheless, even though the Network of Geoparks of the Slovak Republic, providing comprehensive information on the subject matter, is mentioned under the main article devoted to the Novohrad Geopark, no hyperlink is included.

While comparing the geotourism-related content of the websites and the responses of the embedded search engines to the selected search terms (i.e., geotourism, geosite, geopark, geoheritage, geology, landscape, nature), it seems that most of them do not serve their function successfully. 'Geotourism', 'geosite', and 'geoheritage' were not used within the website's content at all; even though there is every reason behind linking them with the phenomenon in question, no results were shown. The most serious problem though seems to be giving preference to information of lesser importance from the point of view of the tourist attractions' rank. For instance, after having typed the word 'geopark', the Polish Tourism Organization's embedded search engine provides information on the Geopark Kielce only, but completely ignores the Muskauer Arch Geopark, which is one of the UNESCO Global Geoparks. Information on the remaining geoparks of the national rank in Poland is also excluded from the website's content. Yet another example from the Polish NTO is that some search words such as 'geology' are not thematically linked with the right articles. To exemplify, using the embedded search engine in order to find information on the above-mentioned Geopark Kielce that abounds in geosites, one will not succeed in doing so. Therefore, the reader does not stand a chance of finding comprehensive information on the Polish attractions within their scope of interest so as to make an informed traveling decision. Meanwhile, examples from Czech Tourism show that the aforementioned phenomenon can also be reversed, and the searching process might result in finding information on the places that are only remotely related to the search terms under discussion. Moreover, the overall number of embedded search engine responses to the words 'nature' and 'landscape' makes it hard for the reader to find geotourism-specific information. A clear preference in most articles is given to the biotic aspects of both nature and landscape, whereas the abiotic ones are either eliminated or diminished. Conducting a similar comparison for the Slovak NTO's website was mostly impossible due to the fact that the website's embedded search engine is linked with the general Google search engine.

Providing limited information in the English version of the NTOs' websites in comparison to their versions written in the mother tongue is understandable, as long as such a limitation is connected with the rank of attractions to being covered. Neglecting information on tourist attractions that are already internationally recognizable, or that stand a chance of becoming ones in the near future, is unjustified and should not take place.

Even though it has become standard to make the mobile versions of the NTOs' websites available for their users, making theme or interest-oriented mobile applications, especially with the geo-prefix in their name, is still a niche to be covered. 'Nature' and 'landscape' seem to be too vast in their character and understanding to serve this function properly from the perspective of geotourists.

With reference to the language accuracy, all the geotourism-related content on the websites of the NTOs under analysis represents native or a native-like level of language proficiency in translation. The most common mistakes are typos or inconsistencies in the translation of the proper names, as indicated in detail in Tables 1–3. Nonetheless, they do not influence the overall level of understanding of the text for the English-speaking recipients.

5. Conclusions

The descriptive content analysis of the selected NTOs' websites enabled the authors to draw the following inferences and improvement suggestions:

1. Information on georesources provided by the NTOs of selected European countries is dispersed and usually of fragmentary character; as long as the generalization of facts is fully understandable and inevitable on national portals, excluding information on the existence of geotourism attractions of an international rank in particular should not take place, especially when they constitute good examples of attractions that stand a chance in raising the tourists' level of awareness on the concepts of geoprotection and sustainability;

2. Geotourism-related information falls under divergent umbrella terms, and therefore, it usually requires much effort from the readers to come across the content of their interest; it could be improved at least by implementing the key word 'geotourism' or one of its derivatives (e.g., 'geosite') to the NTOs' online travel planners;

3. Both animate and inanimate landscape values co-create the inextricably linked tourism space; therefore, it seems unjustified to prioritize one for the sake of the other and focus promotional actions mainly on the animate landscape tourism resources;

4. As it is impossible to draw a clear-cut line between various types of tourism activities relying upon the same resources, online information availability should be ensured by an appropriately working system of search terms that are prepared in cooperation with tourism experts who are aware of their mutual relations, which could in the end result in a more effective use of the websites' content;
5. Even though there is every reason for distinguishing 'geoparks' as a separate bookmark on the websites of NTOs, or at least a subcategory of nature-related attractions, it is not a common practice;
6. Access to further geoheritage information provided via hyperlinks on the websites of the NTOs analyzed is possible only in some cases; usually, it is just a theory due to the issues in translation and/or the information provided is incomplete, out of date, or the reader is led to a different website than the hyperlinks' names suggest. It is most common in reference to the official websites of the national parks; therefore, there seems to be too much lost opportunity and the current situation calls for a change;
7. Making the brochures available as downloadable PDF files focused on the concept of geotourism could be one of the possible solutions for presenting each country's geoheritage potential in a holistic way;
8. Presenting the geotourist offer including the spatial distribution of georesources could also be made available via mobile applications as they are ever gaining in popularity, especially among the younger generations of visitors; moreover, information provided this way does not have a direct physical interference whatsoever with the landscape, and therefore helps to maintain it as untouched by the tourism traffic and its needs as possible;
9. Including information on the availability of geoparks under the UNESCO auspices, at least in a form of a hyperlink, or an active picture directing the reader to further information on UNESCO Global Geoparks is a good promotional opportunity;
10. NTOs present the country's tourist offer in a holistic way, which is in agreement with the general idea for presenting certain content and the overall website design. Nevertheless, there is a need for a think tank enabling expert consultations, especially between the marketing and promotion sector representatives and experts in the tourism industry that would lead to such online presentation of information that, on one hand, would maximize the use of the countries' tourist offer potential in accordance with the rules of sustainable development, and on the other hand, would not neglect or diminish the role of the abiotic aspects of the environment;
11. Promoting geotourism on a national level via NTO's websites with its scientifically and cognitively valuable resources alongside other, well-consolidated forms of tourist activities can be conducive to gaining knowledge on and appreciating the geological component in both natural and urban recipient areas.

To summarize, the actions conducive to the accurate presentation of geotourism resources on the NTOs' websites, as presented in Figure 7, are threefold. All three levels, namely the primary actions focused on the geotourism resources per se, followed by undertaking the cooperation of the experts in three realms of knowledge, and finally, the actions concentrated on the e-aspect of their promotion are of equal importance. It is worth emphasizing that as far as the very first step of actions to be taken is concerned, the scientific value of geosites is of utmost importance. The procedure of their recognition, exploration, and taking an inventory of the sites can vary in different countries. Nevertheless, the elements that the above-mentioned procedure usually have in common are being drawn by teams of academics, most frequently specializing in geology, and financed among others from the ministerial funds. To exemplify, in Poland, the binding documents drawn in this respect in 2006 and 2012 respectively were:

- The Catalogue of Geotourist Sites in Poland (written under the scientific supervision of Słomka, Doktor, Joniec, and Kicińska-Świderska, and published by the AGH University of Science and Technology in Cracow) [113];

- The Catalogue of Geotourist Sites in Nature Reserves and Monuments (written under the scientific supervision of professor Słomka from the AGH University of Science and Technology, Department of General Geology, and Geotourism in Cracow) [114].

Figure 7. Flowchart of actions conducive to the accurate presentation of geotourism resources on the NTOs' websites. Source: Authors' own elaboration.

All three countries whose NTOs' websites underwent analysis possess considerable geotourism potential. Therefore, the main research aim was to check how the potential of georesources is presented online amongst the other tourism resources of each country. The Internet is a tool enabling an instant reaction; hence, this instant reaction can be expected in presenting the newest tourism trends. In other words, the Internet, as quick of a tool as it is, should be equally quick in the hands of the decision makers responsible for the final shape of the regional promotion in tourism. Even more so, taking into consideration that geotourism as a separate category has been growing in popularity since the 1990s, it goes hand in hand with the concept of geoconservation and sustainable development.

The main limitation behind the study, and at the same time a further research perspective refers to the overall number of NTOs' websites that underwent the descriptive content analysis. This paper is the first step in the trial to create a model of a proper online presentation of geotourism resources. At this point in this research, it is unfeasible to clearly state the level of representativeness for other European countries, as no comparative data is available in the academic literature. Notwithstanding, the similarities and differences in the online presentation of georesources shown in the sections of results and conclusions can surely be referred to other Central European countries that underwent the same system changes, and therefore had comparable development and economic growth obstacles to overcome. Similarly, even though well-defined examples of the online presentation of single georesources can be found on the websites of all the analyzed countries (of the Czech Republic in particular), the current picture of the online presentation focusing on the georesources of the remaining European countries, according to their regional division, is yet to be drawn to perceive a holistic 'geotourism online map' of Europe as a tourism destination. Secondly, as the research outcomes have shown, the hyperlinks system within the described websites is far from ideal, and the seemingly more in-depth information provided by them should also become the subject of further analyses. Thirdly, even though the importance of the NTOs from the perspective of tourism marketing cannot be denied, the broadly understood Internet offers much more than the website information; therefore, especially the aspect of using social media as a promotional tool for geotourism is the research area that should attract more attention in the academic literature.

Author Contributions: Conceptualization, A.R.; methodology, A.R., K.W. and Z.J.; formal analysis, A.R., K.W. and Z.J.; resources, K.W. and A.R., writing—original draft preparation, A.R., K.W.; writing—review and editing, A.R., K.W. and Z.J.; visualization, A.R., K.W.; supervision, K.W.; Z.J. All authors have read and agreed to the published version of the manuscript.

Funding: This research received no external funding.

Acknowledgments: The authors would like to express gratitude to the NTOs' authorities for providing consent to use the visual material available on the official websites for the purpose of drawing this article, as well as for being open to cooperation.

Conflicts of Interest: The authors declare no conflict of interest.

References

1. Noti, E. The websites of National Tourism Organisations: A challenge of E-marketing. *Angl. J.* **2013**, *2*, 241–249.
2. Vyas, C. Evaluating state tourism websites using Search Engine Optimization tools. *Tour. Manag.* **2019**, *73*, 64–70. [CrossRef]
3. MacKay, K.; Vogt, C. Information technology in everyday and vacation contexts. *Ann. Tour. Res.* **2012**, *39*, 1380–1401. [CrossRef]
4. Law, R.; Buhalis, D.; Cobanoglu, C. Progress on information and communication technologies in hospitality and tourism. *Int. J. Contemp. Hosp. Manag.* **2014**, *26*, 727–750. [CrossRef]
5. Feng, R.; Morrison, A.M.; Ismail, J.A. East versus West: A comparison of online destination marketing in China and the USA. *J. Vacat. Mark.* **2004**, *10*, 43–56. [CrossRef]
6. Baloglu, S.; Pekcan, Y.A. The website design and Internet site marketing practices of upscale and luxury hotels in Turkey. *Tour. Manag.* **2006**, *27*, 171–176. [CrossRef]
7. Yuan, Y.-L.; Gretzel, U.; Fesenmaier, D.R. The role of information technology use in American convention and visitors bureaus. *Tour. Manag.* **2006**, *27*, 326–341. [CrossRef]
8. Choi, S.; Lehto, X.Y.; Oleary, J.T. What does the consumer want from a DMO website? A study of US and Canadian tourists' perspectives. *Int. J. Tour. Res.* **2007**, *9*, 59–72. [CrossRef]
9. Pai, C. K.; Xia, M.L.; Wang, T. W. A comparison of the official tourism website of five east tourism destinations. *Inf. Technol. Tour.* **2014**, *14*, 97–117. [CrossRef]
10. Malenkina, N.; Ivanov, S. A linguistic analysis of the official tourism websites of the seventeen Spanish Autonomous Communities. *J. Destin. Mark. Manag.* **2018**, *9*, 204–233. [CrossRef]
11. Mariani, M.M.; Baggio, R.; Buhalis, D.; Longhi, C. *Tourism Management, Marketing, and Development*; Palgrave Macmillan US: New York, NY, USA, 2014; ISBN 978-1-349-47471-4.
12. Liang, K.; Law, R. A modified functionality performance evaluation model for evaluating the performance of China based hotel websites. *J. Acad. Bus. Econ.* **2003**, *2*, 193–208.
13. Auger, P. The Impact of Interactivity and Design Sophistication on the Performance of Commercial Websites for Small Businesses. *J. Small Bus. Manag.* **2005**, *43*, 119–137. [CrossRef]
14. Buhalis, D. *ETourism. Information Technology for Strategic Tourism Management*; Financial Times Prentice Hall: New York, NY, USA, 2003; ISBN 0582357403.
15. Buhalis, D.; Law, R. Progress in information technology and tourism management: 20 years on and 10 years after the Internet—The state of eTourism research. *Tour. Manag.* **2008**, *29*, 609–623. [CrossRef]
16. Bastida, U.; Huan, T.C. Performance evaluation of tourism websites' information quality of four global destination brands: Beijing, Hong Kong, Shanghai, and Taipei. *J. Bus. Res.* **2014**, *67*, 167–170. [CrossRef]
17. Del Vasto-Terrientes, L.; Fernández-Cavia, J.; Huertas, A.; Moreno, A.; Valls, A. Official tourist destination websites: Hierarchical analysis and assessment with ELECTRE-III-H. *Tour. Manag. Perspect.* **2015**, *15*, 16–28. [CrossRef]

18. Fernández-Cavia, J.; Huertas-Roig, A. City Brands and their Communication through Web Sites: Identification of Problems and Proposals for Improvement. In *Information Communication Technologies and City Marketing: Digital Opportunities for Cities around the World*; Gascó-Hernandez, M., Torres-Coronas, T., Eds.; Information Science Reference, IGI Global: Hershey PA, USA, 2009; pp. 26–49.

19. Lee, W.; Gretzel, U. Designing persuasive destination websites: A mental imagery processing perspective. *Tour. Manag.* **2012**, *33*, 1270–1280. [CrossRef]

20. Baggio, R. A Websites Analysis of European Tourism Organizations. *Anatolia* **2003**, *14*, 93–106. [CrossRef]

21. Holland, J.; Baker, S.M. Customer participation in creating site brand loyalty. *J. Interact. Mark.* **2001**, *15*, 34–45. [CrossRef]

22. Barwise, P.; Elberse, A.; Hammond, K. Marketing and the Internet: A research review. In *Handbook of Marketing*; Weitz, B.A., Wensley, R., Eds.; SAGE: London, UK, 2002; pp. 527–558. ISBN 978-0761956822.

23. Ilfeld, J.S.; Winer, R.S. Generating Website Traffic. *JAR* **2002**, *42*, 49–61. [CrossRef]

24. Santos, J. The role of tour operators' promotional material in the formation of destination image and consumer expectations: The case of the People's Republic of China. *J. Vacat. Mark.* **1998**, *4*, 282–297. [CrossRef]

25. Wang, X.; Zhang, J.; Gu, C.; Zhen, F. Examining antecedents and consequences of tourist satisfaction: A structural modeling approach. *Tinshhua Sci. Technol.* **2009**, *14*, 397–406. [CrossRef]

26. Tang, L.; Jang, S. Investigating the Routes of Communication on Destination Websites. *J. Travel Res.* **2012**, *51*, 94–108. [CrossRef]

27. Hyun, M.Y.; O'Keefe, R.M. Virtual destination image: Testing a telepresence model. *J. Bus. Res.* **2012**, *65*, 29–35. [CrossRef]

28. Lehto, X.Y.; Kim, D.-Y.; Morrison, A.M. The effect of prior destination experience on online information search behaviour. *Tour. Hosp. Res.* **2006**, *6*, 160–178. [CrossRef]

29. Law, R.; Leung, R. A Study of Airlines' Online Reservation Services on the Internet. *J. Travel Res.* **2000**, *39*, 202–211. [CrossRef]

30. Govers, R.; Go, F. Projected destination image online: Website content analysis of pictures and text. *Inf. Technol. Tour.* **2004**, *7*, 73–89. [CrossRef]

31. Dwivedi, M. Online destination image of India: A consumer based perspective. *Int. J. Contemp. Hosp. Manag.* **2009**, *21*, 226–232. [CrossRef]

32. Chung, N.; Lee, H.; Lee, S.J.; Koo, C. The influence of tourism website on tourists' behavior to determine destination selection: A case study of creative economy in Korea. *Technol. Forecast. Soc. Chang.* **2015**, *96*, 130–143. [CrossRef]

33. Ho, C.-I.; Lee, Y.-L. The development of an e-travel service quality scale. *Tour. Manag.* **2007**, *28*, 1434–1449. [CrossRef]

34. Zins, A.H. Exploring Travel Information Search Behaviour beyond Common Frontiers. *Inf. Technol. Tour.* **2007**, *9*, 149–164. [CrossRef]

35. Yovcheva, Z.; Buhalis, D.; Gatzidis, C.; van Elzakker, C.P.J.M. Empirical Evaluation of Smartphone Augmented Reality Browsers in an Urban Tourism Destination Context. *Int. J. Mob. Hum. Comput. Interact.* **2014**, *6*, 10–31. [CrossRef]

36. Horng, J.-S.; Tsai, C.-T. Government websites for promoting East Asian culinary tourism: A cross-national analysis. *Tour. Manag.* **2010**, *31*, 74–85. [CrossRef]

37. Stepchenkova, S.; Tang, L.; Jang, S.; Kirilenko, A.P.; Morrison, A.M. Benchmarking CVB website performance: Spatial and structural patterns. *Tour. Manag.* **2010**, *31*, 611–620. [CrossRef]

38. Cano, V.; Prentice, R. Opportunities for endearment to place through electronic 'visiting': WWW homepages and the tourism promotion of Scotland. *Tour. Manag.* **1998**, *19*, 67–73. [CrossRef]

39. Gretzel, U.; Yuan, Y.-L.; Fesenmaier, D.R. Preparing for the New Economy: Advertising Strategies and Change in Destination Marketing Organizations. *J. Travel Res.* **2000**, *39*, 146–156. [CrossRef]

40. Hvass, K.A.; Munar, A.M. The takeoff of social media in tourism. *J. Vacat. Mark.* **2012**, *18*, 93–103. [CrossRef]

41. Xiang, Z.; Gretzel, U. Role of social media in online travel information search. *Tour. Manag.* **2010**, *31*, 179–188. [CrossRef]

42. Chung, T.; Law, R. Developing a performance indicator for hotel websites. *Int. J. Hosp. Manag.* **2003**, *22*, 119–125. [CrossRef]

43. Morrison, A.M.; Taylor, J.S.; Douglas, A. Website Evaluation in Tourism and Hospitality. *J. Travel Tour. Mark.* **2004**, *17*, 233–251. [CrossRef]

44. Ting, P.-H.; Kuo, C.-F.; Li, C.-M. What Does Hotel Website Content Say About a Property—An Evaluation of Upscale Hotels in Taiwan and China. *J. Travel Tour. Mark.* **2012**, *29*, 369–384. [CrossRef]

45. Law, R.; Qi, S.; Leung, B. Perceptions of Functionality and Usability on Travel Websites: The Case of Chinese Travelers. *Asia Pac. J. Tour. Res.* **2008**, *13*, 435–445. [CrossRef]

46. Law, R. Evaluation of hotel websites: Progress and future developments (invited paper for 'luminaries' special issue of International Journal of Hospitality Management). *Int. J. Hosp. Manag.* **2019**, *76*, 2–9. [CrossRef]

47. Hose, T.A. Telling the story of stone—Assessing the client base. In *Geological and Landscape Conservation*; O'Halloran, D., Green, C., Harley, M., Stanley, M., Knill, J., Eds.; Geological Society: London, UK, 1994.

48. Hose, T.A. Geotourism—Selling the Earth to Europe. In *Engineering Geology and the Environment*; Marinos, P.G., Koukis, G.C., Tsiambaos, G.C., Stournaras, G.C., Eds.; AA Balkema: Rotterdam, The Netherlands, 1997; pp. 2955–2960.

49. Hose, T.A. European geotourism—Geological interpretation and geoconservation promotion for tourists. In *Geological Heritage: Its Conservation and Management*; Barretino, D., Wimbledon, W.P., Gallego, E., Eds.; Instituto Tecnologico Geominero de Espana: Madrid, Spain, 2000; pp. 127–146.

50. Hose, T.A. Geotourism: Appreciating the deep time of landscapes. In *Niche Tourism: Contemporary Issues, Trends and Cases*; Novelli, M., Ed.; Routledge: London, UK, 2005; pp. 27–37.

51. Hose, T.A. The English Origins of Geotourism (as a Vehicle for Geoconservation) and Their Relevance to Current Studies. *AGS* **2011**, *51*, 343–359. [CrossRef]

52. Newsome, D.; Dowling, R.K. *Geotourism. The Tourism of Geology and Landscape*; Goodfellow: Oxford, UK, 2010.

53. Hose, T.A.; Vasiljević, D.A. Defining the Nature and Purpose of Modern Geotourism with Particular Reference to the United Kingdom and South-East Europe. *Geoheritage* **2012**, *4*, 25–43. [CrossRef]

54. Gray, M. Geodiversity and geoconservation: What, why, and how? In Papers Presented at the George Wright Forum. Santucci, V.L., Ed.; 2005, pp. 4–12. Available online: http://www.georgewright.org/223gray.pdf (accessed on 21 October 2019).

55. Gray, M. *Geodiversity. Valuing and Conserving Abiotic Nature*; John Wiley & Sons: Chichester, UK, 2004.

56. Melelli, L. "Perugia Upside-Down": A Multimedia Exhibition in Umbria (Central Italy) for Improving Geoheritage and Geotourism in Urban Areas. *Resources* **2019**, *8*, 148. [CrossRef]

57. Vuković-Vojnović, D.; Nićin, M. English as a Global Language in the Tourism Industry: A Case Study. In *The English of Tourism*; Rață, G., Petroman, I., Petroman, C., Eds.; Cambridge Scholars Publishing: Newcastle upon Tyne, UK, 2012; pp. 3–18.

58. Polish Tourism Organization. Poland's Highlights 2019/2020 #VisitPoland #Polandtravel. Available online: https://www.poland.travel/attachments/article/26117/Poland%20Handout%202020%20new.pdf (accessed on 21 October 2019).

59. Polish Tourism Organization. New UNESCO Site in Poland. Available online: https://www.poland.travel/en/travel-inspirations/new-unesco-site-in-poland (accessed on 21 October 2019).

60. Lubelskie Marshall's Office. Available online: http://www.turystyka.lubelskie.pl/index.php/lrt/page/135/ (accessed on 16 October 2019).

61. West Pomeranian Marshall's Office. Available online: http://wzp.pl/ (accessed on 18 October 2019).

62. Świętokrzyskie Voivodeship Office. Available online: https://www.kielce.uw.gov.pl/en/swietokrzyskie-province/3,Swietokrzyskie-Province.html (accessed on 18 October 2019).

63. Visit Silesia. Available online: http://www.visitsilesia.eu/ (accessed on 15 October 2019).

64. Polish Tourism Organization. Śląskie—Industrial Adventure. Available online: https://www.poland.travel/en/regions/slaskie-voivodship-industrial-adventure (accessed on 22 October 2019).

65. Polish Tourism Organization. Hiking in Poland—Sudety. Available online: https://www.poland.travel/en/experience/hiking/hiking-in-poland-guide-%E2%80%93-sudety (accessed on 22 October 2019).

66. Erfurt-Cooper, P. The Importance of Natural Geothermal Resources in Tourism. In Proceedings of the World Geothermal Congress 2010, Bali, Indonesia, 25–29 April 2010; Available online: https://www.geothermal-energy.org/pdf/IGAstandard/WGC/2010/3318.pdf (accessed on 15 October 2019).

67. Polish Tourism Organization. Termy w Polsce Sposobem na Jesienną Chandrę. Available online: https://www.polska.travel/pl/zobacz-koniecznie/pomysl-na/termy-w-polsce-sposobem-na-jesienna-chandre (accessed on 22 October 2019).

68. Polish Tourism Organization. Poland: Major Tourist Attractions. Available online: https://pdf.polska.travel/docs/en/hit/Hity_en.pdf?_ga=2.177685733.176097391.1573110931-1079215847.1557493001 (accessed on 23 October 2019).
69. Polish Tourism Organization. Muskauer Park: Fulfilled Vission of Paradise. Available online: https://www.poland.travel/attachments/article/6505/UNESCO_EN.pdf (accessed on 23 October 2019).
70. Polish Tourism Organization. Geoparki. Available online: https://www.polska.travel/pl/poznaj-atrakcje-i-zabytki/zabytki-i-inne-atrakcje/geoparki-2 (accessed on 21 October 2019).
71. Global Geoparks Network. Muskau Arch Geopark—GERMANY/POLAND. Available online: http://www.europeangeoparks.org/?page_id=656 (accessed on 20 October 2019).
72. Ministry of the Environment. Available online: https://www.gov.pl/web/srodowisko (accessed on 21 October 2019).
73. Polish Geological Institute. Geotourism. Available online: https://www.pgi.gov.pl/geoturystyka-606.html (accessed on 21 October 2019).
74. Polish Geological Institute. Geoturystyka. Available online: https://www.pgi.gov.pl/oferta-inst/geoturystyka.html (accessed on 21 October 2019).
75. Polish Tourism Organization. Geopark in Kielce. Available online: https://www.poland.travel/en/heritage/archeological-sites/geopark-in-kielce (accessed on 24 October 2019).
76. Geopark Kielce. Historia Geoparku Kielce. Available online: http://geopark-kielce.pl/o-nas/historia-geoparku-kielce/ (accessed on 20 October 2019).
77. Polish Tourism Organization. Embedded Search Engine Results for the Word 'Geology'. Available online: https://www.poland.travel/en/finder?ordering=newest&searchword=geology (accessed on 24 October 2019).
78. Polish Tourism Organization. Plan Your Trip. Available online: https://travelplanner.poland.travel/planner/selectAttractions?poisPerPage=19 (accessed on 24 October 2019).
79. Czech Tourism. Bohemian Switzerland. Available online: https://www.czechtourism.com/c/bohemian-switzerland-national-park/ (accessed on 25 October 2019).
80. Czech Tourism. Bohemian Paradise. Available online: https://www.czechtourism.com/s/bohemian-paradise-walk/ (accessed on 27 October 2019).
81. Czech Tourism. Bohemian Paradise along the Greenway Jizera. Available online: https://www.czechtourism.com/s/bohemian-paradise-greenway-jizera-cycling/ (accessed on 27 October 2019).
82. Czech Tourism. Bohemian Switzerland. Available online: https://www.czechtourism.com/s/hrensko-soutesky-walk/ (accessed on 28 October 2019).
83. Czech Tourism. Natural Heritage. Available online: https://www.czechtourism.com/a/natural-heritage/ (accessed on 24 October 2019).
84. Czech Tourism. Rock towns. Available online: https://www.czechtourism.com/a/rock-towns/ (accessed on 25 October 2019),
85. Czech Tourism. UNESCO Global Geopark. Available online: http://pdf.czechtourism.com/unesco_en/ (accessed on 24 October 2019).
86. Czech Tourism. The Adventurous Heart of Bohemia. Available online: https://www.czechtourism.com/tourists/stories/cesky-raj/ (accessed on 25 October 2019).
87. Czech Tourism. Návrh geoturistického produktu Geoparku Vysočina. Available online: https://www.czechtourism.cz/getattachment/Pro-media/Tiskove-zpravy/TOP-prace-studentu-Service-design-i-geoturismus/Navrh-geoturistickeho-produktu-Geoparku-Vysocina.pdf.aspx?ext=.pdf (accessed on 30 October 2019).
88. Czech Tourism. Search Results: Geopark. Available online: https://www.czechtourism.com/tourists/search/?searchtext=geopark&searchmode=anyword (accessed on 25 October 2019).
89. Czech Tourism. Search Results for the Search Term "Geology". Available online: https://www.czechtourism.com/tourists/search/?searchtext=geology&searchmode=anyword (accessed on 30 October 2019).
90. Czech Tourism. Brdy Mountains. Available online: https://www.czechtourism.com/c/brdy-mountains/ (accessed on 25 October 2019).
91. Czech Tourism. Pravčická brána: Get to know a European Great: The Pravčická Brána. Available online: https://www.czechtourism.com/c/pravcicka-brana/ (accessed on 31 October 2019).
92. Slovakia Travel. Tatras and Northern Spiš. Available online: http://slovakia.travel/en/places-to-go/the-best-of-slovakia/tatras (accessed on 26 October 2019).

93. Slovakia Travel. You Will Find (Only) in Slovakia. Available online: http://slovakia.travel/en/places-to-go/unique-slovakia#prettyPhoto (accessed on 25 October 2019).
94. Slovakia Travel. Geoparks: Attractive Geoparks Luring for Hiking. Available online: http://slovakia.travel/en/things-to-see-and-do/nature-and-the-countryside/geoparks (accessed on 25 October 2019).
95. Slovakia Travel. Novohrad Geopark. Available online: http://slovakia.travel/en/novohrad-geopark (accessed on 25 October 2019).
96. Bansko Bystricky Geopark. SLOVAKIA—LAG Banská Bystrica Geopark. Available online: http://www.geoparkbb.sk/en (accessed on 26 October 2019).
97. Slovakia Travel. Subterranean Treasures. Available online: http://slovakia.travel/en/things-to-see-and-do/nature-and-the-countryside/caves (accessed on 25 October 2019).
98. Slovakia Travel. Show Caves in Slovakia. List of Accessible Caves in Slovaki. Available online: https://sacr3-files.s3-eu-west-1.amazonaws.com/PDF%20zoznamy/Jaskyne/Cave_EN.pdf (accessed on 27 October 2019).
99. Slovakia Travel. National Park of Slovenský Raj. Available online: http://slovakia.travel/en/national-park-of-slovensky-raj (accessed on 25 October 2019).
100. Slovakia Travel. Map. Available online: http://slovakia.travel/en/travel-essentials/map (accessed on 25 October 2019).
101. Slovakia Travel. Vedúca Skamenelina. Available online: https://trip.slovakia.travel/event/8916bd61-1d6c-4c94-8fb6-a3461e56ff15?lang=en (accessed on 25 October 2019).
102. Slovakia Travel. Planner. Available online: https://trip.slovakia.travel/explore?categories=bb05c2c7-6eeb-4e46-92d4-9ed3972e06d6,33796e39-e303-4848-a975-ea0d2369b9e9&lang=en¢er=19.7011,48.6402&accommodationPersons=2 (accessed on 25 October 2019).
103. Slovakia Travel. Banská Bystrica Geopark. Available online: http://slovakia.travel/en/banska-bystrica-geopark (accessed on 25 October 2019).
104. Slovakia Travel. Results to the Search Term 'Geosite'. Available online: http://slovakia.travel/en/search?q=geosite (accessed on 25 October 2019).
105. Slovakia Travel. Košice Region Tourism (app). Available online: http://slovakia.travel/en/kosice-region-tourism-app (accessed on 31 October 2019).
106. Slovakia Travel. Mobile Applications. Available online: http://slovakia.travel/en/travel-essentials/mobile-applications?tx_viator_list%5Bdata_3012%5D%5Bfilters%5D%5B1%5D%5B%5D=748 (accessed on 1 November 2019).
107. Slovakia Travel. Planner. Available online: https://trip.slovakia.travel/explore?lang=en¢er=19.7011,48.6402&categories=75994246-67ba-49a0-b868-bc33ac58d6c9 (accessed on 25 October 2019).
108. Polish Tourism Organization. Underground Attractions. Available online: https://www.poland.travel/en/discover/underground-attractions (accessed on 15 October 2019).
109. Slovakia Travel. Region of Gemer and Malohont. Available online: http://slovakia.travel/en/region-of-gemer-and-malohont (accessed on 22 October 2019).
110. Slovakia Travel. Small Area but with Nine National Parks. Available online: http://slovakia.travel/en/things-to-see-and-do/nature-and-the-countryside/national-parks (accessed on 22 October 2019).
111. Global Geoparks Network. What is a UNESCO Global Geopark? Available online: http://www.globalgeopark.org/aboutGGN/6398.htm (accessed on 25 October 2019).
112. Czech Tourism. Bohemian Paradise—UNESCO Geopark: The Fairytale Bohemian Paradise Area. Available online: https://www.czechtourism.com/c/bohemian-paradise-unesco-geopark/ (accessed on 23 October 2019).
113. Słomka, T.; Kicińska-Świderska, A.; Doktor, M.; Joniec, A. Katalog obiektów geoturystycznych w Polsce (obejmuje wybrane geologiczne stanowiska dokumentacyjne). Available online: https://archiwum.mos.gov.pl/g2/big/2009_07/8f91b006e724bc23a00b8bb5779d22bb.pdf (accessed on 18 December 2019).
114. Bartuś, T.; Bębenek, S.; Doktor, M.; Golonka, J.; Ilcewicz-Stefaniuk, D. The Catalogue of Geotourist Sites in Nature Reserves and Monuments. Available online: http://www.kgos.agh.edu.pl/download/Katalog_Obiekow_Geoturystycznych_2012.pdf (accessed on 18 December 2019).

resources

Article

Protected Areas: Geotourist Attractiveness for Weekend Tourists Based on the Example of Gorczański National Park in Poland

Krzysztof Widawski [1,*], Piotr Oleśniewicz [2], Agnieszka Rozenkiewicz [1], Anna Zaręba [1] and Soňa Jandová [3,4]

1   Institute of Geography and Regional Development, University of Wrocław, pl. Uniwersytecki 1, 50-137 Wrocław, Poland; agnieszka.rozenkiewicz@uwr.edu.pl (A.R.); anna.zareba@uwr.edu.pl (A.Z.)
2   Faculty of Physical Education, University School of Physical Education in Wrocław, al. Ignacego Jana Paderewskiego 35, 51-612 Wrocław, Poland; piotr.olesniewicz@awf.wroc.pl
3   Faculty of Mechanical Engineering, Technical University of Liberec, Studentská 2, 461 17 Liberec, Czech Republic; sona.jandova@tul.cz
4   Department of Sports and Outdoor Activities, College of Physical Education and Sport PALESTRA, Slovačíkova 400/1, 197 00 Prague 19, Czech Republic
*   Correspondence: krzysztof.widawski@uwr.edu.pl; Tel.: +48-71-37-59-579; Fax: +48-71-343-5184

Received: 1 February 2020; Accepted: 23 March 2020; Published: 25 March 2020

Abstract: The aim of the publication was to assess the geotourist attractiveness of protected areas in Poland among weekend tourists based on the example of Gorczański National Park. The park location near urbanized areas makes it an attractive field for research on weekend tourism development. The tourist potential of the park is presented, starting from geological aspects and geotourist values. Then, the tourist potential was analysed, with a focus on geotourist resources, which include tourist trails and didactic routes. The tourist traffic volume was also examined. On the basis of legal documents, such as nature conservation plans, threats related to tourism development in protected areas were presented as indicated by park managers. In accordance with the Act on Nature Conservation, the threats are divided into four groups: internal existing and potential threats and external existing and potential threats. The tourists' opinion on the geotourist attractiveness of the park was investigated with surveys conducted during selected weekends significant in the context of tourist traffic volume. Thus, a profile of people visiting the park for short stays was obtained, as well as their assessment of the tourist resources of the area, with particular emphasis on geotourist values.

Keywords: geotourism; protected areas; weekend tourism; tourists' opinion; national parks

1. Introduction

The very idea of nature conservation appeared in 1872, when the first national park, Yellowstone, was established [1,2]. Since then, protected areas have been exceptionally attractive for tourist penetration. Notably, the first national park in the world's history is still considered a most appealing destination because of its geotourist character [3]. For many decades, tourism and nature conservation have coexisted in the tourist space, as highlighted by the subject literature [4]. The popularity of tourism based on natural resources, both animated and inanimate, increased at the turn of the last century [5–9]. Such an increase can always have, and usually has, a twofold impact on the area itself, its resources, and its environment. On the one hand, it translates into raising environmental awareness of tourism participants. On the other hand, the uncontrolled presence of tourists itself may pose a threat to any kind of natural resources protected by the national park institution, including the inanimate ones [10]. These two aspects are indicated in many literature sources. Together with the growing tourist awareness of the need to protect inanimate natural resources, the need for geoconservation was pointed at. The main assumption of this idea is the demand for proper management of the geological heritage, substantial for both aesthetic and educational reasons [11,12]. It should be accompanied by appropriate promotion of the values themselves as well as adequate attitudes [13,14]. This standpoint is an essential part of a broader approach to the protection of geoheritage, which should be reflected in properly defined geoconservation strategies [15–18].

However, the role of tourism should not be limited to minimizing the possible losses associated with its presence in a space requiring protection due to the uniqueness of its resources. Tourism can and should contribute to the protection of natural heritage, as pointed out by many authors [1,19–23]. The key to success is proper environmental education, which should bring definite results and give impetus for further actions. One of the basic effects should be an appropriate attitude of the local community when using the natural resources of the region [24]. A sustainable approach is fundamental, so that resources can be utilized in a long perspective [25–27]. To this end, the region should ensure that its tourist infrastructure is properly developed. Correct handling of tourism is often the key to success, as implied by Bushell [28] or Priskin [29]. In their action strategy, regional authorities should take into account the active role of the local community. Only then it is possible to preserve resources in a relatively unchanged form [30,31], especially if the participation of the local community in the protection translates into specific income related to the development of tourism in the protected area, usually located in so-called marginal areas, away from large industrial or service centres [32].

Protected areas are attractive for many contemporary forms of tourism. For years, they have been associated, not without reason, with green tourism, ecotourism, or nature tourism [4,19–23]. This is natural, at least because of the specific tourist resources assigned to these types of tourist spaces. In addition, other forms of active or specialized tourism, characteristic of the natural environment, are also developing. These include skiing, cycling, and horse-riding tourism, as well as mountain climbing and speleological tourism, requiring specialized knowledge and skills. Against this background, the last three decades have seen a relatively new form of tourism, geotourism, also present in protected areas.

The concept was defined in the middle of the last decade of the previous century. Hose [33–37] was among its first investigators. Hose and Vasiljević ([38], p. 27) pointed at geotourism as a form that constitutes "the provision of interpretive and service facilities enabling tourists to acquire knowledge and understanding of the geology and geomorphology of a site (including its contribution to the development of the Earth sciences) beyond the level of mere aesthetic appreciation". The geological aspect in the development of this form of tourism is important [38]. The context of sustainable development, which should accompany this form of tourism, was highlighted by Newsome and Dowling [39] in their understanding of geotourism as drawing its attractiveness from the heritage of the inanimate nature of the tourist region.

A natural place for the development of geotourism seems to be protected areas, by definition offering nature preserved in an unchanged or barely changed form. The subject literature [11,40] indicates nature reserves, landscape parks, natural parks, but above all, national parks. Owing to their resources, their location as usually far from urbanized areas, and their general attractiveness, national parks are visited by a greater number of tourists year after year. Increased tourist traffic can pose a threat to the protected nature in the park area [41–45]. In the case of inanimate nature, usually trampling, collecting, changes in soil structure, and erosion are mentioned [41,46–56]. Therefore, for the proper functioning of the national park and management of its resources, it is crucial to investigate all forms of tourist traffic occurring in its area [15–18,57,58].

The main aim of this article was to present the tourist attractiveness of Gorczański National Park in the opinion of its weekend visitors. For this purpose, a questionnaire was carried out in three selected periods between 2016 and 2017; responses were collected from 244 properly surveyed tourists who visited the park over a weekend (details are described in the Section 3). Since the publication focuses primarily on the geotourist aspect and its evaluation among the park visitors, geotourist values were chosen to be presented first, starting with a brief introduction to the Gorce Mountains geology. Next, we describe the elements of infrastructure that simultaneously constitute inanimate nature resources in a linear form, such as tourist trails and didactic routes. The analysis of the opinions of weekend tourists visiting the park is preceded by a short presentation of the volume of tourist traffic recorded in the second decade of the 21st century. Its increase in recent years clearly shows the potential of Gorczański National Park, one of the most interesting mountain national parks of Poland, as a tourist destination. Its location near one of the country's major agglomerations, Kraków, as well as the Silesian conurbation, makes it an almost natural weekend tourism destination in this part of the country, and thus an interesting field of research on the geotourist attractiveness of this protected area in the opinion of both ordinary tourists and those with the "geo-" prefix.

2. Study Area

2.1. Gorczański National Park Location

Gorczański National Park is entirely located in the Gorce Mts., surrounded by the Beskid Wyspowy Mts. in the north, the Beskid Sądecki Mts. in the east, the Orawa-Nowy Targ Basin in the south, and the Orawa-Jordanów Foothills in the west. The park occupies 14% of the range area, which is about 500 km^2 [59].

The Gorce Mts. is a homogeneous mountain range belonging to the intermediate mountains. The landscape of Gorczański National Park involves primarily forests and glades. Slightly more than half of the range rises above 600 m a.s.l. Relatively few summits exceed 1000 m a.s.l. The highest peaks include Turbacz (1310 m a.s.l.), Jaworzyna Kamienicka (1228 m a.s.l.), and Kiczora (1282 m a.s.l.). The structure of the Gorce Mts. is distinctive. Six ridges, separated by V-shaped river valleys, spread out from Turbacz in different directions. The two longest ridges spread out in opposite directions: south-east and north-west.

The park was founded on 1 January 1981 by the Ordinance of the Council of Ministers of 8 August 1980 on the creation of Gorczański National Park [60]. In 1988, the Minister for Agriculture, Forestry, and Food Economics incorporated 805 ha of farmland forests. The park involves enclaves: the eastern slopes of Bardo; northern slopes of Jaworzyna Ponicka; Kędzierski forest; Mechów forest; and Morgi Czajkowskie forests. The area of the park currently covers 7019 ha [61] (Figure 1).

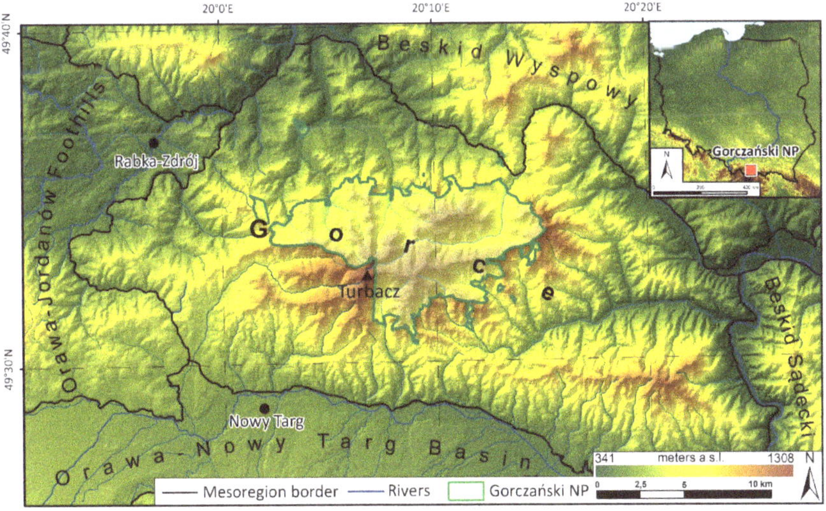

Figure 1. Gorczański National Park.

2.2. Tourist Attractiveness of Gorczański National Park

Tourist attractiveness refers to tourist resources, tourist infrastructure, and communication accessibility [62]. The main role in geotourism is undoubtedly played by the related values generating tourist traffic in specific regions. The Polish subject literature divides tourist resources into two main groups: natural and cultural [61–66]. These constitute a group of sightseeing values [67]. For the purposes of this research, the group of natural values is particularly important; they are a natural reservoir of geotourist values. It is worth emphasizing that not all elements of animated or inanimate nature can be regarded a tourist value. Only those that meet certain conditions may be included in this group:

- they are big enough to be recognizable in the landscape;
- they arouse interest among tourists;
- they are inherently resistant to tourist traffic in their immediate vicinity;
- they are properly adapted to the reception of tourist traffic [67].

A more detailed division of natural values introduces the following three main groups:

- Values that are not influenced by human activity. These are caves and grottoes, groups of rocks, valleys and gullies, erratics, scenic viewpoints presenting geologically and geomorphologically attractive landscapes, and among the elements of an animated nature—peculiarities of fauna and flora.
- Values whose character and shape have been significantly influenced by human activity. This group includes primarily assets of an animated nature, such as historic parks, botanical, and zoological gardens.
- Values of surface character. Human activity did not affect their nature or shape. These are protected areas that constitute a reservoir of many assets of an animated and inanimate nature, protected because of their unique character. In this group, the most important type is national parks, which in Poland have the status of the highest form of nature conservation.

At this point, it is worthwhile to take a brief look at the potential of tourist resources of an inanimate nature in one of the most attractive mountain national parks of Poland: Gorczański National Park.

2.3. Geology of Gorczański National Park

The Gorce Mts. range is a part of the Outer Western Carpathians. The Gorce Mts. are built of flysch rocks of the Magura nappe—sandstones, conglomerates, and shales of the Early and Middle Eocene [68,69]. Structurally, the Magura nappe is the uppermost tectonic element of the Polish Outer Carpathian wedge [70]. Thick-bedded layers of sandstones sedimented in Palaeocene and Eocene (ca. 65–37.5 million years ago) constitute the dominant complex, with a thickness exceeding 2000 m. The central part of the mountain chain is mainly built of Magura sandstones and of conglomerates. These rocks build both the ridge parts of the Gorce Mts. and their highest peak, Turbacz (1310 m a.s.l.), and also form tors of original shapes. The border parts of Gorczański National Park are built of sandstones and shales, with numerous valleys, passes, and depressions resulting from erosion processes [71].

The unique geological character of the Gorce Mts. is typical, especially for its northern part, where Gorczański National Park was established. This is determined primarily by the variety and degree of exposure of the Magura nappe rocks occurring in the Gorce Mts. area. The unique geological character of the Gorce Mts. is typical especially for its northern part, where Gorczański National Park was established [72]. This is determined primarily by the variety and degree of exposure of the Magura nappe rocks occurring in the Gorce Mts. area (Figure 2).

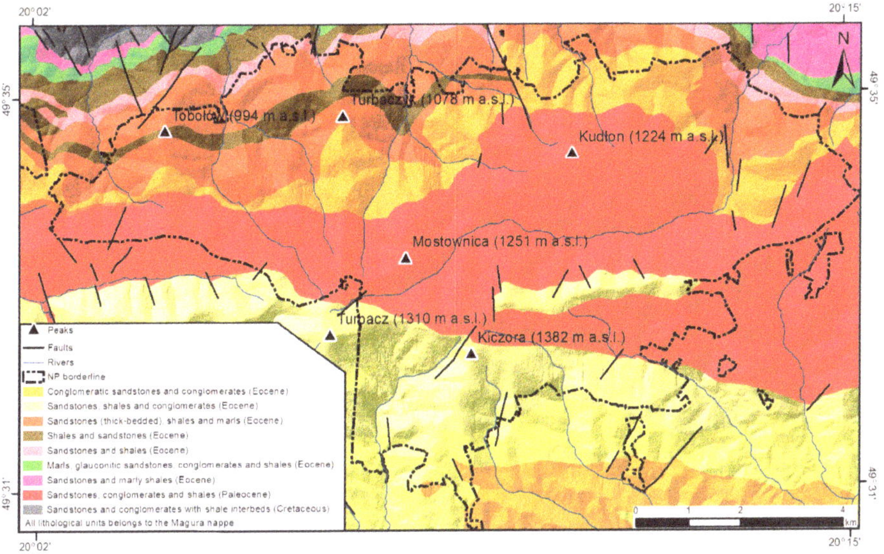

Figure 2. Gorczański National Park geological map. Source: [72].

2.4. Geotourist Values of the Gorce Mts

On the basis of the description of geological scientific values of Gorczański National Park and its surroundings, as suggested by many authors [72,73], the geotourist assets of the indicated area can be divided into three main groups, i.e., tors, mineral and thermal waters, and rock exposures (both natural and artificial).

The tors, most common on the ridges or in the upper parts of the slopes, are extremely varied in terms of shape and size. They consist of thick-bedded sandstones and thin conglomerates. The genesis of tors, either single or in groups, often with accompanying chasms and caves, is associated with several types of natural processes, i.e., with landslide movements, weathering, erosion, and denudation. The northern slopes of the Kudłoń chain are particularly rich in tors, with the most famous attraction being Kudłoński Baca. Zbójecka Jama by the Jaworzyna Kamienicka is considered to be the largest cave in the Gorce Mts. Pucułowski Stawek by the Wysznia glade, an example of a pond formed in a recess of a landslide niche, is an interesting geotourist asset.

The second group of geotourist values of the Gorce Mts. are mineral and thermal waters. Rabka and Szczawa are two health resorts proud of their mineral waters. Rabka (Figure 1) is known for its salt waters ([72] after [74]), including sodium chloride, bromide, iodide, and barium waters. Thermal water was found in deep drillings in Poręba Wielka and Rabka ([72] after [75]).

The rock exposures (both natural and artificial, e.g., quarries) are the third main type of Gorce Mts. assets significant from the point of view of geotourism. A very good degree of flysch rocks exposure, often in a continuous form, occurring especially in the northern part of the chain, mainly in riverbeds and at the banks of rivers and streams, is a distinctive element of the Gorce Mts.

2.5. Other Natural Sightseeing Values (Figure 3)

Sightseeing values important from the point of view of geotourism also include scenic viewpoints, as indicated in the literature. It should be stressed that it is not so much the place that is important here as the panorama that can be admired. Thus, it is a point value in the physical sense, but its essence is a fragment of the landscape which can be appreciated. It is a very significant value, especially for lovers of inanimate nature, who can contemplate and learn about the effects of geological and geomorphological processes on the macroscale.

A total of 24 scenic viewpoints were set up in the park. Their spatial distribution is shown in Figure 3. They are evenly arranged, with a slight concentration in the southern part, so that the tourist visiting the park has a full overview of the most attractive panoramas offered by the park.

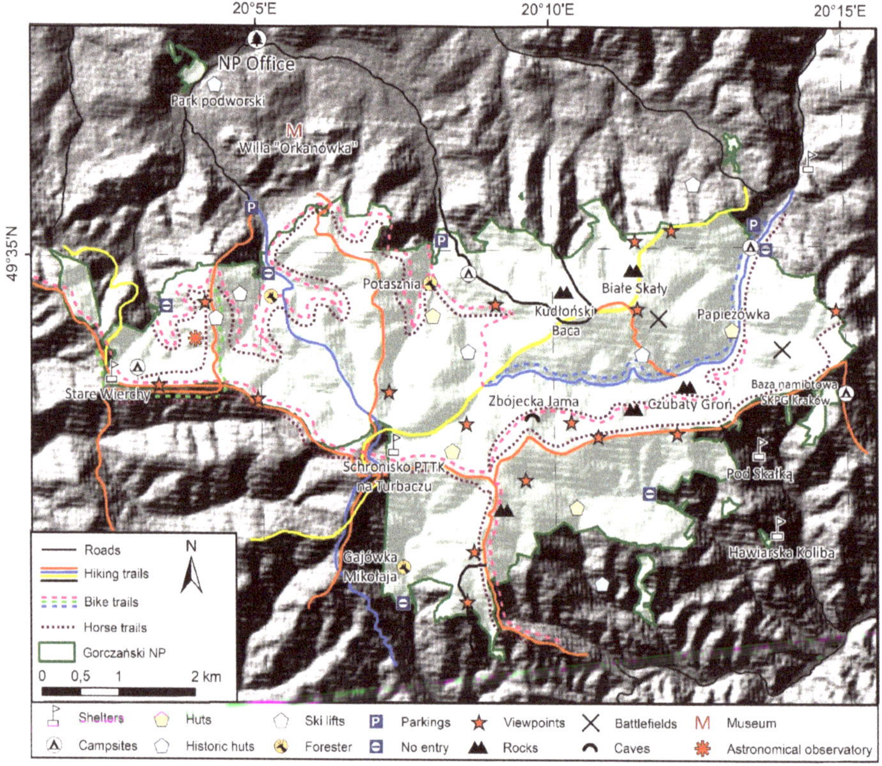

Figure 3. Selected natural and cultural values and tourist infrastructure of Gorczański National Park.

2.6. Accompanying Infrastructure

The accompanying infrastructure includes all facilities, both tourist and paratourist, which facilitate or enable tourists to practise specific forms of tourism. In the case of protected areas, these are usually selected forms of active tourism and cognitive tourism. In this group, geotourism plays an increasingly important role year after year. Among the devices that are present in the tourist space of the park, there are resources important for practising educational tourism. The entire teaching and administrative base is listed in Table 1.

Table 1. Facilities of the accompanying base of Gorczański National Park for the needs of educational tourism.

Facility	Tourist Function
Management building	Educational and conference facilities in Poręba Wielka, conference and exhibition room for temporary exhibitions
Gajówka Mikołaja	The educational and exhibition base in the Łopuszna valley
Astronomical observatory	The highest located observatory in Poland, peak Suhory
The education hall in the park hut in Hala Długa	Educational activities, exhibition of the history of nature and pastoral economy in the Gorce Mts.
PTTK Mountain Tourism Culture Centre at Turbacz	Museum of mountain tourism history in the Gorce Mts. Surrounding area of Turbacz
Papieżówka shelter	The place of education classes, exhibitions; an exhibition about John Paul II
Shepherd's hut	Educational classes

Among other elements of the accompanying base, there are two ski lifts: the Tobołów cableway (a chairlift) and a surface lift at Koninki ski station. In addition, there are 10 forester's lodges in the park, three resting places (Obidowiec and camping fields Oberówka and Trusiówka), a roofed shelter on the route between Tobołczyk and Tobołów, and four car parks (Trusiówka, Potasznia, Koninki–Hucisko, and Parzygnatówka). Moreover, in the area of the park, sanitary facilities, information boards, and waste bins have been placed where tourist traffic is concentrated. There are also two GOPR (Mountain Rescue Service) buildings at Stare Wierchy and Tobołczyk.

2.7. Hiking, Bike, and Horse Tourist Trails

One of the elements of the accompanying base for transport accessibility within the region is tourist trails. There are hiking, bike, and horse trails (Figure 4). In the first group, there are 10 trails of varying degrees of difficulty, time to be devoted to its passage and qualities that are presented on the route. The time of passing, depending on the trail, is from one hour and a quarter to 7.5 h. The thematic scope of the trails is diverse. The blue trail coincides with the Papal Trail, which imitates the course of the favourite routes of the later Pope John Paul II, who often visited the park. The yellow trail, five hours long, passes through 10 glades, where there are huts associated with the pastoral economy of the region.

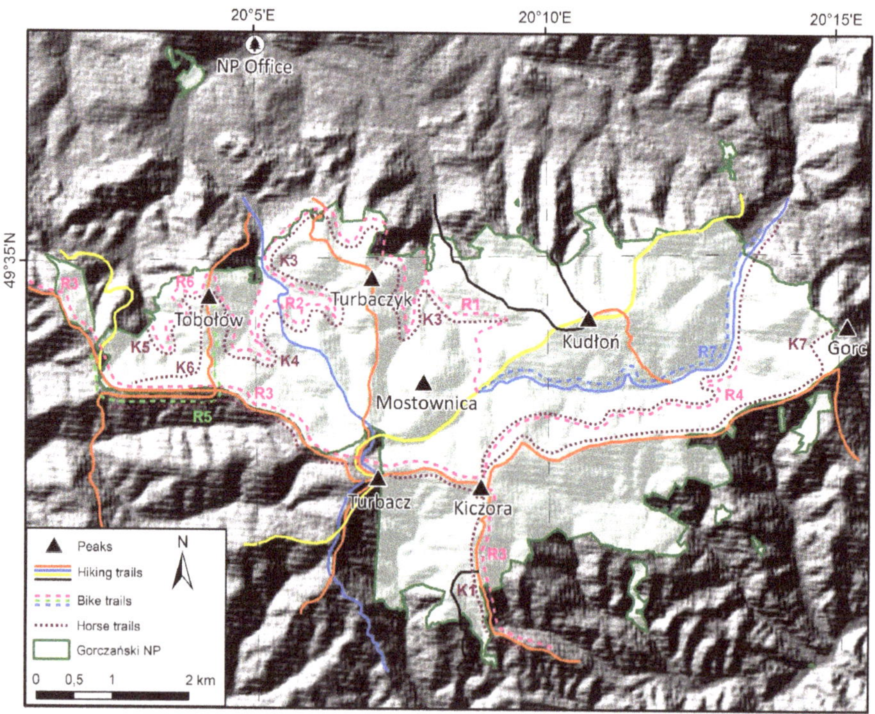

Figure 4. Hiking, bike, and horse trails in Gorczański National Park.

It is worth noting the black trail in the southern part of the park. The walk time is two hours and the theme is the value of the animated and inanimate nature of the park.

For those interested in geology and geomorphology, another black trail, in the northern part of the park, is important. The time of the trail is 4.5 h. On the steep slopes, there are numerous exposures of Carpathian flysch with visible layers of shales and sandstones. There are also several rock outcrops.

Another important geotourism point is the blue trail from the Trusiówka glade through the Kamienica valley to the Borek pass in the eastern part of the park. It will take about 3 h to get there. On the route, we can observe exposures of Carpathian flysch with visible layers of shales, sandstones, and conglomerates. They are a result of water erosion. The effects of the destructive activity of mountain streams can be observed here.

This type of trails is the park's specialty. Their characteristic feature is their low level of difficulty. The main recipients of the offer are families with children.

In total, the trails in the park have 74 km. Bike and horse trails are shorter. The total length of bike trails in the park is 53.5 km. The average length of a trail is 7.6 km, with the longest one—R3—being 11.3 km, and the shortest one—R5—being 3.3 km.

The six horse trails are 47.3 km long. The longest ones are K1—11.3 km and K2—11.4 km. The shortest is K6—only 3.3 km. It should be emphasized that almost 90% of tourists move around the park on foot.

2.8. Didactic Routes

The most important element, also exploited by geotourism, is thematic didactic routes, i.e., hiking trails marked out in a way to include as many objects important from the didactic point of view as possible. Significant components of this kind of routes are stopping points, here referred to as geotourist stops, which expose interesting geological and geomorphological assets. They are accompanied by information boards that present the assets and explain the associated phenomena and processes. There is a total of 10 didactic routes in the park (Figure 5). In almost every case, the leading theme is of an animated and inanimate nature. For two of them, the cultural heritage and history of the region are essential elements. The park authorities put emphasis on developing this offer, as education is one of their priority tasks.

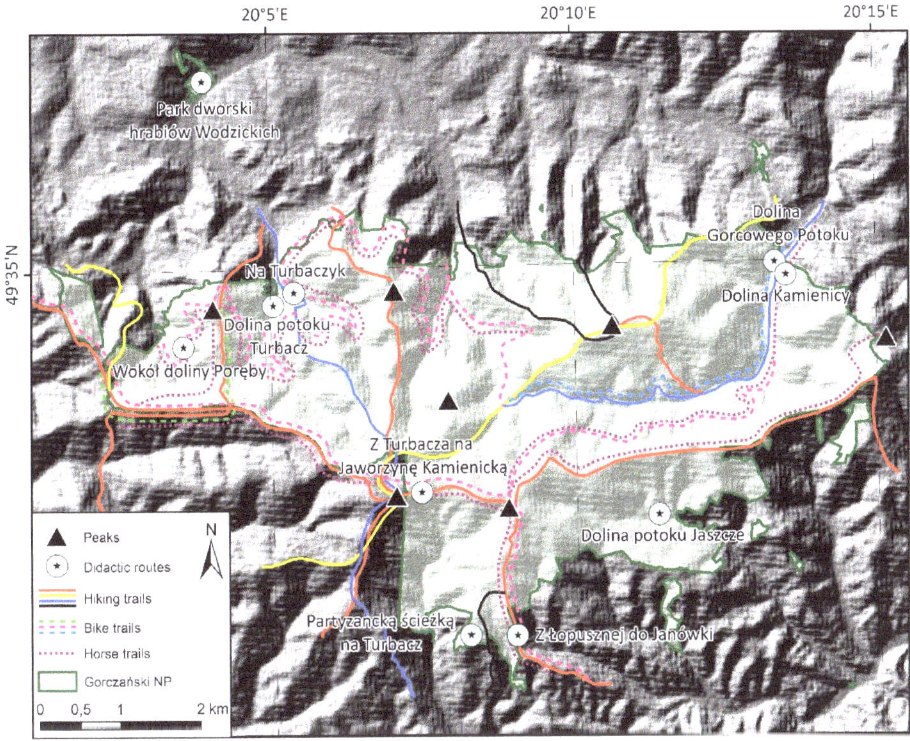

Figure 5. Didactic routes of Gorczański National Park.

Although the educational offer is varied, it is mostly devoted to issues that refer to the animated nature. Despite that, important elements related to the geology and geomorphology of the park are not neglected. Only in the case of two routes, no inanimate nature didactic stop was proposed. In the case of the other routes, there are usually one or two stops, with over 11 learning stops on average (Table 2).

Table 2. Geotourist elements of the didactic routes in Gorczański National Park.

Route	Geotourist Stop
Dolina Gorcowego Potoku	Geomorphology of mountain brooks
Dolina Kamienicy	Geological exposure, rock
Dolina potoku Jaszcze	
Dolina potoku Turbacz	Geological structure, soil, geomorphology of the mountain brook
Na Turbaczyk	
Park dworski hrabiów Wodzickich and Chabówka	Geological structure: tectonic window
Wokół doliny Poręby	
Z Łopusznej na Jankówki	The rock outcrop underneath Wysznia
Z Turbacza na Jaworzynę Kamienicką	Torfowisko wysokie. Trzy Kopce
Partyzancką ścieżką na Turbacz	Mountain brook

It should be emphasized that each of the didactic routes is equipped with a unified system of visual information and small tourist infrastructure of the park, stop posts, signposts and information boards. As intended by the creator, didactic routes are a tourist product allowing visitors to reach the most interesting places in the park.

2.9. Tourist Traffic in Gorczański National Park

In the current decade, we can speak of an increase in tourist traffic in Gorczański National Park that moved from 60,000 visitors in 2010 up to 90,000 visitors in 2017 (Figure 6).

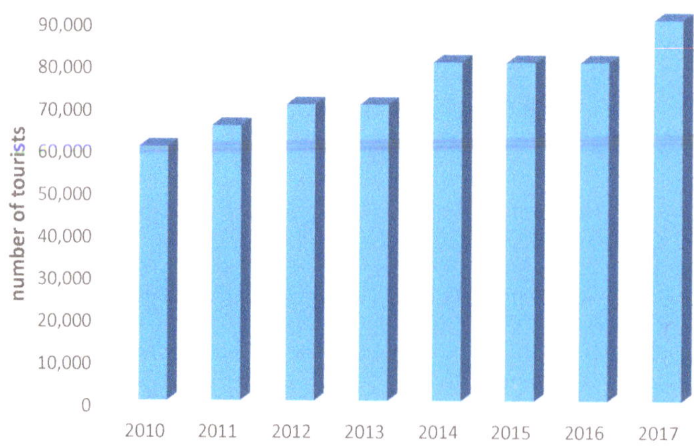

Figure 6. Number of tourists in Gorczański National Park in years 2010–2017. Source: [76–83].

Less than 170 km of trails are traversed by almost 100,000 people [83], which gives 13 people per hectare regarding park area and 533 people per kilometre regarding trails. This puts the park in the 15th place overall when taking into account the number of tourists per kilometre of a trail and the number of tourists per hectare of the park area.

The moderate tourist load of the park paradoxically increases its tourist attractiveness. A relatively small number of tourists (in 2017, the park was the 16th park in the ranking of the volume of tourist traffic in the country, with the number of tourists 42 times lower than the record-breaking Tatrzański National Park) significantly increases the comfort of a tourist holiday, regardless of its character as active or passive. This is also important because of the quality of the park's tourist attractions. With a few exceptions, national parks in Poland look for a sustainable approach between the duty of nature conservation, including inanimate nature, and opening up to the growing tourist traffic and the associated risks every year.

Each national park, also in Poland, is obliged to prepare and consequently implement a nature conservation plan in its area. The same is true for Gorczański National Park, which also implements a nature conservation plan, designed in 2018. The first step is to determine the type and nature of threats to inanimate nature resulting from human activity in the park area, including activities related to the growing tourist traffic every year.

2.10. Threats to the Inanimate Nature Resulting from the Tourist Traffic in Gorczański National Park

The size of tourist traffic in valuable natural areas affects their quality and conservation status. This is a kind of paradox—the more valuable and attractive a natural area is in the opinion of tourists, the greater tourist traffic is observed, which translates into a greater threat to the balance in nature in the protected area. In the literature, a whole set of threats is repeatedly pointed out, of which managers in the protected area are aware. Noise, pollution, or anthropopressure are indicated among them [84]. Partyka [85] draws attention to the excessive attendance of visitors and the thickening of tourist trails in the most popular places, trampling wild paths, damaging root systems, trees, destroying vegetation and soil, noise, disturbing animals, causing fires, littering, changes in the landscape and microclimate, synanthropization of flora and fauna, and changes in the structure of biocenoses. Similarly, Baraniec [86] points to anthropogenic denudation, destruction in vegetation, or littering, which diminishes the aesthetic values of the park and has a negative impact on the animal world. Wieniawska [87] emphasizes the threat related to the development of infrastructure, especially skiing. Hiking, most popular in the summer in mountainous areas, destroys nature on tourist trails. Tourists can destroy the vegetation cover, create shortcuts between paths, destroy the surface within the paths, cause loose material movement, etc. [88].

The overwhelming majority of literature on the subject points to the threats of animated nature, assuming that elements of inanimate nature, as more resistant to the environment, are less threatened [86,88]. It is worth confronting this position with the opinion of institutions that are responsible for the management of protected areas, in this case, national parks.

In accordance with the requirements of the Act on Nature Conservation [89], the threats to the animated and inanimate nature in Gorczański National Park are defined in the conservation plan for the park, developed by the park management. In addition to the identification of threats, ways of eliminating or significantly reducing them are also indicated. The draft plan of 2018 [90] identifies four groups of threats:

- existing internal threats;
- potential internal threats;
- existing external threats;
- potential external threats.

The first group includes 17 types of risks. Erosion of internal roads, routes, and trails as a result of tourist traffic and log-rolling is the most important for inanimate nature in the light of the analysed document. Among the most endangered resources, the authors of the document include inanimate nature, namely soil and bedrock. The way to remedy the problem is, among other things, renovation of roads, trails, and routes, as well as educating tourists and local residents on how to move around the park, combined with tourist traffic control. Other problems are vandalism (littering the ground, burning bonfires, devastation of huts, etc.), forest damage, and poaching. Among the endangered park resources, there are forest ecosystems, selected animal species, cultural assets, and also inanimate nature. One way to prevent potential harm is to educate the local community and tourists accordingly. Another internal threat is tourist traffic, gathering. Inanimate nature is less at risk, and appropriate channelling of this traffic (correction of trails in order to keep the tourists away from the immediate vicinity of assets) combined with education of the traffic participants is indicated as a means to achieve the goal. The last-named threat to inanimate nature, and especially to soils and rock exposures, is inappropriate repairs of internal roads (slope and valley roads), including bridges, culverts, and tourist infrastructure. The answer to the problem is to carry out anti-erosion protection, conduct road repairs with consultation with hydrologists and geomorphologists, and reduce the work in the vicinity of the stream beds to the necessary minimum.

Among the internal threats that are potentially dangerous to inanimate nature are illegal climbing onto rock outcrops, unauthorized cave penetration, and collecting rocks and minerals. As a remedy, it is suggested that the ban on both activities should be maintained, reinforced by an appropriate educational campaign. In the third group, the most important threat to inanimate nature is the loss of ecological connectivity of the park with the neighbouring areas through such activities as development of banks and regulation of water courses, and development of sports and recreation infrastructure. The way to remedy the problem is to monitor investment plans in the buffer zone of the park and to cooperate with local authorities, organizations, and tourist businesses on solutions limiting the unfavourable influence of sport and tourism on nature. No external potential threats are seen.

Thanks to proper management, the natural resources of Gorczański National Park, including geotourist ones, can be an important factor in attracting tourist traffic. The park authorities, trying to protect these resources in accordance with legal requirements, also by ensuring their proper accessibility, have an influence on how tourists assess the attractiveness of the park. It is important to see how the tourist potential of Gorczański National Park is evaluated by tourists who visit this mountain range as part of their weekend tourism and what the function is of the geotourist potential in this group of visitors.

3. Materials and Methods

The study on the tourist attractiveness of the park as an area conserving and presenting valuable natural resources, including those of inanimate nature, was carried out among 252 tourists visiting the park. The survey was conducted in three rounds in the 2016 and 2017 tourist season. In 2016, the survey was held during a so-called long weekend, when a weekend is accompanied by additional days off due to national holidays, as it was in the period of 29 April–3 May 2016; 104 questionnaires were then collected, 5 of which were rejected. At the turn of September and October (30 September–2 October) of 2016, despite good weather conditions, only 39 questionnaires were collected, with only 1 rejected. In 2017, tourists were surveyed in mid-April (12–14 April 2017), when the tourist season practically begins, traditionally marked by the appearance of the park's characteristic crocuses. At that time, 101 questionnaires were collected, while 2 were filled in incorrectly. In total, 252 questionnaires were collected (8 were rejected), so the survey was based on 244 correctly filled-in questionnaires. The main assumption was to determine the attractiveness of Gorczański National Park as a potentially important tourist destination for weekend tourism. This is due to the specific location of the park, about 70 km from the centre of Poland's second largest urban agglomeration, Kraków. It can therefore be assumed that because of the distance and relatively good communication—with national road No. 7, S7 express road, and voivodship road 986—the park is an attractive destination for tourists who wish to actively spend their weekend free time. That is why it was decided to conduct the research in a period traditionally referred to as a long weekend in Poland, when there is an accumulation of holidays: 1 and 3 May are public holidays. In 2016, it was Sunday and Tuesday. Practically then, the weekend tourists had time off from Friday evening 29 April 2016. It is already the full weekend season in the park. The next two research periods were the beginning of the park season, which coincided with the Easter period of 2017, and the end of the season—the weekend at the turn of September and October—when the weekend tourists can still count on good weather, as it was in 2017.

The questionnaire, as in other studies on the tourist attractiveness of protected areas, consisted of three main parts. The first one referred to the characteristics of the group of respondents and included standard questions relating to gender, age, education, and place of residence. It resulted in establishing a socio-demographic profile of tourists visiting Gorczański National Park. Then, the purpose of the visit was determined. In addition to the three most popular objectives proposed in the question, the participants could also indicate another objective if they did not declare any of those contained in the question. The next part focused on the evaluation of the park's tourist offer, which is one of the important ways of presenting natural resources, including geotourist ones, of the protected area. The third, final part was devoted to the general assessment of the tourist attractiveness of Gorczański National Park.

The group of the surveyed respondents comprised all the tourists who appeared on the trail, at the place where the survey was conducted, at the indicated time. The difficulty in determining a representative study sample was highlighted by many authors [10,91,92]. Therefore, we adopted the principle of surveying all tourists who appeared in the park at the time specified in the survey. We also decided to analyse all the questionnaires collected during holiday periods crucial for weekend tourism. The sample size does not differ considerably from that examined during similar studies devoted to the analysis of tourist traffic preferences and assessments of tourists visiting protected areas [10,93–97].

Answers to the survey questions were obtained with the face-to-face interview technique, which ensures a high level of correct replies [94].

The questions were designed carefully, to avoid potential inaccuracies and reduce the number of ambiguous answers. Some of them were closed questions, where the respondent could choose, depending on the question, one or two answers—as in the case of the question about the most attractive of the proposed values located on the tourist trails in the park. In the next group of questions, a 5-point Likert scale was applied. They concerned the respondents' opinion on particular trail elements. The tourists indicated their assessment ranging from "I strongly disagree" to "I strongly agree". The question summarizing the third part of the survey was an open one, so that the respondents could freely express their opinion on the attractiveness of the visited park.

4. Results and Discussion

4.1. The Tourist Attractiveness of Gorczański National Park from the Opinion of Tourists

One of the main objectives to establish national parks was nature conservation, intended to interact with the sustainable use of their natural resources for education, research, and recreation [98,99]. Recreational use of national parks is increasing all over the world [47,100–102]. It is therefore not surprising that the issue of tourism in protected areas, including parks, is an important research subject also in the context of tourist traffic [11,103,104]. The perception of protected areas attractiveness has been explored in many studies [8,92,105–108]. This is all the more significant as the vast majority of environmentally attractive areas are exposed to pressure from tourist traffic [47,109,110], which also impacts on the operation of national parks.

First, it is worth it to present a profile of the tourists visiting Gorczański National Park as it emerges from the information provided by the respondents themselves.

4.1.1. Tourist Profile

Numerous publications list a number of different factors influencing tourism in protected areas. They point at the level of education, place of residence [8,108], satisfaction with the products or services offered [8,92], as well as specific values. The choice of destination is influenced by socio-demographic variables such as age, gender, place of residence, and level of education [21,24,111], as well as other aspects related to recreation [21,112–116].

This applies especially to tourists visiting protected areas [117,118]. Relationships between the examined factors are also indicated. Investigators sometimes demonstrate the correlation between the respondents' age and landscape perception: younger people turn out more sensitive to this element [113]. Other studies emphasize that young and well-educated tourists usually present more ecological attitudes [119–123]. Similar results were obtained in our study on the profile of tourists visiting Gorczański National Park.

The majority of the surveyed visitors to Gorczański National Park were young people aged 15–45 years. Children and visitors over 60 years of age were the least numerous (Figure 7).

Consequently, the park, located relatively close to a large Polish urban agglomeration (Kraków), is an important weekend destination for people of an age that favours an active form of recreation, and the distribution in the three age groups between 16 and 60 was relatively even (Figure 7). The attractiveness of a park as a place of active recreation is also commented on in other studies, which, besides parks' attractiveness resulting from accumulation of valuable natural assets, imply their perfect suitability to practise numerous activities [124,125].

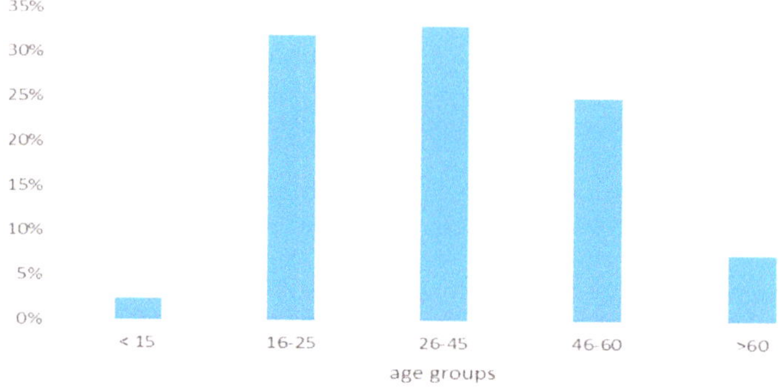

Figure 7. Age structure of respondents.

As far as gender is concerned, the group of women was slightly more populous. Women prevailed in groups up to 25 years of age and in retirement age, preferring the walking trails offer. In the case of men, the park was most popular for weekends in the age groups of 26–45 and 46–60 years. Women show greater interest in understanding nature conservation [23,117] than the surveyed men visiting protected areas. Thus, the potentially greater representation of women among the visitors to Gorczański National Park may translate into a greater understanding of the activities aimed at protecting the park resources, also the ones that may to some extent limit the resources availability to visitors.

It is worth noting that the vast majority of tourists active in the park area were people with higher education. These were 137 people, or 56% of the surveyed population. Every third person had secondary education. Only 11.5% declared vocational education, while only 2.5% had primary education—six girls under 15 years of age. In the group of people with higher and secondary education, women prevailed (55%), while men more often had vocational education (60%) (Figure 8).

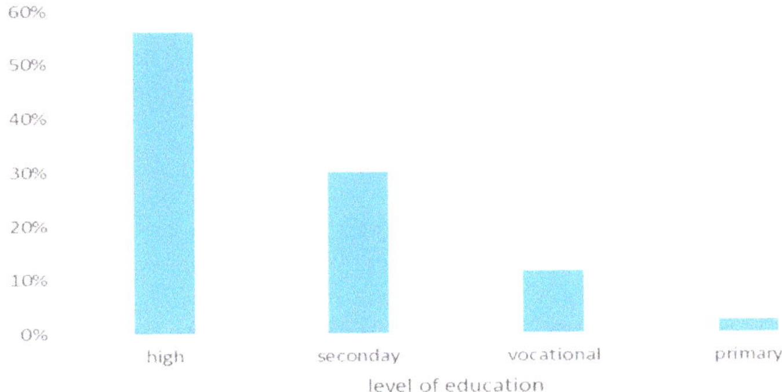

Figure 8. Education structure of respondents.

The results relating to the age or education of visitors to protected areas are in line with a broader trend observed in many national parks with reference to demographic factors [117,118]. Some investigators imply a higher sensitivity to landscape assets among younger tourists [113]; others observe more ecological attitudes in young and well-educated tourists [23,119–123]. Taking these relationships into account and analysing the data from Gorczański National Park, one can risk a thesis that with such a group of tourists, all actions aimed at the sustainable, pro-protective development of the region will be easier, as they will encounter a higher understanding among visitors.

4.1.2. Region of Origin of the Respondents

In the context of the character of weekend tourist trips, the spatial distribution of places from where tourists came to visit Gorczański National Park appears interesting (Figure 9). Not surprisingly, the biggest share was held by people from the Lesser Poland province—the region where the park is located. Almost every third person came from this province (mainly from Kraków and smaller cities in the vicinity). Every fifth tourist arrived from the Silesia province, bordering on the Lesser Poland province. The distribution of the tourists' origin combined with their count indicates that the park, from the tourist's point of view, is a destination of regional importance.

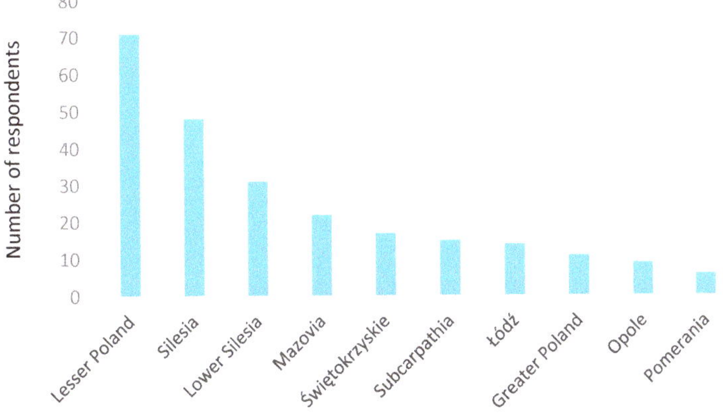

Figure 9. Provinces of origin of respondents visiting Gorczański National Park.

The park attractiveness in the region in which it is located seems to be a common issue for many protected areas studied and described in the literature. Peripheral location together with the attractiveness of the natural values themselves, including geotourist ones, is regarded as an important factor inviting tourist traffic [126]. The relatively close location of the park also increases its attractiveness for weekend tourists and same-day visitors [125].

4.1.3. Purpose of a Visit to Gorczański National Park

Motivations driving tourists to visit areas of natural attractiveness are an important subject of research [127–132]. Butler and Boyd [133] or McCool [134] are not the only ones to stress that the knowledge of tourists' preferences and the purpose of their visits is the foundation to create the tourist offer in protected areas. Wide knowledge of the opinions and preferences of tourists visiting protected areas translates significantly into efficient and coherent management of resources in these areas [135,136]. The research results that refer to the goals and motivations of tourists visiting Gorczański National Park seem interesting.

When asked about the purpose of the trip (Figure 10), the tourists indicated four main answers: a trip to the mountains, recreation, visiting family or friends, a different aim. For 66% of those surveyed, Gorczański National Park was the main tourist destination, while for the remaining 34% it was a secondary one.

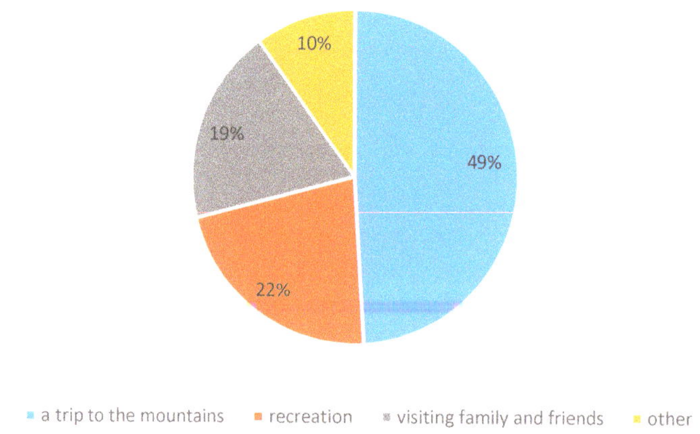

Figure 10. Purpose of a visit to Gorczański National Park.

Although many reasons influence tourists' decisions to visit a protected area, some common points can be identified with other areas described in the literature. Tourists visiting protected areas, including national parks, arrive there to experience positive emotions [137–139]. When analysing the experience of a tourist visiting a park in more detail, one can indicate the categories of tourist experience proposed by Pine and Gilmore [140]. There is an "education" category among them, which includes tourists looking for new experiences by learning based on active participation [141]. This is perfectly suited by the linear tourist product, which will be mentioned below: didactic routes. In the case of protected areas, the "escapism" category is especially significant, referring to the tourist's escape from his daily routine combined with an attempt to learn a new self [130,142–145]. These experiences are usually associated with an active form of space exploration, such as cycling, horse-riding, etc., which is also emphasized by Gorczański National Park visitors. Another important fact is worth noting: The experience of positive memories translates into a positive attitude of the tourist towards the visited region [137,146]. This also refers to protected areas.

The specificity of weekend tourism in Gorczański National Park is rather an individual formula of the trip. This is also true for research on a group of tourists visiting the park. Almost 9 out of 10 people (89%) came here alone (20%) or accompanied by their close family or friends (69%). This is a feature common to national parks visited by weekend tourists [25]. It is important to highlight the declarations of individual visitors, who stressed that they visited the park at any time of year.

4.1.4. Evaluation of Trails and Educational Offer of Gorczański National Park

Tourism should not be assumed to oppose nature conservation. It is worth emphasizing that the broadly understood development of tourism may translate into proper implementation of the area protection goals, including a very important educational goal [147,148].

Research on tourist traffic also affects the management of natural resources in protected areas, including national parks. The knowledge of tourists' preferences, motivations, and behaviours may influence the activities of the managing authorities when creating marketing strategies and specific tourist products offered in protected areas [130,145,149,150]. Examples include the didactic routes of Gorczański National Park.

A well-developed tourist product translates into increased attractiveness of the national park, especially among tourists interested in exploring the natural resources of the region [27,151].

The assessment is all the more important because it shows the natural resources in the park, including geotourist ones, as perceived by the surveyed tourists. The majority of tourists had no objections to the accessibility, safety, or comfort of the trails and presented a positive attitude toward the difficulty of the park trails, finding them rather easy. The tourists were not so unanimous in assessing the cleanliness of the trails: The majority thought that they were clean or rather clean, but almost every third tourist considered them not clean (Figure 11).

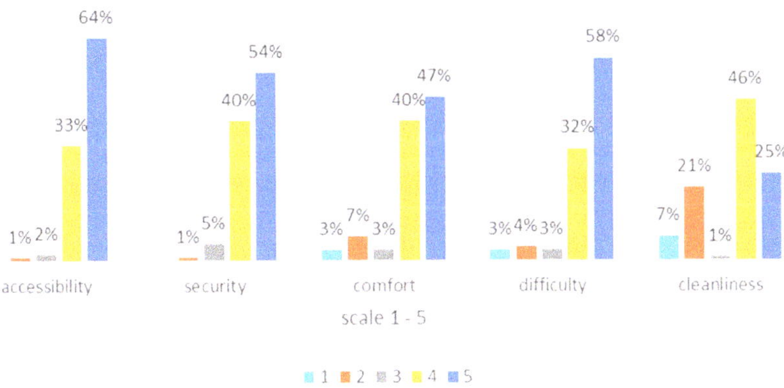

Figure 11. Evaluation of trails in Gorczański National Park.

The respondents were also asked about the attractiveness of particular trail elements. During the preliminary research, the most popular set of elements was determined. These were views/panoramas; water facilities, including streams and ponds; rocks and other geological features; as well as glades with elements of pastoral culture, i.e., huts. Tourists had the opportunity to indicate two of the proposed items that they considered the most attractive. A total of 352 responses were collected out of the possible 488. The panoramas received the biggest number of indications: 54%. The landscape of the protected area with its surroundings was important for tourists. It is not much of a surprise. Many publications point to the decisive role of landscape in the tourism of protected areas [152]. It is therefore not astonishing that knowledge of its reception is important for the proper management of this type of resources and plans for their maintenance and promotion [153]. The quality of the landscape is one of the most important assets of areas of high natural value [154]. This is important for many reasons. The landscape attractiveness can translate into the quality-of-life of the local community and income for the region as it contributes to the quality of the tourist experience [155]. It is important in this context to propose a linear product that would integrate the attractive points of the area visited [156]. One of the ideas are all trails based on a specific type of value, as in the case of didactic routes, discussed below.

Thanks to its location, Gorczański National Park offers one of the most beautiful panoramas, especially the view of the Tatra Mountains—the highest mountain range in Poland. Ranked next in terms of the number of indications were elements of landscape of inanimate nature connected with local culture: glades with huts, which are a hallmark of the park. The third place was taken by other elements of inanimate nature, directly related to the geotourist offer of the park—rocks and rock clusters (Figure 12). The importance of the park's geotourist resources for visitors is worth emphasizing at this point. Although the main protected element in the park is animated nature, the geotourist qualities that constitute inanimate nature resources are considered to be the most attractive to visitors.

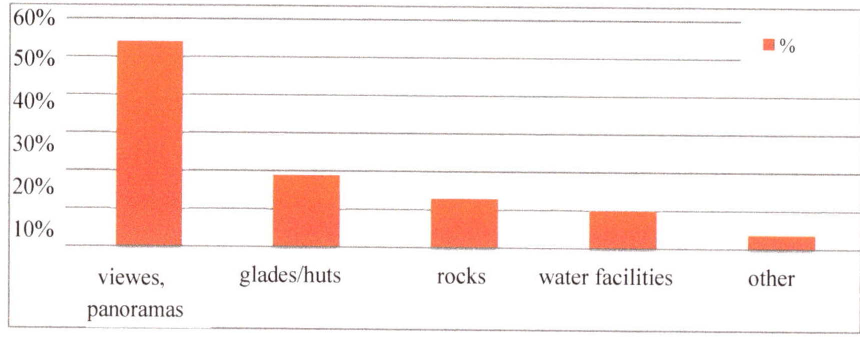

Figure 12. The most attractive elements of trails of Gorczański National Park.

4.1.5. Didactic Routes in the Opinion of Tourists

An important tool for the proper use of a national park values, including geotourist ones, is interpretation. It not only has an educational function, but also supports the protection and maintenance of park resources. After all, it is an important tool for managing the resources. A well applied interpretation may contribute to reducing the negative impact of tourist traffic on the park resources. One of the important channels of interpretation are markings, educational centres, or, which is important in the case of the examined Gorczański National Park, trails, as well as didactic routes [157]. Such an interpretation of a linear nature may help properly distribute the tourist traffic in the park area. It is also worth noting that interpretation facilitates tourists' involvement in the active exploration of the area [158]. Thanks to interpretation, apart from the very knowledge of resources, an understanding arises of the need for nature protection and appropriate behaviour in the natural environment. It also increases the level of tourist satisfaction with visiting the particular region and its resources. The same is true for Gorczański National Park. The type of interpretation that lies behind the didactic routes has been very positively evaluated.

A separate part of the survey was devoted to the evaluation of the educational offer of the park, with geotourist topics prominently present. The offer of didactic routes is the most popular form of education. When asked if they knew about it, the majority of the respondents gave negative answers, despite the park management efforts to promote this kind of education. The surveyed weekend tourists may not be familiar with the offer because it is addressed primarily to organized school groups at different levels of education. Those who knew the educational offer were most frequently familiar with the route Park dworski hrabiów Wodzickich—Chabówka, where one of the 16 stops is dedicated to geotourism (geological structure: tectonic window), then Dolina potoku Turbacz (two geotourist stops), Wokół doliny Poręby, and Dolina Gorcowego Potoku (one geotourist stop).

The majority of people using the offer of the didactic routes had a positive attitude towards the infrastructure accompanying the routes and were satisfied with the number of information boards and their distribution within the park (Figure 13).

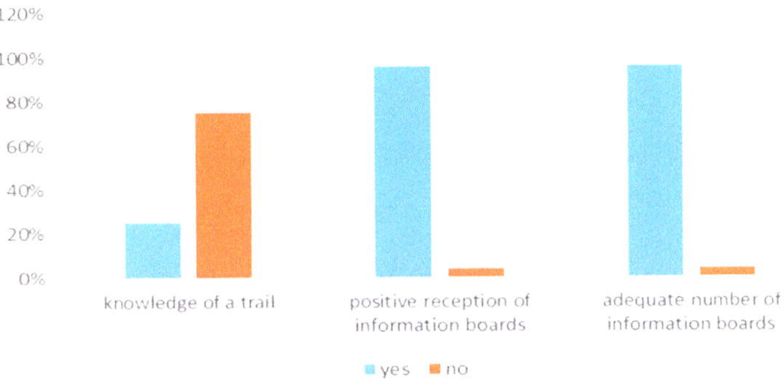

Figure 13. Didactic routes of Gorczański National Park in the opinion of tourists.

4.1.6. General Assessment of Gorczański National Park

Finally, it is worthwhile to look at the problem of tourist satisfaction, which affects the overall assessment of the resources used by the tourist. This is an important issue: customer's satisfaction, its level, as well as knowledge about it are important for tourism management [159,160]. This applies, for example, to the loyalty of the tourist to the destination or the income derived from tourism in the region [161]. After all, it depends on the level of satisfaction whether a given region or product will be recommended to others [160]. Therefore, tourists' satisfaction related to the visited region is crucial for the tourism industry and its economic development [162,163]. The satisfaction level is a derivative of the level of the region tourist development [164]. This is important for protected areas, increasingly dependent on the development of sustainable tourism. A high level of satisfaction translates into a higher probability that the tourist will be more likely to return to a region that brings positive associations [165]. It is interesting to look at the evaluation of Gorczański National Park in this context.

The vast majority of respondents confirmed the validity of the authorities' decision to create a national park and the need for the conservation of animated and inanimate nature in the Gorce Mts., which would emphasize its uniqueness at the same time. The general impression about the tourist attractiveness of the park was very positive among the respondents (Figure 14).

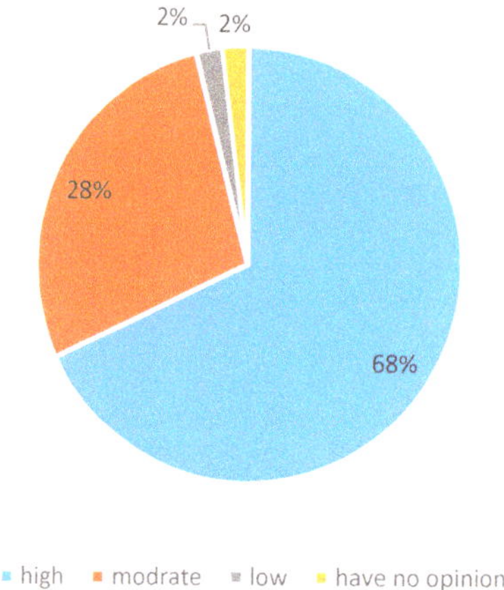

Figure 14. Tourist attractiveness of the park in the opinion of tourists.

Usually, attractiveness results from the possibility to practise active recreation, from attractive landscapes, specific natural values, or a widely understood authentic nature, which can be found in protected areas [111,166]. The opportunity to relax is also important [167]. It is also emphasized that the tourists' perception of park attractiveness is typically affected by more than one factor. It is not different for Gorczański National Park.

Among the positive opinions indicated in the survey, the following advantages appeared most often: sights, panoramas, varied trails, tranquillity, low traffic, favourable density of the trails, and interesting cultural and natural assets. Among the negative ones there were mud, lack of signposting and land development, insufficient care for tidiness on trails, and little variety of views as compared with the respondents' previous experiences from other national parks.

In an open question, the surveyed tourists had the opportunity to freely express their opinion about the park. Not surprisingly, this part contained more negative opinions. People paid attention to the garbage left by tourists on the trails and suggested that the number of trash bins be increased. It is worth reminding that too many trash bins contradict the idea of the park. The management are trying to foster a conscious tourist who takes back with them whatever they brought to the park. People complained about cars and quads running around the park and the buffer zone. Many tourists pointed out the development of the park and its growing popularity. They regretted it because they valued peace and quiet away from a crowd of tourists, which is so characteristic of the most popular parks in Poland.

The above opinions may indicate many of the park's deficiencies, so it is worth emphasizing the answers to the question related to recommending the park to others. The majority would strongly recommend a visit to Gorczański National Park to their friends. Positive perceptions of the visited national park or protected area generally prevail over neutral or negative ones. This is still an indirect evaluation of the park by visitors.

5. Conclusions

Owing to its specificity, Gorczański National Park is particularly attractive for weekend tourists, with all the consequences. First of all, its location—in the vicinity of larger urban areas—makes it a natural destination for their residents who are looking for a place for active recreation. The area of the Gorce Mts. is exactly such a place. Just over 70 km away, there is a region with almost no urbanization, with tranquillity undisturbed by civilization, clean air, and a beautiful landscape.

Moreover, described in the literature as a set of features necessary for recreation [65], the analysed region offers favourable recreation conditions, such as special scenic qualities of the landscape, favourable bio-climatic conditions, or convenient amenities for active recreation [67]. An additional advantage, whose importance cannot be overestimated, is that this is a protected area. It is very important for tourism. As a result, visitors may rest assured that the resource they use (the main attraction, which is well preserved nature) will remain unchanged for many years and decades. The sustainability of the assets significantly increases the attractiveness of a protected area, and a higher form of protection in a region makes it more attractive. In this case, it is a national park, which is the highest form of conservation, regulated by a separate legal act. As a result, the park is a natural destination for many tourists interested in various forms of enjoying nature, ranging from active forms of recreation, such as trekking, cycling, and horse riding, to geotourism, a form of educational and cognitive tourism popular for the previous three decades. In recent years, the park has gained attention owing to its geotourist potential. Currently, 70% of the didactic routes present important geotourist features of the mountains. As part of the accompanying infrastructure, the number of facilities has increased where didactic activities related to animated and inanimate nature characteristic of the park are performed. A large number of scenic viewpoints perfectly introduce the geological and geomorphological context of the area to park visitors. All this translates into a strongly positive opinion on the park's attractiveness. The vast majority of tourists expressed a positive opinion about the park as a place of excellent active recreation. They understood the need for nature conservation in this place and supported activities aimed at promoting nature resources and the ways in which the park management tried to achieve the goals. The geotourist aspect is important for weekend tourists. Those who came here for a short time appreciated the peace, quiet, and adequate distribution of tourist traffic, so there was no issue of a large number of tourists on trails. This is often mentioned as one of the main attractiveness factors. It is coupled with a reduction in the number of investments—the respondents preferred nature preserved in an unchanged form to the development of, for example, ski infrastructure, destroying the attractive inanimate nature. Only 2% of the study subjects regretted the lack of development of this type of infrastructure.

It is also worth pointing at the applicability of the conducted research. It can serve as a reference point for the activities of the authorities managing the park, responsible also for preparing an attractive offer for the tourist traffic.

Generally, the study results confirm the correct direction of development of the tourist offer proposed by the park: The offer of trails and didactic routes was positively received. However, attention was drawn to important details that should be taken into account in order to increase the comfort of tourists visiting the examined protected area. It is worth considering the voices of older participants in the tourist traffic; in their opinion, the trails are difficult. Among other things, mud and stones make it problematic for this group to move. Considering the opinion of the senior group when preparing the offer may increase the overall attractiveness of the park. This is all the more important because the elderly tourist market is becoming an increasingly important segment every year, which should be recognized by all creators of tourist products or managers of regional tourist resources. It is worth emphasizing the importance of inanimate nature for tourists visiting the park. Values of this type were assessed as very attractive, so they can and should constitute a foundation for extending the tourist offer of the park.

On the other hand, significant participation of young people is important and consistent with a wider trend. Young tourists seem natural allies of the park managers in shaping the tourist offer, where conservation of natural resources is a priority. They understand the need of nature conservation even at the expense of their own comfort.

An important conclusion of the research is the educational attractiveness of the geotourist assets of the park, expressed by the high evaluation of the didactic routes set out in its area. As indicated by weekend tourists, geotourism, in addition to its sightseeing advantages, has also a significant educational value.

In conclusion, Gorczański National Park turned out to be an area attractive for weekend tourists. It is an interesting destination in terms of natural resources, but also because of the form of nature conservation, which hopefully makes the resources, including geotourist ones, remain unchanged in their character for many more decades to come.

Finally, one should mention that the study limitations include the sample size. Although three different periods were chosen, which are generally popular among weekend tourists in Poland, the number of correct answers could have been higher. This is a recurring problem in field research. However, considering the fact that the study was survey-based, and that account was taken of the period during which the data were collected, one can assume that the results are representative. Weather conditions constituted another limitation: The timing of the surveys was set relatively early. The research was carried out in spite of the fact that weather conditions were not always favourable for weekend trips, which probably influenced the size of the studied group.

Author Contributions: Conceptualization, K.W.; methodology, K.W., S.J.; validation, K.W.; formal analysis, K.W.; investigation, P.O., A.R.; resources, P.O., A.Z., S.J.; data curation, P.O., A.Z.; writing—original draft preparation, K.W.; writing—review and editing, P.O., A.R., S.J.; supervision, K.W. All authors have read and agreed to the published version of the manuscript.

Funding: This research received no external funding.

Acknowledgments: The authors would like to thank Sabina Łanocha for her contribution to the research and article development.

Conflicts of Interest: The authors declare no conflict of interest.

References

1. Russell, M.A.; Bradford, J.E.; Murphy, L.E.E.C. Waters and development of a turn-of-the-century tourist economy in Yellowstone National Park. *Hist. Archeol.* **2004**, *38*, 96–113. [CrossRef]
2. Hunt, W.J., Jr. A Model of Tourism as Context for Historical Sites: An Example of Historical Archeology at Yellowstone National Park. In *Historical Archeology of Tourism in Yellowstone National Park*; Corbin, A., Russell, M.A., Eds.; Springer: New York, NY, USA, 2010; pp. 1–74.
3. Bergstrom, R.D.; Harrington, L.M.B. Embedded in nature: Challenges to sustainability in communities of the Greater Yellowstone ecosystem. *Sustainability* **2019**, *11*, 1459. [CrossRef]
4. Bushell, R.; Staiff, R.; Eagles, P.F.J. Tourism and Protected Areas: Benefits beyond Boundaries. In *Tourism and Protected Areas: Benefits beyond Boundaries*; Bushell, R., Eagles, P.F.J., Eds.; CABI: Wallingford, UK, 2007; pp. 1–11.
5. Kuenzi, C.; McNeely, J. Nature-Based Tourism. In *Global Risk Governance*; Renn, O., Walker, K.D., Eds.; IUCN Publishing: Gland, Switzerland, 2008; pp. 155–178.
6. Balmford, A.; Beresford, J.; Green, J.; Naidoo, R.; Walpole, M.; Manica, A. A global perspective of trends in nature-based tourism. *PLoS Biol.* **2009**, *7*, e1000144. [CrossRef] [PubMed]
7. Buckley, R. *Ecotourism: Principles and Practises*; CABI: Wallingford, UK, 2009.
8. Newsome, D.; Moore, S.A.; Dowling, R.K. *Natural Area Tourism: Ecology, Impacts and Management*, 2nd ed.; Channel View Publications: Bristol, UK, 2013.
9. Eagles, P.F.J. Research priorities in park tourism. *J. Sustain. Tour.* **2014**, *22*, 528–549. [CrossRef]
10. Garavaglia, V.G.; Diolaiuti, G.; Smiraglia, C.; Pasquale, V.; Pelfini, M. Evaluating tourist perception of environmental changes as a contribution to managing natural resources in glacierized areas: A case study of the Forni Glacier (Stelvio National Park, Italian Alps). *Environ. Manag.* **2012**, *50*, 1125–1138. [CrossRef] [PubMed]
11. Henriques, M.H.; Pena dos Reis, R.; Brilha, J.; Mota, T. Geoconservation as an emerging geoscience. *Geoheritage* **2011**, *3*, 117–128. [CrossRef]
12. Gray, M. Other nature: Geodiversity and geosystem services. *Environ. Conserv.* **2011**, *38*, 271–274. [CrossRef]
13. Burek, C.V.; Prosser, C.D. The history of geoconservation: An introduction. *Geol. Soc. Lond. Spec. Publ.* **2008**, *300*, 1–5. [CrossRef]
14. Gray, M. Geodiversity and Geoconservation: What, Why, and How? In *Papers Presented at the George Wright Forum*; Santucci, V.L., Ed.; Book Concern Printers: Hancock, MI, USA, 2005; pp. 4–12.

15. Serrano, E.; González-Trueba, J.J. Assessment of geomorphosites in natural protected areas: The Picos de Europa National Park (Spain). *Geomorphol. Relief Process. Environ.* **2005**, *11*, 197–208. [CrossRef]

16. Lima, F.; Brilha, J.; Salamun, E. Inventorying geological heritage in large territories: A methodological proposal applied to Brazil. *Geoheritage* **2010**, *2*, 91–99. [CrossRef]

17. Fassoulas, C.; Mouriki, D.; Dimitriou-Nikolakis, P.; Iliopoulos, G. Quantitative assessment of geotopes as an effective tool for geoheritage management. *Geoheritage* **2012**, *4*, 177–193. [CrossRef]

18. Božić, S.; Tomić, N. Canyons and gorges as potential geotourism destinations in Serbia: Comparative analysis from two different perspectives—General tourists' and geotourists'. *Open Geosci.* **2015**, *7*, 531–546. [CrossRef]

19. Dowling, R.K.; Newsome, D. *Geotourism: Sustainability, Impacts and Management*; Elsevier: Oxford, UK, 2006.

20. Honey, M. *Ecotourism and Sustainable Development: Who Own Paradise?* 2nd ed.; Island Press: Washington, DC, USA, 2008.

21. Hose, T.A. Selling the story of Britain's stone. *Environ. Interpret.* **1995**, *10*, 16–17.

22. Panizza, M. Geomorphosites: Concepts, methods and example of geomorphological survey. *Chin. Sci. Bull.* **2001**, *46*, 4–6. [CrossRef]

23. Słomka, T.; Kicińska-Świderska, A. Geoturystyka-podstawowe pojęcia. *Geoturystyka* **2004**, *1*, 5–7.

24. Valentine, P.S. Review: Nature-based tourism. In *Special Interest Tourism*; Weiler, B., Hall, C.M., Eds.; Belhaven Press: London, UK, 1992; pp. 105–127.

25. Tonge, J.; Moore, S. Importance-satisfaction analysis for marine-park hinterlands: A Western Australian case study. *Tour Manag.* **2007**, *28*, 768–776. [CrossRef]

26. McIntosh, R.W.; Goeldner, C.R.; Ritchie, J.R.B. *Tourism: Principles, Practices, Philosophies*; Wiley: New York, NY, USA, 1995.

27. Ceballos-Lascurain, H. *Tourism, Ecotourism and Protected Areas: The State of Nature-Based Tourism around the World and Guidelines for Its Development*; IUCN: Cambridge, UK, 1996.

28. Bushell, B. Global issues for protected areas and nature-based tourism: Case studies of partnership in australia addressing some of these issues. In *International Workshop: Case Studies on Sustainable Tourism and Biological Diversity*; Gunling, L., Korn, H., Specht, R., Eds.; German Federal Agency for Nature Conservation: Bonn, Germany, 1999; pp. 93–114.

29. Priskin, J. Assessment of natural resources for nature-based tourism: The case of the Central Coast region of Western Australia. *Tour Manag.* **2001**, *22*, 637–648. [CrossRef]

30. Walpole, M.J. Feeding dragons in Komodo National Park: A tourism tool with conservation complications. *Anim. Conserv.* **2001**, *4*, 67–73. [CrossRef]

31. Walpole, M.J.; Goodwin, H.J. Local attitudes towards conservation and tourism around Komodo National Park, Indonesia. *Environ. Conserv.* **2001**, *28*, 160–166. [CrossRef]

32. Guyer, C.; Pollard, J. Cruise visitor impressions of the environment of the Shannon-Erne waterways system. *J. Environ. Manag.* **1997**, *51*, 199–215. [CrossRef]

33. Hose, T.A. Telling the story of stone—Assessing the client base. In *Geological and Landscape Conservation*; O'Halloran, D., Green, C., Harley, M., Stanley, M., Knill, S., Eds.; Geological Society: London, UK, 1994; pp. 451–457.

34. Hose, T.A. Geotourism—Selling the earth to europe. In *Engineering Geology and the Environment*; Marinos, P.G., Koukis, G.C., Tsiambaos, G.C., Stournaras, G.C., Eds.; Balkema: Rotterdam, The Netherlands, 1997; pp. 2955–2960.

35. Hose, T.A. Is it any fossicking good? Or behind the signs—A critique of current geotourism interpretative media. In *Unpublished Paper Delivered to the Tourism in Geological Landscapes Conference*; Ulster Museum: Belfast, UK, 1998.

36. Hose, T.A. European geotourism—Geological interpretation and geoconservation promotion for tourists. In *Geological Heritage: Its Conservation and Management*; Barretino, D., Wimbledon, W.P., Gallego, E., Eds.; Instituto Tecnologico Geominero de Espana: Madrid, Spain, 2000; pp. 127–146.

37. Hose, T.A. Towards a history of geotourism: Definitions, antecedents and the future. *Geol. Soc. Lond. Spec. Publ.* **2008**, *300*, 37–60. [CrossRef]

38. Hose, T.A.; Vasiljević, D.A. Defining the nature and purpose of modern geotourism with particular reference to the United Kingdom and South-East Europe. *Geoheritage* **2012**, *4*, 25–43. [CrossRef]

39. Newsome, D.; Dowling, R.K. *Geotourism: The Tourism of Geology and Landscape*; Goodfellow: Oxford, UK, 2010.

40. Santangelo, N.; Romano, P.; Santo, A. Geo-itineraries in the Cilento Vallo di Diano Geopark: A tool for tourism development in southern Italy. *Geoheritage* **2015**, *7*, 319–335. [CrossRef]

41. Liddle, M.J. *Recreation Ecology: The Ecological Impact of Outdoor Recreation and Ecotourism*; Chapman and Hall: London, UK, 1997.

42. Newsome, D.; Moore, S.A.; Dowling, R.K. *Natural Area Tourism: Ecology, Impacts and Management*; Channel View Publications: Sydney, Australia, 2002.

43. Buckley, R. Impacts positive and negative: Links between ecotourism and environment. In *Environmental Impacts of Ecotourism*; Buckley, R., Ed.; CABI: New York, NY, USA, 2004; pp. 1–14.

44. Cole, D.N. Impacts of hiking and camping on soils and vegetation: A review. In *Environmental Impacts of Ecotourism*; Buckley, R., Ed.; CABI: Oxford, UK, 2004; pp. 41–60.

45. Newsome, D.; Cole, D.N.; Marion, J. Environmental impacts associated with recreational horse-riding. In *Environmental Impacts of Ecotourism*; Buckley, R., Ed.; CABI: New York, NY, USA, 2004; pp. 61–82.

46. Kelly, C.L.; Pickering, C.M.; Buckley, R.C. Impacts of tourism on threatened plant taxa and communities in Australia. *Ecol. Manag. Restor.* **2003**, *4*, 37–44. [CrossRef]

47. Pickering, C.M.; Hill, W. Impacts of recreation and tourism on plant biodiversity and vegetation in protected areas in Australia. *J. Environ. Manag.* **2007**, *85*, 791–800. [CrossRef]

48. Ballantyne, M.; Gudes, O.; Pickering, C.M. Visitor trails are an important cause of fragmentation in endangered urban forests. *Landsc. Urban Plan.* **2014**, *130*, 112–124. [CrossRef]

49. Godefroid, S.; Koedam, N. The impact of forest paths upon adjacent vegetation: Effects of the path surfacing material on the species composition and soil compaction. *Biol. Conserv.* **2004**, *119*, 405–419. [CrossRef]

50. Hill, W.; Pickering, C.M. Vegetation associated with different walking track types in the Kosciuszko Alpine area, Australia. *J. Environ. Manag.* **2006**, *78*, 24–34. [CrossRef]

51. Wimpey, J.; Marion, J.L. The influence of use, environmental and managerial factors on the width of recreational trails. *J. Environ. Manag.* **2010**, *91*, 2028–2037. [CrossRef]

52. Kutiel, P.; Zhevelev, H.; Harrison, R. The effect of recreational impacts on soil and vegetation of stabilised Coastal Dunes in the Sharon Park, Israel. *Ocean Coast. Manag.* **1999**, *42*, 1041–1060. [CrossRef]

53. Marion, J.L.; Leung, Y.F. Trail resource impacts and an examination of alternative assessment techniques. *J. Park Recreat. Adm.* **2001**, *19*, 17–37.

54. Monz, C.A.; Cole, D.N.; Leung, Y.F.; Marion, J.L. Sustaining visitor use in protected areas: Future opportunities in recreation ecology research based on the USA experience. *Environ. Manag.* **2010**, *45*, 551–562. [CrossRef] [PubMed]

55. Marion, J.L.; Wimpey, J.F.; Park, L.O. The science of trail surveys: Recreation ecology provides new tools for managing wilderness trails. *Park Sci.* **2011**, *28*, 60–65.

56. Fenu, G.; Cagoni, D.; Ulian, T.; Bacchetta, G. The impact of human trampling on a threatened coastal Mediterranean plant: The case of Anchusa littorea Moris (Boraginaceae). *Flora* **2013**, *208*, 104–110. [CrossRef]

57. McCool, S.F. Constructing partnerships for protected area tourism planning in an era of change and messiness. *J. Sustain. Tour.* **2009**, *17*, 133–148. [CrossRef]

58. Panizza, M.; Piacente, S. *Geomorfologia Culturale*; Pitagora Editrice: Bologna, Italy, 2003.

59. Kondracki, J. *Geografia Regionalna Polski*; PWN: Warszawa, Poland, 2000.

60. Ordinance of the Council of Ministers of August 8, 1980 on the creation of the Gorczański National Park. *Journal of Laws*, 22 August 1980.

61. Kruczek, Z. *Atrakcje Turystyczne. Fenomen, Typologia, Metody Badań*; Proksenia: Kraków, Poland, 2011.

62. Lijewski, T.; Mikułowski, B.; Wyrzykowski, J. *Geografia Turystyki Polski*; Polskie Wydawnictwo Ekonomiczne: Warszawa, Poland, 1998.

63. Warszyńska, J.; Jackowski, A. *Podstawy Geografii Turyzmu*; PWN: Warszawa, Poland, 1978.

64. Rogalewski, O. *Zagospodarowanie Turystyczne*; WSiP: Warszawa, Poland, 1979.

65. Lijewski, T.; Mikułowski, B.; Wyrzykowski, J. *Geografia Turystyki Polski*; Polskie Wydawnictwo Ekonomiczne: Warszawa, Poland, 2002.

66. Kowalczyk, A. *Geografia Turyzmu*; PWN: Warszawa, Poland, 2001.

67. Lijewski, T.; Mikułowski, B.; Wyrzykowski, J. *Geografia Turystyki Polski*, 5th ed.; Polskie Wydawnictwo Ekonomiczne: Warszawa, Poland, 2008.

68. Cieszkowski, M.; Chodyń, R.; Szczęch, M. Gorce—Góry fliszowe. In *Gorczański Park Narodowy. Przyroda i Krajobraz Pod Ochroną*; Czarnota, P., Stefanik, M., Eds.; Gorczański Park Narodowy: Poręba Wielka, Poland, 2015; pp. 39–51.

69. Pawlik, Ł.; Šamonil, P.; Malik, I.; Kroh, P.; Ślęzak, A.; Daněk, P. Geomorphic edge effects in response to abiotic and anthropogenic disturbances in forest ecosystems of the Gorce Mountains, Western Carpathians. *Catena* **2019**, *177*, 134–148. [CrossRef]

70. Andreucci, B.; Castelluccio, A.; Jankowski, L.; Mazzoli, S.; Szaniawski, R.; Zattin, M. Burial and exhumation history of the Polish Outer Carpathians: Discriminating the role of thrusting and post-thrusting extension. *Tectonophys* **2013**, *608*, 866–883. [CrossRef]

71. Gorczański Park Narodowy. Przyroda Nieożywiona. Available online: https://www.gorczanskipark.pl/page,art,id,11,kategoria,Przyroda_nieozywiona.html (accessed on 15 November 2019).

72. Cieszkowski, M. Geologiczne walory naukowe Gorczańskiego Parku Narodowego i jego otoczenia. *Ochr. Besk. Zach.* **2006**, *1*, 45–57.

73. Szczęch, M.; Cieszkowski, M.; Chodyń, R.; Loch, J. Geotouristic values of the Gorce National Park and its surroundings (the Outer Carpathians, Poland). *Geotourism* **2016**, *1–2*, 27–44. [CrossRef]

74. Chowaniec, J.; Poprawa, D.; Witek, K. Stop E8—Rabka: Wody mineralne. In *Carpatian Paleogeography and Geodynamics: A Multidisciplinary Approach. Field Trip Guide, Proceedings of the 12th Meeting of the Association of European Geological Societies, Kraków, Poland, 10–15 September 2001*; Birkenmajer, K., Krobicki, M., Eds.; Państwowy Instytut Geologiczny: Kraków, Poland, 2001; pp. 121–125.

75. Chowaniec, J.; Poprawa, D.; Witek, K. Występowanie wód termalnych w polskiej części Karpat. *Przegl. Geol.* **2001**, *49*, 728–742.

76. Statistics Poland. *Ochrona Środowiska 2018*; Statistics Poland: Warsaw, Poland, 2018.

77. Statistics Poland. *Ochrona Środowiska 2011*; Statistics Poland: Warsaw, Poland, 2011.

78. Statistics Poland. *Ochrona Środowiska 2012*; Statistics Poland: Warsaw, Poland, 2012.

79. Statistics Poland. *Ochrona Środowiska 2013*; Statistics Poland: Warsaw, Poland, 2013.

80. Statistics Poland. *Ochrona Środowiska 2014*; Statistics Poland: Warsaw, Poland, 2014.

81. Statistics Poland. *Ochrona Środowiska 2015*; Statistics Poland: Warsaw, Poland, 2015.

82. Statistics Poland. *Ochrona Środowiska 2016*; Statistics Poland: Warsaw, Poland, 2016.

83. Statistics Poland. *Ochrona Środowiska 2017*; Statistics Poland: Warsaw, Poland, 2017.

84. Kruczek, Z. Frekwencja w polskich atrakcjach turystycznych. Problemy oceny liczby odwiedzających. *Ekon. Probl. Turyst.* **2016**, *3*, 25–35. [CrossRef]

85. Partyka, J. Ruch turystyczny w polskich parkach narodowych. *Folia Turist.* **2010**, *22*, 9–23.

86. Baraniec, A. Turystyka w Babiogórskim Parku Narodowym. In *Użytkowanie Turystyczne Parków Narodowych*; Partyka, J., Ed.; Wydawnictwo Ojcowski Park Narodowy: Ojców, Poland, 2002; pp. 155–165.

87. Wieniawska, B. Turystyka a ochrona przyrody w Karkonoskim Parku Narodowym. In *Użytkowanie Turystyczne Parków Narodowych*; Partyka, J., Ed.; Wydawnictwo Ojcowski Park Narodowy: Ojców, Poland, 2002; pp. 259–270.

88. Fidelus, J. Turystyka piesza i jej wpływ na przemiany rzeźby terenu w otoczeniu hal Gąsienicowej i Kondratowej w Tatrach. In *Stan i Perspektywy Rozwoju Turystyki w Tatrzańskim Parku Narodowym*; Pociask-Karteczka, J., Matuszyk, A., Skawiński, P., Eds.; AWF: Kraków, Poland, 2007; pp. 261–272.

89. Act of April 16, 2004 on nature conservation. Journal of Laws, No. 92, Item 880, Art. 5. Available online: http://prawo.sejm.gov.pl/isap.nsf/DocDetails.xsp?id=WDU20040920880 (accessed on 10 December 2019).

90. Conservation Plan for Gorczański National Park of 2018. Available online: https://www.gorczanskipark.pl/UserFiles/File/2018/BIP%20i%20ogloszenia/PROJEKT%20%20z%2015-01-2018%20Planu%20Ochrony%20GPN.pdf (accessed on 27 November 2019).

91. Beerli, A.; Martín, D. Tourists characteristics and the perceived image of tourists destination: A quantitative analysis of a case study of Lanzarote, Spain. *Tour Manag.* **2004**, *25*, 623–636. [CrossRef]

92. Togridou, A.; Hovardas, T.; Pantis, J.D. Determinants of visitors willingness to pay for the National Marine Park of Zakynthos, Greece. *Ecol. Econ.* **2006**, *60*, 308–319. [CrossRef]

93. Ceurvorst, R.; Lamborn, C. Visitor attitudes and value orientations for a proposed national monument. *J. Outdoor Recreat. Tour.* **2018**, *23*, 33–43. [CrossRef]

94. Mmopelwa, G.; Kgathi, D.L.; Molefhe, L. Tourists' perceptions and their willingness to pay for park fees: A case study of self-drive tourists and clients for mobile tour operators in Moremi Game Reserve, Botswana. *Tour Manag.* **2007**, *28*, 1044–1056. [CrossRef]

95. Jones, N.; Panagiotidou, K.; Spilanis, I.; Evangelinos, K.; Dimitrakopoulos, P. Visitors' perceptions on the management of an important nesting site for loggerhead sea turtle (*Caretta caretta L.*): The case of Rethymno coastal area in Greece. *Ocean Coast. Manag.* **2011**, *54*, 577–584. [CrossRef]

96. Lili, P.; Lijuan, C.; Ming, W. Tourist behaviors in Wetland Park: A preliminary study in Xixi National Wetland Park, Hangzhou, China. *Chin. Geogr. Sci.* **2010**, *20*, 66–73.

97. Beh, A.; Bruyere, B. Segmentation by visitor motivation in three Kenyan national reserves. *Tour Manag.* **2007**, *28*, 1464–1471. [CrossRef]

98. Ehrenfeld, D. The conservation of non-resources. *Am. Sci.* **1976**, *64*, 648–656.

99. Ferreira, S.; Harmse, A. Kruger National Park: Tourism development and issues around the management of large numbers of tourists. *J. Ecotour.* **2014**, *13*, 16–34. [CrossRef]

100. Adams, C.; Seroa da Motta, R.; Ortiz, A.; Reid, J.; Aznar, E.; Sinisgalli, A. The use of contingent valuation for evaluating protected areas in the developing world: Economic valuation of Morro do Diabo State Park, Atlantic Rainforest, São Paolo State (Brazil). *Ecol. Econ.* **2008**, *66*, 359–370. [CrossRef]

101. Duzgunes, E.; Demirel, O. Determining the tourism potential of the Altindere Valley National Park (Trabzon/Turkey) with respect to its conservation value. *Int. J. Sustain. Dev. World Ecol.* **2013**, *20*, 358–368. [CrossRef]

102. Mayer, M. Can nature-based tourism benefits compensate for the costs of national parks? A study of the Bavarian Forest National Park, Germany. *J. Sustain. Tour.* **2014**, *22*, 561–583. [CrossRef]

103. Reimann, M.; Lamp, M.-L.; Palang, H. Tourism impacts and local communities in Estonian National Parks. *Scand. J. Hosp. Tour.* **2011**, *11*, 87–99. [CrossRef]

104. Strickland-Munro, J.; Moore, S. Indigenous involvement and benefits from tourism in protected areas: A study of Purnululu National Park and Warmun Community, Australia. *J. Sustain. Tour.* **2013**, *21*, 26–41. [CrossRef]

105. Muller, M.; Job, H. Managing natural disturbance in protected areas: Tourists' attitude towards the bark beetle in a German national park. *Biol. Conserv.* **2009**, *142*, 375–383. [CrossRef]

106. Oh, C.O.; Draper, J.; Dixon, A.W. Comparing resident and tourist preferences for public beach access and related amenities. *Ocean Coast. Manag.* **2010**, *53*, 245–251. [CrossRef]

107. Peters, H.; Hawkins, J.P. Access to marine parks: A comparative study in willingness to pay. *Ocean Coast. Manag.* **2009**, *52*, 219–228. [CrossRef]

108. Petrosillo, I.; Zurlini, G.; Corliano, M.E.; Zacarrelli, E.; Dadamo, M. Tourist perception of recreational environment and management in a marine protected area. *Landsc. Urban Plan.* **2007**, *79*, 29–37. [CrossRef]

109. Buultjens, J.; Ratnayake, I.; Gnanapala, A.; Aslam, M. Tourism and its implications for management in Ruhuna National Park (Yala), Sri Lanka. *Tour Manag.* **2005**, *26*, 733–742. [CrossRef]

110. Cunha, A.A. Negative effects of tourism in a Brazilian Atlantic forest National Park. *J. Nat. Conserv.* **2010**, *18*, 291–295. [CrossRef]

111. Haukeland, J.V.; Grue, B.; Veisten, K. Turning national parks into tourism attractions: Nature orientation and quest for facilities. *Scand. J. Hosp. Tour.* **2010**, *10*, 248–271. [CrossRef]

112. Fleishman, L.; Feitelson, E.; Salomon, I. The role of cultural and demographic diversity in crowding perception: Evidence from nature reserves in Israel. *Tour. Anal.* **2004**, *9*, 23–40. [CrossRef]

113. Oku, H.; Fukamachi, K. The differences in scenic perception of forest visitors through their attributes and recreational activity. *Landsc. Urban Plan.* **2006**, *75*, 34–42. [CrossRef]

114. Arnberger, A.; Brandenburg, C. Past on-site experience, crowding perceptions, and use displacement of visitor groups to a periurban national park. *Environ. Manag.* **2007**, *40*, 34–45. [CrossRef]

115. Arnberger, A.; Haider, W. A comparison of global and actual measures of perceived crowding of urban forest visitors. *J. Leis. Res.* **2007**, *39*, 668–685. [CrossRef]

116. Leujak, W.; Ormond, R.F.G. Visitor perceptions and the shifting social carrying capacity of south Sinai's coral reefs. *Environ. Plan.* **2007**, *39*, 472–489. [CrossRef]

117. Uysal, M.; Jurowski, C.; Noe, F.P.; McDonald, C.D. Environmental attitude by trip and visitor characteristics—US Virgin Islands National Park. *Tour Manag.* **1994**, *15*, 284–294. [CrossRef]

118. Puhakka, R. Environmental concern and responsibility among nature tourists in Oulanka PAN Park, Finland. *Scand. J. Hosp. Tour.* **2011**, *11*, 76–96. [CrossRef]

119. Dunlap, R.E.; Van Liere, K.D.; Mertig, A.G.; Jones, R.E. Measuring endorsement of the new ecological paradigm: A revised NEP scale. *J. Soc. Issues* **2000**, *56*, 425–442. [CrossRef]

120. Akama, J.S.; Land, C.L.; Burnett, G.W. Conflicting attitudes toward state wildlife conservation programs in Kenya. *Soc. Nat. Resour.* **1995**, *8*, 133–144. [CrossRef]

121. Infield, M. Attitudes of rural communities towards conservation and local conservation areas in Natal, South Africa. *Biol. Conserv.* **1988**, *45*, 21–46. [CrossRef]

122. Newmark, W.D.; Leonard, N.L.; Sariko, H.I.; Gamassa, D.M. Conservation attitudes of local people living adjacent to five protected areas in Tanzania. *Biol. Conserv.* **1993**, *63*, 177–183. [CrossRef]

123. Trakolis, D. Local people's perceptions of planning and management issues in Prespes Lakes National Park, Greece. *J. Environ. Manag.* **2001**, *61*, 227–241. [CrossRef]

124. Hall, T.; Shelby, B. Temporal and spatial displacement: Evidence from a high-use reservoir and alternative sites. *J. Leis. Res.* **2000**, *32*, 435–456. [CrossRef]

125. Manning, R.E.; Valliere, W.A. Coping in outdoor recreation: Causes and consequences of crowding and conflict among community residents. *J. Leis. Res.* **2001**, *33*, 410–426. [CrossRef]

126. Balmford, A.; Green, J.M.H.; Anderson, M.; Beresford, J.; Huang, C.; Naidoo, R.; Manica, A. Walk on the wild side: Estimating the global magnitude of visits to protected areas. *PLoS Biol.* **2015**, *13*, e1002074. [CrossRef]

127. Kretchaman, J.; Eagles, P. An analysis of the motives of ecotourists in comparison to the general Canadian population. *Soc. Leis.* **1990**, *13*, 499–508. [CrossRef]

128. Eagles, P.F. The travel motivations of Canadian ecotourists. *J. Travel Res.* **1992**, *31*, 3–7. [CrossRef]

129. Wright, P.A. North American ecotourism markets: Motivations, preferences, and destinations. *J. Travel Res.* **1996**, *35*, 3–10.

130. Palacio, V.; McCool, S.F. Identifying ecotourists in Belize through benefit segmentation: A preliminary analysis. *J. Sustain. Tour.* **1997**, *5*, 234–243. [CrossRef]

131. Chan, J.K.L.; Baum, T. Motivation factors of ecotourists in ecolodge accommodation: The push and pull factors. *Asia Pac. J. Tour. Res.* **2007**, *12*, 349–364. [CrossRef]

132. Dodds, R.; Graci, S.R.; Holmes, M. Does the tourist care? A comparison of tourists in Koh Phi Phi, Thailand and Gili Trawangan, Indonesia. *J. Sustain. Tourism.* **2010**, *18*, 207–222. [CrossRef]

133. Butler, R.; Boyd, S. (Eds.) *Tourism and National Parks: Issues and Implications*; Wiley: Chichester, UK, 2000.

134. McCool, S.F. Managing for visitor experiences in protected areas: Promising opportunities and fundamental challenges. *PARKS Int. J. Prot. Areas Manag.* **2006**, *16*, 3–9.

135. Arabatzis, G.; Grigoroudis, E. Visitors' satisfaction, perceptions and gap analysis: The case of Dadia-Lefkimi-Souflion National Park. *Landsc. Urban Plan.* **2010**, *97*, 229–238. [CrossRef]

136. Suckall, N.; Fraser, E.D.G.; Cooper, T.; Claire, Q. Visitor perceptions of rural landscapes: A case study in the Peak District National Park, England. *J. Environ. Manag.* **2009**, *90*, 1195–1203. [CrossRef]

137. Andersson, T.D. The tourist in the experience economy. *Scand. J. Hosp. Tour.* **2007**, *7*, 46–58. [CrossRef]

138. Larsen, S. Aspects of a psychology of the tourist experience. *Scand. J. Hosp. Tour.* **2007**, *7*, 7–18. [CrossRef]

139. Ritchie, J.R.; Hudson, S. Understanding and meeting the challenges of consumer/tourist experience research. *Int. J. Tour. Res.* **2009**, *11*, 111–126. [CrossRef]

140. Pine, B.J.; Gilmore, J.H. Welcome to experience economy. *Harv. Bus. Rev.* **1998**, *74*, 97–105.

141. Bansal, H.; Eiselt, H.A. Exploratory research of tourists' motivations and planning. *Tour Manag.* **2004**, *25*, 387–396. [CrossRef]

142. Kim, J.H.; Ritchie, B.W. Motivation-based typology: An empirical study of golf tourists. *J. Hosp. Tour. Res.* **2012**, *36*, 251–280. [CrossRef]

143. Luo, Y.; Deng, J. The new environmental paradigm and nature-based tourism motivation. *J. Travel Res.* **2008**, *46*, 392–402. [CrossRef]

144. Kim, H.; Borges, M.C.; Chon, J. Impacts of environmental values on tourism motivation: The case of FICA, Brazil. *Tour Manag.* **2006**, *27*, 957–967. [CrossRef]

145. Yuan, S.; McDonald, C. Motivational determinants of international pleasure time. *J. Travel Res.* **1990**, *29*, 42–44. [CrossRef]

146. Oh, H.; Fiore, A.M.; Jeoung, M. Measuring experience economy concepts: Tourism applications. *J. Travel Res.* **2007**, *46*, 119–132. [CrossRef]

147. Hovardas, T.; Poirazidis, K. Evaluation of the environmentalist dimension of ecotourism at the Dadia Forest Reserve (Greece). *Environ. Manag.* **2006**, *38*, 810–822. [CrossRef]

148. Hall, C.M. Tourism and biodiversity: More significant than climate change? *J. Herit. Tours* **2010**, *5*, 253–266. [CrossRef]

149. Wahlers, R.G.; Etzel, M.J. Vacation preferences as a manifestation of optimal stimulation and lifestyle experience. *J. Leis. Res.* **1985**, *17*, 283–295. [CrossRef]

150. Ross, E.D.; Iso-Ahola, S.E. Sightseeing tourists motivation and satisfaction. *Ann. Tour. Res.* **1991**, *18*, 226–237. [CrossRef]

151. Sickle Van, K.; Eagles, P.F.J. Budgets, pricing policies and user fees in Canadian parks' tourism. *Tour. Manag.* **1998**, *19*, 225–235. [CrossRef]

152. Scott, D.; Jones, B.; Konopek, J. Implications of climate and environmental change for nature-based tourism in the Canadian Rocky Mountains: A case study of Waterton Lakes National Park. *Tour. Manag.* **2007**, *28*, 570–579. [CrossRef]

153. Jacobsen, J.; Steen, K. Use of landscape perception methods in tourism studies: A review of photo-based research approaches. *Tour. Geogr.* **2007**, *9*, 234–253. [CrossRef]

154. Zube, E.H. Local and extra-local perceptions of national parks and protected areas. *Landsc. Urban Plan.* **1986**, *13*, 11–17. [CrossRef]

155. Hsu, T.-K.; Tsai, Y.-F.; Wu, H.-H. The preference analysis for tourist choice of destination: A case study of Taiwan. *Tour. Manag.* **2009**, *30*, 288–297. [CrossRef]

156. Little, C. *Greenways for America*; Johns Hopkins University Press: Baltimore, MD, USA, 1990.

157. Tilden, F. *Interpreting Our Heritage*, rev. ed.; University of North Carolina Press: Chapel Hill, NC, USA, 1977.

158. Goldman, T.L.; Chen, W.J.; Larsen, D.L. Clicking the icon: Exploring the meanings visitors attach to three national capital memorials. *J. Interpret. Res.* **2001**, *6*, 3–30.

159. Füller, J.; Matzler, K. Customer delight and market segmentation: An application of the three-factor theory of customer satisfaction on life style groups. *Tour. Manag.* **2008**, *29*, 116–126. [CrossRef]

160. Hui, T.K.; Wan, D.; Ho, A. Tourists' satisfaction, recommendation and revisiting Singapore. *Tour. Manag.* **2007**, *28*, 965–975. [CrossRef]

161. Matzler, K.; Füller, J.; Faullant, R. Customer satisfaction and loyalty to Alpine ski resorts: The moderating effect of lifestyle, spending and customers' skiing skills. *Int. J. Tour. Res.* **2007**, *9*, 409–421. [CrossRef]

162. Petrick, J.F. Measuring cruise passengers' perceived value. *Tour. Anal.* **2003**, *7*, 251–258. [CrossRef]

163. Sadeh, E.; Asgari, F.; Mousavi, L.; Sadeh, S. Factors affecting tourist satisfaction and its consequences. *J. Basic Appl. Sci. Res.* **2012**, *2*, 1557–1560.

164. Gursoy, D.; Mccleary, K.W.; Lepsito, L.R. Propensity to complain: Effects of personality and behavioral factors. *J. Hosp. Tour. Res.* **2007**, *31*, 358–386. [CrossRef]

165. Tsiotsou, R.; Vasioti, E. Satisfaction: A segmentation criterion for "short term" visitors of mountainous destinations. *J. Travel Tour. Market.* **2006**, *20*, 61–73. [CrossRef]

166. Raadik, J.; Cottrell, S.P.; Fredman, P.; Ritter, P.; Newman, P. Understanding recreational experience preferences: Application at Fulufjället National Park, Sweden. *Scand. J. Hosp. Tour.* **2010**, *10*, 231–247. [CrossRef]

167. Yoon, Y.; Uysal, M. An examination of the effects of motivation and satisfaction on destination loyalty: A structural model. *Tour. Manag.* **2005**, *26*, 45–56. [CrossRef]

 *resources*

Article

Regional Geotourist Resources—Assessment and Management (A Case Study in SE Poland)

Wojciech Zgłobicki [1,*], **Sylwia Kukiełka** [1] **and Bogusława Baran-Zgłobicka** [2]

[1]   Institute of Earth and Environmental Sciences, Maria Curie-Skłodowska University, Kraśnicka Av. 2d,
     20-718 Lublin, Poland; lublinms@gmail.com
[2]   Institute of Social and Economic Geography and Spatial Management, Maria Curie-Skłodowska University,
     Kraśnicka Av. 2d, 20-718 Lublin, Poland; bbaran@umcs.pl
*   Correspondence: wojciech.zglobicki@umcs.pl; Tel.: +48-81-537-6884

Received: 20 January 2020; Accepted: 17 February 2020; Published: 19 February 2020

Abstract: The appropriate identification of the geoheritage resources of a specific area is the sine qua non of the development of geotourism. The identification of tourists' perception of abiotic nature sites of high scientific value is also particularly important because it determines the possibility of using the potential of geosites. In the study, a detailed analysis was carried out of the assets of geological, geomorphological and hydrologic sites in the Central Roztocze region (SE Poland) comprising the central part of the proposed Geopark "Stone Forest in Roztocze". Data from the Polish Central Register of Geosites, the results of a geotourist assessment and questionnaire surveys were used in the analysis. These data indicate a high potential for geotourism development and consistency between scientific assessments and ratings from tourists. However, this potential is not used to a sufficient degree, while actions aimed at developing geotourism and establishing the Geopark are not appreciated by local authorities and institutions responsible for tourism development. The idea of geoparks and geotourism development is not supported by the State either, whether institutionally or financially. Based on the studies conducted, we propose practical measures that should be implemented to increase the use of the region's geotourist assets.

Keywords: geoheritage; geopark; geosites; regional development

1. Introduction

Geotourism is a form of on-site cognitive tourism focused on geological and geomorphological (landscape) assets [1,2]. It promotes visits to geosites and preservation of geodiversity [3]. Geoparks have a special role in the development of geotourism—*"geographical areas where geological heritage sites are part of a holistic concept of protection, education, and sustainable development"* [4]. The appropriate use of geoheritage resources of a particular area and the development of the geotourist function should be preceded by a diagnosis of the current state in this respect [5–7]. This enables the planning of the kind of measures aimed at transforming geoheritage resources into geotourist attractions. Such studies constitute an indispensable element of establishing geosites and geoparks. An assessment of geotourist assets can be carried out by means of an expert evaluation [8–12] or questionnaire surveys among tourists and residents [13–17]. Both methods of assessing geoheritage resources have their advantages and limitations. Questionnaire surveys seem to be particularly interesting, owing to the possibility of obtaining information directly from persons who are the actual recipients of a specific geotourist offer. Understanding the profiles, motivation and preferences of tourists is indicated as one of the most significant directions of geotourism studies [6].

The problem of Poland's geotourist assets attracts more and more attention in the scientific literature (e.g., References [18–24], and many others). Three national geoparks have been established, and designs of new ones are being prepared. The scientific basis for the development of geotourism is getting better and more complete. However, there is a lack of reliable, popular science information on the geotourist resources of the particular regions, lack of promotion, and a shortage of developed tourist products [11,24,25].

The primary objective of the study was to assess the geotourist resources of the Central Roztocze Region located in south-eastern Poland, for which the design of "The Stone Forest in Roztocze" Geopark was prepared [26]. Using two assessment methods made it possible to compare the perspectives of specialists (scientists) and users (tourists) on the assets of geoheritage sites. Analyses of this kind are carried out relatively rarely. Another objective of the study was to determine whether local authorities and institutions responsible for tourism development appreciate the existence of geotourist assets in this area, as well as possibilities of regional development related to the Geopark project. Our comprehensive studies enabled the identification of the strengths and weaknesses of the region, as well as actions that should be taken to make better use of its geotourist potential.

Until now, three areas in Poland received a national geopark certificate: The Polish part of the cross-border Geopark Łuk Mużakowa (Muskau Arch Geopark) (2009), Geopark Góra Św. Anny (St. Anne's Mountain Geopark) (2010) and Geopark Karkonosze (Karkonosze Mountains Geopark) (2010). The certificates were issued by the Minister of the Environment, but after the establishment of the geoparks above, the State lost interest in this initiative. The Muskau Bend Geopark now has the status of a global UNESCO geopark, and the "Geopark Łuku Mużakowa" association works in its territory (http://geopark.muzakowski.pl/). The legal status of geoparks in Poland is undefined: There are no formal grounds for the functioning and management of geoparks. There is no organisation or association dedicated to the development of geotourism on the national scale. Geoparks are isolated projects that currently receive no support from the State. The geopark area partially overlaps with the national park area (Karkonosze Mountains Geopark) or nature reserve area (St. Anne's Mountain Geopark). The protection of geoheritage assets in Poland is based on the protection of areas and objects: National parks, nature reserves, natural phenomena. In the case of a geopark, the location within a protected area with a strict conservation regime may be an obstacle to the geopark's development because the supremacy of protecting outstanding natural assets, including geodiversity precludes the full achievement of the goals for which geoparks are established. Only the cooperation of nature protection services, local authorities and tourist organisations can intensify the development of geotourism.

The study encompassed detailed investigations of the central part of the proposed geopark in Central Roztocze, characterised by a high intensity of tourist traffic and concentration of major geotourist attractions. Tourists are attracted mainly by the region's high diversity of landscapes. The landforms and elements of the geological structure of this meso-region become the destination for more and more tourists who begin to appreciate its geotourist assets. One can also observe an improvement in tourism infrastructure, which increases the accessibility of assets not only along general tourist trails, but also geotourist trails [27,28]. There are many indications that geotourism has a high potential for development but, for the time being, this potential is not fully used.

2. Materials and Methods

2.1. Geoheritage Resources of Central Roztocze

Roztocze is an elongated morphological ridge extending from the NW to the SE, between the Lublin Upland and the Volhynia–Podolia Upland in the north and north-east and the Sandomierz Basin in the south (Figure 1). It is a series of elevations reaching 300–400 m a.s.l., rising 100–150 m above the surrounding areas [26]. The morphological escarpments of Roztocze correspond to the fault lines separating it from the depressed neighbouring areas; hence, it has the character of a horst dissected by several displacements into numerous blocks [29]. The Roztocze ridge is composed of

Upper Cretaceous opoka (silica-calcareous marine sedimentary rock), gaize (glauconitic sandstone) and marl that form a horizontally distributed, thick lithological complex [30].

Patches of younger Tertiary deposits lying on the Cretaceous rock include sandstone, conglomerates, *Serpula* and *Lithothamnium* limestone. Tertiary limestone overlying Upper Cretaceous rock was eroded, due to its low resistance to destructive factors. They survived only on residual hills. The surface deposits of Central Roztocze are primarily Quaternary sandy and silty deposits formed as a result of periglacial processes [29].

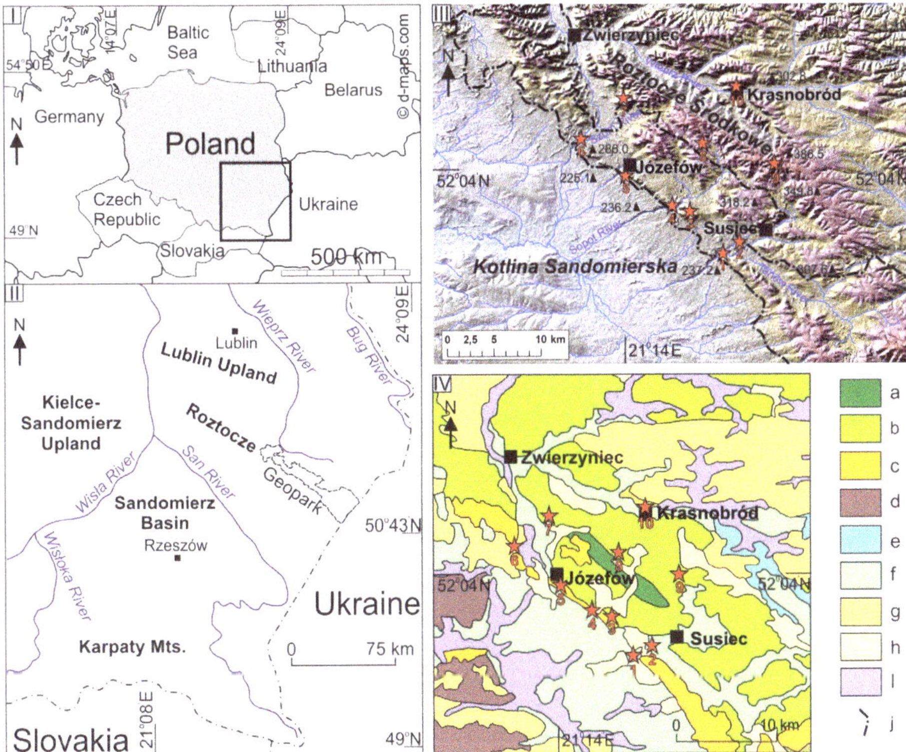

Figure 1. Location of the studied area in Central Europe, rectangle indicates the location of Figure 1.II; (**I**), SE Poland (**II**). Location of the studied geosites against: (**III**) Geomorphology of the Central Roztocze (**III**), (**IV**) Geological Map of Central Roztocze (based on [31]). (a) opoka, marl, limestone (Campanian), (b) limestone, chalk, opoka (Maastrichtian), (c) organogenic limestone (Miocene), (d) glacial till, sand and gravel (Pleistocene), (e) lacustrine sand and loam (Pleistocene), (f) fluvial sand, gravel and loam (Pleistocene), (g) loess (Pleistocene), (h) aeolian sand (Pleistocene), (i) fluvial sand, gravel and alluvial deposits (Holocene), (j) border of the proposed Geopark. 1—Nad Tanwią nature reserve, 2—Waterfall on the Jeleń river, 3—Quarry in Nowiny, 4—Sopot river gap valley in the Czartowe Pole nature reserve, 5—Quarry in Józefów, 6—Szum river gap valley, 7—"Piekiełko" rock forms, 8—Underground quarry in Senderki, 9—Wapielnia Hill, 10—Quarry in Krasnobród.

Flat plateaus built of Cretaceous opoka and dissected by vast valleys are the main landform. Limestone residual hills tower above them, Wapielnia being the highest (386.2 m a.s.l.). In the south, Central Roztocze is delimited by a distinct escarpment zone consisting of several morphological elements parallel to each other. In the west, it is composed of a chain of hills formed by Tertiary rocks.

The land relief features deep and narrow gap valleys of the Sopot, Tanew and Szum rivers. Series of cascades in Cretaceous or Tertiary rocks are visible in the river channels [32].

The proposed Geopark "Stone Forest in Roztocze" is 65 km long (NW-SE) and 2–18 km wide, covering an area of about 640 km^2. It encompasses a considerable piece of the western part of Eastern Roztocze, the southern piece of Central Roztocze (its escarpment zone), and a small piece of Western Roztocze. The name of the Geopark is related to the pieces of fossil trunks and branches of *Taxodioxylon taxodii* Gothan occurring in this area [26]. Within the proposed geopark, about 150 geosites that can form the basis of geotourism development were indicated, including 85 located within Central Roztocze (the central part of the Geopark). Most of the geosites in this area are geological outcrops and exposures (40 sites), water sites, i.e., springs, waterfalls and lakes (23), geomorphological sites (16), and sites related to cultural heritage (6). Below is a description of the most important geosites of the proposed Geopark, located within parts of Central Roztocze intensively visited by tourists (Figure 2). Seven of them are located within the Central Roztocze Geotourist Trail [33]. The waterfall on the Jeleń river, the Szum river gap valley, and "Piekiełko" rock forms are situated off the trail.

Nad Tanwią nature reserve (1)—Numerous cascades forming small waterfalls are visible in the gap valley of the Tanew river. The biggest waterfalls reach a height of 1.2 m. The formation of the cascades is linked to the vertical uplifting movements of Roztocze occurring since the Neogene until the present [32]. The most noteworthy geotourist assets of the nature reserve include the deeply incised valley of the Tanew river, the cascades revealing the geological structure of the area, the meandering river channel, and numerous outcrops of Upper Cretaceous, Miocene and Holocene rocks [21].

Waterfall on the Jeleń river (2)—The highest waterfall in Roztocze (1.5 m) formed on a cascade in the Jeleń river channel built of Cretaceous gaize.

Quarry in Nowiny (3)—Cretaceous gaize covered by Miocene limestone can be observed in this inoperative quarry. The quarry walls reach the height of 12 m. Outcrops of Upper Badenian deposits—calcarenite—occur in the walls. Miocene limestone was quarried here for many years. The site is located in the vicinity of two tourist trails [33].

Sopot river gap valley in the Czartowe Pole nature reserve (4)—A unique geotourist asset in the nature reserve is the deeply incised valley featuring cascades forming riffles between 25 and 90 cm high (Figure 2a). They formed on layers of Miocene limestone [33]. An educational path was designated along the river.

Quarry in Józefów (5)—It is one of the few where it is possible to admire the sedimentary features of the Miocene deposits. The outcrops show organogenic limestone, mainly from algae, along with silty deposits. The quarry walls are up to 15 m tall. This is the biggest quarry in Central Roztocze, and the ongoing quarrying activity is very limited (Figure 2b). A 19-metre observation tower was built at the quarry, offering views of the entire site and the surrounding landscape [33].

Szum river gap valley (6)—The river flows across the escarpment zone of Roztocze, along a winding, narrow and deeply incised valley. The steep slopes of the valley formed in Miocene limestone, quartz sand and Cretaceous gaize. Riffles and small cascades formed along a 300-metre stretch of the river, within gaize outcrops. The relatively steep gradient of the channel results in fast flow rates leading to the formation of cascades. Eleven waterfalls, up to 0.5 m high, can be found here [33].

"Piekiełko" rock forms (7)—The hill is built of organogenic limestone and Badenian sandy limestone. Several dozen rock forms were created by weathering and mass movements. The largest ones are two "towers", 4.5–5 m tall. Alongside large rocks, there are also smaller forms: Rock ledges, pulpits and ridges [34]. A waymarked educational path with interpretation panels about the origins of the rocks leads to the nature reserve.

Underground quarry in Senderki (8)—The underground galleries, 1 to 1.5 m high were built in order to obtain millstones. At present, six openings lead to the excavation galleries. Miocene rock occurs here in the form of layers of organogenic limestone, sandy limestone, sandstone and sand. Inoperative quarries, depressions and waste heaps can be seen on the surface. Although the underground quarry is not accessible to visitors, an educational path enables the observation of the surface traces of quarrying activity [35].

Wapielnia Hill (9) is a residual hill (subjected to denudation), built of rocks that were uplifted by about 70 m during the Alpine tectonic movements. Outcrops of reef limestone and *Lithothamnium -Bryozoa* limestone constituting various rock forms occur at the top of the hill. The limestone obtained here was used in industry and local construction work [21].

Figure 2. There are two general types of geosites of planned Geopark: River channels with small waterfalls (**a**) and quarries (**b**). (**a**) Sopot river gap valley in Czartowe Pole nature reserve (**b**) Quarry in Józefów.

Quarry in Krasnobród (10)—Upper Cretaceous opoka was quarried here. The main ingredients of opoka are calcium carbonate and silica formed mainly by various organisms, i.e., *Porifera*, *Bryozoa*, *Bivalvia*, *Echinoidea*, coccoliths. A small number of fossils can still be seen in the outcrops. The wall of the quarry is about 150 m long and 25 m tall. An observation tower stands at the top of the quarry wall [33].

2.2. Methods

In the assessment of the geotourist assets of Central Roztocze, a four-stage procedure was adopted. First, the results of the assessment of the scientific, educational and tourist value of the sites located in the region and included in the Polish Central Register of Geosites were analysed (http://geoportal.pgi.gov.pl/portal/page/portal/geostanowiska/). Data concerning the sites located in Central Roztocze were introduced into the register as part of the preparation of the geopark plan. The assessment in the Register is quite subjective because it is not based on specific formal criteria. However, it offers a general image of the value of the sites based on the assessment of experts knowing the region and its geoheritage features. Among the highest rated geosites, ten different sites were selected for further assessment (four river valleys with waterfalls, three quarries, one underground mine and two hills). At the same time, they are sites frequently visited by tourists, widely recognized as the region's largest natural tourist attractions. A detailed assessment of their geotourist assets was then carried out by means of a method developed by the authors (second stage). The attractiveness of the same geosites was also assessed based on the survey of tourists visiting Central Roztocze (four stage). The next step consisted of analyzing strategic documents prepared at various levels of administration—from province to district level—in terms of the presence of information on the possibility of using geoheritage assets for tourism and regional development.

Many methods of assessing the geotourist assets of geoheritage components have been developed so far (see, for example, reviews by Kubaliková [9] and Brilha [12]). The assessment used in the present study includes the criteria used in most evaluations of this kind. It is a modification of the method proposed by Warowna et al. [11]. The assessment comprises 12 criteria belonging to four groups:

(i) Scientific value, (ii) educational value, (iii) functional value, (iv) tourist value (Table 1). Each criterion was assessed according to a three-degree scale—(i) 0.0, (ii) 0.5, (iii) 1.0.

Table 1. Categories and geotourist value assessment criteria.

Type of the Value	Scores
	Scientific
Scientific knowledge	0.0—unknown, lack of publications, 0.5—less than 10 scientific publications, 1.0—more than 10 publications
Rarity	0.0—not among the top 3 most important sites, 0.5—one of the three most important sites, 1.0—the only occurrence
Diversity	0.0—only one abiotic feature (process), 0.5—2 visible abiotic features, 1.0—three and more abiotic features
	Educational
Representativeness	0.0—low representativeness/poor visibility of features, 0.5—medium (primarily for experts), 1.0—high (also for non-experts)
Exposure (visibility)	0.0—obstacles to observation all year round (e.g., vegetation), 0.5—obstacles to observation in some seasons of the year, 1.0—no obstacles to observation
Educational products	0.0—none, 0.5—one product related to geoheritage, 1.0—two or more products related to geoheritage
	Functional
Accessibility (getting there)	0.0—by car and more than 1000 m walking distance, 0.5—by car and less than 1000 m walking distance, 1.0—by public transport and less than 500 m walking distance
Accessibility (difficulty of reaching the site)	0.0—difficult access (vegetation, the character of substratum), 0.5—minor obstacles to access or accessibility in certain seasons of the year, 1.0—no obstacles to access
Presence of tourist trails and educational paths	0.0—path or trail more than 3 km away, 0.5—path or trail up to 3 km away, 1.0—the site lies on a trail or path
	Tourist
Biotic or cultural value	0.0—no additional value, 0.5—moderate additional value, 1.0—significant additional value
Landscape beauty	0.0—small aesthetic value, 0.5—moderate aesthetic value, 1.0—landmark of the region
Viewpoints	0.0—no viewpoints in the vicinity (3–4 km), 0.5—viewpoint 1–2 km away 1.0—viewpoint up to 1 km away (or the site is the viewpoint)

Besides a general assessment, various ranks were assigned to the individual criteria depending on the degree of the potential tourists' interest in the assets of geoheritage, according to the approach used in such assessments [24,36,37]. Two groups of tourists were distinguished-geotourists for whom the scientific and educational values of geosites are key when assessing sites, and tourists who pay more attention to tourist and functional values. The number of points obtained in the particular groups of criteria was then multiplied by the appropriate coefficients (Table 2).

Table 2. Ranks of criteria groups for different audiences (multiplication coefficient).

	Scientific	Educational	Functional	Tourist
Geotourists	2.0	1.5	1.0	1.0
Tourists	1.0	1.0	1.5	2.0

A survey questionnaire was the second tool used to obtain information from tourists. The respondents were selected randomly, and the answers were anonymous. The respondents were asked to complete a paper questionnaire with 14 questions concerning the geotourist assets of Central Roztocze (Appendix A). The survey was conducted in the following locations—in Zwierzyniec, Górecko Kościelne, Józefów and Krasnobród—in 2014 and 2015. The respondents came from seven provinces: Lubelskie, Mazowieckie, Małopolskie, Śląskie, Podkarpackie, Lubuskie and Świętokrzyskie. A total of 311 survey questionnaires were collected from tourists visiting the areas above.

3. Results

3.1. Geotourist Values

The synthetic evaluation of 85 geosites in Central Roztocze listed in the Polish Central Register of Geosites indicates that their value is moderate. It should be noted that each value in the Register (scientific, educational and tourist) is assigned a rating from 1 to 10. Among them, there are sites with a score of 10 points (maximum value) in the individual categories, as well as sites with a score of 3 or 4 points (Figure 3, Table 3). The average scores of the particular groups of values are as follows: (i) Scientific value—4.8; (ii) educational value—4.7; (iii) tourist value—4.9. The correlations (correlation coefficient) between the types of values are the following: Scientific and educational—0.78; scientific and tourist—0.31, educational and tourist—0.6. Geological sites received higher ratings of their scientific value, while geomorphological and hydrologic sites obtained higher scores for their educational and tourist value.

Table 3. Assessment of the value of 10 selected geosites in Central Roztocze according to data from the Polish Central Register of Geosites (maximum score—30).

Geosite	Scientific	Educational	Tourist	Total Score
"Piekiełko" rock forms (7)	7	9	9	25
Quarry in Nowiny (3)	8	8	8	24
Szum river gap valley (6)	8	8	8	24
Nad Tanwią nature reserve (1)	8	8	7	23
Quarry in Józefów (5)	8	8	7	23
Underground quarry in Senderki (8)	5	7	10	22
Sopot river gap valley in Czartowe Pole nature reserve (4)	6	7	8	21
Waterfall on the Jeleń river (2)	6	7	7	20
Quarry in Krasnobród (10)	3	5	7	15
Wapielnia Hill (9)	4	4	6	14

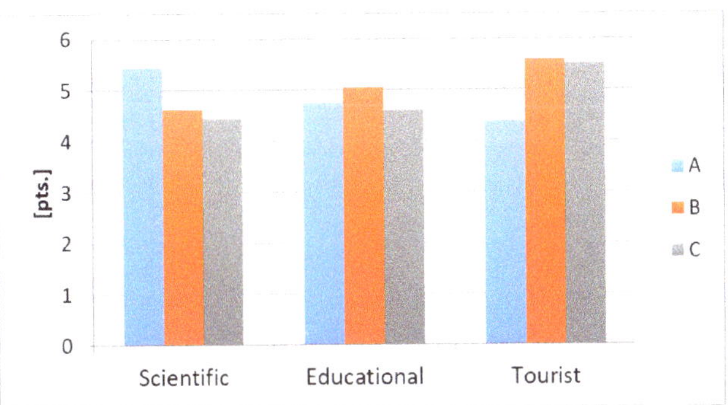

Figure 3. Mean geotourist value rating for geosites in Central Roztocze based on the Polish Central Register of Geosites (rating scale from 1 to 10). A—geological sites (quarries, outcrops), B—geomorphological sites (sand dunes, gullies, hills), C—hydrologic sites (springs, peat bogs, waterfalls).

On average, the sites received 60% of the maximum score. It can be, thus, concluded that their geotourist value is moderately high (Table 4). The lowest score was given to the group of tourist values—30% of the maximum score. The highest mean scores were obtained by the following criteria: (i) Presence of tourist trails and educational paths, (ii) scientific knowledge, (iii) rarity, (iv) visibility. The lowest mean scores occurred in the case of the following criteria: (i) Presence of viewpoints, (ii) scenic beauty, (iii) additional biotic or cultural assets. Two quarries, in Józefów and Krasnobród, received the highest scores—87% and 83% of the maximum rating, respectively. The lowest rating was obtained by the underground quarry in Senderki and the Wapielnia Hill (33% and 41%, respectively).

The synthetic assessment indicates the occurrence of sites with high main values and moderate additional values (Figure 4). With regard to the main values, the highest score was achieved by the Nad Tanwią nature reserve, quarry in Józefów, Sopot river gap valley in the Czartowe Pole nature reserve, and "Piekiełko" rock forms. On the other hand, the quarries in Józefów and Krasnobród, as well as the Sopot river gap valley in the Czartowe Pole nature reserve, were rated as having the highest additional values.

The geoutourist value assessments taking into account various audience groups indicated relatively small differences (Table 5). The differences primarily concern the order of geosites with the highest score even though the top three sites are the same in the case of both assessments. The group of sites with the lowest score also contains the same geosites.

Table 4. Results of the own geotourist assessment of 10 selected geosites in Central Roztocze (maximum score—12).

Geosite	Total Score	Scientific Value	Educational Value	Functional Value	Tourist Value
Quarry in Józefów (5)	10.5	2.5	2.5	3.0	2.5
Quarry in Krasnobród (10)	10.0	2.0	2.5	3.0	2.5
Nad Tanwią nature reserve (1)	9.5	3.0	3.0	2.5	1.0
Sopot river gap valley in Czartowe Pole nature reserve (4)	9.5	3.0	2.5	2.5	1.5
"Piekiełko" rock forms (7)	7.5	2.5	2.5	1.5	1.0
Quarry in Nowiny (3)	7.5	2.0	2.0	3.0	0.5
Szum river gap valley (6)	6.5	1.5	2.0	2.5	0.5
Waterfall on the Jeleń river (2)	6.5	1.5	3.0	1.5	0.5
Wapielnia Hill (9)	5.0	1.0	1.5	1.0	1.5
Underground quarry in Senderki (8)	4.5	1.5	1.0	1.0	1.0
Mean	7.3	1.9	2.0	2.0	1.0

Table 5. The variation of value assessment results among geotourists and tourists.

Geosite	Geotourist Score	Tourist Score
Nad Tanwią nature reserve	14.5	11.25
Quarry in Józefów	14.25	15.5
Sopot river gap valley in Czartowe Pole nature reserve	13.75	11.25
Quarry in Krasnobród	13.25	14.5
"Piekiełko" rock forms	11.25	9.25
Quarry in Nowiny	10.0	9.5
Waterfall on the Jeleń river	10.0	7.75
Szum river gap valley	8.0	8.25
Wapielnia Hill	6.25	7.0
Underground quarry in Senderki	6.0	5.5

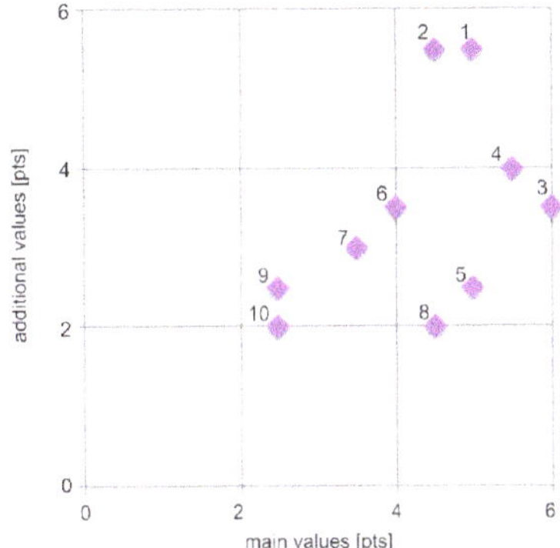

Figure 4. Total main (scientific and educational) and additional values (functional and tourists) for the 10 geosites. 1—Quarry in Józefów, 2—Quarry in Krasnobród, 3—Nad Tanwią nature reserve, 4—Sopot river gap valley in the Czartowe Pole nature reserve, 5—"Piekiełko" rock forms, 6—Quarry in Nowiny, 7—Szum river gap valley, 8—Waterfall on the Jeleń river, 9—Wapielnia Hill, 10—Underground quarry in Senderki.

3.2. Questionnaire Survey Results

The possibility of direct contact with nature and of admiring the landscapes and diverse landforms was the main motivation for tourists to come to Central Roztocze (36%) (Figure 5a). Many respondents were also motivated by the possibility of active leisure (24%) and visiting interesting historic sites (20%). The respondents associated the region primarily with forests and fresh air (36%) and attractive landscapes (31%) (Figure 5b).

When rating the tourist attractiveness of a given site, area or phenomenon, 24% of the respondents take into account mainly aesthetic values, and 22%—good access to a given tourist attraction (Figure 6). 20% pay attention to a large number of tourist attractions in the vicinity. 16% of the respondents regard free admission or low cost of admission as very important.

The respondents looked for information about the geotourist assets of Central Roztocze mostly on the Internet (43%), in brochures, flyers, tourist guides (22%), and among family and friends (20%). Other sources of information were rarely used (Figure 7).

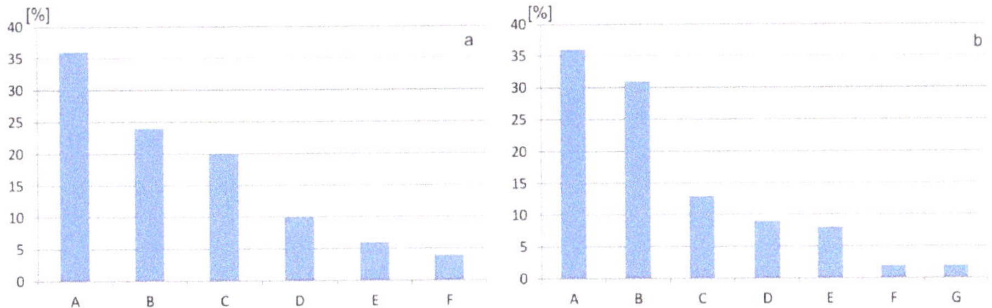

Figure 5. (**a**) Motivation for coming to Central Roztocze. A—nature tourism, B—active tourism, C—cultural tourism, D—family visit, E—event tourism, F—education. (**b**) The tourist assets of Central Roztocze. A—forests and fresh air, B—beautiful landscapes, C—active recreation, D—cultural values, E—agritourism, F—regional cuisine, G—tourist infrastructure.

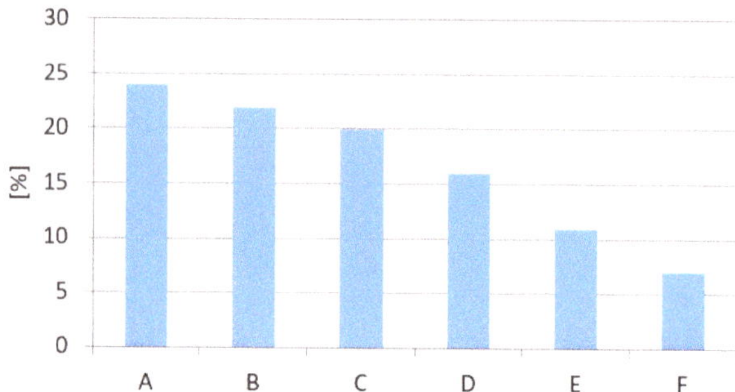

Figure 6. Most significant values for tourists. A—aesthetic value, B—good access, C—many attractions in the vicinity, D—cost of entrance, E—location against tourist trails, F—scientific value.

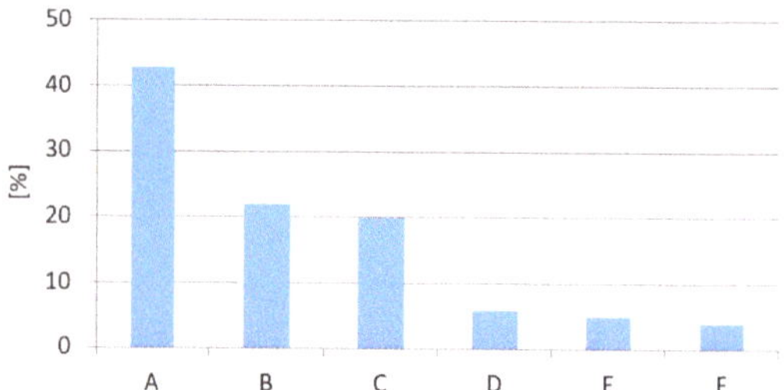

Figure 7. Sources of information about the region and its tourist offer. A—Internet, B—booklets, leaflets, C—friends, family, D—tourist information, E—do not look for information, F—scientific publications.

The term "geotourism" is known to 43% of the respondents (Figure 8). However, when attempting to define geotourism, 64% of the tourists gave an incorrect answer. Geotourism was described as tourism focused on natural environment assets, including biotic and abiotic nature. Respondents usually associated geotourism with geography, which was influenced by the prefix geo-. About 36% of the respondents connected geotourism with geoheritage: "Geological tourism", "tourism related to visiting geological sites" (Table 6). Only 12% of the tourists were familiar with the term Geopark "Stone Forest in Roztocze". In this case, most of the respondents who declared their familiarity with the term provided a correct answer (Table 7).

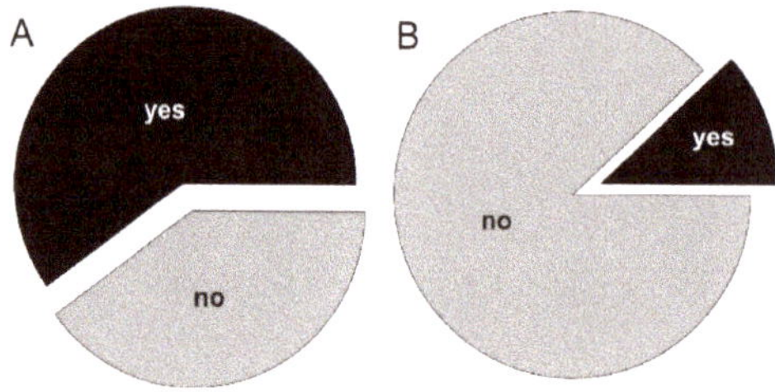

Figure 8. The familiarity of tourists with the term "geotourism" (**A**), geopark "Stone Forest in Roztocze" (**B**).

Table 6. Selected definitions of the term "geotourism" according to respondents.

	Definitions of Geotourism Provided by Tourists
1.	Tourism related to geology, geological tourism
2.	Tourism related to visiting geological sites
3.	Type of tourism related to the protection of abiotic nature
4.	Visiting forest sites
5.	Discovering the past
6.	Tourism within a specific region
7.	Visiting an area
8.	Discovering a natural phenomenon created without human impact
9.	Discovering the values of a specific natural site
10.	Visiting and discovering geological sites, history of mining, geological history
11.	Discovering the origin of the Earth, its shape, structure, history
13.	Something related to the Earth
14.	Multifunctional form of tourism—visiting and discovering nature, history of a region, geological processes
15.	Discovering landscape assets
16.	Form of cognitive tourism based on discovering geological sites
17.	Science related to tourism and nature
18.	Term related to the Earth and tourism
19.	Tourism and taking pictures
21.	Tourism related to archaeology, studying the Earth, rocks, caves
23.	Cognitive tourism
24.	Visiting and discovering "green" nature
25.	Tourism and geography
26.	Visiting geographical sites
27.	Tourism related to geographical assets
28.	Tourism related to visiting rocky landforms

Table 7. Definitions of the "Stone Forest" Geopark according to the respondents.

Definitions of the "Stone Forest" Geopark Provided by Tourists
Located within the Roztocze region
Stone Forest created as a result of geological processes
An area located within the Roztocze region where one can observe petrified wood
An area with defined borders where sites of high geological value are located
Fossils created as a result of natural processes
Located within the Roztocze region, petrified wood and other fossils
An area important from the geological perspective
An area where one can see petrified wood

A vast majority of the respondents correctly selected four geotourist assets from the prepared set: Fossils (74% of the respondents), quarries (68%), springs (64%), sand dunes (60%). It should be noted, however, that 46% of the respondents indicated sites belonging to the category of cultural and biotic assets (Figure 9). The respondents were allowed to choose more than one answer. The respondents mentioned the Kraków-Częstochowa Upland (Jura Krakowsko-Częstochowska) and Central Roztocze as areas of high geotourist value, and the Kaczawskie Mountains and the Suwałki Region as areas of the lowest geotourist value in Poland (Table 8).

The geotourist sites received a relatively high rating from the respondents (Table 9). In the assessment, each site was given a particular rating without any particular criteria (1—lowest rating, 5—highest rating). The five biggest attractions of Central Roztocze include two geotourist sites: Waterfalls on the Tanew river in the Nad Tanwią nature reserve (second place) and the gap valley of the Szum river (fifth place). The synthetic assessment for six geosites was lower than the mean. The lowest scores were received by the underground quarry in Senderki and rock forms in the "Piekiełko" nature reserve. No site received a score lower than 3.5 in a 5-degree scale (1—least attractive, 5—most attractive). This means that, in the opinion of visitors, the geotourist assets of Central Roztocze can be regarded as attractive. Religious sites turned out to be the best rated attractions that surpassed 80% of all geotourist attractions: The church "On the Island" in Zwierzyniec, the church in Krasnobród, and Saint Roch's Chapel.

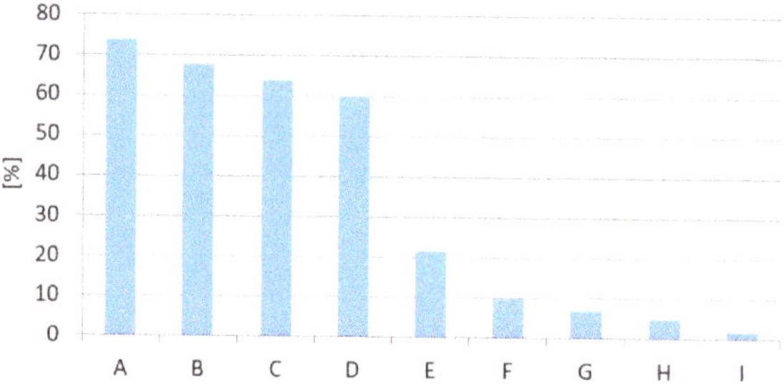

Figure 9. Kinds of geotourist assets (according to tourists). A—fossils, B—quarries, C—springs, D—dunes, E—botanical gardens, F—battlefields, G—manor houses, H—museums, I—festivals.

Table 8. Assessment of the geotourist value of the selected regions of Poland.

Region	Respondents' Assessment	Actual Scientific Value
Kraków-Częstochowa Upland	High	High
Central Roztocze	High	High
Białowieża Forest	Moderate	Low
Biebrza Valley	Moderate	Low
Nałęczów Plateau	Low	Moderate
Kaczawskie Mountains	Low	High
Suwałki Region	Low	High

Table 9. Respondents' assessment of the value of tourist sites (maximum value—5). Bolded names refer to the geosites studied in the paper.

Tourist Site	Mean Score	Lack of Knowledge about the Site (%)
Church "On the Island" in Zwierzyniec	4.40	17
Nad Tanwią nature reserve	4.37	16
Saint Roch's Chapel	4.30	21
Church in Krasnobród	4.27	17
Sopot river gap valley in the Czartowe Pole nature reserve	4.24	39
Polish Horse sanctuary	4.22	33
Szum river gap valley	4.11	23
Wapielnia Hill	4.11	51
Quarry in Józefów	4.02	34
Quarry in Nowiny	4.00	51
Quarry in Krasnobród	4.00	42
Waterfall on the Jeleń river	4.00	41
Regional Museum in Krasnobród	4.00	31
"Piekiełko" rock forms	3.89	45
Homestead in Guciów	3.69	20
Underground quarry in Senderki	3.56	69

4. Discussion

The questionnaire survey results indicate that a portion of tourists (less than 30%) visiting Central Roztocze can be regarded as more or less conscious geotourists. This is confirmed by their correct answers to questions concerning terms, such as geotourism, geotourist attractions, geotourist assessment of regions. At the same time, the assessment of the value of 10 selected geosites, carried out based on the survey results, is consistent with the results of geotourist assessment results—Polish Central Register of Geosites and the assessment developed by the authors. Particularly in the latter case, a large similarity of the assessments occurred, the correlation coefficient is 0.57 (Figure 10). Of course, one can hardly assume that the respondents' assessment was determined by the scientific value of the sites. However, the high synthetic assessment of their tourist value indicates that abiotic nature sites arouse the interest of visitors and can be the basis for the development of geotourism. The favourable perception of geoheritage sites by tourists and residents is of key importance from the perspective of spatial management and preservation of abiotic nature assets [17].

Questionnaire surveys also indicate that there is a group of tourists who do not regard geoheritage assets as important and interesting. Familiarity with the assessed sites plays a very significant role in this respect. A low rating of geosites very often results from the lack of information about them: The correlation coefficient between the assessment results and the respondents' lack of knowledge about geosites is 0.62. The underground quarry in Senderki is an example of such a site. One can suppose that the situation was similar in the case of the low score of geotourist assets of regions, such as the Kaczawskie Mountains or the Suwałki Region which are actually unique on a national scale in terms of geological structure (remnants of volcanic phenomena and forms) and geomorphology (young glacial landscape). The lack of knowledge about the studied geosites was indicated by 16 to 69% (average of 38%) of the respondents. It should be stressed, however, that the Kaczawskie Mountains and the Suwałki Region are among the most valuable and generally quite frequently visited areas (with few exceptions). In the case of other geosites in Central Roztocze, the level of knowledge about them is clearly lower. Zgłobicki, Baran-Zgłobicka [15] mentioned the significant variation of familiarity with valuable geotourist sites in the area of the potential Małopolska Vistula Gap Geopark.

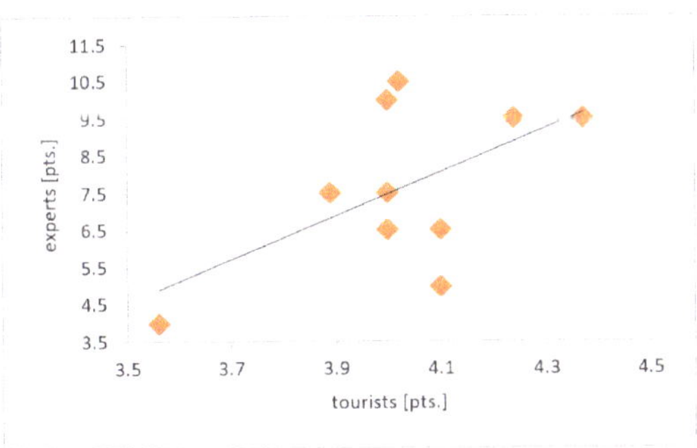

Figure 10. Correlation between tourists' assessment (questionnaire survey) and author's geotourist assessment (data from Tables 4 and 9).

A high degree of consistency between assessments occurs in the case of the following geosites—Nad Tanwią nature reserve, Waterfall on the Jeleń river, Sopot river gap valley in the Czartowe Pole nature reserve (Table 10). This applies both to sites of undoubtedly high scientific and tourist value (Nad Tanwią nature reserve, Sopot river gap valley in the Czartowe Pole nature reserve) and those of lower value. A moderate degree of consistency occurs for the Wapielnia Hill and the Quarry in Józefów. In the case of some sites, however, quite a considerable variation of assessments was found, namely—Underground quarry in Senderki, quarry in Nowiny, "Piekiełko" rock forms, Quarry in

Krasnobród. These differences result primarily from the tourists' poor familiarity with these sites despite their high scientific value.

Table 10. Comparison of the assessment results of selected geosites (their rank).

Geosites	A	B	C
Quarry in Józefów	3	1	5
Quarry in Krasnobród	7	2	7
Nad Tanwią nature reserve	3	3	1
River gap in Czartowe Pole nature reserve	5	3	2
"Piekiełko" rock forms	1	4	9
Quarry in Nowiny	2	4	6
Szum river gap valley	2	5	3
Waterfall on the Jeleń river	6	5	8
Wapielnia Hill	8	6	4
Underground quarry in Senderki	4	7	10

A—assessment based on the Polish Central Register of Geosites, B—geotourist assessment developed by the authors, C—tourists' assessment.

The gap valleys of the rivers in Central Roztocze are definitely popular and frequented by tourists. They are among the best rated geomorphological sites in the Lublin Region [15]. They arouse the interest among tourists in other regions of the world as well [29,38].

One of the reasons behind the high rating given by tourists to such geosites is their unquestionable visual (landscape) value. The significance of landscape beauty in the perception, and consequently, assessment of natural components and entire landscapes are indicated by the results of previous studies [17,39]. There is no doubt that geotourism can also develop based on the assets of the cultural landscape [40]. In this respect, a special role can be played by a large number of quarries in Central Roztocze [28,41,42]. Some of them are located close to important tourist localities, which is a significant advantage. In addition, their tourist and geotourist value are enhanced by the existence of observation towers [43]. The quarries also make it possible to actively learn about geological assets by observing and collecting rocks and fossils. It is particularly important from the perspective of the educational function of geosites, particularly with respect to children and youth [24]. Thus, it seems that the first promotional measures and development of geotourist products in Central Roztocze should be based on greater use of the selected quarries.

Very few people have heard about the proposed geopark, which is not surprising, given that this term is mostly used in the scientific literature. Studies conducted in Kielce (central Poland) at the turn of the 21st century show that even geography students were not familiar with this concept: Only 40% of them were able to provide a correct definition [14]. At the same time, actions aimed at establishing the geopark seem to be the only possible direction of geotourism development in this area. Therefore, as other authors indicate, it is necessary to conduct in-depth studies on the motivation of tourists visiting abiotic nature sites in Central Roztocze [16].

Tourists indicated certain deficiencies in the geotourism development of Central Roztocze. The most frequently mentioned issues include the lack or inadequate marking of tourist trails, the lack of or insufficiently detailed interpretation panels, poor condition of geotourist sites, lack of information and maps with marked geotourist sites (Table 11). According to tourists, the most important elements are absent, namely the marking of trails and interpretation panels. The only geotourist product developed so far is the Geotourist Trail of Central Roztocze [33], but, since the map of the trail is no longer available for purchase, its impact is limited. Seven among the 10 geosites under study are located on this trail and are "equipped" with geotourist interpretation panels (Figure 11). It seems, however, that these boards need some modification because their current content is sometimes too scientific.

Table 11. Tourism in the development strategies of districts in Central Roztocze.

The Strategy Part of Documents	
Directions of Actions/Strategic Objectives	**Operational Objectives**
Urban-rural district Józefów	
Tourism development and improving the district's tourist attractiveness through the effective use of the cultural potential and heritage	Improving the district's image and tourist attractiveness through the proper marking of sites of historical and tourist interest Supporting the residents of rural areas in the creation of agri-tourist farms by means of promotion, information and consultancy
Improving and developing the district's promotional activities	Improving the district's image through the proper marking of roads, localities and sites of interest Creating a coherent system for the promotion of the district's tourist and recreational attractions
Urban-rural district Krasnobród	
Increasing the competitiveness of the town and district of Krasnobród as a centre of tourist and recreational traffic	Supporting the development of small and medium-sized enterprises, including businesses serving tourists Increasing the tourist and health resort attractiveness and strengthening the natural environment assets
Rural district Susiec	
Economic stimulation of the district through the use of its tourist assets	Increasing the tourist attractiveness of the district
Rural district Tereszpol	
Caring for the natural environment and ensuring tourism development in the district	Developing and implementing an Integrated Tourist Product in cooperation with other local government units as part of the "Active Roztocze" initiative Building, extension and modernisation of tourist infrastructure Tourist promotion of the district
Urban-rural district Tomaszów Lubelski	
Using the natural environment assets, cultural heritage and public space of the district in the process of its development	Promoting the district by highlighting the natural environment assets and historic monuments within its territory Building tourist and tourism-related infrastructure
Urban-rural district Zwierzyniec	
Developing and improving the quality of infrastructure and the social sphere	Developing tourist and recreational infrastructure Promotion of the district
Districts belonging to the "Active Roztocze" initiative	
Integrated development of tourism	Increasing the accessibility of areas of natural and cultural value by integrating local tourist trails with supra-regional cycling trails Preserving natural and cultural heritage sites Developing tourism specialisation through the integration of tourist products

Figure 11. Elements of tourist infrastructure. (**A**) Observation tower at the quarry in Józefów, (**B**) Walkways and panels along the educational path in the Czartowe Pole nature reserve.

The peripheral location of the studied area at the Polish-Ukrainian border and in relation to larger cities, is a challenge to the further development of tourism here. There is a lack of good roads to arrive in the area although their condition is improving (Figure 12).

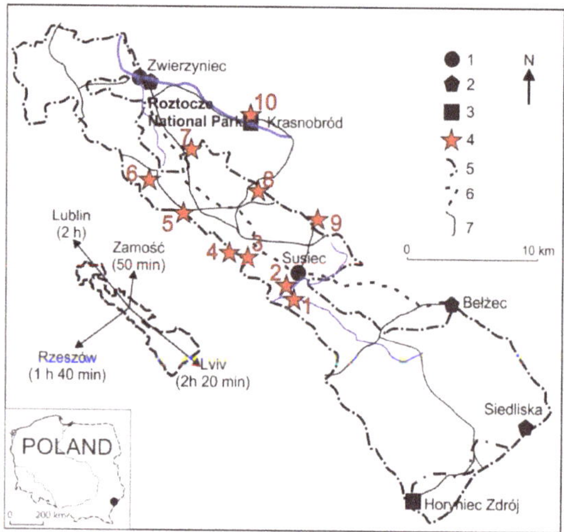

Figure 12. The location of the studied geosites within the proposed geopark and the distance (travel time by car) from larger cities. (1) Main tourist localities, (2) cultural assets—museums, (3) health resorts, (4) geosites, (5) boundaries of the proposed geopark, (6) railway line, (7) main roads. Border of proposed Geopark after Krąpiec et al. (2012) [26]

The way of presenting scientific knowledge to tourists is a key challenge to the development of geotourism [44,45]. Studies show that the tourists' knowledge about abiotic nature is distinctly more limited than in the case of the fauna and flora [46]. The same also applies to students of natural science study programmes [14,25]. It is necessary to prepare popular science products presenting geoheritage information in an accessible way. Even in the case of sites of outstanding value, such as the Giant's Causeway, tourists indicate the need for the improvement of interpretation panels with regard to format, size and content [46]. Questionnaire surveys conducted in the vicinity of the Iguazu Falls National Park show that about 70% of tourists do not read the information presented on the interpretation panels [43]. This may result from the inadequate location of the panels, the poor finishing of the panels, or long-winded, hermetic language. On the other hand, individuals who *have* read the information presented on the panels give them a favourable rating. Well-prepared interpretation panels can, thus, be an important means for tourists to acquire knowledge about the geological and geomorphological assets of a given region.

In modern times, the Internet is the most common and the fastest source of information. The Polish Central Register of Geosites is the database providing information about the geological assets. However, this information is often incomplete, and the descriptions are not easily digestible for ordinary tourists. Efforts should be made to create and launch a geoportal—a database of natural assets (components of biotic and abiotic nature) at the regional level. It could be complemented by a mobile application providing the basic information on the region's natural environment, geosites and geotourist paths. However, Rozenkiewicz et al. [47] indicate that still in Central Europe *"information on geotourism resources available online is rather dispersed"*.

All actions aimed at establishing geoparks and developing geotourism should be based on the involvement of local communities in the process [4]. Therefore, it is necessary to persuade residents that using landscape and preserving geoheritage assets for geotourism purposes generates social and economic benefits [48]. The influence of geoparks on local economy and regional development can occur by means of [49]: (i) Production of geoproducts (souvenirs, handicrafts, geofood), (ii) involving local business in geotourism marketing, (iii) development of recreational and sports activities related to geomorphology and geology (geokayaking, geohiking, climbing). In the case of Central Roztocze, Brzezińska-Wójcik [27] mentions several decorative geoproducts, geofood, and themed towns associated with masonry that can help in the education, interpretation and promotion of geoheritage.

Based on examples from Iceland, Australia and Malaysia, Dowling [48] indicates that the immediate benefits that local communities can derive from the development of sustainable geotourism include increased opportunities for employment and generation of additional revenue. The bottom-up strategy is the most efficient for the promotion and development of geotourism, as well as actions aimed at establishing geoparks. This is exemplified by the activities of the Local Action Group *Land of Extinct Volcanoes*, encompassing 14 districts in SW Poland, aimed at promoting and educating the public about geoheritage assets [50]. Unfortunately, the attempts to establish such an organisation in Central Roztocze have not been successful so far.

The appropriate positioning of the preservation and tourist use of abiotic nature assets in the spatial planning and economic development system is of fundamental importance to the establishment of the Geopark and the development of geotourism, including its new forms. At the central planning level, in the 2030 National Spatial Development Concept [51], geoparks are merely mentioned in the spatial development vision as elements (nodes) of Poland's ecological network. In the draft National Raw Material Strategy [52], the establishment of geoparks is an action within the preservation of geodiversity.

At the regional level, in the Development Strategy for Lubelskie Province for the years 2014–2020 [53], and in the Spatial Development Plan for Lubelskie Province [54], the Central Roztocze meso-region is in a separate area of strategic intervention—area of the economic use of natural and cultural assets—in *"the Roztocze and Solska Forest functional area and areas with identified health resort assets."* The Strategy [53] defines the operational objective that encompasses the rational and effective use of natural environment resources for economic and recreational needs, while preserving and conserving the natural environment assets. The Plan [54] also sets out the framework for the conservation of abiotic nature, within which it proposes the establishment of geoparks and placing the most valuable assets of abiotic nature under the protection of the law. It indicates the proposed geopark "Stone Forest in Roztocze" as an area of geological heritage protection that should be used for the extensive promotion of geological sciences and propagation of the educational and tourist functions of geosites, while ensuring their effective protection. According to the Plan, nature and sightseeing tourism using the assets of abiotic nature requires special developmental stimuli.

The Programme for Tourism Development in Lubelskie Province until 2020 with a perspective until 2030 [55] mentions the problem of geotourism only when indicating the need for tailoring specialised products to specific audiences as part of nature tourism. The lack of a broader approach to geoparks and geotourism in a sectoral problem at the regional level does not create a climate favourable to their development. What is more, the Polish legal system does not provide any regulations concerning the formal establishment, rules of functioning and management of geoparks.

The situation is not much better at the local level. In the development strategies of local governments of six districts within which the sites included in the questionnaire surveys are located, only the abiotic resources and a very general framework of tourism development are presented to a varying extent (Table 11). In some cases, specific measures are indicated, but they do not apply directly to geotourism either. Issues related to tourism are covered most broadly in the Development Strategy of the Town and District of Krasnobród. Districts located in the functional area of Roztocze and Solska Forest [56] established a group called "Active Roztocze", for which a strategic territorial-functional plan was prepared up to the year 2023. The Geopark is indicated as a significant element of the integrated tourist product.

The area of the planned geopark has valuable abiotic nature assets but, at the same time, there are several unfavourable internal factors that pose a serious challenge to the development of geotourism (Table 12). Making the most of the existing potential can be facilitated or impeded by external determinants. Developmental stimuli arising from the formal and legal regulation of the functioning of geoparks, along with the institutional and financial support from central and regional governments, are particularly important.

Table 12. Development of geotourism in Central Roztocze—expert-based SWOT analysis.

	HELPFUL	HARMFUL
	Strengths	**Weaknesses**
INTERNAL	- scientific research results and documentation of the potential geopark confirming the valuable geoheritage assets	- lack of identification and appreciation of the geotourist potential in strategy documents
	- numerous geosites of high geotourist value	- peripheral location of the area and its poor accessibility in terms of transport
	- high level of tourism development	- lack of qualified tour-guides
	- numerous additional tourist attractions (including cultural ones)	- lack of extensive promotion of geotourist assets
	- existence of the Central Roztocze Geotourist Trail	- poor involvement of local communities in the development of geotourism
	Opportunities	**Threats**
EXTERNAL	- establishment of the geopark	- the lack of support for the idea of establishing geoparks in Poland
	- systemic regulation of the status and principles of geopark management	- the lack of understanding of the idea of a geopark as an opportunity for regional development
	- intensive tourist traffic within some geosites	- the lack of financial support for actions focused on geotourism
	- development of regional geotourist products	- the lack of understanding among local governments with regard to geotourism development
	- growing interest in active and alternative tourism	- competition from other tourist regions

Spatial planning does not offer any practical solutions with regard to the tourist use of geoparks because the planning of spatial development is merely a segment of spatial planning. The problems of the location and rules of the functioning of tourist infrastructure are discussed in planning documents on a par with problems related to housing and economy, social and technical infrastructure, environmental protection, etc. These problems as a whole are regulated by the Act on Spatial Planning and Development [57]. At the same time, the formal links between strategic and spatial planning are quite loose at the local level. It should be stressed that the geoparks are not discussed in the current Act on Nature Conservation [58]. Since they do not exist in the legal system, including the laws on nature conservation, it is very difficult to take the initiative to establish and manage them. The lack of a formal framework makes it impossible to obtain organizational and financial support for the process of geopark creation. They are not included in the spatial planning system either. Only the proposed National Raw Materials Policy mentions the protection of geodiversity through the establishment of geoparks, among other measures. Geoparks were mentioned only in the bill on the Polish Geological Agency with the attached Council dealing with the certification of geoparks and responsible for information, education and promotion. At the same time, there are no regulations imposing an obligation to prepare master plans for geoparks.

Geotourism, as a form of the sustainable development of rural areas and using land resources of high natural value, requires the involvement of local small business, as well as appropriate instruments and management [48]. Local governments in Poland are not actively initiating the creation of geoparks or supporting the actions of local communities [59]. The lack of systemic regulation of the formal status and management rules of geoparks is currently the biggest challenge. The need to protect geoheritage, on the one hand, and appropriate tourist development, on the other, forces the conflict-free coexistence of various functions in the use of space. The proper functioning of geoparks, stimulating local development, requires appropriate provisions in strategy and planning documents at various levels of government. Particularly significant is the local level where, due to the autonomy of districts with regard to planning, it is necessary to introduce detailed provisions into planning documents. In the context of tourism (including geotourism) planning and management, it is necessary to develop strategies of tourism development. In the case of geoparks, it is indispensable to prepare professional master plans [60] that are the basis for the operation of managing bodies, typically established by way of agreements of local governments, and enable the practical implementation of the idea of a geopark as part of business [61]. A geopark should be managed by a clearly defined structure, operating in accordance with the provisions of the law that would also enable the preservation of the assets and sustainable development. The body managing a geopark could play an active role in the economic development of a region. It should cooperate with local businesses in order to promote and support the creation of new products associated with geological heritage [60–65].

5. Conclusions

Central Roztocze is one of the most important potential areas for the development of geotourism in south-eastern Poland. The relatively large tourist traffic offers opportunities for building an offer based on geoheritage assets. The number of geotourist products existing here is distinctly greater than in another potential geopark in the region—the Małopolska Vistula Gap.

Some tourists can be regarded as conscious geotourists, able to accurately identify the assets of abiotic nature and for whom these assets are an important factor in making the decision to visit an area. However, the knowledge of the idea of establishing the Geopark in Central and Eastern Roztocze is very poor, which results from the total lack of promotion of the idea.

The present study found a strong consistency between the respondents' assessments and geotourist assessments. What is particularly important, sites of significant scientific value received high ratings, which creates the possibility of using the educational potential of the geosites.

The fundamental challenge to the development of geotourism and establishment of the Geopark is the total lack of activity among local governments and local communities in this respect. They do not regard the Geopark and geotourism as an opportunity for the economic development of the area. This may result from a lack of familiarity with the idea of geoparks, which shows the need for research on the perception of geotourism among residents.

The main obstacle to the creation of geoparks and their inclusion in the economic development of the region through the development of geotourism is the lack of legal regulations concerning the rules of their functioning and management. These sites are not integrated into the system of economic and spatial development planning at a more detailed level. At the same time, the idea of geoparks will not be able to develop without the institutional and organizational support from the State.

Author Contributions: Conceptualization, W.Z. and S.K.; methodology, W.Z. and S.K.; investigation, W.Z., S.K., B.B.-Z.; writing—original draft preparation, W.Z. and B.B.-Z.; writing—review and editing, W.Z. and B.B.-Z.; visualization, W.Z. All authors have read and agreed to the published version of the manuscript.

Funding: This research received no external funding.

Acknowledgments: The authors wish to thank the Editors and anonymous reviewers for their valuable comments and suggestions to improve the quality of this paper.

Conflicts of Interest: The authors declare no conflict of interest.

Appendix A

Geotourist values of Central Roztocze—questionnaire survey
Faculty of Earth Sciences and Spatial Management, Maria Curie-Sklodowska University
Al. Kraśnicka 2cd 20-718 Lublin
The aim of the study is to gather information on the motivation of visitors to Central Roztocze and to determine geotouristic values of the region. The survey is anonymous, and the information collected will be for scientific purposes only.

1. Which of the above motives prompted or will lead you to come to Central Roztocze?

Please select, at most, two answers
- ☐ Cultural tourism (architectural monuments)
- ☐ Nature tourism (the possibility of communing with nature, admiring landscapes, topography)
- ☐ Active tourism (biking, walking, Nordic walking, kayaking)
- ☐ Family visit
- ☐ Event tourism (imprezy kulturalne, rozrywkowe)
- ☐ Educational

2. What do you associate with Central Roztocze most?

Please select, at most, two answers
- ☐ Forests and fresh air
- ☐ Cultural values
- ☐ Well-developed tourist infrastructure
- ☐ Agrotourism
- ☐ Active tourism
- ☐ Regional cuisine
- ☐ Beautiful landscapes

3. What do you pay attention to when assessing the tourist attractiveness of a given object, area or phenomenon?

Please select, at most, three answers
- ☐ Good access to tourist attraction
- ☐ Cost of entrance
- ☐ Aesthetic value
- ☐ Many atractions in the vicinity
- ☐ Scientific value
- ☐ Location against network of tourist trails

4. Where are you looking for or would you like to find information about the geotouristic values of Central Roztocze?

    ☐   Booklets, leaflets
    ☐   Internet
    ☐   Opinions of friends and family
    ☐   Scientific publications
    ☐   I am not looking for such information

5. Do you know the term "geotourism"?

    ☐   Yes
    ☐   No

6. Please try to explain the concept of geotourism

… …
… …
… …

7. Do you know the concept of the Stone Forest Geopark in Roztocze?

    ☐   Yes
    ☐   No

8. Please try to explain the concept of Stone Forest Geopark in Roztocze

… …
… …
… …

9. Choose geotourist values from the list below

Please select only one answer
    ☐   Fossils
    ☐   Festivals
    ☐   Quarries
    ☐   Battlefields
    ☐   Dunes
    ☐   Botanical gardens
    ☐   Springs
    ☐   Manor houses
    ☐   Museums

10. Which of the regions of Poland do you consider to be the most geotouristically attractive?

    ☐   Białowieżą Forest
    ☐   Central Roztocze
    ☐   Biebrza River Valley
    ☐   Nałęczów Plateu
    ☐   Suwałki region
    ☐   Kaczawskie Mountains
    ☐   Kraków-Częstochowa Upland

11. How do you assess the attractiveness of individual tourist attractions in Central Roztocze?

1—no attractive 5—very attractive
Please mark only one answer per line

	1	2	3	4	5	I don't know
1. Church "On the Island" in Zwierzyniec	□	□	□	□	□	□
2. Polish Horse sanctuary	□	□	□	□	□	□
3. Nad Tanwią nature reserve	□	□	□	□	□	□
4. Saint Roch's Chapel	□	□	□	□	□	□
5. Church in Krasnobród	□	□	□	□	□	□
6. Szum river gap valley	□	□	□	□	□	□
7. Underground quarry in Senderki	□	□	□	□	□	□
8. "Piekiełko" rock forms	□	□	□	□	□	□
9. Waterfall on the Jeleń river	□	□	□	□	□	□
10. Quarry in Józefów	□	□	□	□	□	□
11. Regional Museum in Krasnobród	□	□	□	□	□	□
12. Quarry in Krasnobród	□	□	□	□	□	□
13. Wapielnia Hill	□	□	□	□	□	□
14. Homestead in Guciów	□	□	□	□	□	□
15. Sopot river gap valley in the Czartowe Pole nature reserve	□	□	□	□	□	□
16. Quarry in Nowiny	□	□	□	□	□	□

12. Please indicate deficiencies in the geotourism development of Central Roztocze

… …
… …
… …

13. Age

- □ 18–25 years
- □ 26–35 years
- □ 36–45 years
- □ 46–60 years
- □ more 60 years

14. Education

- □ Primary
- □ Secondary
- □ Higher

15. Place of stay (province)

… …
… …

Thank you for completing the survey! ☺

References

1. Dowling, R.K. Geotourism's global growth. *Geoheritage* **2011**, *3*, 1–13. [CrossRef]
2. Hose, T.A. 3 G's for modern geotourism. *Geoheritage* **2012**, *4*, 7–24. [CrossRef]
3. Newsome, D.; Dowling, R.K. Setting an agenda for geotourism. In *Geotourism: The Tourism of Geology and Landscape*; Newsome, D., Dowling, R., Eds.; Goodfellow Publishers Limited: Oxford, UK, 2010; pp. 1–12.
4. Global Geoparks Network. Guidelines and Criteria for National Geoparks Seeking UNESCO's Assistance to Join the Global Geoparks Network (GGN). Available online: http://www.unesco.org/new/fileadmin/MULTIMEDIA/HQ/SC/pdf/sc_geoparcs_2010guidelines.pdf (accessed on 31 December 2019).
5. Ólafsdóttir, R.; Tverijonaite, E. Geotourism: A Systematic Literature Review. *Geosciences* **2018**, *8*, 234. [CrossRef]
6. Kubalíková, L. Assessing Geotourism Resources on a Local Level: A Case Study from Southern Moravia (Czech Republic). *Resources* **2019**, *8*, 150. [CrossRef]

7. Selmi, L.; Coratza, P.; Gauci, R.; Soldati, M. Geoheritage as a Tool for Environmental Management: A Case Study in Northern Malta (Central Mediterranean Sea). *Resources* **2019**, *8*, 168. [CrossRef]
8. Pralong, J.P. A method for assessing tourist potential and use of geomorphological sites. *Geomorphol. Relief Process. Environ.* **2005**, *11*, 189–196. [CrossRef]
9. Kubaliková, L. Geomorphosite assesment for geotourism purposes. *Czech J. Tour.* **2013**, *2*, 80–104. [CrossRef]
10. Kubalíková, L.; Kirchner, K. Geosite and geomorphosite assessment as a tool for geoconservation and geotourism purposes: A case study from Vizovicka vrchovina highland (eastern part of the Czech Republic). *Geoheritage* **2016**, *8*, 5–14. [CrossRef]
11. Warowna, J.; Zgłobicki, W.; Kołodyńska-Gawrysiak, R.; Gajek, G.; Gawrysiak, L.; Telecka, M. Geotourist values of loess geoheritage within the planned Geopark Małopolska Vistula River Gap, E Poland. *Quat. Int.* **2016**, *399*, 46–57. [CrossRef]
12. Brilha, J. Inventory and quantitative assessment of geosites and geodiversity sites: A review. *Geoheritage* **2016**, *8*, 119–134. [CrossRef]
13. Kim, S.S.; Kim, M.; Park, J.; Guo, Y. Cave tourism: Tourists' characteristics, motivations to visit, and the segmentation of their behavior. *Asia Pac. J. Tour. Res.* **2008**, *13*, 299–318. [CrossRef]
14. Wójtowicz, B. The attitude and expectation of the local community toward development of geological tourism in protected areas near the city of Kielce and the Świętokrzyski region. *Probl. Ekol. Kraj.* **2011**, *29*, 123–132. (In Polish)
15. Zgłobicki, W.; Baran-Zgłobicka, B. Geomorphological Heritage as a Tourist Attraction. A Case Study in Lubelskie Province, SE Poland. *Geoheritage* **2013**, *5*, 137–149. [CrossRef]
16. Fung, C.K.; Jim, C. Segmentation by motivation of Hong Kong Global Geopark visitors in relation to sustainable nature-based tourism. *Int. J. Sustain. Dev. World Ecol.* **2015**, *22*, 76–88. [CrossRef]
17. Martins, B.; Pereira, A. Residents' Perception and Assessment of Geomorphosites of the Alvão—Chaves Region. *Geosciences* **2018**, *8*, 381. [CrossRef]
18. Alexandrowicz, Z. Geoparki-nowe wyzwanie dla ochrony dziedzictwa geologicznego. *Przegląd Geol.* **2006**, *54*, 36–41.
19. Solarska, A.; Hose, T.A.; Djordjije, A.; Vasiljević, D.A.; Mroczek, P.; Jary, Z.; Marković, S.B.; Widawski, K. Geodiversity of the loess regions in Poland: Inventory, geoconservation issues, and geotourism potential. *Quat. Int.* **2013**, *296*, 68–81. [CrossRef]
20. Koźma, J. Transgraniczny Geopark Łuk Mużakowa. *Przegląd Geol.* **2011**, *59*, 276–290.
21. Słomka, T. (Ed.) *Katalog Obiektów Geoturystycznych w Obrębie Pomników i Rezerwatów Przyrody Nieożywionej*; AGH: Kraków, Poland, 2012.
22. Migoń, P.; Pijet Migoń, E. Overlooked geomorphological component of volcanic geoheritage: Diversity and perspectives for tourism industry, Pogórze Kaczawskie Region, SW Poland. *Geoheritage* **2015**, *8*, 333–350. [CrossRef]
23. Alexandrowicz, Z.; Alexandrowicz, W.P.; Buczek, K. Conservation of the Natura 2000 Areas in the Context of Environmental Changes in Past and Present: A Case from the Polish Carpathians Geoheritage. *Geoheritage* **2019**, *11*, 517. [CrossRef]
24. Gajek, G.; Zgłobicki, W.; Kołodyńska-Gawrysiak, R. Geoeducational Value of Quarries Located Within the Małopolska Vistula River Gap (E Poland). *Geoheritage* **2019**, *11*, 1335–1351. [CrossRef]
25. Zgłobicki, W.; Kołodyńska-Gawrysiak, R.; Gawrysiak, L. 2015: Gully erosion as a natural hazard: The educational role of geotourism. *Nat. Hazards* **2015**, *79* (Suppl. 1), 159–181. [CrossRef]
26. Krąpiec, M.; Jankowski, L.; Margielewski, W.; Urban, J.; Krąpiec, P. Geopark "Kamienny Las na Roztoczu" i jego walory geoturystyczne. *Przegląd Geol.* **2012**, *60*, 468–479.
27. Brzezińska-Wójcik, T. Strategia hands-on activity w kreowaniu geoproduktów w kontekście edukacji, interpretacji i promocji geodziedzictwa na Roztoczu (Środkowowschodnia Polska). *Ekon. Probl. Tur.* **2015**, *1*, 169–193.
28. Brzezińska-Wójcik, T.; Skowronek, E.; Świeca, A. Nature and culture heritage in the tourist offer of the border region of Roztocze. *Sci. Rev. Phys. Cult.* **2017**, *7*, 129–142.
29. Buraczyński, J. *Roztocze, Budowa–Rzeźba–Krajobraz*; Wydawnictwo UMCS: Lublin, Poland, 1997.
30. Leszczyński, K. Rozwój litofacjalny późnej kredy Niżu Polskiego. *Biul. Państw. Inst. Geol.* **2010**, *443*, 33–54.
31. Marks, L.; Ber, A.; Gogołek, W.; Piotrowska, K. (Eds.) *Geological Map of Poland 1:500 000*; Polish Geological Institute-National Research Institute: Warsaw, Poland, 2006.

32. Brzezińska-Wójcik, T. Aktywność tektoniczna w strefie krawędziowej Roztocza Tomaszowskiego w świetle wskaźników morfometrycznych. *Ann. UMCS B* **1997**, *52*, 57–75.
33. Brzezińska-Wójcik, T.; Grabowski, T.; Moskal, A.; Pawłowski, A.; Wiechowska, I. *Część Opisowa do Mapy "Szlak Geoturystyczny Roztocza Środkowego", Informator–Mapa Turystyczna 1:50,000*; Kartpol: Lublin, Poland, 2011.
34. Złonkiewicz, Z. Zabytki skałkowe Roztocza. *Ochr. Przyr.* **1990**, *47*, 309–333.
35. Gazda, L.; Ruska, A. Górnictwo kamieni młyńskich w Senderkach. In *Historia i Współczesność Górnictwa na Terenie Lubelszczyzny*; Krzowski, Z., Ed.; Wydawnictwo Politechniki Lub.: Lublin, Poland, 2005; pp. 54–62.
36. Božić, S.; Tomić, N. Canyons and gorges as potential geotourism destinations in Serbia: Comparative analysis from two perspectives–general geotourists' and pure geotourists'. *Open Geosci.* **2015**, *7*, 531–546. [CrossRef]
37. Różycka, M.; Migoń, P. Customer-Oriented Evaluation of Geoheritage–on the Example of Volcanic Geosites in the West Sudetes, SW Poland. *Geoheritage* **2018**, *10*, 23–37. [CrossRef]
38. Cocean, G.; Cocean, P. An assessment of gorges for purposes of identifying geomorphosites of geotourism value in the Apuseni Mountains (Romania). *Geoheritage* **2017**, *9*, 71–81. [CrossRef]
39. Zgłobicki, W.; Baran-Zgłobicka, B.; Ziółek, M.; Ziółek, G. Atrakcyjność wizualna krajobrazu polskich parków narodowych a ich wartości przyrodnicze. *Parki Nar. Rezerw. Przyr.* **2005**, *24*, 135–151.
40. Gordon, J.E. Geoheritage, Geotourism and the Cultural Landscape: Enhancing the Visitor Experience and Promoting Geoconservation. *Geosciences* **2018**, *8*, 136. [CrossRef]
41. Brzezińska-Wojcik, T. 2017, Zasoby przyrody nieożywionej Roztocza jako podstawa kreowania obszarowego produktu turystycznego. *Zeszyty Nauk. Wyższej Szkoły Tur. Języków Obcych Warszawie* **2017**, *2*, 5–22.
42. Brzezińska-Wójcik, T.; Skowronek, E. Tangible Heritage of the Historical Stonework Centre in Brusno Stare in the Roztocze Area (SE Poland) as an Opportunity for the Development of Geotourism. *Geoheritage* **2020**, *12*, 10. [CrossRef]
43. Migoń, P.; Pijet-Migoń, E. Viewpoint geosites-Values, conservation and management issues. *Proc. Geol. Assoc.* **2017**, *128*, 511–522. [CrossRef]
44. Moreira, J.C. Interpretative panels about the geological heritage—A case study at the Iguassu Falls National Park (Brazil). *Geoheritage* **2012**, *4*, 127–137. [CrossRef]
45. Migoń, P.; Pijet-Migoń, E. Interpreting Geoheritage at New Zealand's Geothermal Tourist Sites—Systematic Explanation Versus Storytelling. *Geoheritage* **2017**, *9*, 83–95. [CrossRef]
46. Crawford, K.R.; Black, R. Visitor understanding of the geodiversity and the geoconservation value of the Giant's Causeway world heritage site, Northern Ireland. *Geoheritage* **2012**, *4*, 115–126. [CrossRef]
47. Rozenkiewicz, A.; Widawski, K.; Jary, Z. Geotourism and the 21st Century–NTOs' Website Information Availability on Geotourism Resources in Selected Central European Countries: International Perspective. *Resources* **2020**, *9*, 4. [CrossRef]
48. Dowling, R.K. Global geotourism—An emerging form of sustainable tourism. *Czech J. Tour.* **2013**, *2*, 59–79. [CrossRef]
49. Farsani, N.; Coelho, C.; Costa, C. Geotourism and geoparks as novel strategies for socio-economic development in rural areas. *Int. J. Tour. Res.* **2011**, *13*, 68–81. [CrossRef]
50. Pijet-Migoń, E.; Migoń, P. Promoting and Interpreting Geoheritage at the Local Level—Bottom-up Approach in the Land of Extinct Volcanoes, Sudetes, SW Poland. *Geoheritage* **2019**, *11*, 1227–1236. [CrossRef]
51. *Koncepcja Przestrzennego Zagospodarowania Kraju 2030, Monitor Polski*; Uchwała Rady Ministrów nr 239; Warszawa, Polska, 2012; p. 252. Available online: http://prawo.sejm.gov.pl/isap.nsf/DocDetails.xsp?id=WMP20120000252 (accessed on 18 February 2020).
52. Ministerstwo Środowiska. *Projekt Polityki Surowcowej Państwa*; Ministerstwo Środowiska: Warszawa, Poland, 2018.
53. Urząd Marszałkowski Województwa Lubelskieg. *Strategia Rozwoju Województwa Lubelskiego na Lata 2014–2020 (z Perspektywą do 2030 Roku)*; Urząd Marszałkowski Województwa Lubelskiego: Lublin, Poland, 2014.
54. Zarząd Województwa Lubelskiego. *Plan Zagospodarowania Przestrzennego Województwa Lubelskiego*; Zarząd Województwa Lubelskiego: Lublin, Poland, 2015.
55. Urząd Marszałkowski Województwa Lubelskieg. *Program Rozwoju Turystyki w Województwie Lubelskim do 2020 Roku z Perspektywą do 2030*; Urząd Marszałkowski Województwa Lubelskiego: Lublin, Poland, 2016.
56. Tomaszów Lubelski: Lublin. *Strategiczny Plan Terytorialno-Funkcjonalny w Ramach Inicjatywy "Aktywne Roztocze" do 2023 r*; Tomaszów Lubelski: Lublin, Poland, 2016.

57. Ustawa z dnia 27 marca 2003, r.o planowaniu i zagospodarowaniu przestrzennym. Dziennik Ustaw nr 80/2003 poz. 717 (tekst jedn. Dz. U. 2018; p. 1945). Available online: http://prawo.sejm.gov.pl/isap.nsf/DocDetails.xsp?id=WDU20030800717 (accessed on 18 February 2020).

58. Ustawa z dnia 16 kwietnia 2004 r. o ochronie przyrody. Dziennik Ustaw nr 92/2004 poz. 880 (tekst jedn. Dz. U.2018; p. 1614). Available online: http://prawo.sejm.gov.pl/isap.nsf/DocDetails.xsp?id=WDU20040920880 (accessed on 18 February 2020).

59. Alexandrowicz, Z.; Miśkiewicz, K. Geopark–od idei do realizacji, ze szczególnym uwzględnieniem Polski. *Chrońmy Przyr. Ojczystą* **2016**, *72*, 243–253.

60. Mc Keever, P.J.M.; Zouros, N. Geoparks: Celebrating Earth Heritage. Sustaining Local Communities. *Episodes* **2005**, *28*, 274–278. [CrossRef] [PubMed]

61. Baran-Zgłobicka, B.; Harasimiuk, M. Znaczenie badań geograficznych w poprawie jakości planowania w jednostkach samorządu terytorialnego różnego szczebla. *Barom. Reg.* **2016**, *14*, 127–140.

62. Zouros, N.; Mc Keever, P.J.M. European Geoparks: Geoconservation and Sustainable Local Development. In Proceedings of the International Conference "Studying, Modeling and Sense Making of Planet Earth", Lesvos, Greece, 1–6 June 2008; University of the Aegean: Lesvos, Greece, 2008.

63. Zouros, N.; Valiakos, I. Geoparks management and assessment. *Bull. Geol. Soc. Greece* **2010**, *43*, 965–977. [CrossRef]

64. Gravis, I.; Németh, K.; Twemlow, C.; Németh, B. The Case for Community-Led Geoheritage and Geoconservation Ventures in Māngere, South Auckland, and Central Otago, New Zealand. *Geoheritage* **2020**, *12*, 19. [CrossRef]

65. Muntoni, F.; Balvis, T.; Rizzo, R.; Loru, P. Territorial Planning of Geological Mining Historical and Environmental Park of Sardinia. *Geoheritage* **2020**, *12*, 22. [CrossRef]

MDPI

St. Alban-Anlage 66

4052 Basel

Switzerland

Tel. +41 61 683 77 34

Fax +41 61 302 89 18

www.mdpi.com

Resources Editorial Office

E-mail: resources@mdpi.com

www.mdpi.com/journal/resources